Inorganic and Organometallic Macromolecules

Design and Applications

Alaa S. Abd-El-Aziz • Charles E. Carraher, Jr.
Charles U. Pittman, Jr. • Martel Zeldin
Editors

Inorganic and Organometallic Macromolecules

Design and Applications

Alaa S. Abd-El-Aziz
The University of
British Columbia, Okanagan
3333 University Way
Kelowna, British Columbia
Canada V1V 1V7
alaa.abd-el-aziz@ubc.ca

Charles E. Carraher, Jr.
Department of Chemistry & Biochemistry
Florida Atlantic University
777 Glades Rd.
Boca Raton, Florida 33431
USA
carraher@fau.edu

Charles U. Pittman, Jr.
Department of Chemistry
Mississippi State University
Mississippi State, Mississippi 39762
USA
cpittman@chemistry.msstate.edu

Martel Zeldin
Department of Chemistry
University of Richmond
28 Westhampton Way
Richmond, Virginia 23173
USA
mzeldin@richmond.edu

Library of Congress Control Number: 2007934536

ISBN: 978-0-387-72946-6 e-ISBN: 978-0-387-72947-3

Printed on acid-free paper.

9 8 7 6 5 4 3 2 1

springer.com

Preface

Polymeric materials of the 21st century often contain atoms that are not present in traditional polymers. Polymers containing nontraditional atoms are now of interest because of their unique properties. This book demonstrates the breadth of these properties and some of the specialized analytical techniques that have been developed to characterize them.

Chapters 1, 2, 3, 4 and 7 emphasize the emerging special properties of materials dealing with the transmission of light for the purpose of communication, as well as other efforts. Later chapters deal with the use of materials in treating a variety of disease-causing microbes—including viruses responsible for pandemic herpes and the common cold (Chapter 8), cancers (Chapter 11), and bacterial infections (Chapter 17). The interaction of these materials for future biological investigations is investigated in Chapters 5 and 6.

Chapter 12 provides a comprehensive review of the application of Mossbauer spectroscopy to metal-containing polymers and Chapter 13 reviews the application of a new mass spectrometry technique. The use of metal-containing polymers as catalysts is described in Chapters 1, 9, and 10. Their use as precursors for advanced ceramics (Chapter 14), high temperature materials (Chapter 15), and flame retardants (Chapter 16) is also discussed. The unusual property of selected materials to spontaneously form fibers is described in Chapter 18.

This book includes a cross-section of novel polymeric materials containing nontraditional atoms and emphasizes current chemical, biological, engineering, ceramic, and optical areas of application. It is intended for those interested in the general areas of biomedicine, catalysis, electronics and light, thermal stability, and analysis of materials. The polymers reported in this volume represent early research but are inidicative of future application.

Preface

Contents

Contributors

Alaa S. Abd-El-Aziz
Department of Chemistry, University of British Columbia Okanagan, Kelowna, British Columbia, Canada V1V 1V7

Francis E. Appoh
Department of Chemistry, University of Saskatchewan, Saskatoon, Saskatchewan, Canada S7N 5C9

Girish Barot
Department of Chemistry and Biochemistry, Florida Atlantic University, Boca Raton, FL 33431, USA

Amitabh Battin
Florida Atlantic University, Department of Chemistry and Biochemistry, Boca Raton, FL 33431, USA and Florida Center for Environmental Studies, Palm Beach Gardens, FL 33410, USA

Samuel Bernard
Laboratorie des Multimateriaux et Interfaces, CNRS-Universite Claude Bernard Lyon 1, 69622 Villeurbanne, France

Christopher W. Bielawski
Department of Chemistry and Biochemistry, University of Texas at Austin, Austin, TX, 78712, USA

A. M. Bochkin
Institute of Problems of Chemical Physics, Russian Academy of Sciences, Chernogolovka, Moscow oblast 142432, Russia

Andrew J. Boydston
Department of Chemistry and Biochemistry, University of Texas at Austin, Austin, TX 78712, USA

Charles E. Carraher, Jr.
Department of Chemistry and Biochemistry, Florida Atlantic University, Boca Raton, FL 33431, and Florida Center for Environmental Studies, Palm Beach Gardens, FL 33410, USA

Cara L. Carraher
Department of Biomedicine, Princeton University, Princeton, NJ 08540, USA

David Cornu
Laboratorie des Multimateriaux et Interfaces, CNRS-Universite Claude Bernard Lyon 1, 69622 Villeurbanne, France

Hongchen Dong
Department of Chemistry, The Hong Kong University of Science and Technology, Clear Water Bay, Kowloon, Hong Kong, China

Sylvain Duperrier
Laboratorie des Multimateriaux et Interfaces, CNRS-Universite Claude Bernard Lyon 1, 69622 Villeurbanne, France

G. I. Dzhardimalieva
Institute of Problems of Chemical Physics, Russian Academy of Sciences, Chernogolovka, Moscow oblast 142432, Russia

Daming Fan
Center for Applications in Polymer Science, Central Michigan University, Mount Pleasant, MI 48859, USA

Tiziana Fiore
Dipartimento di Chimica Inorganica e Analitica "Stanislao Cannizzaro", Universita di Palermo, Viale delle Scienze, Parco d'Orleans, 90128 Palermo, Italy

N. D. Golubeva
Institute of Problems of Chemical Physics, Russian Academy of Sciences, Chernogolovka, Moscow oblast 142432, Russia

Pierre D. Harvey, Department de Chimie, Universite de Sherbrooke, Sherbrooke, PQ, Canada J1K 2R1

Matthias Häußler
Department of Chemistry, The Hong Kong University of Science and Technology, Clear Water Bay, Kowloon, Hong Kong, China

Bobby A. Howell
Center for Applications in Polymer Science and Department of Chemistry, Central Michigan University, Mount Pleasant, MI 48859, USA

Teddy M. Keller
Chemistry Division, Naval Research Lab., Washington, DC, 20375, USA

Manoj K. Kolel-Veetii
Chemistry Division, Naval Research Lab., Washington, DC 20375, USA

Heinz-Bernhard Kraatz, Department of Chemistry, University of Saskatchewan, Saskatoon, Saskatchewan, Canada, S7N 5C9

Philippe Miele
Laboratorie des Multimateriaux et Interfaces, CNRS Universite Claude Bernard Lyon 1, 69622 Villeurbanne, FR

Yoshihiro Mori
Okayama University of Science, Research Institute of Technology, Okayama 700-0005, Japan

Kazutaka Nagao
Research Institute of Technology, Okayama University of Science, Okayama 700-00005, Japan

Yoshinobu Naoshima
Department of Computer Simulation, Okayama University of Science, Okayama 700-000005, Japan

Claudia Pellerito
Dipartimento di Chimica Inorganica e Analitica "Stanislao Cannizzaro", Universita di Palermo, Viale delle Scienze, Parco d'Orleans, 90128 Palermo, Italy

Lorenzo Pellerito
Dipartimento di Chimica Inorganica e Analitica "Stanislao Cannizzaro", Universita di Palermo, Viale delle Scienze, Parco d'Orleans, 90128 Palermo, Italy

A. D. Pomogailo
Institute of Problems of Chemical Physics, Russian Academy of Sciences, Chernogolovka, Moscow oblast, 142432, Russia

S. I. Pomogailo
Institute of Problems of Chemical Physics, Russian Academy of Sciences, Chernogolovka, Moscow oblast 142432, Russia

Leela Rakesh
Center for Applications in Polymer Science, Central Michigan University, Mount Pleasant, MI 48859, USA

Michael R. Roner
Department of Biology, University of Texas at Arlington, Arlington, TX 76010, USA

A. S. Rozenberg
Institute of Problems of Chemical Physics, Russian Academy of Sciences, Chernogolovka, Moscow oblast 142432, Russia

Theodore S. Sabir
Department of Chemistry, Providence Christian College, Ontario CA 91761, USA

Michelangelo Scopelliti
Dipartimento di Chimica Inorganica e Analitica "Stanislao Cannizzaro", Universita di Palermo, Viale delle Scienze, Parco d'Orleans, 90128 Palermo, Italy

Patrick O. Shipman
Department of Chemistry, University of British Columbia Okanagan, Kelowna, BC, Canada V1V 1V7

Ben Zhong Tang
Department of Chemistry, The Hong Kong University of Science and Technology, Clear Water Bay, Kowloon, Hong Kong, China

Bérangère Toury
Laboratorie des Multimateriaux et Interfaces, CNRS-Universite Claude Bernard Lyon 1, 69622 Villeurbanne, France

Wai-Yeung Wong
Department of Chemistry and Centre for Advanced Luminescence Materials, Hong Kong Baptist University, Kowloon Tong, Hong Kong, P. R. China

Anna Zhao
Department of Chemistry and Biochemistry, Florida Atlantic University, Boca Raton, FL 33431, USA

Synthetic Versatility and Structural Modularity in Organometallic Polymers

Andrew J. Boydston and Christopher W. Bielawski

1 Background

With regard to tunability within a functional material, there are two primary areas of general discussion: (1) versatile synthetic strategies and (2) breadth of compatible structural features within the monomeric scaffold. These two issues rarely avoid some degree of overlap, yet a universal solution to both within any polymer design is non-trivial. Synthetic versatility can be further broken down into having either multiple access routes to obtaining the general monomer template or having a versatile and multifunctional monomer that can partake in more than one type—or in mechanistically distinct—polymerizations (e.g., copolymerizations) with high control. Structural modularity is inherently dependent on several factors: The polymerization method, the stability of the metal center, and the location of the metal's center (i.e., whether main- or side-chain metal incorporation). However, assuming general compatibility of the reaction conditions with the functional groups desired, the monomer design should accommodate installation of said groups.

Synthetically, there are a handful of methods for preparing main-chain organometallic polymers. This chapter provides an introduction to the methods that involve homo- and copolymerization of organometallic monomers, copolymerization of organometallic with organic monomers, and copolymerization of inorganic reagents with organic monomers such that those bonds formed to the metal involved are the ones that lead to polymer formation. There are multiple strategies for each type

A.S. Abd-El-Aziz et al. (eds.), *Inorganic and Organometallic Macromolecules: Design and Applications.*

of polymerization method and additional details can be found in later chapters. Each method of polymer formation has been studied for years. Although many intricate details could be elaborated, for the purposes of this chapter only a brief overview will be given.

1.1 Polymerizations of Organometallic Monomers

Synthesis of functionalized organometallic compounds is often an entry point into structurally simple metal-containing polymers. With the vast body of knowledge available for small-molecule synthesis of metal complexes, it is of no surprise that this method is widely utilized and spans multiple subclasses of macromolecules. Many groups have had success in designing ligands with reactive sites either distal to the point of contact with the metal or proximal, such as arenes bearing halogens poised for substitution. The stability of the organometallic polymer is often determined by the binding affinity between the ligand and the transition metal incorporated within the polymer chain.

1.1.1 Olefin and Alkyne Polymerization

The polymerization of vinyl ferrocene (**1**) by Arimoto and Haven (Scheme 1.1) is regarded as the birth of organometallic polymers [1]. Since that report, the surge of additional methods and monomer structures suitable for alkene polymerizations has strengthened considerably. Typically, olefin polymerization approaches are used to obtain side-chain organometallic polymers, although many examples of main-chain systems have also been achieved. An attraction of this approach is that, assuming reasonable stability of the organometallic moieties, virtually any robust alkene or alkyne metathesis reaction compatible with organic monomers is compatible with organometallic variants as well. Whereas less focus has been placed on alkyne metathesis, one of the key features of this method is its ability to generate an organometallic polymer with a fully conjugated all-carbon backbone [2].

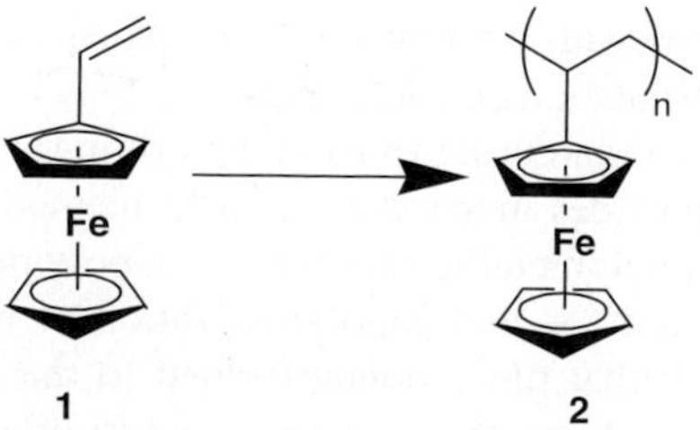

Fe
1
n
Fe
2

Scheme 1.1

1.1.2 Substitution and Condensation Reactions

The S_NAr approach and polycondensation reactions are excellent methods for generating organometallic polymers. Most of the examples in these areas involve use of an organic moiety as a comonomer and are highlighted in Section 1.2. With regard to structural complexity and control, perhaps the most exemplary organometallic polymers are alternating bimetallic polymers. As shown in Scheme 1.2, isolation of Fe-complex (**3**) (Scheme 1.3) and subsequent reaction with cationic Ru complex (**4**) gave an alternating bimetallic polymer with excellent control [3]. This route has several key advantages. First, each organometallic monomer can be constructed and characterized independently. Second, the use of a heterocoupling copolymerization reaction gives perfect control over the alternating positioning of each metal-containing moiety within the polymer chain. Finally, the number of metal combinations within metallocene chemistry is vast, thus modularity in this system should be high.

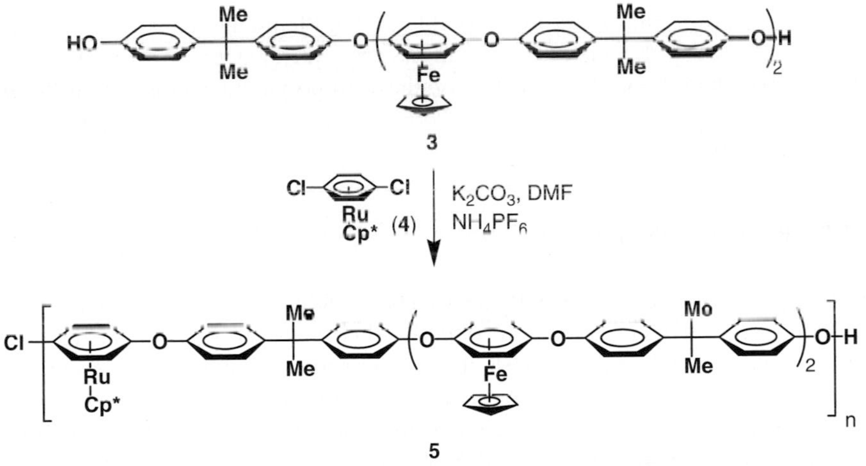

Scheme 1.2

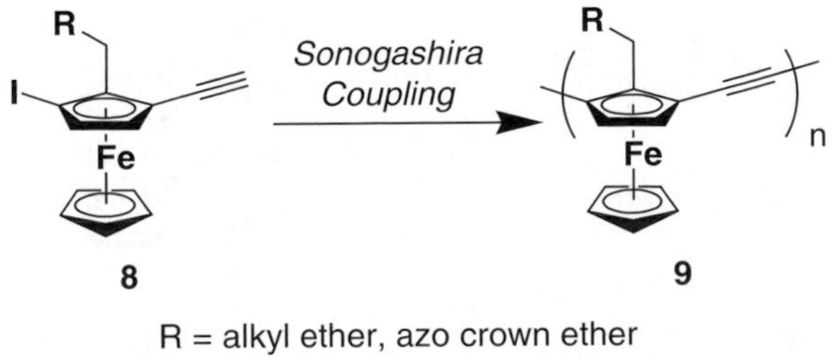

Scheme 1.3

1.1.3 Electropolymerization

Electrochemical polymerization [4] is another attractive route to synthesizing metal-containing polymers from discrete organometallic monomers. Most often, the polymers obtained are not main-chain, but rather side-chain organometallics (Figure 1.1, **6**) [5]. There are, however, some examples of main-chain organometallic polymers obtained by electropolymerization. Constable, for example, made use of a functionalized Ru(terpy)$_2$ complex bearing electropolymerizable thiophenes on the periphery to achieve polymers such as Figure 1.1, **7** [6].

1.1.4 Alkyne Cross-Coupling Reactions

The examples of homopolymerizations discussed so far are of AA-type monomers. A convenient method for controlled AB-type monomer polymerizations is to use alkyne cross-coupling methodologies. For example, highly functionalized monomers (**8**) were prepared by Plenio [7] that featured an aza crown ether as well as an iodo and ethynyl group poised for homopolymerization (Scheme 1.3). It is at the heart of our discussion to point out that Plenio and coworkers had previously reported polymers showing interesting optical activity with structures also based on **9** (Scheme 1.3) [8]. The synthesis of polymers having very different potential applications, yet stemming from a common synthetic design demonstrates the importance of modularity.

1.1.5 Ring-Opening Polymerization

Ring-opening polymerization (ROP) has seen broad utility for synthesizing main-chain organometallic polymers. Initially reported by Rauchfuss [9], and thoroughly developed by Manners and coworkers [10], the transformation from **10** to **11** (Scheme 1.4) has been optimized to include various conditions for polymerization such as thermal, anionic, photo, and metal-mediated polymerizations; Both solution and solid-state polymerization have also been reported. Molecular weights on the order of 10^6 Da have been achieved and the ability to prepare monomers of varying functionality has assisted

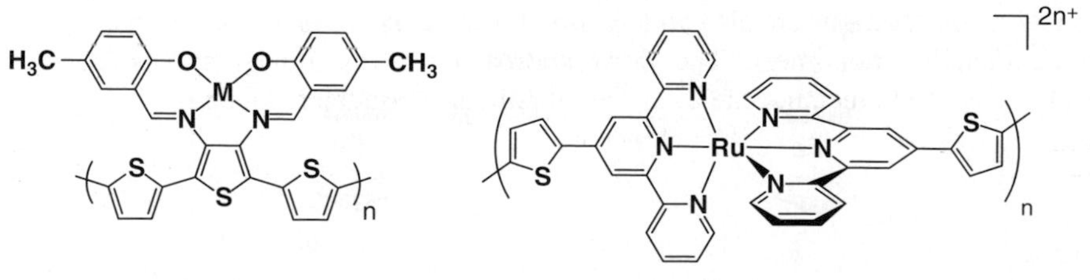

Figure 1.1 Examples of polymers obtained via electropolymerization of organometallic monomers

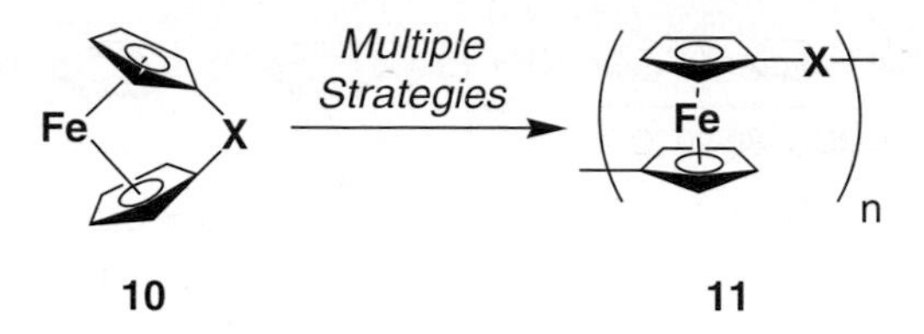

Scheme 1.4

in overcoming any inherent solubility limitations. Bridging groups have included hydrocarbon, sulfur, boron, tin, germanium, phosphorus, and silicon bridges as well as various segments obtained from block copolymerizations.

1.2 Copolymerization of Organometallic with Organic Monomers

1.2.1 Alkene Polymerizations

The majority of reactions available for alkene polymerization can be used to copolymerize organometallic and organic substrates. There has been substantial progress in this area since Pittman first described the radical polymerization of acrylate functionalized metal-arene complexes in the early 1970s [11]. Although some monomers underwent homopolymerization smoothly, typically an organic comonomer was necessary to obtain macromolecular products of significant molecular weight. It is noteworthy that in addition to organic comonomers, organometallic substrates have also been used to obtain mixed metal copolymers. This method is also compatible with contemporary, highly controlled polymerization methods. For example, Frey and coworkers reported the living free radical polymerization of vinyl ferrocene with styrene [12].

1.2.2 Substitution and Condensation Reactions

Similar to the all-organometallic polymers discussed previously, substitution and condensation reactions are also widely used for the copolymerization of organic and organometallic monomers. The S_NAr approach utilizes almost exclusively the application of the metallocene as an electrophile because the halogen is activated by metal complexation to the arene. Since the report by Segal on the polymerization of CpRu complexed with *p*-dichlorobenzene (**12**) along with various bis(phenate) ions (Scheme 1.5) [13], this method has been developed to include a wide range of structures. The greatest structural variations are found in the organic comonomers (Figure 1.2); However, polyhalogenation of the metallocene offers an avenue for structural tuning as well. Extending from their work with Cp^*Ru-dichlorobenzene complexes [14], Dembek and coworkers have demonstrated the ability to generate

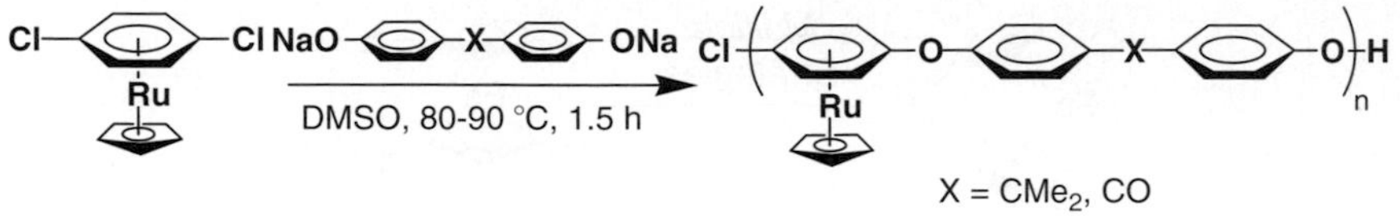

Scheme 1.5

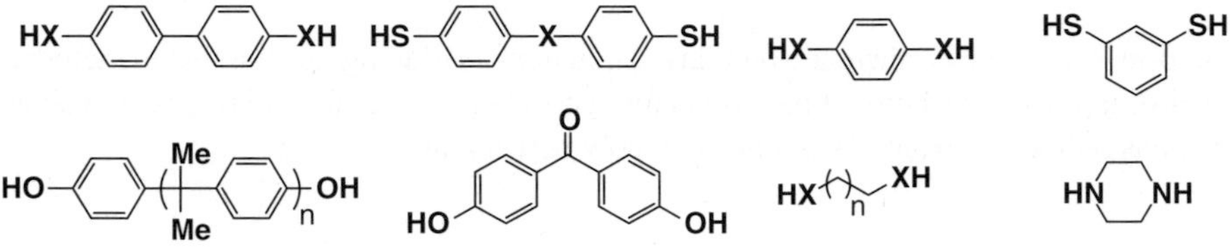

Figure 1.2 Commonly used organic spacers for S_NAr copolymerization with di- or polyhalogenated metallocenes (X = O, S)

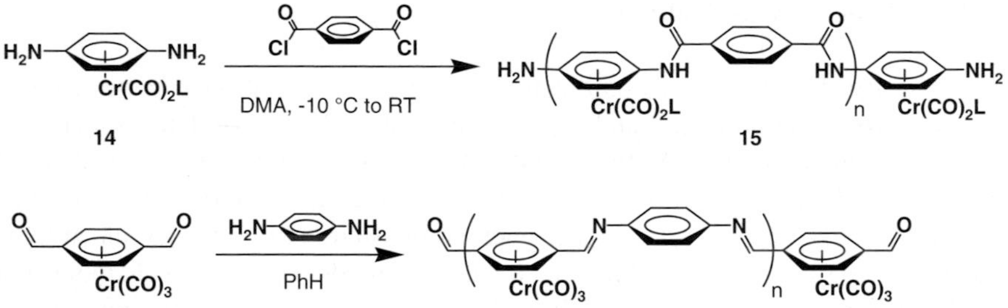

Scheme 1.6

highly branched materials using tri- and tetrachlorobenzene complexes [15]. This latter example illustrates the molecular and architectural complexity that can be achieved with this route while requiring only simple synthetic manipulations such as polyhalogenation of arenes followed by metal complexation.

Many examples exist of metal-arene complexes bearing nucleophiles poised for polycondensation to give polyesters, amides, imines, etc. As early as 1961, polycondensation reactions using diacid chloride variations of metallocenes [16] have been studied with various linkers such as 1,4-hydroquinone and *p*-phenylenediamine. Polycondensation reactions have the added versatility of using the organometallic monomer as either the nucleophilic or electrophilic partner. Examples of each are depicted in Scheme 1.6. Jin and Kim showed the use of phenylenediamine-$Cr(CO)_3$ complex (**14**) with terephthaloyl chloride to give the corresponding copolymer (**15**) in good yields (Scheme 1.6) [17]. Complimentary to this example, dehydration reactions to form polyimines were performed by Wright and Lowe-Ma using a Cr complex of terephthalaldehyde (**16**) and *m*-phenylenediamine (Scheme 1.6).

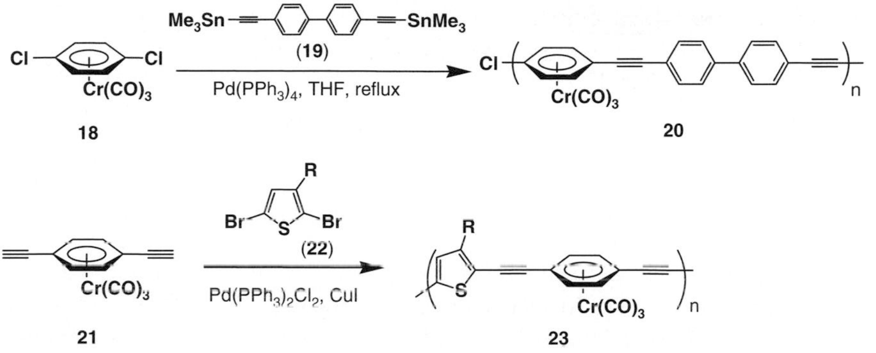

Scheme 1.7

Although this particular polymer showed limited solubility, the approach could be easily adapted to include monomers with increased solubilizing ability.

1.2.3 Cross-Coupling Reactions

Others have achieved an all-carbon backbone using standard cross-coupling techniques to install conjugated linkers. In these cases, the metal-arene complex has been employed in both roles (individually and dually) of the cross coupling. As one partner, Wright used the metallocene (**18**) as the electrophile in combination with Stille reagent **19** under standard conditions to give the corresponding highly conjugated organometallic polymer (Scheme 1.7) [18]. As might have been expected from the linear rigid framework, these polymers displayed poor solubility despite a relatively low molecular weight (ca 7.8 kDa). Alternatively, Chujo and coworkers prepared a functionalized thienylene containing copolymer **23** using 1,4-diethynylbenzene chromium complex (**21**) in combination with dibromothiophenes (**22**) under Sonogashira conditions (Scheme 1.7). In these studies, polymers with molecular weights ranging from 13.5 to 24.4 kDa (PDIs = 3.2 – 3.6) were obtained and extensive π-delocalization was confirmed from comparative UV-Vis spectroscopy. Notably, the polymers were found to be semiconducting when undoped ($\lambda_s = 10^{-6}$ S/cm) [19].

1.3 Polymerizations Involving Metal-Binding Events During Polymerization

1.3.1 Metal-Containing Polyynes

The metal-containing polyynes are an interesting class of main-chain organometallic polymers that have been under investigation since the 1970s. These systems are known for their rigid-rod structures and electronic communication over the extensively

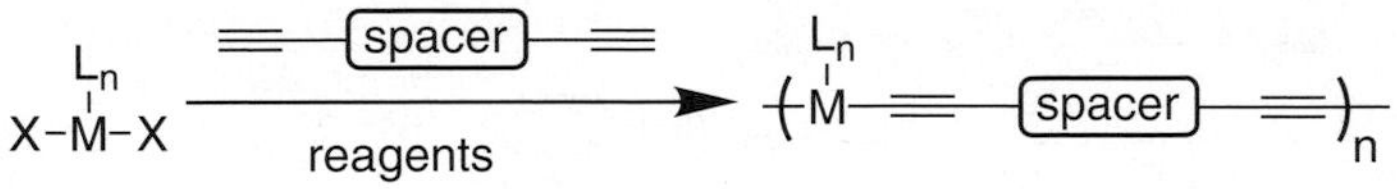

Scheme 1.8

delocalized π-system leading to interesting optical properties. Since Hagihara's initial report on Pd- and Pt-containing polymers [20], much optimization and development has followed. Synthetic avenues typically involve use of a metal(II) halide (**24**) in combination with an α,ω-diyne (**25**). As implied in Scheme 1.8, various linkers have been used and many transition metals have been explored as well as ligand effects at the metal center. Electronic and solubility tuning is also achieved through simple functionalization of the arene linkers. Mixed-metal bimetallics have also been obtained using metal-alkyne linkage in the polymerization step [21].

1.3.2 Coordination Polymers

When one considers the vast body of knowledge that is nearly taken for granted regarding neutral donor ligands in metal complexes, it is not surprising that many researchers have used these moieties in the design of organometallic polymers. The task put forward would seem to simply be one of attaching two known ligands at either end of an extended and/or rigid linker. Whereas this is essentially the overall scheme, achieving macromolecular materials in this way is simple in theory, but nontrivial in its execution. Common donor moieties have historically been phosphines, mono-, bidentate-, or tridentate amines, ethers, imines, nitriles, and thio compounds. High binding affinities are necessary to generate high-molecular-weight materials, and often characterization of the polymers is hampered by the inherent tendency toward depolymerization, especially in dilute solutions typical of gel permeation chromatography (GPC), UV-Vis, and mass spectroscopy.

High-molecular-weight materials have been obtained using difunctional bis(phosphines) as in the reports by Sijbesma (Figure 1.3, **26**) [22]. In these studies, it was found that simply combining either Pd(II) or Pt(II) salts with a bis(phosphine) produced macromolecular materials. Although a common concern with coordination polymers of labile ligands is there inherent lack of structural integrity and strength, the polymers reported by Sijbesma were sufficiently stable to form fibers. Studies were conducted to establish the nature of the metal center and conditions to control linear versus cyclic oligomerization. Bis(phosphine)s of conjugated linkers have also been used. Examples reported by Puddephatt [23] use Au(I) capped bis(acetylides) in combination with

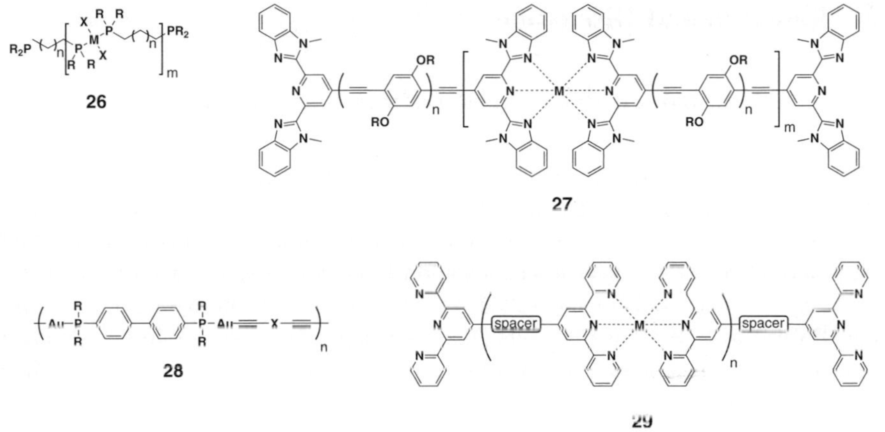

Figure 1.3 Various scaffolds of coordination polymers synthesized from copolymerization of an organic linker with a transition metal salt

bis(phosphine)s to produce macromolecular coordination polymers (Figure 1.3, **28**). Interestingly, they demonstrated that coordination of AuCl to the bis(phosphine) followed by reaction of the bis(acetylide), in a manner analogous to those described in Section 1.3.1, also produced to polymeric materials. Coordinating amines are also widely used. Prior to Puddephatt's report, Takahashi had used pyridines linked through hydrocarbon chains to coordinate between metal centers providing cationic metallo-polyynes of interesting structure and properties [24]. However, when amines are used it is more common that each binding pocket is made up of a di- or triamine. Difunctional linkers derived from terpyridyl (terpy) ligands, for example, offer a very high binding affinity and many structural derivatives used for polymer formation (Figure 1.3, **29**) are known [25]. In these cases, a range of conjugated and otherwise functionalized spacers have been used to connect two terpy moieties. Rowan and Weder have used a pyridine-based chromophore functionalized with two benzimidazoles to form metal binding sites on the ends of *p*-phenyleneethynylene oligomers. In the presence of Zn(II) or Fe(II), supramolecular polymers such as **27** were obtained [26]. Closely related are various diamine linkers which have also been used to make organometallic coordination polymers. Rehahn's work [27] exemplifies some of the key features of polymers in this subclass. Using phenanthrolines connected through rigid conjugated spacers, polymers containing Cu(I) or Ag(I) were synthesized. The properties of these polymers are quite interesting because macromolecular structure can be solvent-controlled. In noncoordinating solvents, a linear structure having a "classical" polymeric form is obtained, and concentration dependence of molecular weight and structure is not observed. In more coordinating solvents, aggregates are assumed to be formed which, at high dilution, appeared as cyclic oligomers.

2 Research and Discussion

2.1 New Approach to Modular Difunctional Monomers

The various methods described so far for achieving organometallic polymers offer several key features. Multiple transition metals can be incorporated and various polymerization protocols have been presented. Coordination polymers, although often too labile for practical implementation in devices, display dynamic behavior that may be optimized for controlled reversible polymerizations. Metal-arene and metallocene polymers have the versatility of being either main-chain or side-chain organometallic macromolecules, with occasional hybrids of the two being studied. Incorporation of the metal moieties can be accomplished pre-, post-, and during polymerization (or copolymerization) with equal diversity in the mode of polymerization. From a survey of the body of work available on organometallic polymers, it seems apparent that a universal monomer scaffold of high tunability would offer the particular features: (1) Monomer synthesis should be simple and straight-forward with high overall yield, and the general structure should have multiple access routes to facilitate modification; (2) the steric environment around the metal should be easily manipulated, predictable, and modular; (3) a broad range of transition metals should be compatible without the need to significantly alter the polymerization method; (4) sites for functionalization should be apparent and easily manipulated to control both physical and electronic features, preferably independent of one another; (5) the polymers should display "bench-stability" toward moisture and air while dually exhibiting controllable dynamic behavior at the metal center; (6) the polymerization protocol should be robust, proceed under mild conditions, and not require the need for inert atmosphere or dry solvents, and (7) high molecular weights should be obtainable as well as control over molecular weight and end groups.

Many of the most efficient strategies for organometallic polymer synthesis (with regard to controllability and molecular weight) make use of highly developed polymerization reactions optimized initially for all-organic substrates. Many of the polymers synthesized under these criteria are inherently side-chain type and offer restricted communication between metal centers. Additionally, polymerization protocols often require the need for inert atmosphere and dry solvents due to the sensitivity of either the organometallic moiety or the reactivity of the comonomers. With regard to metallocene-based systems, it is important to note that because the point of attachment, and synthetic focus, is primarily on functionalization of the arene, the same sites used for electronic and steric tuning are often coincident with those used for attachment of polymerizable functional groups. Although for an S_NAr approach there are plenty of handles for structural tuning, one exception is that the organometallic complex must almost always be designated as the electrophilic partner because the substitution reaction relies on activation of the arene-halide bond after metal complexation. This last issue is resolved for polycondensation reactions and when precise stoichiometric control and pristinely pure reagents are used, excellent control over molecular weight can be achieved.

Though the list of criteria for a highly modular design seems daunting, there appeared to be an organometallic scaffold that would potentially lend itself to a solution to such extraordinary demands. We envisioned the use of heterocyclic carbenes in a new fashion to achieve organometallic polymers exhibiting all of the features listed above. Prior to our contributions, there were very few reports of bis(carbene)s used in the synthesis of macromolecular organometallic materials beyond labile Ag-based aggregates [28]. This was surprising considering the features of heterocyclic carbenes that are desirable for polymer formation such as broad structural diversity and tunable affinities toward virtually every transition metal. Our design would require the construction of a monomer bearing two facially opposed heterocyclic carbene moieties linked through a rigid organic framework that would give control over the site of metallation and shepherd complexation away from intramolecular chelation. Years of effort produced stable and isolable N-heterocyclic carbenes. The task of a bis(carbene) target, however, seemed loftier, especially considering the precedence for challenges in preparing such a class of difunctional ligands [29]. Our current research remains focused on the design and synthesis of new structures and incorporation of stable bis(carbene) moieties into macromolecules that are engineered to execute useful tasks.

2.2 *Difunctional Heterocyclic Carbenes as Linkers*

Our original series of bis(carbene) structures focused on a practical synthesis that could demonstrate modularity primarily of the conjugated linker between the metal centers. This was accomplished using tetraamino arenes bearing various aromatic frameworks including benzo, biphenyl, and dioxin-based chromophores (Scheme 1.9). Cyclization of the tetraamines (**30**) with formic acid gave each bis(imidazole) (**31**) in excellent yield and high purity. Fourfold alkylation was accomplished by first treating the bis(imidazoles) each with NaH in refluxing $PhCH_3$ followed by introduction of the electrophile. Addition of dimethylformamide (DMF) as cosolvent at this point in the reaction facilitated dissolution of partially alkylated intermediates and the final products (**32**) were precipitated from the cooled reaction mixture cleanly in good to excellent yields. This protocol was both rapid and high yielding, however the installation of N-substituents was limited to primary halides. Expanding the possible N-substituents involved finding additional pathways to the monomer template.

Installation of larger substituents was accomplished using a two-step high-yielding protocol (Scheme 1.10) [30] involving a fourfold aryl amination to give tetraamines (**34**); using a modified version of Harlan's procedure [31]. Our initial efforts focused on bulky aliphatic amines, but were later extended to include functionalized arylamines as well. Interestingly, although tetraamino arenes are typically plagued by oxidative instability, we found that large alkyl groups (e.g., ^{t}Bu, ^{t}Oct, Ad) significantly suppressed the rate of oxidation. These compounds can be stored for days under ambient atmosphere. Alternatively, more reactive tetraamines were isolated as their respective hydrochloride salts and found to be highly resistant to

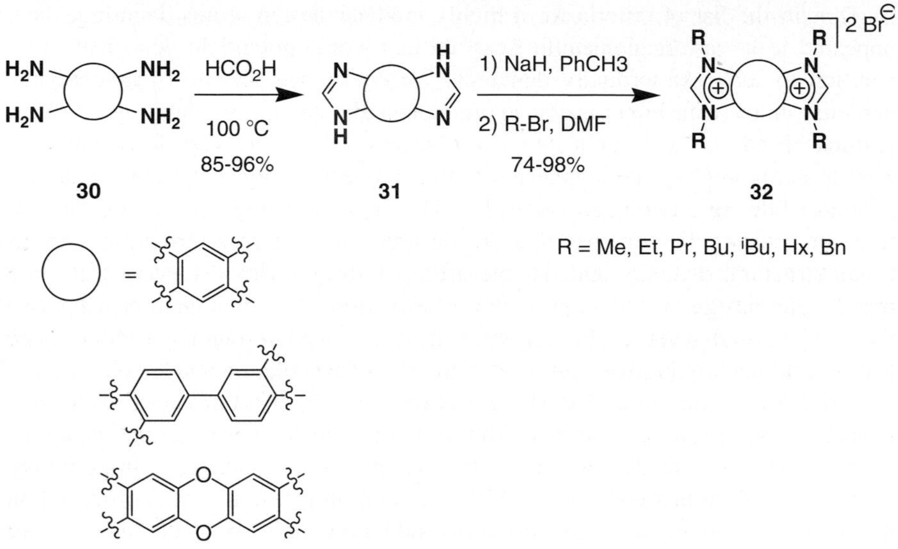

Scheme 1.9

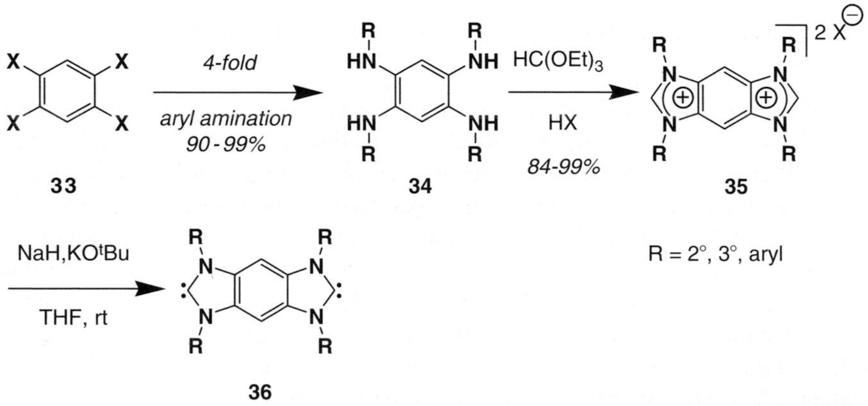

Scheme 1.10

oxidation. Successful ring-closure was accomplished using triethylorthoformate and HCl to provide the bis(azolium) salts (**35**). In cases where increased solubility was desired, simply performing the double ring closure in the presence of HBF_4 in place of HCl provided the more soluble tetrafluoroborate salts. In cases involving very large N-substituents, deprotonation gives the stable bis(carbene)s (**36**) which can be isolated and stored indefinitely.

As mentioned previously, desymmetrization of monomers is often a focal point for accomplishing structural complexity. We have found that in addition to symmetric bis(azolium) salts, a short sequence was available to obtain high yields of asymmetric monomers as well. Ultimately, we have achieved a level of structural diversity arriving at bis(carbene) scaffolds bearing two or three different R-groups, and varying heteroatoms, with complete regiocontrol. This was accomplished using a double-S_NAr reaction, or two sequential S_NAr substitutions, with dichlorodinitrobenzene (**37**) and virtually any primary amine of choice. This effectively installs the first two R-groups regiospecifically. Following an *In situ* reduction-cyclization protocol optimized in our laboratories, the *N,N'*-disubstituted benzimidazoles are obtained in excellent overall yield and high purity. Alkylation is accomplished in high yield to give the bis(azolium) salts with varying substitution patterns.

In the first route shown in Scheme 1.11, substitution of two equivalents of an amine in refluxing ethanol provides the diamine (**38**) as it precipitates from the reaction mixture. This method is best employed when installing R-groups not compatible with alkylation via nucleophilic attack on alkyl halides. Following reduction and cyclization to give **39**, alkylation can be performed to give the bis(azolium) salt with two different R-groups regiospecifically. In further studies, we found that mono-alkylation of **39** could be controlled to give high yields of monoazolium salt (**41**). This compound proved to be a useful asymmetric building block in preparing alternating bimetallic polymers as will be discussed below. A second alkylation of **41** yielded the bis(azolium) salt (**42**) bearing three different R-groups. Three different R-groups can also be installed in cases where the desymmetrizing substituents are not compatible with alkylation. This is accomplished through temperature control during the S_NAr reaction. Treatment of **37** with an amine in EtOH at room temperature gave a nearly quantitative yield of the monoamine product. The second substitution was performed

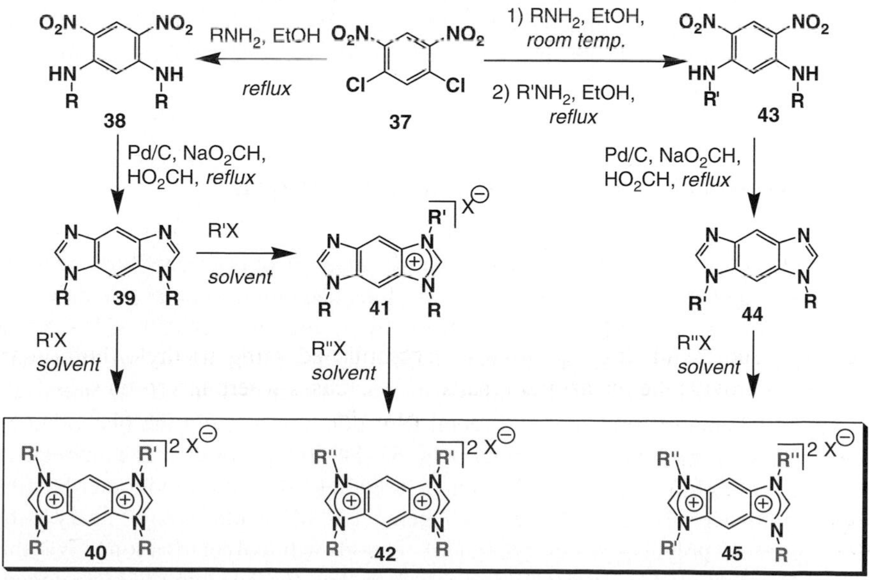

Scheme 1.11

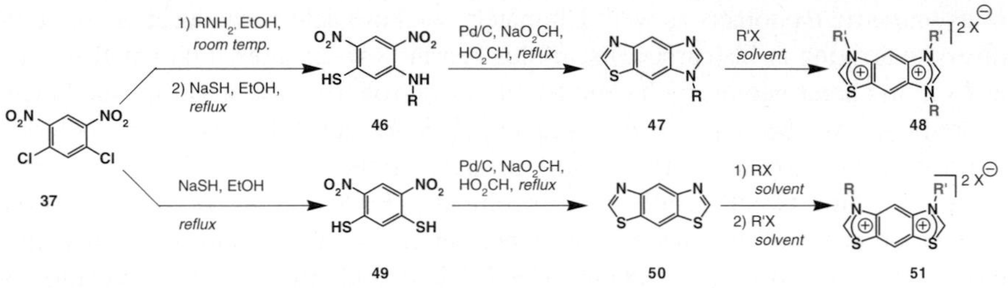

Scheme 1.12

cleanly and in high yield which effectively desymmetrized the monomer template (**43**). Reduction and cyclization to give **44**, followed by alkylation, provided the bis(azolium) (**45**) that is complimentary in the R-group structure to **42**.

Heteroatom variation has been accomplished using NaSH as a nucleophile in the S_NAr reactions (Scheme 1.12). Treatment of **37** with an amine, followed by reaction with NaSH furnished asymmetric dinitroarene (**46**). Reduction and cyclization gave a hybrid structure (**47**) which was cleanly alkylated to give **48**. Use of excess NaSH with **37** yielded **49** which was used en route to bis(thiazole) (**50**). Alkylation could be performed stepwise, as in the synthesis of **42**, to ultimately yield asymmetric structures, such as **51**, bearing two thiazolium moieties.

In summary, in monomer syntheses there is broad flexibility with regard to heteroatom content, N-substituent functionality, electronic asymmetry, steric asymmetry (through judicious placement of different N-substituents), conjugation through arene linker, solubility, and overall function. Each synthetic route is streamlined to be short, high-yielding, and pose minimal technical difficulty. In the next section we will discuss how these attributes were used to obtain polymers of desired functionality.

2.3 Bis(carbene)-Based Organometallic Polymers

All of the bis(azolium) salts presented above undergo smooth copolymerizations with Pd(II) and Pt(II) salts in the presence of acetate anion in polar solvents (e.g. dimethyl sulfoxide [DMSO], DMF, N-methylpyrrolidone [NMP], CH_3CN) as described in our original report in this area [32]. Treatment of our bis(azolium) salts with $Pd(OAc)_2$ or $PtCl_2$/NaOAc in polar solvents between 50 and 110 °C effectively executed the copolymerization and metal incorporation to yield the bis(carbene)-based main-chain organometallic polymers (**52–54**) in high yield after precipitation into MeOH or H_2O (Scheme 1.13). Our first studies focused on monomer bearing aliphatic alkyl groups as described in Scheme 1.8. Molecular weight analysis by GPC revealed polydispersities typical of step-growth polymerizations. For this series, M_n was higher for Pt-containing polymers than for the structurally analogous Pd-containing systems. Notably, very high M_n (up to 1.8×10^6 Da) were obtained

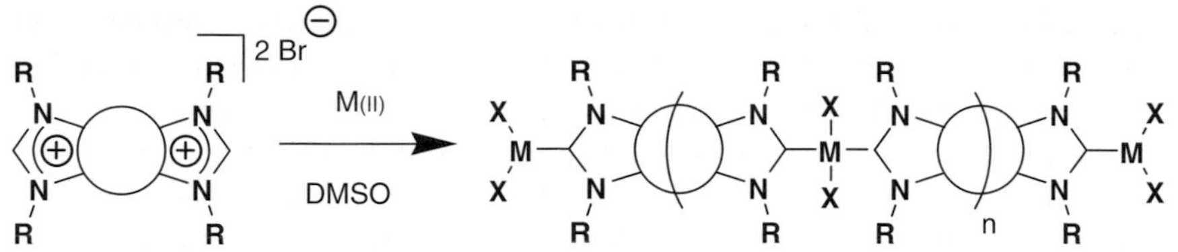

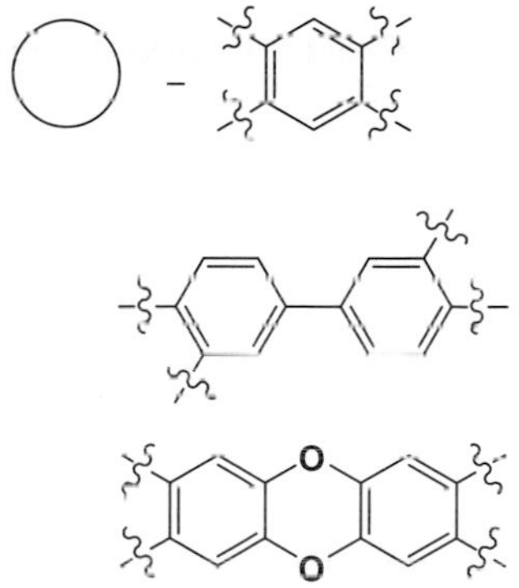

52a R = Bn, M(II) = $Pd(OAc)_2$, X = Br
52b R = Bu, M(II) = $Pd(OAc)_2$, X = Br
52c R = Bn, M(II) = $PtCl_2$/NaOAc, X = Cl
52d R = Bu, M(II) = $PtCl_2$/NaOAc, X = Cl

53a R = Bn, M(II) = $Pd(OAc)_2$, X = Br
53b R = Bu, M(II) = $Pd(OAc)_2$, X = Br

54a R = Bn, M(II) = $Pd(OAc)_2$, X=Br
54b R = Bn, M(II) = $PtCl_2$/NaOAc, X=Cl

Scheme 1.13

with this method. The electronic absorption spectra revealed a fairly narrow range of absorption maxima (308–329 nm). Thermal stability was measured by thermal gravimetric analysis (TGA) under nitrogen atmosphere and T_d was consistently found to be between 280 and 300 °C for polymers of this structure.

One key limitation in this synthetic route was the incompatibility of using metals other than Pd and Pt. Of specific interest initially was the failure to polymerize using Ni(II) salts, which have also demonstrated nonreactivity in attempts to generate analogous small molecule Ni-NHC complexes via Herrmann's procedure [33]. To verify the stability of the resultant Ni-based polymers, we generated the free bis(carbene), which reversibly formed the corresponding homopolymer [34], and subsequently added anhydrous $NiCl_2$. This method provided the desired polymer as a stable macromolecule. Another alternative route utilized Lin's Ag-mediated NHC transfer reaction. Treatment of the bis(azolium) salts with Ag_2O produced thick gels when solvated which, when reacted with divalent metal halides, led to the corresponding organometallic polymers. This method, however, produces a stoichiometric amount of metal waste and results in difficulty confirming complete transmetallation. Procedures for small molecule NHC-Ni complexes are known [35] that involve monomeric azolium salts with predried $Ni(OAc)_2$ at high temperature under vacuum either neat [36] or in an ionic liquid [33].

In general, the polymers in this series were poorly soluble in most common solvents (THF, CH_2Cl_2, dioxane, CH_3CN), but readily dissolved in more polar solvents such as DMF, DMSO, and NMP. Subsequent changes have involved simply using longer alkyl chains for the N-substituents (e.g., hexyl) which resulted high-molecular-weight polymers that exhibited good solubility in solvents such as THF, $CHCl_3$, CH_2Cl_2, and dioxane. Alternatively, when increased solubility is desired in combination with relatively small N-substituents, installation of additional functionality on the arene is useful. For example, we are working toward producing bis(azolium) salts with varying chain-length alkyl groups on the arene (Scheme 1.14). Fourfold electrophilic aromatic substitution was performed on 1,4-dialkyl benzenes (**55**) to give either tetrahalo (**56**) or tetranitro (**57**) products. Reductive cyclization of the tetranitro followed by alkylation, or four-fold aryl amination of the tetrahalo followed by cyclization, should afford the corresponding bis(azolium) salts (**58**) which should exhibit markedly improved solubilities.

One key feature of the bis(NHC) organometallic polymers is their reversibility, or "dynamicity." The dynamic nature of the copolymerization prompted us to investigate the use of chain-transfer agents to modulate both polymer molecular weight and end-group structure (Scheme 1.15). Copolymerization with $Pd(OAc)_2$ in

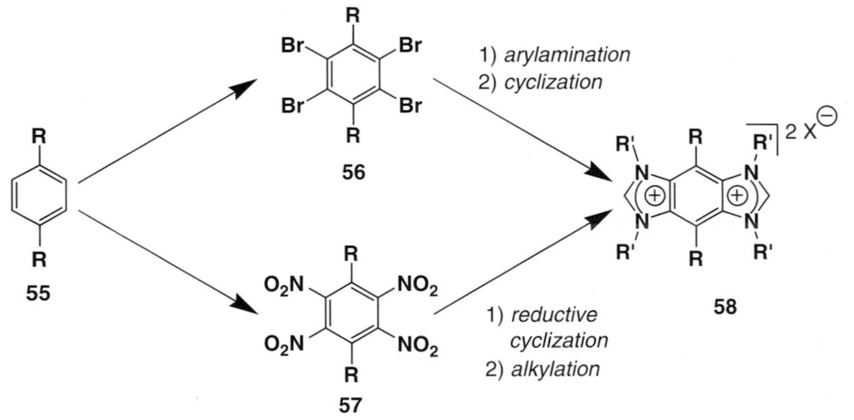

Scheme 1.14

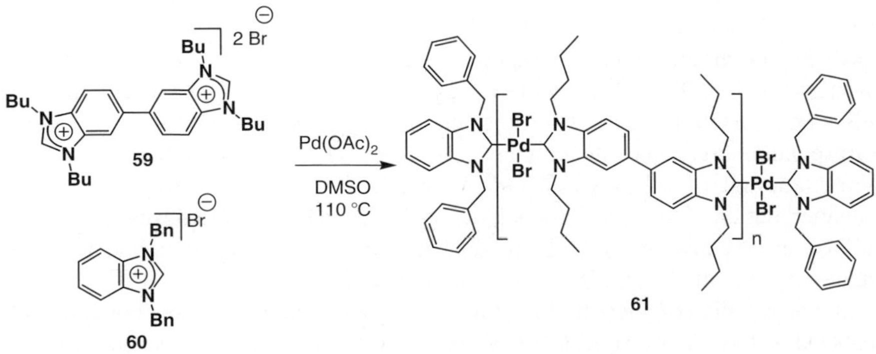

Scheme 1.15

the presence of monfunctional benzimidazolium bromide (**60**) as a CTA produced end-capped polymers (**61**). In these experiments, excellent agreement was observed between the theoretical DP based on loading ratios of **60/59** and experimental DP determined by ^{1}H nuclear magnetic resonance (NMR) analysis. Later studies have focused on using the reversible polymer formation to design dynamic copolymers of varying monomer structure and self-healing networks.

Postpolymerization modification is yet another method of tailoring polymer properties and can be considered a modular attribute of a macromolecule [37]. Given that the metals in the bis(NHC) polymers are coordinatively unsaturated, the use of an exogenous ligand added postpolymerization was expected to bind to the metal thus altering the physical and electronic properties of the polymer. Addition of PPh_3 or PCy_3 to a suspension of polymer **53a** in THF quickly affected complete dissolution of the phosphine-bound polymer. Ligation was confirmed by ^{1}H and ^{31}P NMR spectroscopy.

Generation of bis(NHC)-based organometallic polymers containing varying transition metals of choice was a key breakthrough. Unfortunately, metals typically showing poor hydrolytical stabilities as NHC complexes (e.g., Cu) were impractical using our first generation of monomer scaffolds. To address this issue, we targeted a monomer that would increase the affinity of the NHC ligand for the metal. Considering examples from similar small-molecule organometallic complexes, we focused on using phenolic imidazoles to generate an additional (ionic) bond with the metal. Using similar chemistry to that described previously, dichloro-dinitrobenzene (**37**) was reacted with 2-aminophenol, followed by reduction-cyclization to yield **62** (Scheme 1.16) [40]. Alkylation arrived at the desired monomer (**63**) in high overall yield. Optimization studies revealed that polymerization was most successful with the addition of an exogenous weak base to level the mineral acid generated from reaction of the phenol with the metal halide. After successful formation of both Pd- and Pt-containing polymers with the new monomer design, we targeted transition metals such as

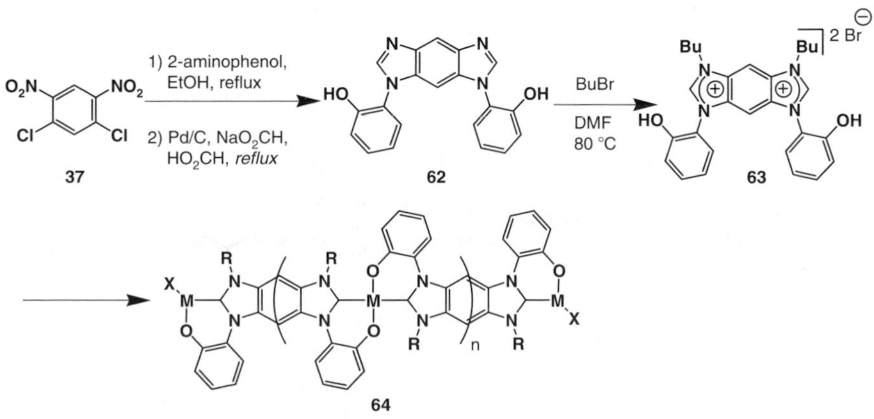

Scheme 1.16

Ni and Cu. Typically, small-molecule analogs are synthesized via a full deprotonation of both the phenol and the azolium under inert atmosphere, followed by introduction of a soluble (ligated) metal salt [38]. Alternatively, Hoveyda has successfully employed Lin's NHC transfer reaction to obtain naphtholic NHC-metal complexes [39]. Additionally, no studies on benzimidazolium salts functionalized with phenol substituents had been reported. Subjecting Ni(II) and Cu(II) salts to our polymerization conditions in the presence of stoichiometric NaOAc produced excellent yields of the corresponding polymers. Incorporation of these transition metals could be done directly under ambient atmosphere in high yields. Increased thermal stability of the systems was also observed. All of the phenolic polymers are air and moisture stable, and TGA analysis under nitrogen atmosphere revealed T_d ranging from 340 to 362 °C. In comparison to the analogous Pd- and Pt-based polymers bearing N-alkyl groups (**52**), thermal stabilities increased by approximately 50 °C via incorporation of phenoxide substituents. Varying the transition metal had a small, but discernable, impact on the electronic absorption spectra. Each absorbs in the infrared with λ_{max} ranging from 287 (Ni) to 319 nm (Cu). The increased binding affinity appeared to be tailored independently of the overall electronic nature. That is, the phenolic Pd and Pt systems exhibit λ_{max} values nearly identical to the related *N*-alkyl polymers (**52**).

Two independently impressive displays of polymer design lie in areas involving directionality (head-to-tail selectivity) and mixed-metal systems. We are working toward accomplishing these goals using our bis(carbene) approach. Further use of the phenol ligand, in combination with "left–right" desymmetrization, led to a designer monomer with the potential to form a discrete directional polymer via head-to-tail ordering of the bis(carbene) linker. Specifically, sequential temperature-controlled S_NAr reactions using an aliphatic amine followed by introduction of 2-aminophenol gave an asymmetric diamine of structure **43**. Carrying through to the corresponding organometallic polymer gave materials under study for the potential to exist as depicted in the form shown in Figure 1.4. Use of **41** to form a transition metal complex with terminal imidazole functionality poised to ultimately generate an alternating bimetallic polymer with excellent control over metal placement could lead to polymers of the structure **66** (Figure 1.4). The broad combinations of transition metals compatible with our polymerization protocols should bring about new opportunities for electronic fine-tuning.

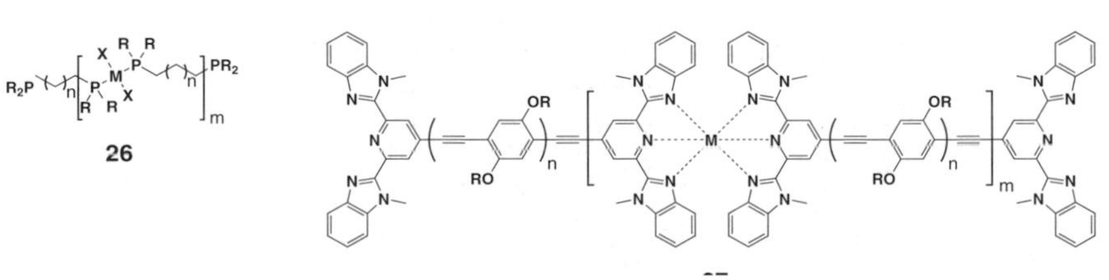

Figure 1.4 Directional polymers and mixed-metal polymers from bis(carbene) monomers

3 Further Considerations and Outlook

We have demonstrated the high level of modularity that is achieved by using bis(carbene) scaffolds as monomer for organometallic polymers. Many handles exist that allow tuning of nearly every desirable feature of the polymers including solubility, thermal stability, metal compatibility, dynamicity, and electronic communication. The frontier of this approach to main-chain organometallic polymers will be pushed further and faster with the addition of structurally variable monomer architectures and will span many areas of material science. The use of cross-linked networks of our systems are already being investigated for application in conductive self-healing materials. Other areas of research involve dynamic polymers as "heat-activated" catalysts. Conductivity optimization through redox matching of the heterocyclic carbenes and the transition metals will ultimately lead to improvements on the already semiconducting (undoped) properties observed from our polymers. All of these areas are facilitated by the common theme of a modular design that not only allows versatility in the metal involved in the polymer, but also in the steric and electronic features of the organic moieties. The changes in design have been brought about quickly and efficiently due to a monomer template that has many avenues for synthesis and ultimately new applications.

References

1. Arimoto FS, Haven AC. (1955) J Am Chem Soc. 77:6295.
2. Buchmeiser M, Schrock RR. (1995) Macromolecules. 28:6642.
3. de Denus CR, Hoffa LM, Todd EK, Abd-El-Aziz AS. (2000) J Inorg Organomet Polym. 10:189.
4. Roncali J. (1999) J Mater Chem. 9:1875.
5. Holliday BJ, Swager TM. (1995) Chem Commun. 23.
6. Constable EC, Hagfeldt A. (2002) Chem Commun. 284.
7. Plenio H, Hermann J, Sehring A. (2000) Chem Eur J. 6:1820.
8. Plenio H, Hermann J, Leukel J. (1998) Eur J Inorg Chem. 12:2063.
9a. Brandt PF, Rauchfuss TB. (1992) J Am Chem Soc. 114:1926.
9b. Galloway CP, Rauchfuss TB. (1993) Angew Chem. Int Ed. 32:1319.
10. Nguyen P, Gómez-Elipe P, Manners I. (1999) Chem Rev. 99:1515.
11. Abd-El-Aziz AS. (2002) Macromol Rapid Commun 23:995.
12. Baumert M, Frohlich J, Stieger M, Frey H, Mülhaupt R, Plenio H. (1999) Macromol Rapid Commun. 20:203.
13. Segal JA. (1985) J Chem Soc. Chem Commun. 1338.
14. Dembek AA, Fagan PJ, Marsi M. (1993) Macromolecules. 26:2992.
15. Dembek AA, Fagan PJ, Marsi M. (1994) Polym Mater Sci Eng 71:158.
16. Knobloch FW, Rauscher WH. (1961) J Polym Sci. 54:651.
17. Jin J-I, Kim R. (1987) Polym J. 19:977.
18. Wright ME, Lowe-Ma CK. (1995) Inorg Chim Acta. 232:223.
19. Morisaki Y, Chen H, Chujo Y. (2002) Polym Bull. 48:243.
20. Fujikura Y, Sonogashira K, Hagihara N. (1975) Chem Lett. 1067.
21. Long NJ, White AJP, Williams DJ, Younus M. (2002) J Organomet Chem. 649: 94.
22. Paulusse JMJ, Huijbers JPJ, Sijbesma RP. (2995) Macromolecules 38: 6290.

23. Jia G, Puddephatt RJ, Scott JD. (1993) J Vittal Organometallics 12: 3565.
24. Onitsuka K, Ogawa H, Joh T, Takahashi S. (1988)Chem Lett. 1855.
25. Andres PR, Schubert US, Adv Mater. 16:1043.
26. Knapton D, Rowan SJ, Weder W. (2006) Macromolecules 39:651.
27. Lahn B, Rehahn M. (2002) e-Polym. 1:1.
28a. Guerret A, Solé S, Gornitzka H, Teichert M, Trinquier H, Bertrand G. (1997) J Am Chem Soc. 119:6668.
28b. Chiu PL, Chen CY, Zeng JY, Lu CH, Lee HM. (2005) J Organomet Chem. 690:1682.
29. Kim H-J, (2004) Doctoral Dissertation, University of Alabama.
30a. Khramov DM, Boydston AJ, Bielawski CW. (2006) Org. Lett. 8:1831.
30b. Boydston AJ, Khramov DM, Bielawski CW. (2006) Tetrahedron Lett. 47:5123.
31. Wenderski T, Ligh KM, Ogrin D, Bott SG, Harlan CJ. (2004) Tetrahedron Lett. 34:6851.
32. Boydston AJ, Williams KA, Bielawski CW. (2005) J Am Chem Soc. 127:12,496.
33. Huynh HV, Holtgrewe C, Pape T, Koh LL, Hahn E. (2006) Organometallics 25:245.
34. Kamplain JW, Bielawski CW. (2006) Chem Commun10.1039/b518246h.
35. Herrmann WA, Schwarz J, Gardiner MG, Spiegler ML. (1999) Organomet Chem. 575:80.
36. McGuinness DS, Mueller W, Wasserscheid P, Cavell KJ, Skelton BW, White AH, Englert U. (2002) Organometallics 21:175.
37. Boaen NK, Hillmyer MA. (2005) Chem Soc Rev. 34:267.
38. Waltman AW, Grubbs RH. (2004) Organometallics 23:3105.
39. Larsen AO, Leu W, Oberhuber CN, Campbell JE, Hoveyda AH. (2004) J Am Chem Soc. 126:11,130.
40. Boydston AJ, Rice JD, Sanderson MD, Dykhno OL, Bielawski CW. (2006) Organometallics 25:6087.

Hyperbranched Polymers Containing Transition Metals: Synthetic Pathways and Potential Applications

Matthias Häußler, Hongchen Dong, and Ben Zhong Tang

1 Background

Incorporation of transition metals into organic monomers and polymers has been thoroughly examined over the past five decades in light of the promising electrical, magnetic, optical, sensing and catalytic properties that these organometallic materials possess [1–4]. Thanks to their intriguing properties, which are often inaccessible by their pure organic parents, these organometallic polymers have found applications in the coating, pharmaceutical, and aerospace industries.

Whereas many of these studies focused on the synthesis of linear polymers with either transition metals integrated into the main chain or attached as pendant groups at the side chains, the preparation and study of highly branched three dimensional (3-D) macromolecular architectures—such as dendrimers and hyperbranched polymers containing organometallic complexes—has only recently received greater attention. Depending on their position, metal centers have been shown to act as cores, simulating artificial models of biological systems such as metalloenzymes, as well as connectors, branching points, and terminal (surface) units distributed throughout the whole structure with potential applications in the field of sensors, catalysts, and as light-harvesting antennas [5,6]. Despite their structural beauty, dendrimer synthesis needs to be carried out in a thoughtful manner involving multistep reaction and purification protocols in order to construct the various tree-like generations, which will in many cases restrict their potential applications to academic interests only [7]. Moreover, recent advances in mass spectrometric techniques have revealed depictions of dentrimers showing them to be highly idealized and that the real samples indeed exhibit imperfections and structural defects [8].

A.S. Abd-El-Aziz et al. (eds.), *Inorganic and Organometallic Macromolecules: Design and Applications.*

Compared with their "perfect" dentrimer congeners, "imperfect" hyperbranched polymers often exhibit similar, or even comparable, molecular properties despite their random and polydisperse structures. Nevertheless, they can be readily prepared by single-step polymerization procedures, allowing access to large-scale production and thus widening their potential uses and applications.

Our group is particularly interested in the synthesis of hyperbranched conjugated organic and organometallic polymers [9–14]. We have developed different synthetic routes toward high-metal loaded materials by either polymerizing metal containing monomers or by postfunctionalizing the hyperbranched scaffolding with organometallic complexes [15–20]. This chapter review our results along with the relevant work of expert groups on this young but promising research field.

2 Research and Discussion

2.1 Theoretical Background

Hyperbranched polymers exhibit tree-like molecular structures and have been the focus of active research since the groundbreaking work of Kim and Webster [21,22]. Throughout the structure three main units can be identified (Chart 1): Dendritic or branching units (D), linear units (L), and terminal units (T).

Different synthetic strategies have been employed for the preparation of the pure organic hyperbranched polymers [23–25]. The most commonly adopted approach is self-condensation of AB_n-type monomers with $n \geq 2$ [26–29]. This type of polymerization can be carried out in a concurrent mode or by slow addition of the monomer or even in the presence of a core molecule of B_f ($f \geq 3$), which allows various structural control over the growing polymer [30–32]. Another approach is copolymerizations of A_2 monomers with B_n comonomers ($n \geq 3$) [33–35]. However, the stoichiometric requirements between the pairs of the functional comonomers

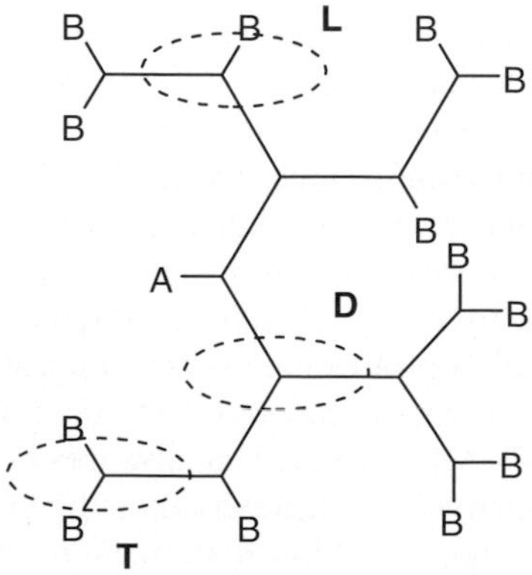

Chart 1

and the potential risk of gelation are severe drawbacks. As an alternative concept, Frechet has reported the synthesis of hyperbranched polymers by self-condensing vinyl polymerization (SCVP), which has recently been further applied to various other types of living/controlled polymerization, such as nitroxide-mediated radical polymerization, atom transfer radical polymerization, group transfer polymerization, and ring-opening polymerization [32,36–46].

Different from their dentrimer counterparts, hyperbranched polymers contain not only dendritic and terminal repeating units but also linear ones, which can be expressed in the degree of branching (DB). The DB is an important structural parameter of hyperbranched polymers and can be described in the following Eq. 2.1 [47].

$$\mathrm{DB} = (D + T) / (D + L + T) \tag{2.1}$$

where D is the number of dendritic units, T is the number of terminal units, and L is the number of linear units. Frey has suggested a modified definition of DB that is based on the direction of growth, as shown in Eq. 2.1 [48]

$$\mathrm{DB} = 2D / (2D + L) = (D + T - N) / (D + T + L - N) \tag{2.2}$$

where N is the number of molecules. For the hyperbranched polymers with high molecular weights N is negligible and Eqs. 2.1 and 2.2 are almost the same. Frey also pointed out that the DB statistically approaches 0.5 in the case of the polymerization of AB_2 monomers, which is much lower than their dendrimer counterparts whose DB is unity.

2.2 *Synthetic Pathways*

In order to directly synthesize organometallic hyperbranched polymers, all the above described established methods could be utilized to knit metal-containing monomers together, provided they are stable under the applied polymerization conditions and do not interfere with the reaction mechanism. As an alternative, suitable pure organic hyperbranched polymers can be functionalized with organometallic complexes. Both approaches have been utilized by other research groups and will be briefly reviewed at the beginning of Chapters 3 and 4.

2.2.1 Incorporation of Transition Metals through the Building Block

Although a wide range of methodologies exist for the preparation of hyperbranched polymers, examples of metal-containing materials—which are directly synthesized from organometallic monomers—are very limited. Reinhoudt and coworkers reported the preparation of hyperbranched polymers via self-assembly of an

AB_2-type monomer (**1**) composed of organopalladium complexes, sandwiched in between SCS pincer ligands and attached labile acetonitrile molecules (Scheme 2.1) [49]. Ligand exchange through solvent removal leads reversibly to spherical assemblies, as confirmed by atomic force microscopy (AFM) and transmission electron microscopy (TEM). The size of the spheres is controllable through manipulation of the substituents on the pincer ligand as well as by exchanging the counter anions [50,51]. Interestingly, linear analogs did not show any globular structures, confirming the necessity of the branching units.

Lewis et al. attempted hyperbranched organometallic polymers via $A_2 + B_3$ protocol by reacting $Pt(PBu_3)Cl_2$ with 1,3,5-triethynylbenzene in a molar ratio of 3:2. However, the resulting product was insoluble in common organic solvents and only the addition of excess amounts of *p*-1,4-diethynylbenzene (triyne:diyne = 1:50) could depress the involved cross-linking reactions [52]. In an alternative approach by Takahashi and coworkers, a formally similar hyperbranched polymer was constructed by self-polycondensation reaction from the AB_2-type analog of Lewis' monomers (Scheme 2.2) [53]. The resultant organometallic polymer was soluble in

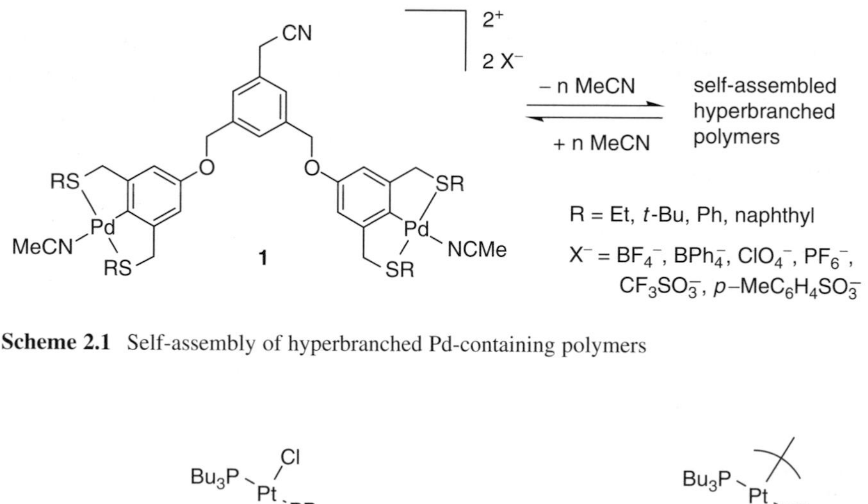

Scheme 2.1 Self-assembly of hyperbranched Pd-containing polymers

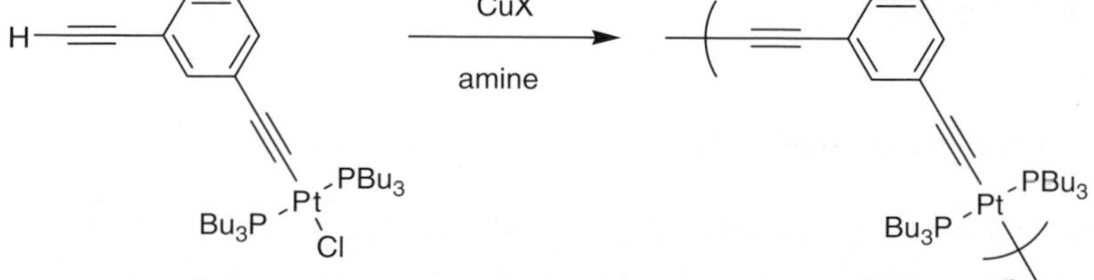

Scheme 2.2 Synthesis of hyperbranched platinum-containing polyyne from AB_2-type monomers

common organic solvents and could be characterized by means of spectral analyses and gel permeation chromatography (GPC).

Recently, Abd-El-Aziz and colleagues reported different hyperbranched poly(arylethers) (**4,5**) and poly(arylthioethers) (**6**) containing cyclopentadienyliron moieties, which were successfully furnished by nucleophilic substitution of $A_2 + B_3$ type monomers (Scheme 2.3) [54]. The polymers were thoroughly characterized by standard spectroscopic analysis techniques, they exhibited generally low viscosities, and the organometallic complexes were stable up to 230 °C, as evaluated by thermal gravimetric analysis (TGA).

Ferrocene is an attractive building block for the preparation of highly branched materials. Galloway and Rauchfuss reported the synthesis of high-molecular-weight poly (ferrocenylenepersulfides) by desulfurization-induced ring-opening polymerization (ROP) (Scheme 2.4) [55]. Whereas polymer **9** was insoluble, the attached bulky *t*-butyl group of **10** kept the polymer network soluble.

Similarly, our group has utilized ferrocene as a metal-containing building block and prepared hyperbranched poly(ferrocenylsilanes) through salt eliminative polycoupling of 1,1-dilithioferrocene with alkyltrichlorosilanes (Scheme 2.5) [56,57]. The solubility as well as the molecular weight increased with increasing spacer length from methyl to *n*-dodecyl-substituted polymers. Spectroscopic analyses revealed that the polymers possess rigid skeleton structures with extended conjugations, with their absorption spectra tailing into the infrared region (>700 nm). This

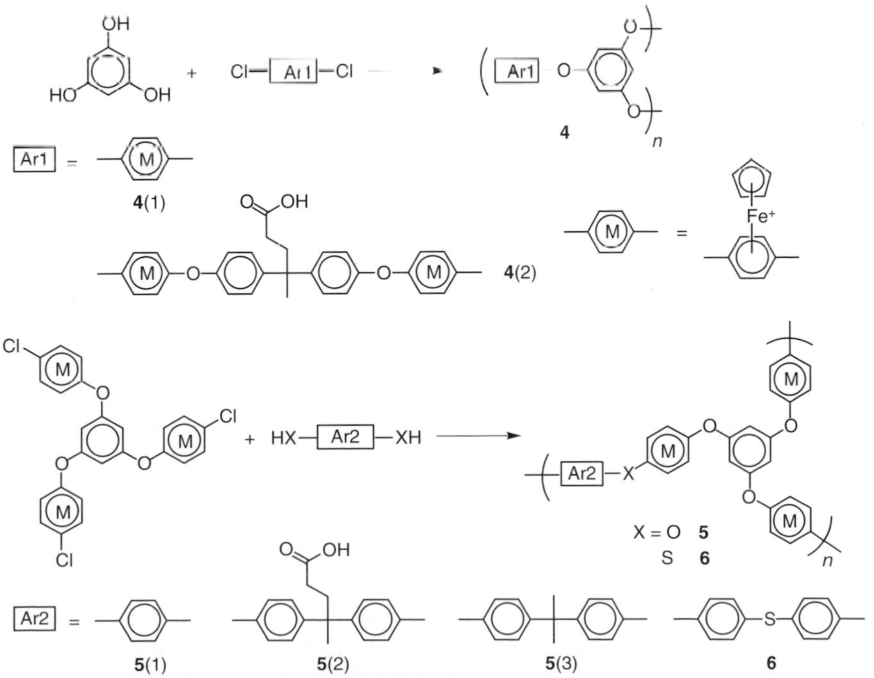

Scheme 2.3 Synthesis of cyclopentadienyliron-containing polymers via $A_2 + B_3$ method

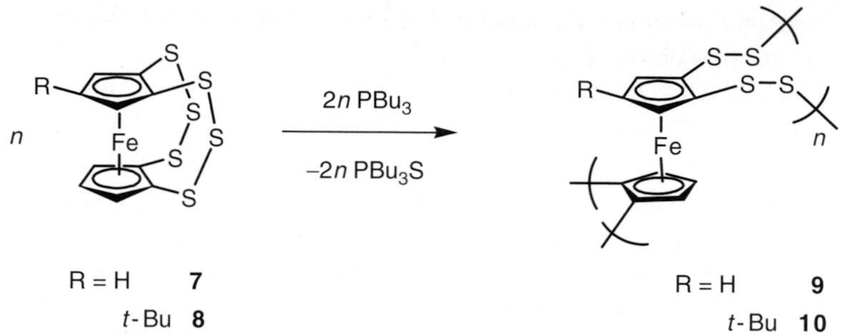

Scheme 2.4 Synthesis of ferrocene-containing polymers via ROP

E	R	polymer
Si	$-CH_3$	**11**(1)
Si	$-C_8H_{17}$	**11**(8)
Si	$-C_{12}H_{25}$	**11**(12)
Si	$-C_{16}H_{33}$	**11**(16)
Si	$-C_{18}H_{37}$	**11**(18)
Si	$-(CH{=}CH_2)$	**11**(Vi)
Si	$-C_6H_5$	**11**(Ph)
P		**11**(P)
P =O		**11**(PO)
Sb		**11**(Sb)

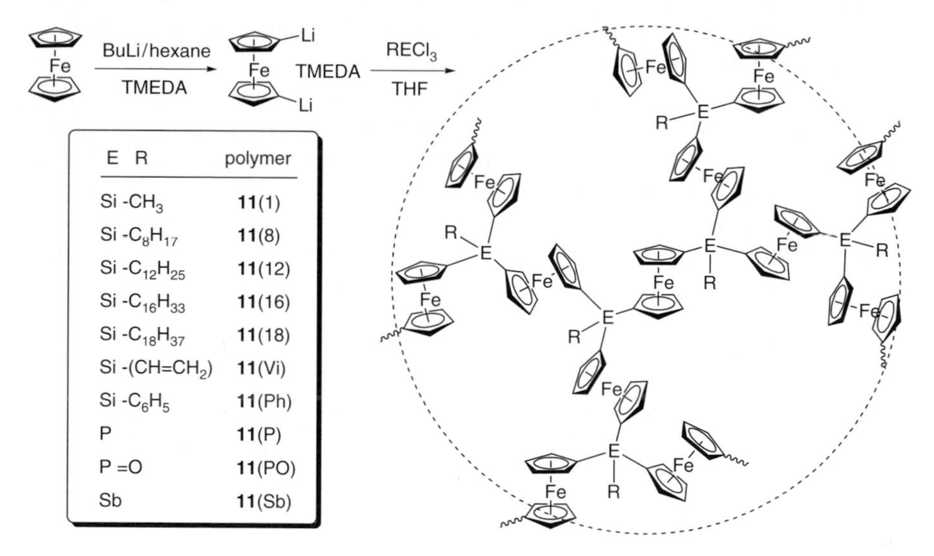

Scheme 2.5 Syntheses of hyperbranched poly(ferrocene)s by desalt polycoupling of dilithioferrocene with trichlorides of silicon, phosphorus and antimony

methodology was recently extended to other group 14 and 15 elements including germanium, phosphor, and antimony [58]. However, most of the polymers showed only limited solubility, making a detailed structural analysis difficult. Nevertheless, all the hyperbranched polymers served as excellent precursors for the preparation of metal-containing ceramics by heating under an inert gas atmosphere. Generally, the pyrolytic yields were found to be higher than their corresponding linear analogs. Whereas calcinations of the Si-containing polymers at 1,000 °C under nitrogen gave ceramics containing mostly α-Fe nanoparticles, those of Ge- and Sb-containing polymers were completely transformed into their iron-alloys. The ceramics from the P-containing polymers showed diffraction patterns of iron phosphides. Interestingly, iron silicide nanocrystals of larger sizes were obtained when the pyrolysis of the methyl-substituted hyperbranched poly(ferrocenylsilane) **11**(1) was conducted at a higher temperature of 1,200 °C under argon. This ceramic was highly magnetizable with magnetic saturations (M_s) up to 51 emu/g and showed near-zero remanence and coercivity.

Recently, our ongoing investigations led to the development of a new protocol for the synthesis of hyperbranched poly(aroylarylene)s by amine-catalyzed regioselective polycyclotrimerization of bis(aroylacetylene)s containing ferrocene moieties (Scheme 2.6) [59]. The incorporation of the ferrocene motif was achieved by either homopolycyclotrimerization of diyne **12** or by copolycyclotrimerization of **14** with monoyne **15** in yields up to ~70% (M_w 9 100, M_w/M_n = 2.8–3.1). Furthermore, the two synthetic methodologies, homo- and copolycyclotrimerization, allow a tailoring of the molecular structure with the ferrocene building blocks either well-distributed throughout the whole hyperbranched polymer or mainly located on the outside as terminal units. The polymers are equipped with numerous benzophenone and triaroylbenzene functionalities, which are known to readily cross-link upon exposure to ultraviolet (UV) light or other high-energy sources [60]. Figure 2.1a shows an example of an optical micrograph of **16** after UV irradiation. Here, the spin-coated hyperbranched polymer functions as negative photoresist and the unexposed parts were completely removed by the organic solvent, leaving behind well-resolved patterns with sharp edges. The resolution might reach submicron to nanometer scale as already demonstrated by the nonmetallic counterparts [59].

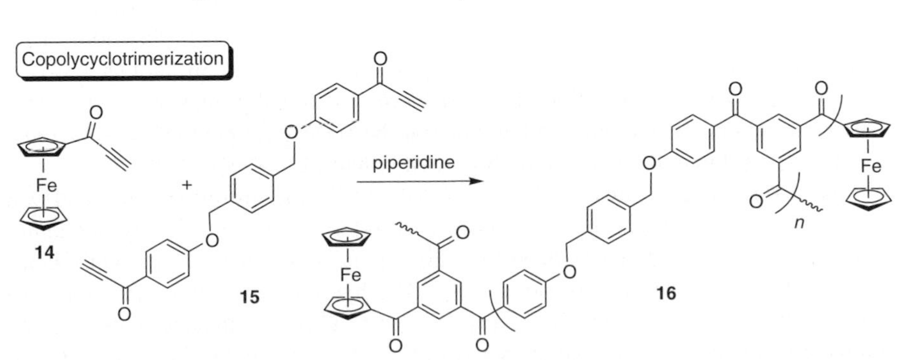

Homopolycyclotrimerization

piperidine

12

13

Copolycyclotrimerization

piperidine

14

15

16

Scheme 2.6 1,3,5-Regioselective homo- and copolycyclotrimerization of ferrocene-containing aroylacetylenes

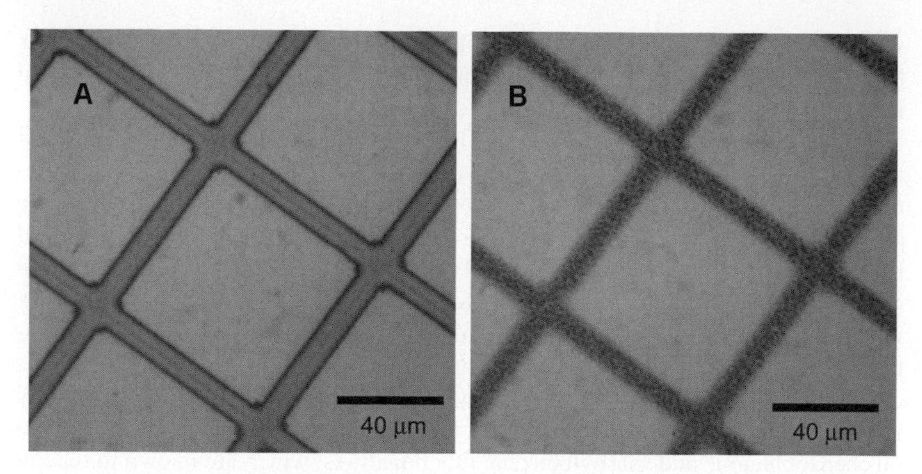

Figure 2.1 Optical micrograph of (**a**) **16** fabricated by UV photo-lithography using a Cu-negative mask and (**b**) ceramic pattern of **16** pyrolyzed under nitrogen at 1,000 °C for 1 h

Ceramization of the silicon wafers of **16** in a tube furnace at 1,000 °C for 1 hour under a steam of nitrogen gave a ceramic pattern with excellent shape retention with respect to their polymer precursors (Figure 2.1b). Close inspection of the ceramic pattern under higher magnification revealed a morphology transformation from a uniform thin-film into congeries of tiny ceramic clusters. The composition of these ceramic patterns consists of Fe and Fe_2O_3 nanoparticles embedded in a carbon matrix.

2.2.2 Incorporation of Transition Metals Through Postfunctionalization

Besides the direct synthesis from their metal-containing monomer building blocks, organometallic polymers are also accessible through postfunctionalization reactions, where suitable chelating groups inside the molecular architecture serve as macroligands for metal complexes and nanoparticles. With this concept in mind, the hyperbranched scaffoldings might serve as homogeneous nanoreactors for the incorporation of catalytically active metal species [61]. Such macrocatalysts might furthermore easily be recovered by precipitation or filtration methods and reused in another reaction cycle. Depending on the position of the chelating functional groups, the metal complexes can be introduced either in the core, on the surface (along the many terminal units), or throughout the whole hyperbranched structure.

Pioneering works in this field were performed by the groups of Frey and van Koten, who functionalized hyperbranched carbosilane polymers (**17**) by palladium complexes (Scheme 2.7) and utilized **18** as homogeneous organometallic catalysts for a standard aldol condensation reaction [62]. The hyperbranched polymer-supported metal catalysts showed reactivities very similar to those of analogous dentrimers, implying that structural perfection is not always required. Recently, this nanocapsules-concept was extended to amiphiphilic hyperbranched polyglycerols (**19**), which selectively immobilized pincer-platinum(II) complexes (**20**) and palladium(II) salts (**21**) into the hydrophilic core [63,64]. The catalytic activity of the Pt-containing

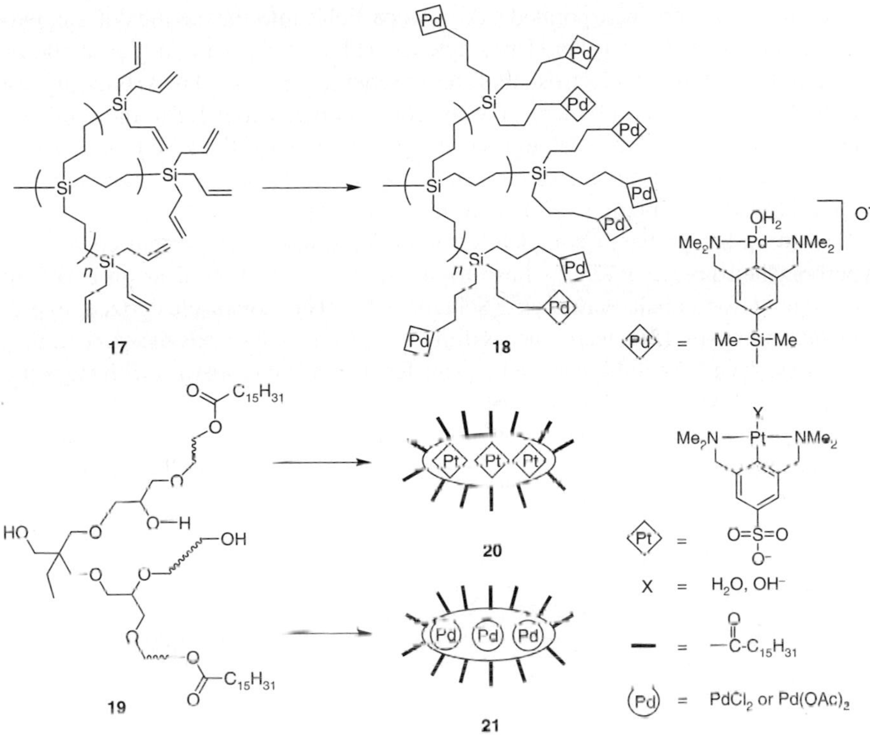

Scheme 2.7 Different motifs of metal encapsulation

nanocapsules was tested in a double Michael addition reaction, which was lower compared with the respective unsupported catalyst possibly because of decreased accessibility of the catalyst in the interior of the nanocapsule. Furthermore, the optically active hyperbranched analog produced no enantiomeric excess from the asymmetric model addition between methyl vinyl ketone and α-cyanopropionate, revealing that the chiral nanocapsule backbone had no influence on the resulting product [65]. The Pd salts of **21** could be reduced to metallic nanoparticles, which could be stabilized by the hyperbranched scaffold. The resulting Pd-colloids were probed as homogeneous catalysts for the hydrogenation of cyclohexene and were found to exhibit higher activity than commercial available Pd/activated charcoal catalysts.

A hyperbranched polymer, structurally similar to **19**, but terminated with 1,2-dimethylimidazolium instead of palmitoyl end groups, were functionalized with monosulfonated triphenylphosphine via counterion exchange. Immobilization of $[Rh(acac)(CO)_2]$ onto the hyperbranched surface successfully furnished polymer-bound complexes, which showed moderate activity in the hydroformylation of 1-hexene in methanol [66].

Other groups reported in a similar way the formation of metal-nanoparticles such as Ag, Au, Cu, Pt, and Pd stabilized through different kinds of hyperbranched polymers including poly(ethyleneimides) [67,68], poly(amidoamines) (structurally similar to PAMAM dentrimers) [69], poly(amine-esters) [70], and aromatic poly(amides) [71–73] and evaluated their activity towards various chemical reactions.

Bai and coworkers incorporated CdS nanoparticles into the cavities of a hyperbranched conjugated poly(phenylenevinylenes) (PPV) comprising different alkoxy side chains (**22**) through electrostatic forces (Scheme 2.8) [74]. The results suggest that the hyperbranched polymers showed efficient energy-transfer to the semiconductor nanoparticles, while at the same time decreasing their self-aggregation problem. Enhanced dispersion of the nanoparticles was achieved through the introduction of water-solublizing sulfonic groups (**23**) [75].

Our group has prepared completely soluble high molecular weight conjugated hyperbranched polyynes (**24**), whose triple bond scaffold served as an excellent macroligand for cobalt carbonyls (Scheme 2.9). The cobalt-clusterized hyperbranched polyynes (**25**) were successfully transformed into advanced ceramics, which were found to exhibit outstanding soft-ferromagnetic properties with magnetic saturations as high as ~118 emu/g [76].

Unfortunately, the high metal-loaded polyyne becomes partly soluble or even insoluble as a result of the formation of supramolecular aggregates once precipitated in poor

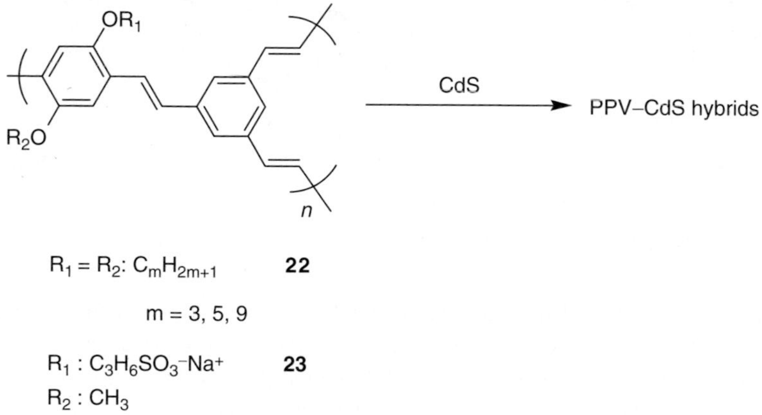

OR_1 R_2O CdS → PPV–CdS hybrids

$R_1 = R_2$: C_mH_{2m+1} **22**

m = 3, 5, 9

R_1 : $C_3H_6SO_3^-Na^+$ **23**

R_2 : CH_3

Scheme 2.8 Preparation of conjugated hyperbranched polymer/CdS nanoparticle hybrids

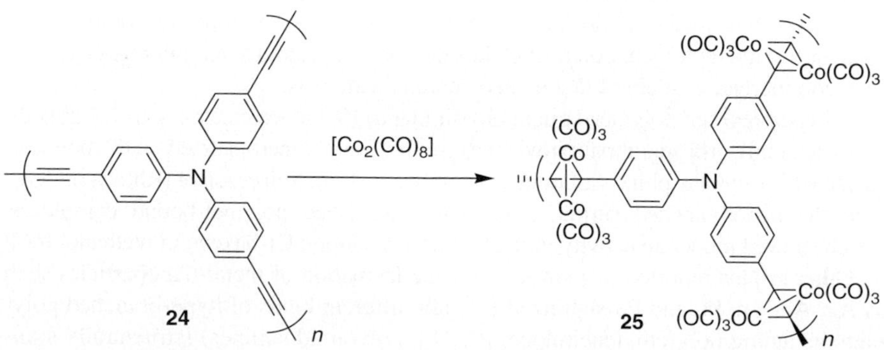

24 $[Co_2(CO)_8]$ → **25**

$(OC)_3Co$ $Co(CO)_3$ $(CO)_3$ Co N Co $(CO)_3$ $Co(CO)_3$ $(OC)_3OC$

Scheme 2.9 Synthesis of cobalt-containing hyperbranched polyynes

solvent. Delightfully, thin films of good quality (thickness: ~1000 nm, as determined by ellipsometry) could still be obtained by spin-coating the freshly prepared solutions of the organometallic polymers onto silicon wafers. Photopatterning of the films through a Cu-negative mask resulted in a photobleaching and features in the size range of ~10 to 100 μm could easily obtained (Figure 2.2).

Wavelength-dependent refractive index measurements of the unexposed and exposed thin films of **25** revealed interesting results. Whereas the metallified polymer, similar to its nonmetallated parent (**24**), showed very high refractive indexes (n = 1.813–1.714) in the spectral region of 600 to 1600 nm, the refractive index dropped after UV irradiation significantly (n = 1.777–1.667) possibly resulting from the decomposition of the organometallic moieties (Figure 2.3). Materials with such a high refractive index change are promising for photonic applications: For example, it may function as photorefractive material in holographic devices [77] or work as high refractive index optical coatings [78].

Cobalt complexes among other metals such as iron and nickel are known to catalyze the growth of carbon nanotube (CNT) by chemical vapor deposition (CVD) [79-81]. Thanks to the thermal stability of the hyperbranched polyyne backbone, spin-coated films of the organometallic polymer were successfully probed to function as catalyst and arrays of CNT bundles could be prepared (Figure 2.4). This preliminary result already suggests potential application in the field of patternable tailor-made catalysts.

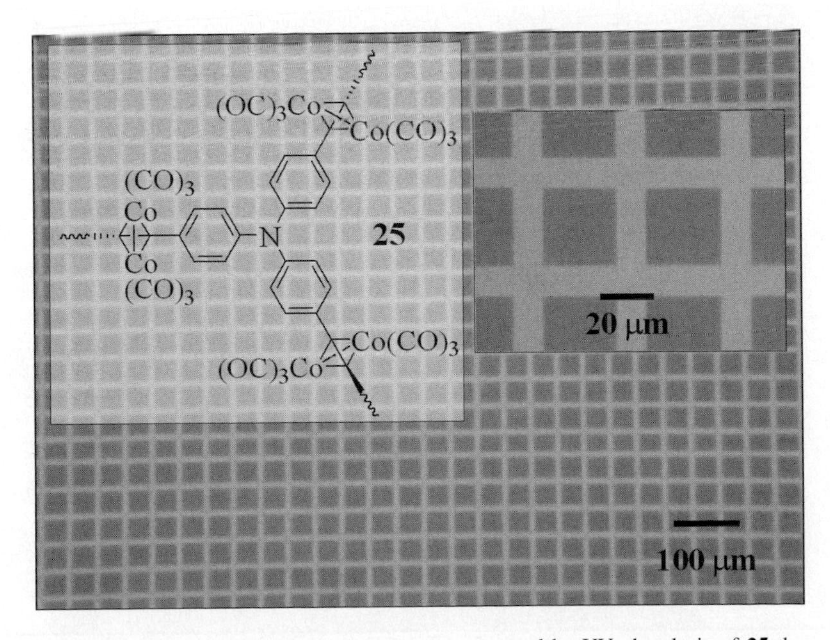

Figure 2.2 Optical micrograph of the photopattern generated by UV photolysis of **25** through a Cu-negative mask. Insets: molecular structure of **25** and enlarged fraction of the optical micrograph

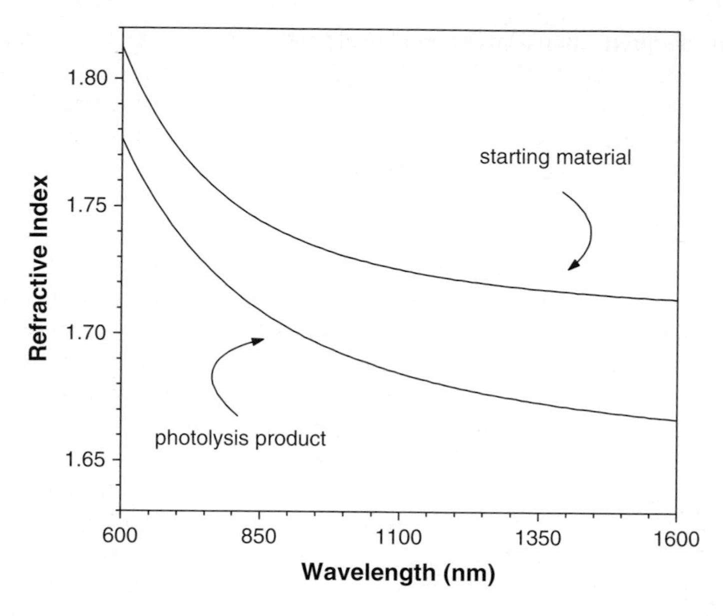

Figure 2.3 Wavelength dependences of the refractive indexes of a thin film of **25** and its photolysis product after UV irradiation for 30 minutes

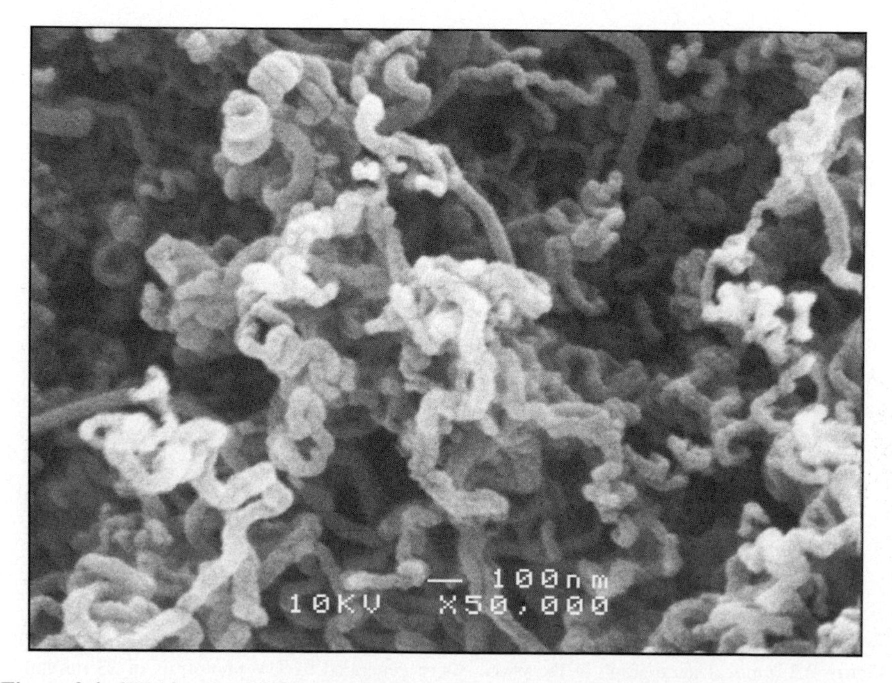

Figure 2.4 SEM image of CNT prepared by CVD from spin-coated thin films of **25**

3 Conclusion and Future Considerations

In this contribution, we have summarized the latest developments in the field of hyperbranched organometallic polymers. Two main synthetic pathways, namely synthesis of hyperbranched polymers from metal containing building blocks and postfunctionalization of hyperbranched architectures, have been reviewed, showing interesting inorganic–organic hybrid materials. Existing results already suggest that structural perfection, as it can be found from dentrimers, is not always required. In particular, future applications in the field of homogeneous reusable catalysis, various kinds of adhesives and coatings, photoresists, as well as processable precursor for advanced materials are already on the horizon, which will advance this young but promising research area.

References

1. Abd-El-Aziz AS. (2002) Macromol Rapid Commun. 23:995.
2. Neuse EW, Rosenburg H. (1970) Metallocene polymers. Marcel Dekker, New York.
3. Manners I. (1996) Angew Chem Int Ed Engl. 35:1602.
4. Nguyen P, Gomez-Elipe P, Manners I (1999) Chem Rev. 99:1515.
5. Newkome GR, He E, Moorefield CN. (1999) Chem Rev. 99:1689.
6. Cuadrado I, Morán M, Casado CM, Alonso B, Losada. (1999) J Coordi Chem Rev. 395:193.
7. Froehlich PJ. (2004) Polym Sci Part A: Polym Chem. 42:3110.
8. Manners I. (2004) Synthetic metal-containing polymers. Wiley, Weinheim, p. 237.
9. Häußler M, Lam JWY, Zheng R, et al. (2003) C R Chimie 6:833.
10. Dong H, Lam JWY, Häußler M, et al. (2004). Curr Trends Polym Sci. 9:15.
11. Lam JWY, Peng H, Häußler M, Zheng R, Tang BZ. (2004) Mol Cryst Liq Cryst. 415:43.
12. Peng H, Lam JWY, Tang BZ. (2005) Macromol Rapid Comm 26:673.
13. Zheng R, Dong H, Peng H, Lam JWY, Tang BZ. (2004) Macromolecules 37:5196.
14. Chen J, Peng H, Law CCW, Dong YP, Lam JWY, Williams ID, Tang BZ. (2003) Macromolecules 36:4319.
15. Xu K, Peng H, Sun Q, et al. (2002) Macromolecules 35:5821.
16. Peng H, Cheng L, Luo J, et al. (2002) Macromolecules 35:5349.
17. Zheng R, Dong H, Tang BZ. (2005) In: Abd-El-Aziz A, Carraher C, Pittman C, Sheats J, Zeldin M (eds) Macromolecules containing metal- and metal-like elements, Vol. 4. Wiley, New York, p 7.
18. Häußler M, Lam JWY, Dong H, Tong H, Tang BZ. (2006) In: Newkome GR, Manners I, Schubert US (eds) Metal-containing and metallo-supramolecular polymers and materials, ACS Symposium Series, American Chemical Society, Washington, DC, p. 244.
19. Sun Q, Peng H, Xu K, Tang BZ. (2003) In: Abd-El-Aziz A, Carraher C, Pittman C, Sheats J, Zeldin M (eds) Macromolecules containing metal- and metal-like elements, Vol. 2. Wiley, New York, p 29.
20. Häußler M, Dong H, Lam JWY, Zheng R, Qin A, Tang BZ. (2005) Chin J Polym Sci. 23:567.
21. Kim YH, Webster, OW. (1988) Polym Prep. 29:310.
22. Kim YH, Webster OW. (1992) Macromolecules 25:5561.
23. Voit BJ. (2000) Polym Sci Polym Chem. 38:2505.
24. Inoue K. (2000) Prog Polym Sci. 25:453.
25. Hawker CJ. (1999) Curr Opin Coll Interf Sci. 4:117.
26. Gao C, Yan D. (2001) Macromolecules 34:156.

27. Sendijarevic I, McHugh AJ, Markoski LJ, Moore JS. (2001) Macromolecules 34:8811.
28. Wang F, Wilson MS, Rauh RD, Schottland P, Reynolds JR. (1999) Macromolecules 33:4272.
29. Hölter D, Frey H. (1997) Acta Polym. 48:298.
30. Bernal DP, Bedrossian L, Collins K, Fossum E. (2003) Macromolecules 36:333.
31. Bharathi P, Moore JS. (2000) Macromolecules 33:3212.
32. Sunder A, Hanselmann R, Frey H, Mulhaupt R. (1999) Macromolecules 32:4240.
33. Chang Y-T, Shu C-F. (2003) Macromolecules 36:661.
34. Jikei M, Chon S-H, Kakimoto M, Kawauchi S, Imase T, Watanebe J. (1999) Macromolecules 32:2061.
35. Emrick T, Chang H-T, Frechet JMJ. (1999) Macromolecules 32:6380.
36. Frechet JMJ, Henmi M, Gitsov I, Aoshima S, Leduc MR, Grubbs RB. (1995) Science 269:1080.
37. Muller AHE, Yan D, Wulkow M. (1997) Macromolecules 30:7015.
38. Yan D, Muller AHE, Matyjaszewski K. (1997) Macromolecules 30:7024.
39. Radke W, Litvinenko GI, Muller AHE. (1998) Macromolecules 31:239.
40. Yan D, Zhou Z, Muller AHE. (1999) Macromolecules 32:245.
41. Litvinenko GI, Muller AHE. (2002) Macromolecules 35:4577.
42. Cheng K-C, Chuang T-H, Chang J-S, Guo W, Su W-F. (2005) Macromolecules 38:8257.
43. Simon PFW, Radke W, Muller AHE. (!997) Macromol. Rapid Commun. 18:865.
44. Simon PFW, Muller AHE. (2004) Macromolecules 37:7548.
45. Sakamoto K, Aimiya T, Kira M. (1997) Chem. Lett. 1245.
46. Gao C, Yan D. (2004) Prog Polym Sci. 29:183.
47. Hawker CJ, Lee R, Frechét JMJ. (1991) J Am Chem Soc. 113:4583.
48. Hölter D, Burgath A, Frey H. (1997) Acta. Polym. 48:30.
49. Huck WTS, vanVeggel FCJM, Kropman, BL, et al. (1995) J Am Chem Soc 117:8293.
50. Huck WTS, vanVeggel FCJM, Reinhoudt DNJ. (1997) Mater Chem. 7:1213.
51. Huck WTS, SnellinkRuel BHM, Lichtenbelt JWT, vanVeggel FCJM, Reinhoudt DN. (1997) Chem Commun. 9.
52. Khan MS, Schwartz DJ, Pasha NA, Kakkar AK, Lin B, Raithby PR, Lewis JZ. (1992) Anorg Allg. Chem. 616:121.
53. Onitsuka K, Ohshiro N, Fujimoto M, Takei F, Takahashi S. (2000) Mol Cryst Liq Cryst 342:159.
54. Abd-El-Aziz AS, Carruthers SA, Aguiar PM, Kroeker SJ. (2005) Inorg Organomet Polym Mater. 15:349.
55. Galloway P, Rauchfuss TB. (!993) Angew Chem Int Ed Engl 32:1319.
56. Sun QH, Lam JWY, Xu KT, et al. (2000) Chem Mater. 12:2617.
57. Sun Q, Xu K, Peng H, Zheng R, Häussler M, Tang BZ. (2003) Macromolecules 36:2309.
58. Haussler M, Sun QH, Xu KT, Lam JWY, Doug HC, Tang BZ. (2005) J Inorg Organomet Polym Mater. 15:67.
59. Dong H, Zheng R, Lam JWY, Häussler M. Tang BZ. (2005) Macromolecules 38:6382.
60. Hasegawa, M, Horie K. (2001) Prog Polym Sci. 26:529.
61. Vriezema DM, Aragonès MC, Elemans JAAW, Cornelissen JJLM, Rowan AE, Nolte RJM. (2005) Chem Rev. 105:1445.
62. Schlenk C, Kleij AW, Frey H, van Koten G. (2000) Angew Chem Int Ed. 19:3445.
63. Slagt MQ, Stiriba S-E, Klein Gebbink RJM, Kautz H, Frey H, van Koten G. (2002) Macromolecules 35:5734.
64. Mecking S, Thomann R, Frey H, Sunder A. (2000) Macromolecules 33:3958.
65. Slagt MQ, Stiriba SE, Kautz H, Gebbink RJMK, Frey H, van Koten G. (2004) Organometallics 23:1525.
66. Schwab E, Mecking S. (2005) Organometallics 24:3758.
67. Aymonier C, Schlotterbeck U, Antonietti L, Zacharias P, Thomann R, Tiller JC, Mecking S. (2002) Chemical Communications. 3018.
68. Krämer M, Pérignon N, Haag R, et al. (2005) Macromolecules. 38:8308.

69. Pérignon N, Mingotaud A-F, Marty J-D, Rico-Lattes I, Mingotaud C. (2004) Chem Mater. 16: 4856.
70. Wei X, Zhu B, Xu Y. (2005) Colloid Polym Sci. 284:102.
71. Tabuani D, Monticelli O, Chincarini A, Bianchini C, Vizza F, Moneti S, Russo S. (2003) Macromolecules. 36:4294.
72. Tabuani D, Monticelli O, Komber H, Russo S. (2003) Macromol Chem Phys. 204:1576.
73. Tabuani D, Monticelli O, Chincarini A, Chini CB, Vizza F, Moneti S, Russo S. (2004) Macromolecules. 37:680.
74. Yang J, Lin H, He Q, Ling L, Zhu C, Bai F. (2001) Langmuir. 17:5978.
75. Peng QG, Yang JL, He QG, Bai FL. (2003) Synth Met. 135:163.
76. Häussler M, Zheng R, Lam JWY, Tong H, Dong H, Tang BZ. (2004) J Phys Chem B. 108:10645.
77. Hendrickx E, Engels C, Schaerlaekens M, Van Steenwinckel D, Samyn C, Persoons A. (2002) J Phys Chem B. 106:4588.
78. Lu C, Guan C, Liu Y, Cheng Y, Yang B. (2005) Chem Mater. 17:2448.
79. Chatterjee AK, Sharon M, Banerjee R, Neumann-Spallart M. (2003) Electrochimica Acta. 48:3439.
80. Huang ZP, Wang DZ, Wen JG, Sennett M, Gibson H, Ren ZF. Appl Phys A. 74:387.
81. Deck CP, Vecchio K. Carbon. 44:267.

Transition Metal σ-Acetylide Polymers Containing Main Group Elements in the Main Chain: Synthesis, Light Emission and Optoelectronic Applications

Wai-Yeung Wong

1 Introduction

Incorporation of transition metal elements into macromolecular organic structures allows the combination of the interesting physical characteristics of metals such as electronic, optical, and magnetic properties with the solubility and processability of traditional carbon-based polymers [1–5]. Transition metal centers which can possess a large variety of structural geometries, ligand environments, and oxidation states may offer novel physical, optoelectronic, and structural properties on these purely organic polymeric systems. Within the framework of synthetic metal-containing polymers, polymers with metal–carbon σ-bonds in the main chain represent one of the most common and important categories of these materials [1–4].⁻ Rigid-rod transition metal acetylide polymers or polymetallaynes currently constitute one of the most extensively studied areas of these metallopolymers in the literature [6].

The development of synthetic methodologies towards transition metal-acetylide oligomers and polymers has shown tremendous progress following the initial reports by Hagihara et al. in the 1970s on the synthesis of polymeric Pt and Pd acetylides [7–10], and the interest has been largely stimulated from their wide range of applications in molecular electronics and materials science (Figure 3.1) [1,2]. In the early 1990s, there were also a number of reports on the synthesis of rigid-rod metal-containing polyynes of groups 8 and 10 by Lewis and coworkers based on bis(trimethylstannylacetylide) precursors [11–14]. The linear geometry of the alkynyl unit and its unsaturated character

A.S. Abd-El-Aziz et al. (eds.), *Inorganic and Organometallic Macromolecules: Design and Applications.*

have made metal alkynyls versatile structural motifs for molecular wires and organometallic oligomeric and polymeric materials which can exhibit unique properties, such as electrical conductivities, rich luminescence properties, nonlinear optical properties, liquid crystallinity, and photovoltaic behavior [6]. The general structure for the polymer is shown in Figure 3.2, in which the polymer has a linear backbone comprised entirely of the metal center M, the spacer group R and the auxiliary ligand L. The prototypical polymer for much of this work is *trans*-$[-Pt(PBu_3)_2C\equiv C(p\text{-}C_6H_4)C\equiv C-]_n$ (**I**) and more recently, a large series of derivatives incorporating various conjugated carbocyclic and heteroaromatic ring systems have been prepared [1,6]. All these materials are generally soluble in organic solvents and the solubility and polymer length can be modified by an appropriate selection of M, R, or L units. The advance in chemical synthesis has led to an extremely versatile route for the production of a vast range of application-tailored conjugated polymers of this kind that can display diverse optical features and photophysics.

Figure 3.3 illustrates the typical energy level diagram for a simple photoluminescence (PL) system in a molecular assembly. Upon absorption of photons by the molecule, there are two possible radiative decay pathways, namely, fluorescence

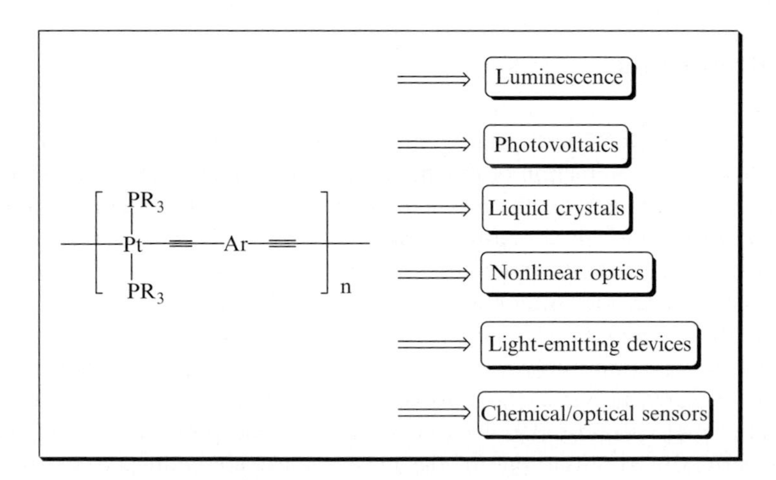

Figure 3.1 Different areas of applications for polyplatinaynes

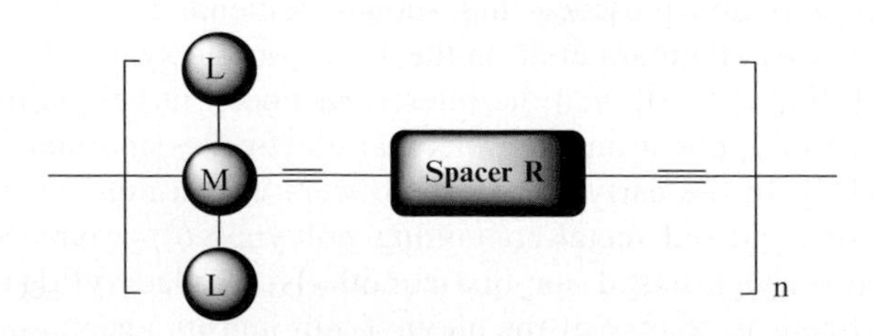

Figure 3.2 General skeleton of transition metal polyyne polymers

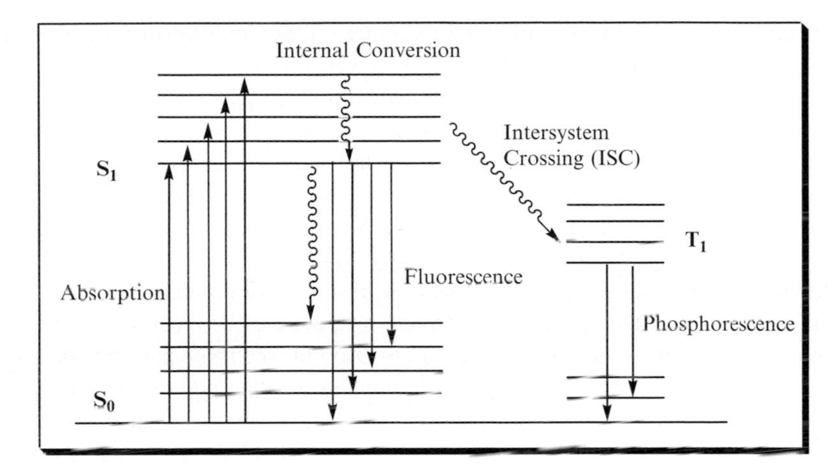

Figure 3.3 The Jablonski diagram for a simple PL system

($S_1 \rightarrow S_0$) and phosphorescence ($T_1 \rightarrow S_0$) (Figure 3.3). The relative positions of the singlet and triplet excited states strongly affect the intersystem crossing (ISC) rate into the triplet manifold. For organic systems, this provides a major nonradiative decay mechanism and reduces the luminescence efficiency. Intensive studies of conjugated polymers over the past few decades have already clarified some fundamental issues about the nature of the singlet excited states [15–17], yet still little is known about the nature of triplet excited states in conjugated polymers. Much of the recent work has shown that triplet states play an important role in optical and electrical processes within conjugated polymers with direct implications for their technological applications [18–26]. It has been demonstrated that the ultimate efficiency of light-emitting diodes (LEDs) is controlled by the fraction of triplet states generated or harvested [27–31]. Therefore, a thorough understanding of triplet photophysics is essential if a full picture of basic excitations in conjugated polymers is to be understood. With this in mind, there has been increasing experimental and theoretical attention focusing on the energy levels of singlet and triplet states in conjugated polymers with the aim of designing highly efficient luminescent materials. Conjugated ethynylated materials, such as arylacetylenes [32–34], poly (aryleneethynylenes)s [35], and metal-containing acetylide complexes and polymers [5,6], have been extensively studied in this regard.

Although it is important to control the relative energy levels of the singlet and triplet excitons for the chemical tuning of the singlet–triplet gap, a vast body of transition metal-containing polyynes has been extensively employed to study the triplet excited states in conjugated polymers directly in which the presence of heavy metal centers can increase the spin-orbit coupling, thus enabling the spin-forbidden triplet emission (phosphorescence) partially allowed [36–51]. In accordance with the energy gap law, the ISC rate depends exponentially on the energy gap to the nearest triplet excited state [53]. In fact, the triplet emission is extremely efficient, approaching 100% efficiency at low temperatures. This would render them promising materials for high-efficiency LEDs [20,53,54].

The synthetic modification of the organometallic acetylide polymers to produce materials of variable bandgaps can be categorized according to the types of metal groups, central spacer units, and ligands on the metals. To our knowledge, little research effort has been devoted to the study of transition metal polyyne polymers functionalized with main group elements in the main chain. With an appropriate design of the main group elements, metals, and organic components in these systems, this research should not only represent a solid contribution to our fundamental knowledge of this class of materials, but can also lead to the development of new advanced materials that could be used in electronic and optoelectronic industries. All the metal polyynes and their molecular diynes and oligoynes reported in this chapter were fully characterized by satisfactory elemental analyses such as fast-atom bombardment (FAB), infrared (IR), and nuclear magnetic resonance (NMR) spectroscopies. Generally, the metal diynes can be considered as building blocks for the high-molecular-weight polymers and valuable insights for the polymer properties can be obtained through the studies of these model compounds. Purification of compounds was achieved by preparative thin layer chromatography (TLC) or column chromatography on silica using appropriate solvents. The weight-average (M_w) and number-average (M_n) molecular weights of all the polymers in our studies were determined by gel permeation chromatography (GPC) in tetrahydrofuran (THF) against polystyrene calibration (Table 3.1). Whereas GPC does not give absolute values of molecular weights but provides a measure of hydrodynamic volume, the hydrodynamic properties of rigid-rod type polymers in solution are quite different from those for flexible polymers. Because of this, calibration of the GPC with polystyrene standards is likely to inflate the molecular weight values of the metal polyynes. The lack of discernible NMR resonances attributable to the end groups supports a high degree of polymerization (DP) for these polyynes in most cases. The thermal stability data of the polymers studied by thermogravimetric analyses (TGA) and their optical energy gaps (E_g) as estimated from the onset wavelength of the solid-state absorptions are also summarized in Table 3.1.

2 Research and Discussion

2.1 Metallopolyynes Containing Group 14 Main Group Elements

2.1.1 Background

With the realization of polyacetylenes and other conjugated heteronuclear polymers as potentially important electronic and optically active materials, research on electron delocalized organic and metalloorganic polymers has received much

Table 3.1 GPC and thermal stability data for polyplatinaynes containing main group elements

Pt polyyne	M_w	M_n	DP	T_{decomp} (onset) (°C)	E_g (eV)	Ref.
1	78,700	52,000	63	381	3.70	85
4	146,000	82,700	80	366	3.10	90
7	42,970	13,580	11	410	3.05	91
8	22,950	11,150	12	414	3.04	91
9	22,790	16,150	10	407	3.04	91
10	23,970	10,400	15	404	3.00	91
11	28,350	12,830	13	407	3.00	91
12	22,600	17,550	6	418	3.00	91
13	37,390	12,750	46[a]	363	3.40	96
15	11,430	7,020	14[a]	361	3.26	115
16	11,100	6,550	13[a]	354	3.28	115
17	15,480	7,930	18[a]	335	3.18	115
18	17,530	9,930	18[a]	348	2.10	115
27	219,940	[a]	301[a]	[b]	2.80	120
28	181,900	56,180	69	275	2.55	120
29	82,860	64,560	72	281	2.40	120

[a] Calculated from M_w

[b] Not reported

attention [1,17,26]. The cross-fertilization between organic and main group element chemistry is growing rapidly in the scientific community [55,56]. Among these, organosilicon polymers are becoming important in many aspects of device technology and play an increasingly vital role in the electronics industry today [56–58]. Moreover, the use of chemical approaches to improve the processing properties and performance of advanced ceramic materials is another rapidly growing area of research and development. In this context, it was found that polysilanes, polycarbosilanes, and poly(silylacetylene)s could be pyrolyzed to silicon carbides (SiC) and such a discovery has been commercialized for the production of SiC fibers [59]. More recently, the chemistry of main group element–alkyne molecular systems has drawn increasing attention. In particular, the molecular design and synthesis of linear or dendritic systems composed of silicon atoms and acetylene units have attracted enormous research interest within the field of materials science [55,60–64]. For instance, all polymers **II–V** (Figure 3.4) exhibit semiconductivity in the range of 10^{-6} to 10^{-3} S cm^{-1} after doping [55], while **II–IV** can show second (χ^2)[65] and third harmonic (χ^3) nonlinear optical properties [66]. These substances possess good potential as functional materials such as semiconductors [67], hole-transporting [68,69] and heat resistant materials [70,71], photoresists [72,73], and ceramic precursors [59]. The role of the cross-linking process in the ceramization of **V** has been demonstrated [55,74]. π-Conjugated organosilicon systems incorporating phenylene, anthrylene, or oligothienylene bridges have been extensively studied regarding their applications to electro- and optical devices [75–80]. As shown in Table 3.2, the lower

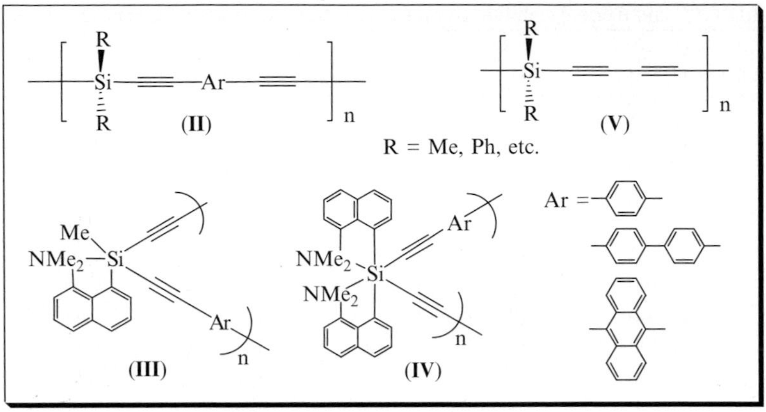

Figure 3.4 Some examples of poly(aryleneethynylenesilylene)s

Table 3.2 The first ionization energies (I.E.) of some group 14 elements

Group 14 element	First I.E. (kJ mol^{-1})
C	1,090
Si	791
Ge	762
Sn	704

ionization energy (I.E.) of the silicon atom and its higher congener compared with carbon is expected to enhance the through-bond interaction of ethynyl units along the backbone [81].

To set the future direction of the rapidly advancing field of novel conjugated materials, a program was initiated in our laboratory to investigate the capacity of main group elements to alter the chemical and physical properties of π-conjugated metal alkynyl oligomeric and polymeric materials. Based on the energy gap law, the radiative rate of triplet emission can be optimized by raising the energy level of the triplet excited state of the material [51,52]. Although the sp^3-silicon unit is generally not as good as an sp- or sp^2-carbon structural motif in facilitating electronic conjugation, recent studies have demonstrated that deliberate inclusion of conjugation-interrupting units shows good potential in improving the resulting optical and photophysical properties of these conjugated compounds [82–84]. We therefore embarked on a project aimed at preparing and characterizing new molecular and polymeric materials through incorporation of main group 14 elements and alkyne functionalities with transition metals.

2.1.2 Synthesis and Chemical Characterization

Although considerable effort has been devoted to the use of (hetero)aromatic or carbon-rich spacer units, much less is known about the silicon-linked system and little research has been performed in the area of metal polyyne chemistry. Along these lines, the first synthesis, characterization, and luminescence behavior of a series of novel platinum-containing oligo- and poly(silylacetylenes) (**1–3**) was described based on the use of dichlorodiethynylsilane (**L_1**) [85]. The relevant synthetic pathways are outlined in Scheme 3.1. Complexes **2** and **3** provide the best finite models for the polymeric analog **1**. These new compounds were prepared using the CuI-catalyzed dehydrohalogenating route at room temperature (r.t.) and isolated as off-white to light yellow solids. Complex **2** was structurally characterized by X-ray crystallography (Figure 3.5) and was the first example of a silylacetylene-linked diplatinum complex in the literature. Two discrete mononuclear group 8 metal complexes of ruthenium and osmium(II) containing the ligand **L_1** were also known, which proved useful as synthons to afford other homo- and heterometallic alkynyl compounds [86].

All these Si-bridged platinum complexes were air-stable and had good solubility in common organic solvents. The number of repeat units per chain calculated from M_n was lower for **1** than that for *trans*-$[-Pt(PBu_3)_2C\equiv CC\equiv CC\equiv C-]_n$ [87,88]. The single resonance in the ^{31}P-$\{^1H\}$ NMR spectra flanked by platinum satellites for **1** and **2** confirmed the *trans* arrangement of the phosphine ligands in a square-planar

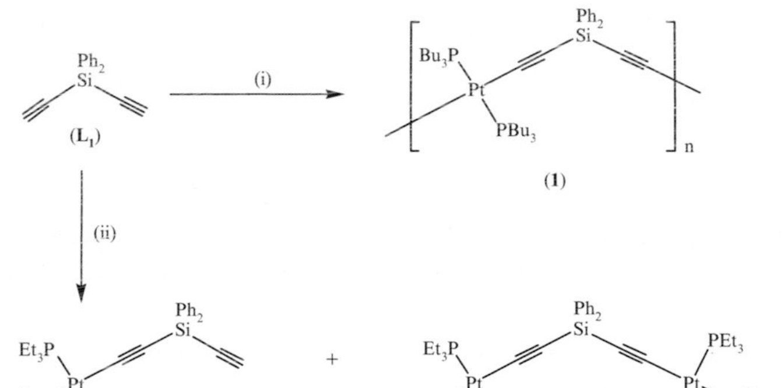

Reagents and conditions: (i) *trans*-$[PtCl_2(PBu_3)_2]$ (1equiv.), CuI, $^iPr_2NH$–CH_2Cl_2, r.t.; (ii) *trans*-$[PtCl(Ph)(PEt_3)_2]$ (1equiv.), CuI, $^iPr_2NH$–CH_2Cl_2, r.t.; (iii) *trans* -$[PtCl_2(PBu_3)_2]$ (0.5 equiv.), CuI, $^iPr_2NH$–CH_2Cl_2, r.t..

Scheme 3.1 Formation of **1–3**

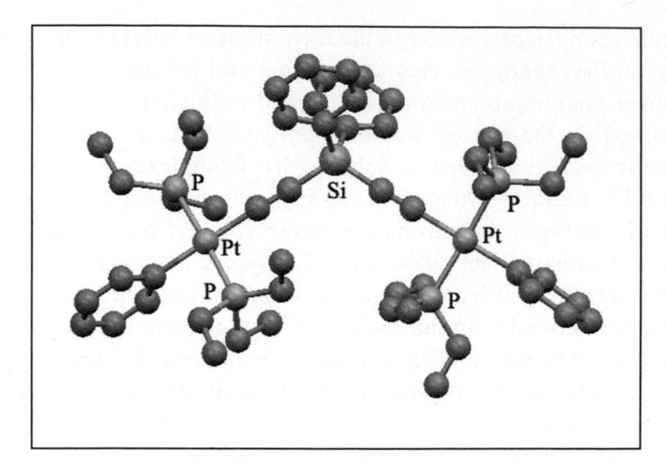

Figure 3.5 X-ray crystal structure of **2**

Table 3.3 ^{29}Si NMR data for oligo- and polyplatinaynes containing alkynylsilane ligands

Compound	^{29}Si NMR (ppm)	Compound	^{29}Si NMR (ppm)
$\mathbf{L_1}$	–48.42	$\mathbf{L_2}$	–48.49
1	–57.50	**4**	–48.16
2	–58.10	**5**	–48.33
3	–57.75	**6**	–49.32

geometry. Two distinct ^{31}P-{^{1}H} signals were observed for the terminal and central phosphine groups in **3**. The ^{29}Si NMR chemical shifts of **1–3** ($\delta = -57.50$ to -58.10) appeared at further downfield positions in comparison with $\mathbf{L_1}$ ($\delta = -48.42$) (Table 3.3) and do not differ significantly from those for related polymers consisting of silylacetylene and π-conjugated organic groups [66]. Polymer **1** was thermally stable up to (~380 °C and the onset decomposition temperature was higher than those in related polymers *trans*-[–Pt(PBu_3)$_2$C ≡ CRC ≡ C–]$_n$ (R = 1,4-phenylene, 2,5-thienylene, 2,5-pyridyl, 9-substituted 2,7-fluorenylene) [36], suggesting that the silylene unit can help increasing the thermal stability of Pt(II) polyynes. The decomposition step was found to result from loss of two PBu_3 groups from the polymer.

2.1.3 Optical and Light-Emitting Properties

For **1–3**, the first electronic absorption band was mainly based on π–π^* transition in the organic system and π-conjugation of the ligand was preserved through the Pt site. The apparent lack of an energy shift in the S_0–S_1 transition among them showed that the lowest singlet excited state was confined to a single repeat unit. When compared with the other conjugated spacers, the degree of electronic conjugation roughly followed the order $C \equiv C \geq$ 2,5-thienyl > 1,4-C_6H_4 > 2,5-pyridyl > $SiPh_2$ for both diynes and polyynes (Figure 3.6), implying that the energy of the S_1 singlet state was highest for the Si-bridged compounds. These results were also consistent with the density functional theory (DFT) calculations at the B3LYP level for **2, 30** (vide infra), and *trans*-[Pt(Ph)$(Et_3)_2C \equiv CC \equiv CC \equiv C$Pt(Ph)$(PEt_3)_2$]. The highest occupied molecular orbitals (HOMOs) are π-bonding orbitals in the conjugate organic structural units that are mixed with the metal *d* orbitals. The lowest unoccupied molecular orbitals (LUMOs) are largely related to the vacant $p_\perp$ orbitals of the Pt center mixed with π^* orbitals of the conjugate organic moieties. Here, $p_\perp$ stands for the *p* orbitals perpendicular to the square planes around Pt center. From the contour plots shown in Figure 3.7, the calculated HOMO–LUMO gaps were in line with the extent of π-delocalization in the conjugation chain as depicted in Figure 3.6.

The temperature dependence of the PL spectra of **1** is shown in Figure 3.8. There was a strong increase in triplet emission intensity with decreasing temperature. At 11 K, the emissive peak occured at about 508 nm for **1**. The luminescence lifetime (τ_p) values for **1** were determined to be about 11.0 μs at 290 K and 40.2 μs at 11 K (versus τ_p = 30 μs at 10 K for **I**), which was in line with the π–π^* character of the phosphorescent triplet state of the emission. The reduced conjugation in **1** shifted the phosphorescence to the blue by 0.06 and 0.38 eV, respectively, as compared with the related Pt(II) polyynes **I** and **30**.

Attempts were also made to extend the work to some robust and stable silylacetylene derivatives functionalized with aromatic entity such as L_3 [89], which can then afford new Pt polyynes **4** and oligoynes **5** and **6** [90]. Scheme 3.2 depicts the synthetic routes to these materials. Polymer **4** was prepared as a structural analog to *trans*-[–Pt$(PBu_3)_2C \equiv C$(p-C_6H_4)$C \equiv CPt(PEt_3)_2C \equiv C$($p$-$C_6H_4$)$C \equiv C$-$]_n$ (**VI**) in order to evaluate the effect of

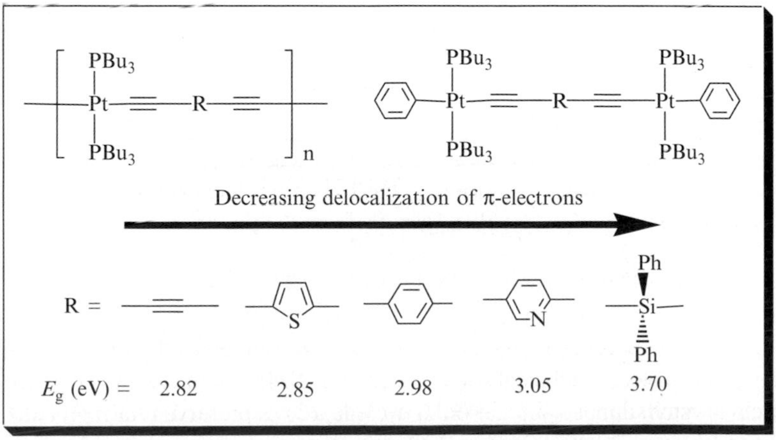

Figure 3.6 Optical bandgaps (E_g) for platinum polyynes and diynes with different spacer groups

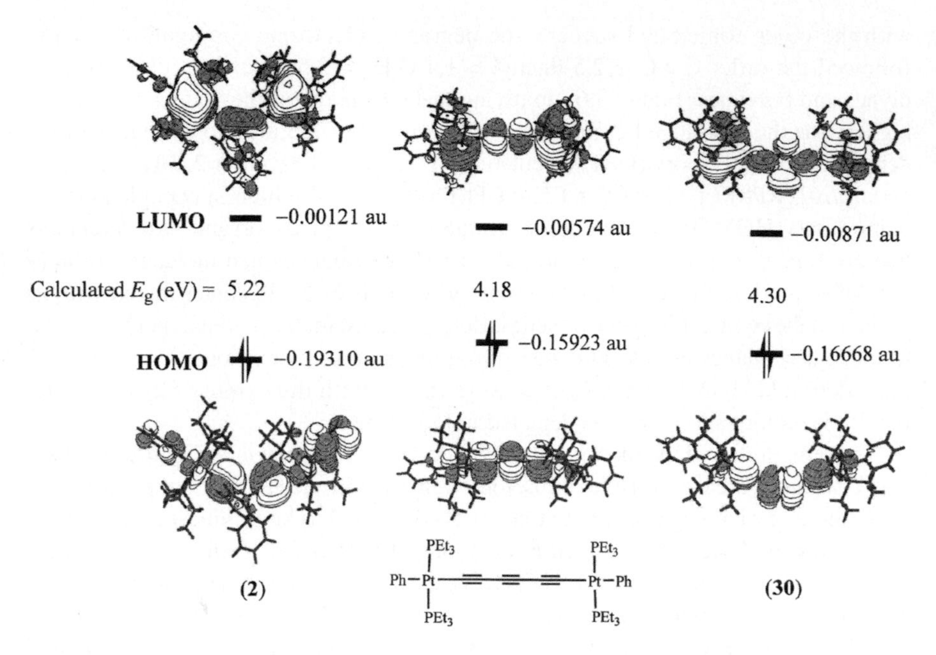

Figure 3.7 The DFT-estimated HOMO and LUMO levels for three platinum diynes

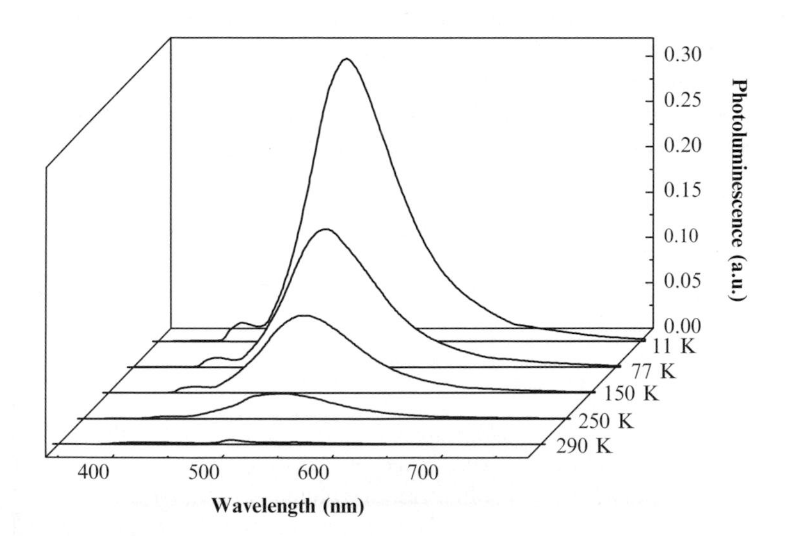

Figure 3.8 Temperature dependence of the PL spectra of **1**

$SiPh_2$ in the metal polyyne system. Compounds **4–6** were generally organic-soluble in aromatic, chlorinated, or ethereal solvents. The ^{29}Si NMR chemical shifts of these platinum-rich alkynylsilanes (δ = −48.16 to −49.32) remained relatively unshifted relative to $\mathbf{L}_2$ (δ = −48.49) (Table 3.3). Once again, it was shown that addition of $SiPh_2$ units in **4** could enhance the thermal stability of these Pt(II) polyynes.

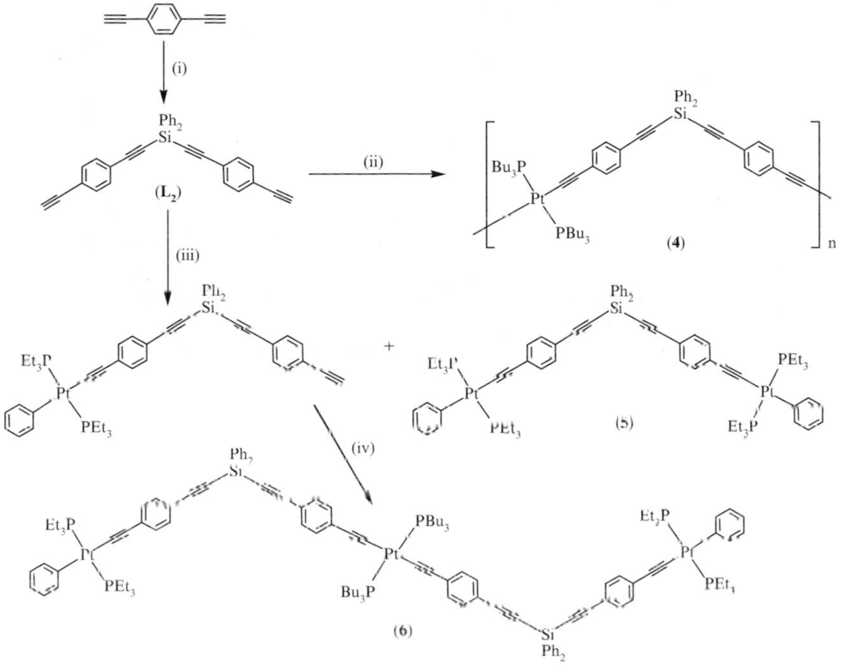

Reagents and conditions: (i) ^{n}BuLi, 195 K, Ph_2SiCl_2, THF; (ii) *trans*-$[PtCl_2(PBu_3)_2]$ (1 equiv.), CuI, $^{i}Pr_2NH$–CH_2Cl_2, r.t.; (iii) *trans*-$[PtCl(Ph)(PEt_3)_2]$, CuI, $^{i}Pr_2NH$–CH_2Cl_2, r.t.; (iv) *trans*-$[PtCl_2(PBu_3)_2]$ (0.5 equiv.), CuI, $^{i}Pr_2NH$–CH_2Cl_2, r.t.

Scheme 3.2 Synthesis of **4–6**

For **4–6**, we assigned the lowest energy absorption peak to the (0,0) vibronic peak of a $S_0 \rightarrow S_1$ transition from the HOMO to the LUMO, which were essentially delocalized π and π^* orbitals. The observation of a red shift in the bands upon introducing Pt units indicated π-conjugation of the ligands through metal sites. Polymer **4** possessed a well-extended singlet excited state. The absorption maximum followed the order **I** (380 nm) ≈ **VI** (383 nm) > **4** (355 nm), showing that $SiPh_2$ was a less effective spacer than the C_6H_4 and $Pt(PEt_3)_2$ units in electronic conjugation and was thus associated with the highest S_1 level. Moreover, we found that the E_g values gradually decreased in the order **5** > **6** > **4** with increasing oligomer chain length.

Two characteristic emission bands were observed at 11 K for **4** in which the higher energy weak emission peak near 400 nm corresponded to fluorescence and the lower-lying emission beyond 500 nm was attributed to the spin-forbidden phosphorescence. In addition, the emissions did not change in dilute solutions, excluding an excimer origin. The temperature dependence of the PL spectra for **4** is displayed in Figure 3.9. For **4**, the T_1 state appeared to be strongly localized. Compounds **5** and **6** also showed similar PL features. Based on the spectroscopic data, the energy level diagram shown in Figure 3.10 was obtained for the polymers **4**, **I**, and **VI** and their corresponding

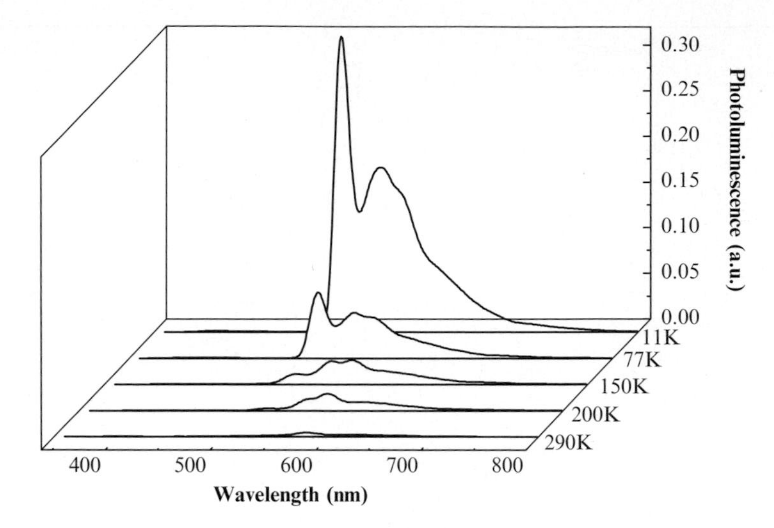

Figure 3.9 Temperature dependence of the PL spectra of **4**

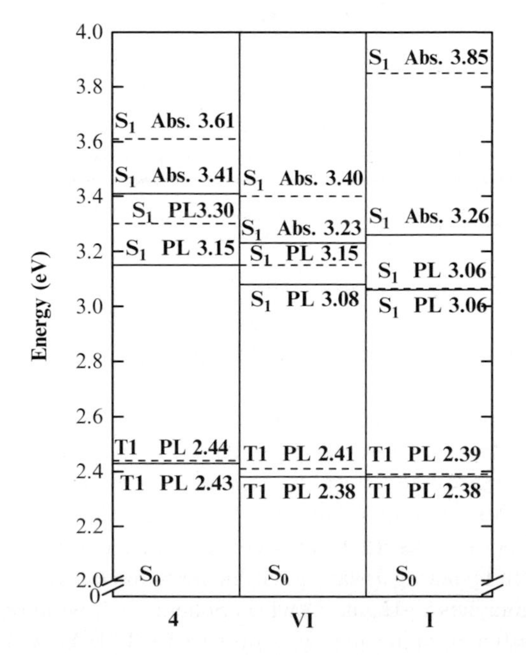

Figure 3.10 Electronic energy level diagram of **4**, **I**, and **VI** *(solid lines)*. The levels for the corresponding Pt diynes are shown by the *dashed lines*. All energy values are quoted in eV

dimeric model compounds. We noted a very high phosphorescence rate efficiency for **4**, and the order of S_1–T_1 crossover efficiency was **4** > **VI** > **I**. This agreed with the solution PL data in which the ratio of integrated intensities of phosphorescence to fluorescence was greater than 1 for **4** but was less than 1 for **VI** (Figure 3.11). The

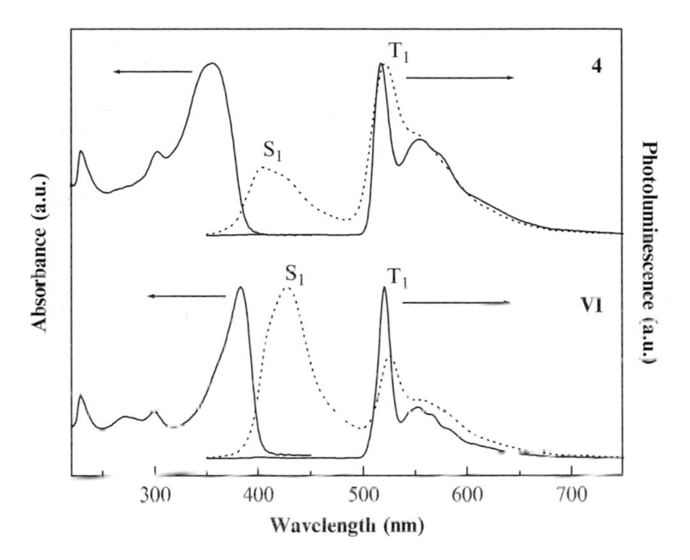

Figure 3.11 Solid-state absorption and PL (11 K) spectra of **4** and **VI** *(solid line)*. The PL spectrum in CH_2Cl_2 solution at room temperature is represented by the *dotted line* in each case

intensity of phosphorescence thus reduces with energy of the T_1–S_0 emission. The use of a conjugation-interrupting silyl segment in **4** fine-tuned the effective conjugation length of metal polyynes and gave rise to efficient crossover between the S_1 and T_1 states. The S_1–T_1 energy gap for **4** was found to lie within the constant range of 0.7 ± 0.1 eV, typical of other analogous polymetallaynes [36].

Extension of these synthetic methodologies to other group 14 elements also warranted examination. Although study of oligo- and poly(aryleneethynylenesilylene)s represents a topical research area nowadays, little is understood on related systems with the heavier group 14 element germanium of even lower I.E.. Knowing this, the first examples of Ge-containing Pt(II) metallopolymers, derived from the blue-light-emitting oligo(fluorenyleneethynylenegermylene)s L_3–L_8, are described which can display very fast phosphorescence decay rate [91]. Inclusion of a conjugation-interrupting sp^3-Ge linker can limit the effective conjugation length and trigger the triplet light emission by taking advantage of the heavy-atom effect of Ge atoms.

Stable polyplatinaynes **7–12**, as shown in Scheme 3.3, were obtained by dehydrohalogenative polymerization between *trans*-$[PtCl_2(PBu_3)_2]$ and the germanium-based oligomers. The excellent film-forming properties and GPC data suggested high-molecular-weights for these materials. Polymers **7–12** displayed excellent thermal stability in the TGA curves. The onset decomposition temperatures were almost invariant of the chain length and the R unit. Relative to the parent polymer *trans*-$[-Pt(PBu_3)_2C\equiv CRC\equiv C-]_n$ (R = 9,9-dihexylfluorene-2,7-diyl), which decomposes at 349 °C [40], the presence of the GeR_2 unit between the fluorene-acetylene segment significantly increased the thermal stability of the Pt polyynes.

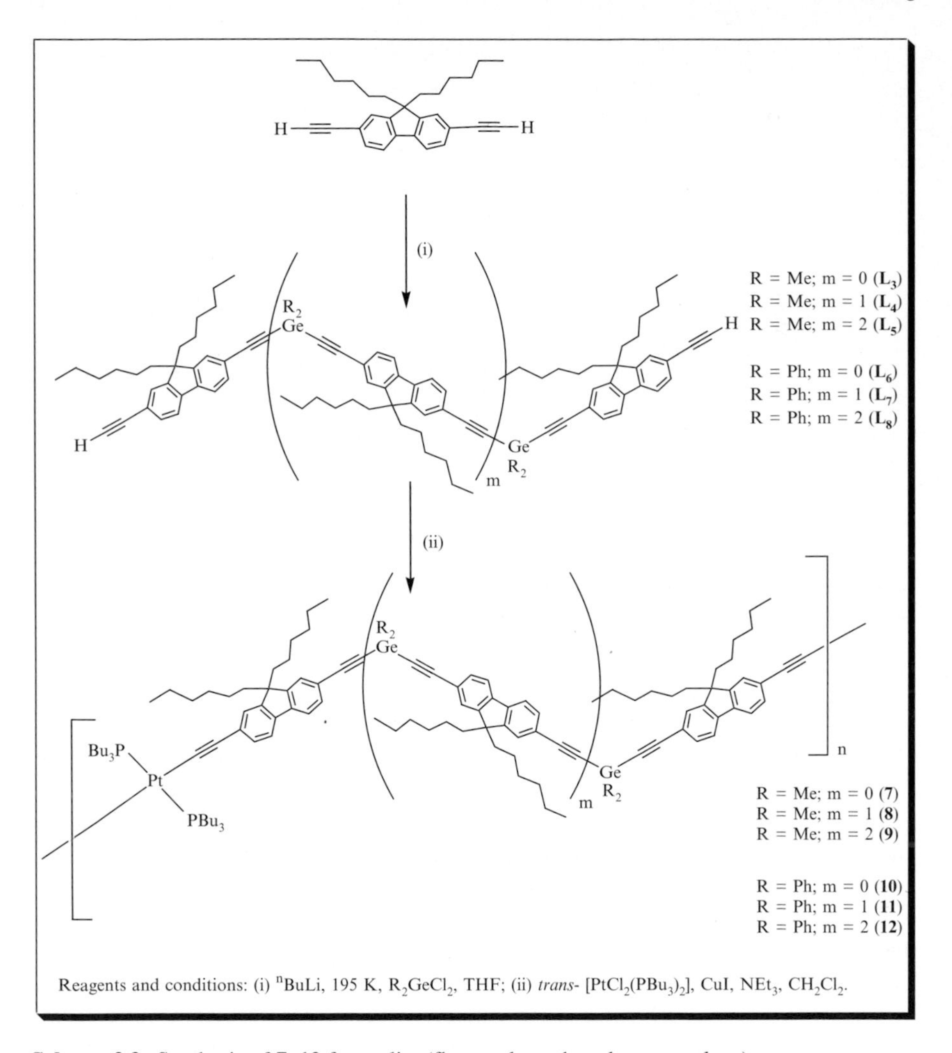

Scheme 3.3 Synthesis of **7–12** from oligo(fluorenyleneethynylenegermylene)s

The transition energies of **7–12** were lowered relative to those of the metal-free oligoacetylenic germanes, suggesting a well-extended singlet excited state in the polymers. Examination of their absorption behavior in both CH_2Cl_2 and in the solid state precluded the formation of solid-state aggregates in thin films. For **7–12**, both fluorescent and phosphorescent emissions arose from ligand-centered $(\pi\pi)^*$ transitions. In dilute fluid solutions, we observed an intense $^1(\pi\pi^*)$ fluorescence near 400 nm, which did not show a large energy shift at high concentrations. The presence of an excimeric state is therefore unlikely. At low temperatures, another phosphorescence band emerged at a wavelength $(\lambda_{em})_P$ between 545 and 548 nm. The substantial Stokes shifts of these peaks from the dipole-allowed absorptions, coupled with the long emission lifetimes in the microsecond range, are suggestive of their triplet parentage. The triplet energy did not vary much with oligomer chain length. In other words, the lowest T_1 state wass confined to a single repeat unit. Variation of the R group did not appear to alter this strong confinement. Insertion of conjugation-hindered GeR_2 group in these polymers shifted the phosphorescence peaks to the blue relative to *trans*-$[-Pt(PBu_3)_2C{\equiv}CRC{\equiv}C-]_n$ (R = 9,9-dihexylfluorene-2,7-diyl). Values of S_0–T_1 separation were found to be about 2.27–2.28 eV for both the Me and Ph series. This corresponded to S_1–S_0 gaps of ~ 3.0 eV. The S_1–T_1 gaps were between 0.73 and 0.75 eV, falling into the usual range for other metal polyynes of group 10–12 elements.

For systems with third-row transition metal chromophores in which the ISC efficiency is close to 100% [92–94], the phosphorescence radiative $(k_r)_P$ and nonradiative $(k_{nr})_P$ decay rates were related to the measured lifetime of triplet emission (τ_P) and the phosphorescence quantum yield (Φ_P) by Eqs. 3.1 and 3.2:

$$(k_{nr})_P = (1 - \Phi_P)/\tau_P \tag{3.1}$$

$$(k_r)_P = \Phi_P / \tau_P \tag{3.2}$$

Using Eqs. 3.1 and 3.2 τ_P, Φ_P, $(k_r)_P$, and $(k_{nr})_P$ values at 20 K were calculated for **7–12** (Table 3.4). The measured Φ_P values were relatively insensitive to the value of m but they changed with the ER_2 group (E = Si, Ge). The phosphorecence of the $GeMe_2$ systems was twice as efficient as the $GePh_2$ congeners. However, substituting $SiPh_2$ for $GePh_2$ reduced Φ_P by almost half (Φ_P ~ 10–13% for the $SiPh_2$ assembly) which was attributed to the heavy-atom effect resulting from the presence of Ge groups in the former case. The $(k_r)_P$ values at 20 K were $(2.1–3.5) \times 10^5$ s^{-1} for **7–9** and $(1.3–1.7) \times 10^5$ s^{-1} for **10–12**. Relative to *trans*-$[-Pt(PBu_3)_2C{\equiv}CRC{\equiv}C-]_n$ (R = 9,9-dihexylfluorene-2,7-diyl) ($(k_r)_P \sim 4.4 \times 10^4$ s^{-1}), insertion of the germylene component increased $(k_r)_P$ by about 10 times. Remarkably, it is possible to get comparable orders of magnitude for $(k_{nr})_P$ and $(k_r)_P$, which was previously unprecedented in the literature for polymetallaynes. Hence, heavy-atom derivatization using Pt and Ge atoms together with conjugation interruption by the latter can greatly boost $(k_r)_P$ values by about. 5 orders of magnitude [95]. It is probable that the high energy benzene stretching modes of Ph group are efficient at promoting $(k_{nr})_P$ for **10–12**, making Φ_P and $(k_r)_P$ values smaller than those observed in **7–9**.

Table 3.4 Photophysical data of **7–12** at 20 K

Pt polyyne	$(\lambda_{em})_P$ (nm)	Φ_P	τ_P (μs)	$(k_{nr})_P$ (s^{-1})	$(k_r)_P$ (s^{-1})
7	547	0.45	1.27	4.3×10^5	3.5×10^5
8	544	0.43	2.08	2.7×10^5	2.1×10^5
9	544	0.45	1.41	3.9×10^5	3.2×10^5
10	548	0.17	1.32	6.3×10^5	1.3×10^5
11	545	0.18	1.21	6.8×10^5	1.5×10^5
12	545	0.20	1.16	6.9×10^5	1.7×10^5

We also recently reported using the electron-poor silole unit in the preparation of a soluble donor-acceptor polyplatinayne **13** and its model compound **14** (Figure 3.12) [96]. A silole ring is of great interest by virtue of its low-lying LUMO level because of the σ^*–π^* conjugation in the ring and so 2,5-difunctional silole ring systems were frequently employed as a prominent structural unit of π-conjugated polymers of low E_g values [91–101]. The incorporation of electron-accepting silole units in the metallopolymer main chain created a new π-conjugated system with unique donor–acceptor characteristics that reduced the HOMO–LUMO gap of **13**. The ultraviolet (UV)-vis absorption spectra of both compounds **13** and **14** in CH_2Cl_2 exhibited strong π–π^* transitions in the near UV region and relatively low-energy shoulder bands in the visible range that tailed off between 570 and 590 nm. Compared with the absorption band for 2,5-dibromo-1,1-diethylsilole (λ_{max} = 326 nm), the positions of the low-lying shoulder bands (λ_{max} = 504 (**13**) and 470 (**14**) nm) were remarkably red-shifted by about 178 and 144 nm for **13** and **14**, respectively, after the inclusion of heavy transition metal chromophores. This likely resulted from an intramolecular interaction between metal ethynyl units and silole rings that notably perturbed the electronic structure of the silole unit. The E_g value for **13** was 2.10 eV and it was significantly lowered by about 0.6 to 1.0 eV relative to **1** (3.10 eV) [90] and the electron-rich thienyl analog (2.70 eV) [102]. Metallaynes **13** and **14** also showed broad and unstructured PL bands peaking between 537 and 525 nm, respectively. The small Stokes shifts and the measured lifetimes in the nanosecond time region for **13** and **14** identified S_1 as the emitting state. Unlike other Pt polyynes of higher bandgaps, we observed no room temperature emission from a T_1 state over the measured wavelength range for both compounds.

Complexes **13** and **14** displayed a low-lying LUMO energy level ranging from −3.52 to −3.54 eV. These LUMO levels were lower than those of 2-(4-biphenylyl)-

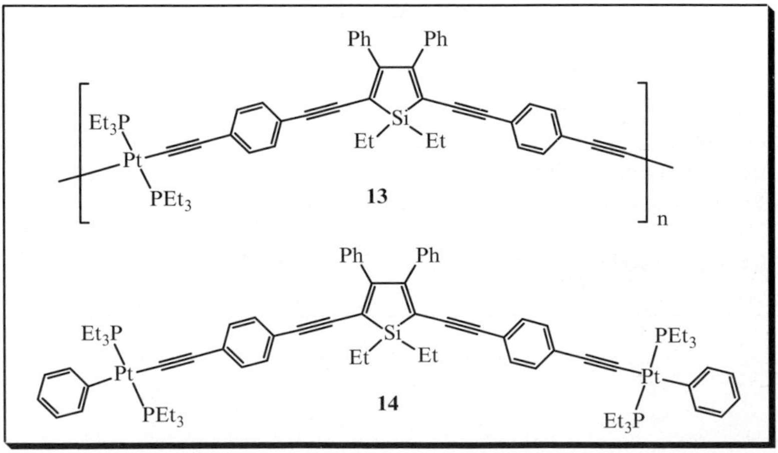

Figure 3.12 Chemical structures of **13** and **14**

5-(4-*tert*-butylphenyl)-1,3,4-oxadiazole (−2.4 eV) and other oxadiazole-based materials [103,104] and that of poly(*p*-phenylenecyanovinylene) (−3.10 eV) [104], which featured good electron-transporting ability in polymeric LEDs (PLEDs). Such energy levels match quite well with the work function of Al when it is used as the electron-transporting material in PLEDs. The HOMO energy levels of **13** and **14** ranged from −5.62 to −5.72 eV which made them attractive candidates as the hole-blocking unit in PLEDs.

2.2 Metallopolyynes Containing Group 16 Main Group Elements

2.2.1 Background

We were interested in the synthesis of oligoacetylenic compounds containing main group elements. The chemistry of some oligoacetylenic sulfides, such as **VII** and **VIII**, has recently been described (Figure 3.13) [105]. In the literature, work on the nonradiative decay of triplet states in conjugated polymers is still in its infancy. Yet radiationless transitions are more common than radiative transitions, so that knowledge of triplet state decay mechanisms is necessary for a full understanding of the photochemistry of conjugated aryleneethynylene systems. A number of literature reports have shown that the nonradiative decay of the triplet states in a series of platinum-containing conjugated polymers and their metal model complexes can be described by the energy gap law [52]. Work should therefore be focused on materials with high-energy triplets in order to avoid competition with nonradiative decay processes.

We continued to work on novel systems to circumvent the problem of the triplet state being nonemissive by using a model system consisting of d^8 Pt-containing ethynylenic conjugated complexes and polymers containing some non-π-conjugated group 16 chalcogen units. Also, although numerous research reports have appeared on the syntheses and applications of purely organic polymeric systems of arylene ether, sulfide, sulfoxide, and sulfone [106–109], work on

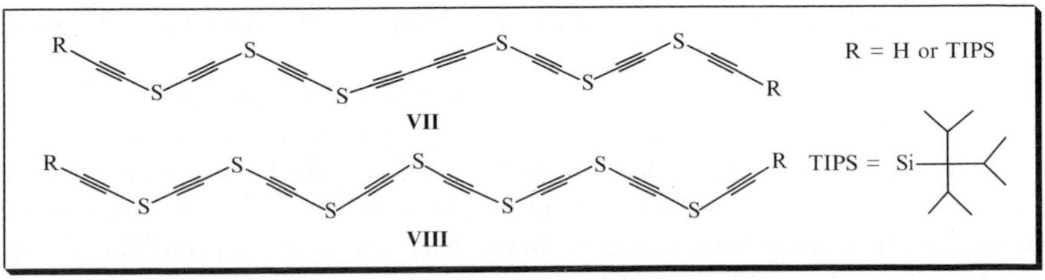

R = H or TIPS

VII

TIPS = Si

VIII

Figure 3.13 Some examples of oligoacetylenic sulfides

rigid-rod chalcogen-derived metallaynes and their corresponding polyynes remains unexplored. Whereas such heteroatoms can act as the conjugation-breaking unit to interrupt the electronic π-conjugation akin to the silyl fragment described in Section 2.1, these materials can also offer an attractive combination of physical, optical, and mechanical properties. In fact, sulfone-based poly(arylene ether)s have gained considerable commercial and industrial importance. Aromatic polymers that contain aryl ether or aryl sulfone linkages generally have lower glass-transition temperatures and greater chain flexibility than their corresponding polymers that do not have these groups in the chain [110–112]. In addition, it is well documented that aromatic sulfoxides are photochemically active molecules and characterization of the triplet states of simple organic molecules such as PhEPh (E = S, SO, SO_2) have played a major role in the development of their photochemistry and photophysics [113,114].

We have widened our scope in the current work to form a new series of platinum aryl-acetylide polymers and their dinuclear counterparts containing O, S, SO, and SO_2 functional groups. These are remarkable model systems that can be used to evaluate how chalcogen-based conjugation-interrupters would limit the effective conjugation length of polymetallaynes.

2.2.2 Synthesis and Chemical Characterization

The syntheses of bis(alkynylated) precursors bearing 4,4′-diphenyl ether, 4, 4′-diphenyl sulfide, 4,4′-diphenyl sulfoxide, and 4,4′-diphenyl sulfone $\mathbf{L_9}$–$\mathbf{L_{12}}$ are outlined in Scheme 3.4. The as-prepared diethynyl synthons were then used to form a series of platinum(II) metallopolymers **15–18** and their model complexes **19–22** by adaptation of the classical dehydrohalogenation protocols (Scheme 3.5) [115]. The feed mole ratios of the platinum chloride precursors and the diethynyl ligands were 2:1 and 1:1 for the diyne and polyyne syntheses, respectively. All the new complexes and polymers were air-stable and could be stored without any special precautions. They were soluble in chlorinated solvents such as CH_2Cl_2 but were generally insoluble in hydrocarbons. The three-dimensional molecular structures of **20–22** were confirmed by single-crystal X-ray diffraction studies (Figure 3.14). All of these molecules adopted angular geometry at the sp^3 hybridized heteroatom. The coordination geometry at each Pt center was square-planar with the two PEt_3 groups *trans* to each other and the metal capping groups connected by the three diethynyl chalcogenide ligands. None of these structures exhibited π–π stacking of the arene rings in the solid state.

It is well-documented that aromatic polymers containing aryl ether or sulfone linkages have excellent thermal and mechanical properties [107]. Analysis of the TGA traces for **15–18** revealed that they all exhibit good thermal stability. They retained more than 95% of their mass in the range of 335 to 363 °C. Their onset decomposition temperatures appeared higher than that in *trans*-[–Pt$(PBu_3)_2$C ≡ C (p-$C_6H_4)_2$C ≡ C–$]_n$ (~300 °C) [116]. Thus, the introduction of a heteroatom can enhance the thermal stability of this class of metal polyynes. We observed a sharp

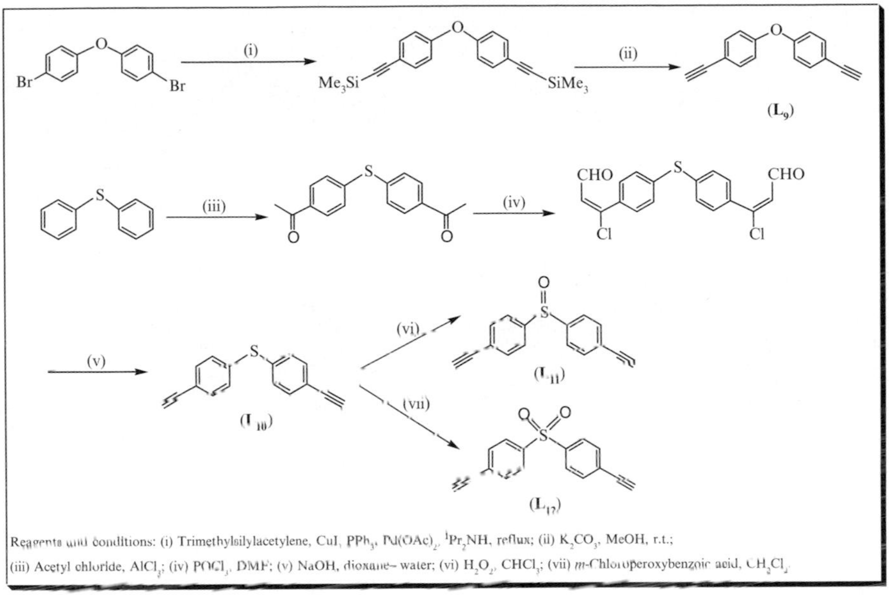
(i)
(ii)
(iii)
(iv)
(v)
(vi)
(vii)
(L_9)
(L_{10})
(L_{11})
(L_{12})
Reagents and conditions: (i) Trimethylsilylacetylene, CuI, PPh_3, $Pd(OAc)_2$, $^iPr_2NH$, reflux; (ii) K_2CO_3, MeOH, r.t.;
(iii) Acetyl chloride, $AlCl_3$; (iv) $POCl_3$, DMF; (v) NaOH, dioxane–water; (vi) H_2O_2, $CHCl_3$; (vii) *m*-Chloroperoxybenzoic acid, CH_2Cl_2

Scheme 3.4 Synthesis of the diethynyl chalcogenide ligand precursors L_9–L_{12}

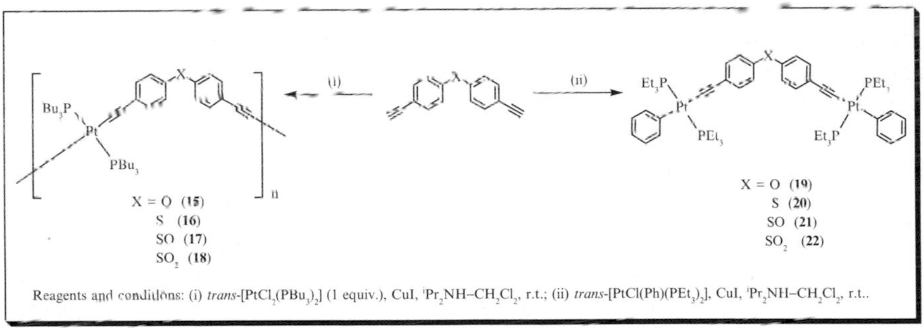
X = O (15)
S (16)
SO (17)
SO_2 (18)
X = O (19)
S (20)
SO (21)
SO_2 (22)
Reagents and conditions: (i) *trans*-$[PtCl_2(PBu_3)_2]$ (1 equiv.), CuI, $^iPr_2NH$–CH_2Cl_2, r.t.; (ii) *trans*-$[PtCl(Ph)(PEt_3)_2]$, CuI, $^iPr_2NH$–CH_2Cl_2, r.t..

Scheme 3.5 Synthetic pathways to compounds **15–22**

weight loss of 28% for **15** whereas 40% of the weight was lost for **16–18**. The decomposition step was ascribed to the removal of one PBu_3 group from **15** and the loss of six Bu groups from **16–18**. These platinum polymers also revealed no discernible glass transitions in the DSC curves.

2.2.3 Optical and Light-Emitting Properties

Compounds **15–22** all displayed similar structured absorption bands in the near UV region and the dependence of the absorption energies on the ligand type suggested

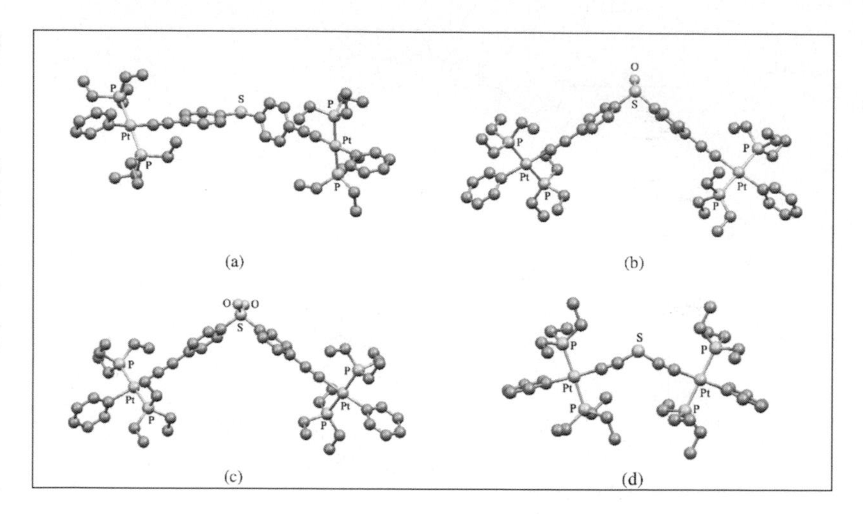

Figure 3.14 X-ray crystal structures of **20** (**a**), **21** (**b**), **22** (**c**), and **25** (**d**)

that the absorption pattern was dominated by intraligand $^1(\pi-\pi^*)$ transitions, possibly mixed with some contributions from metal orbitals. We noted a red shift of the lowest energy absorption upon the insertion of metal fragments. π-Conjugation was thus preserved through the metal site through mixing of the frontier orbitals of the metal and ligand. A well extended singlet excited state in the Pt polyynes was also indicated by the transition energies of the platinum polymers, which were lower than those of the diplatinum complexes. The absorption energy of the oxygen-linked metal compound was the largest whereas the sulfone-containing counterpart showed the lowest absorption energy between **15–18** and **19–22** (Figure 3.15). Relative to *trans*-[Pt(Ph)$(PEt_3)_2C \equiv C(p\text{-}C_6H_4)_2C \equiv CPt(Ph)(PEt_3)_2$] and *trans*-[–Pt$(PBu_3)_2C \equiv C(p\text{-}C_6H_4)_2C \equiv C$–$]_n$ (λ_{abs} = 346 and 372 nm, respectively), we found that the non-π-conjugated chalcogen unit between the two phenyl rings hindered conjugation and causeed a blue shift in the absorption wavelength (325–345 and 343–361 nm) for our Pt diynes and polyynes, respectively. In other words, the energy of the S_1 state varied with the electronic properties of the spacer group, which goverened the efficiency of triplet-state emission (vide infra). Upon photoexcitation, we observed an intense purple-blue $^1(\pi-\pi^*)$ fluorescence peak near 400 nm in dilute fluid solutions at 290 K for each. As the temperature was cooled to 20 K, there was almost no fluorescence band but rather a prominent phosphorescence band from the central ligand chromophore. At 20 K, the intense PL peaks in the low-lying region exhibited large Stokes shifts from the absorption spectra (Figure 3.15) and possessed long emission lifetimes. The emission maxima were dependent on the nature of the acetylide ligand; thus, the lowest emissive states in these compounds can tentatively be assigned as metal-perturbed ligand-localized $^3(\pi-\pi^*)$ transitions. The emission assignment was also supported by the observed temperature dependence of the emission data. Although we have had a long-term

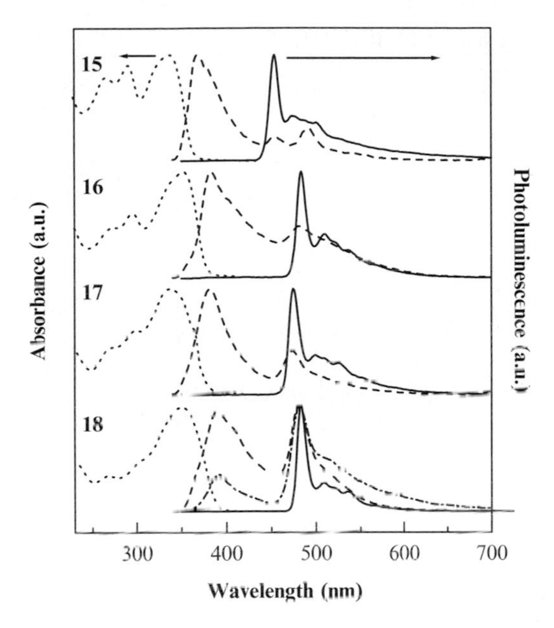

Figure 3.15 The absorption and PL spectra of **15–18**. The absorption spectra are the higher energy dotted lines measured at 290 K. PL spectra were taken at both 290 *(dashed lines*, in CH_2Cl_2), and 20 K *(solid lines*, as thin films). For the polyyne **18**, the 290 K PL spectrum *(dash dotted line)* as thin film is also shown

interest over the past decade in rendering the observation of triplet emission under ambient conditions, this work was attractive in that room-temperature phosphorescence was clearly identified for our platinum polymers (Figure 3.15). This is not well documented for related systems containing extensively conjugated spacers. Presumably because of the internal heavy-atom effect of the sulfone group, polyyne **18** showed the most intense phosphorescence band at 290 K among the four polymers and we even observed very strong solid-state triplet emission for **18** in which the ratio of integrated intensities of phosphorescence to fluorescence was greater than unity.

Table 3.5 lists some photophysical parameters for **15–18** and Figure 3.16 displays the energy level scheme for their lower-lying excitations; the data for the silylene-spaced congener **4** is also included for comparison. The lowest T_1 state essentially remained strongly localized, and can be inferred from the small energy difference between triplet emissions in the Pt diynes and polyynes. All polymers exhibit T_1–S_0 gaps of 2.5 eV or above, which correspond to S_1(PL)–S_0 gaps in the range of 3.13 to 3.57 eV. The measured S_1–T_1 gaps are constant at around 0.7 ± 0.1 eV, corresponding closely to the S_1–T_1 energy gaps estimated for some organic conjugated polymers [117,118]. All of the sulfides had triplet energies lower than the corresponding sulfoxides, whereas those of the sulfones were typically higher than those of the sulfoxides. These findings are in agreement with some simple organic

Table 3.5 Photophysical data of **15–18** at 20 K

Pt polyyne	$(\lambda_{em})_P$ (nm)	Φ_P	τ_P (μs)	$(k_r)_P$ (s^{-1})	$(k_{nr})_P$ (s^{-1})
15	454	0.46	6.38	7.18×10^4	8.50×10^4
16	484	0.49	6.02	8.08×10^4	8.54×10^4
17	475	0.51	7.27	7.06×10^4	6.69×10^4
18	484	0.46	8.57	5.41×10^4	6.26×10^4

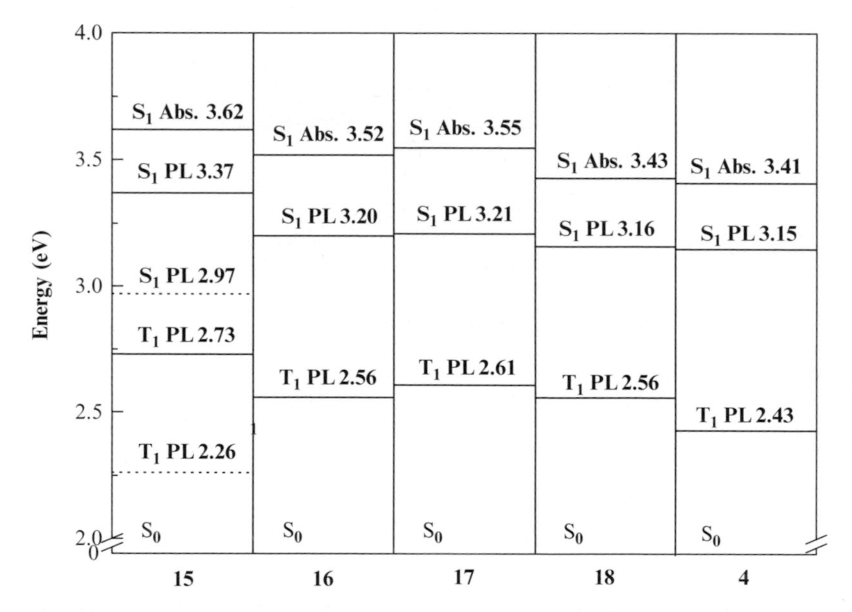

Figure 3.16 Electronic energy level diagram of Pt diynes and polyynes containing group 16 main group elements determined from absorption and PL data. *Dashed lines* represent the levels for the Pt diynes and polyyne containing bi(*p*-phenylene) spacer. The S_0 levels are arbitrarily shown to be of equal energy and all energy values are quoted in eV

aromatic molecules ArEAr (Ar = aryl, E = S, SO, SO_2). It was observed that the T_1–S_0 gap generally changed with the E unit according to the order O > SO_2 > SO > S for the metal diynes. However, the T_1 level for **18** appeared to be slightly lower than those of **16** and **17** probably because of its higher degree of polymerization (DP~18) than the latter two (~13–14). Hindered conjugation through use of different group 16 element-derived units shifted the phosphorescence to the blue as compared with the conjugated biphenyl-spaced counterparts whose largest blue shift was encountered for the diphenyl ether derivatives (ca. 0.44–0.48 eV).

Generally, we observed that the Pt polyynes gave more efficient phosphorescence than the Pt diynes. But, Φ_P did not vary much with the type of E groups in this study. Comparing $(k_r)_P$ and $(k_{nr})_P$ values for the individual metal polyynes and diynes, it was noted that the values for $(k_{nr})_P$ were comparable to those for $(k_r)_P$. Incorporation of the group 16 conjugation-interrupters in the Pt polyyne systems resulted in $(k_r)_P$ values of $(5.4–8.1) \times 10^4\ s^{-1}$ which were more efficient relative to **I** ($(k_r)_P = 6 \times 10^3\ s^{-1}$ at 20 K) by about 1 order of magnitude [95]. This implied that high-energy triplet states (and concurrently high optical gaps) intrinsically led to more efficient phosphorescence in these metal-containing aryleneethynylenes.

In a related context, a series of platinum(II) acetylide complexes of oligoacetylenic sulfides of various chain lengths **23–26** were prepared as molecular models for their long-chain organometallic polymers (Figure 3.17) [119]. The crystal structure of **25** was determined (Figure 3.14). Compounds **23** and **25** with monoacetylenic sulfide units were shown to be air-stable in the solid state, but the oily diacetylenic sulfide species **24** and **26** turned dark within several days when exposed to light at room temperature. The optical properties of these molecules were briefly examined and will likely act as a benchmark for future polymer characterization. Compounds **24** and **26** showed significant red shifts in the emission wavelengths with respect to the monoacetylenic counterparts **23** and **25**, respectively.

Extension to the oligothiophene system was equally appealing as a versatile synthetic step toward sulfur-containing metallaynes and a series of conjugated platinum(II) acetylide polymers containing oligothiophene units **27–29**, and their corresponding dimeric complexes **30–32**, were synthesized according to Scheme 3.6. The solid-state structures of **30** and **31** were established by X-ray crystallography (Figure 3.18). The photophysical, optoelectronic and structural properties of these metallaynes have been extensively studied in terms of the number of oligothienyl units within the bridging ligand [120,121].

The absorption and PL spectra of **27–29** at room temperature are shown in Figure 3.19. The optical gap energy of **27–29** decreased with an increase in the

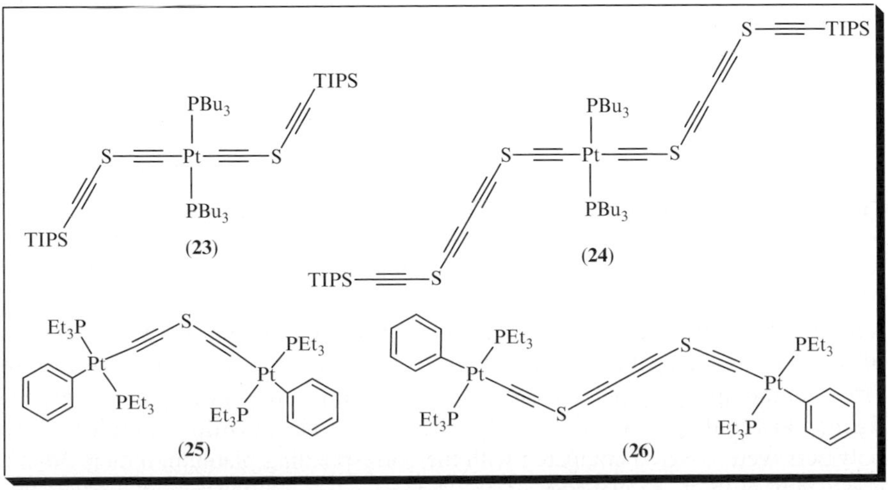

Figure 3.17 Chemical structures of **23–26**

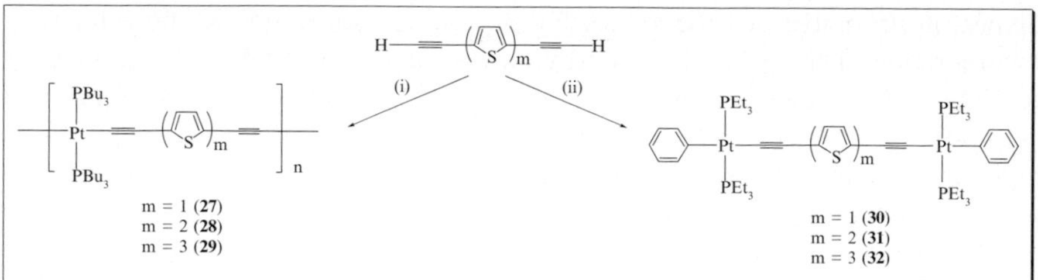

m = 1 (**27**)
m = 2 (**28**)
m = 3 (**29**)

m = 1 (**30**)
m = 2 (**31**)
m = 3 (**32**)

Reagents and conditions: (i) *trans*-[$PtCl_2(PBu_3)_2$] (1 equiv.), CuI, $^iPr_2NH-CH_2Cl_2$, r.t.; (ii) *trans*-[$PtCl(Ph)(PEt_3)_2$], CuI, $^iPr_2NH-CH_2Cl_2$, r.t..

Scheme 3.6 Preparations of compounds **27–32**

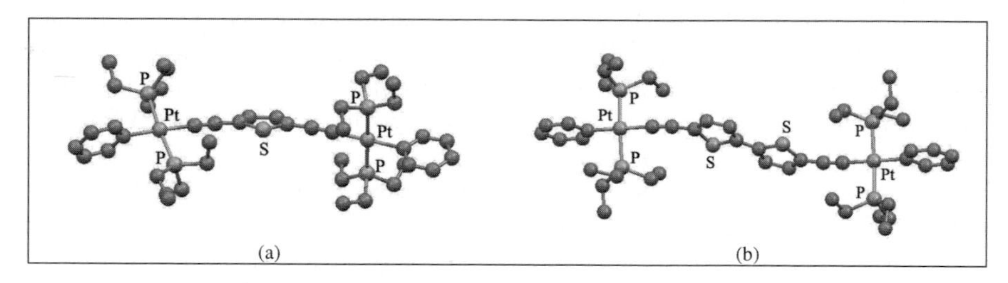

Figure 3.18 X-ray crystal structures of **30** (**a**) and **31** (**b**)

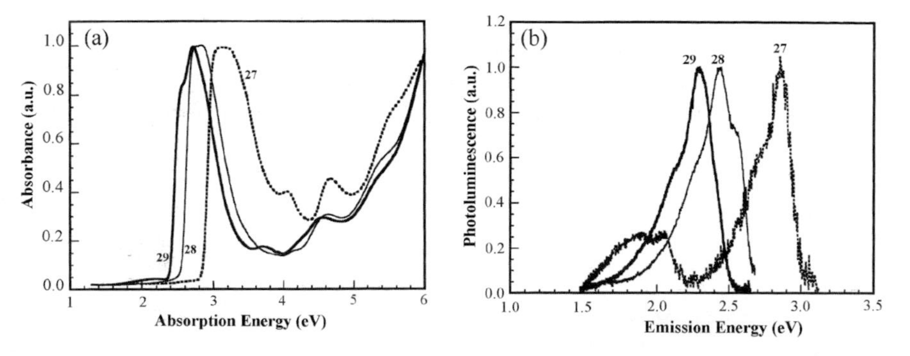

Figure 3.19 Absorption (**a**) and PL spectra (**b**) of **27–29** at room temperature

number of thienyl rings (m). We attributed this to an increased delocalization of π-electrons along the polymer chain. As the m value increased, the overall effect on the optical gap decreased. Moreover, addition of Pt fragments into the ligands lowered the transition energy. Likewise, the transition energies of the polymers were lowered compared with the corresponding platinum dimers **30–32**, consistent with π-conjugation of the ligands through the metal center. The emission features were shifted to lower energy with increasing chain length of the thiophene segment. We ascribed the peaks at 2.85, 2.44, and 2.28 eV for **27–29**, respectively, to arise from the singlet state. Such singlet emissive transition involved the mixed ligand–metal orbitals, dominated either by the intraligand HOMO–LUMO π–π* transition or charge-transfer type transition. The broad PL feature within 1.6 to 2.2 eV for **27** can be assigned to the phosphorescence emission. The triplet emission peak at 2.05 eV in **27** strongly increased in intensity as the temperature was lowered. This emission intensity enhancement revealed a triplet exciton that was quenched by thermally activated diffusion to dissociation sites at room temperature. For **28**, there was a weak triplet emission shoulder centered at about 1.67 eV. There was no room temperature triplet emission for **29** over the measured range (1.2–3.0 eV). At 18 K, the triplet emissive bands became more apparent at 2.05, 1.67, and 1.53 eV for **27–29**, respectively, and they all ranged from 0.80 to 0.88 eV below the S_1 energy level (Figure 3.20). The intensity of this triplet emission decreased rapidly with increasing m value. This was anticipated because the higher m value in the ligand reduced the influence of the heavy metal center, which was mainly responsible for the ISC process. Also, in oligothiophene systems themselves, ISC is reduced with an increasing m value, as the energy of the singlet excited state drops below the corresponding resonance state for ISC in those systems [122]. The energy of the triplet emission also experienced a red shift when additional thiophene rings were added to the ligand, suggesting that the triplet excited state should be extended over several thiophene rings (i.e., three or more) in that system. This shift of triplet emission energy observed in **27–29** was in agreement with the calculations by Beljonne et al. on the evolution of the triplet excitation energy in purely organic oligothiophene systems [122,123].

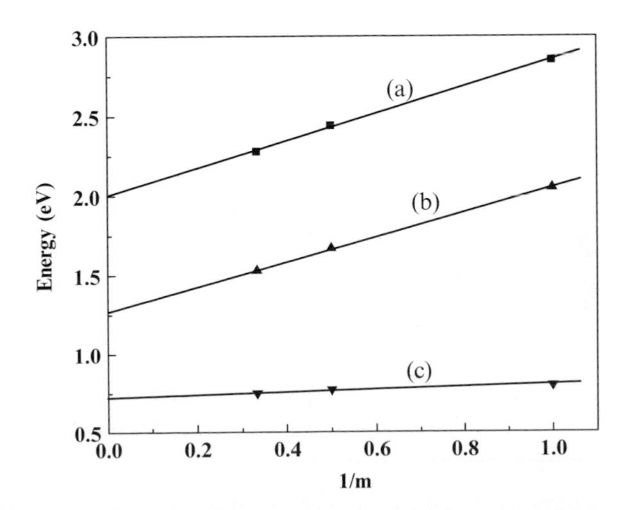

Figure 3.20 Energy versus chain length dependence of **27–29** for S_1–S_0 (**a**), T_1–S_0 (**b**), and S_1–T_1 (**c**) gaps

2.3 Potential Optoelectronic Applications

A study of these new metal alkynyl compounds in suitable electronic and optoelectronic structures forms an essential part of our research program [124–129]. Some examples are given below. The photoconductivity of polymer **4** was examined, and a photocell in a single-layer structure of ITO/**4**/Al was fabricated (Figure 3.21). The quantum efficiency detected was about 0.01%, a typical value for single-layer devices. The photocurrent was found to increase with an increase in bias voltage (Figure 3.21b) [90]. Similar to polyyne **I**, the excited states were mainly strongly bound triplet excitons in such Pt polyynes upon photoexcitation and were confined to one monomer unit [124]. Alternate substitution of $Pt(PBu_3)_2$ units in **I** by $SiPh_2$ did not appear to significantly alter the photoconducting properties of this class of polymetallaynes.

It was shown that **27–29** were also good photoconductors in suitably fabricated configurations [120,126]. The photocurrent spectra of the Au/**27**/Al, ITO/**28**/Al, and ITO/**29**/Al photocells showed two peaks, one at the onset of absorption [2.92 (**27**), 2.64 (**28**), and 2.43 eV (**29**)] and one at higher energies [3.81 (**27**), 3.56 (**28**), and 3.38 eV (**29**)] (Figure 3.22). The second photocurrent peaks were likely resulted from absorption into the higher-lying absorption bands. Polymers **27–29** showed a short-circuit quantum efficiency of about 0.04% at the first photocurrent peak, which is a common value for single-layer devices. We observed no great difference in quantum efficiency with variation of the thiophene content in these Pt-containing polymers. The quantum efficiency of the second peak was different among **27–29** and was very sensitive to air exposure [126]. The overall photocurrent increased when exposed to air and was reduced after annealing under vacuum. The I–V characteristics at the first peak in the spectral response resulted in open-circuit voltages of 0.50, 0.75, and 0.47 V and fill-factors of 0.32, 0.35, and 0.30 for **27–29**, respectively (Figure 3.22b).

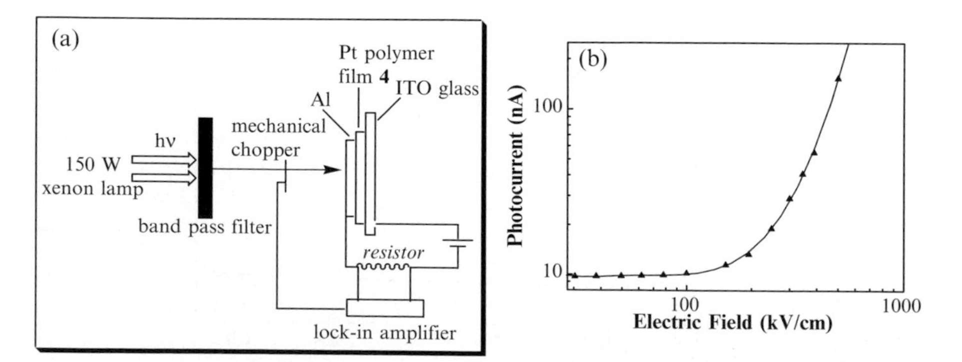

Figure 3.21 Device structure for photoconductivity measurements based on a ITO/**4**/Al photocell (**a**) and the corresponding logarithmic plot of the photocurrent against the applied electric field (**b**)

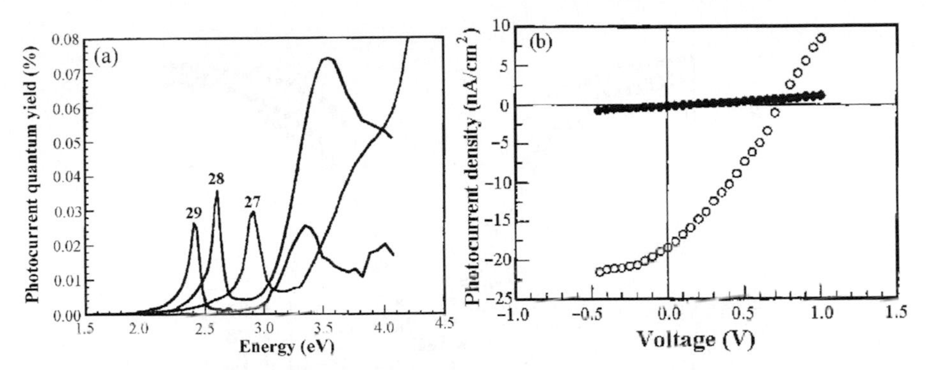

Figure 3.22 Photocurrent spectra for **27–29** (**a**) and I–V curve for the ITO/**28**/Al photocell in the dark (•) and under light illumination (o) (**b**)

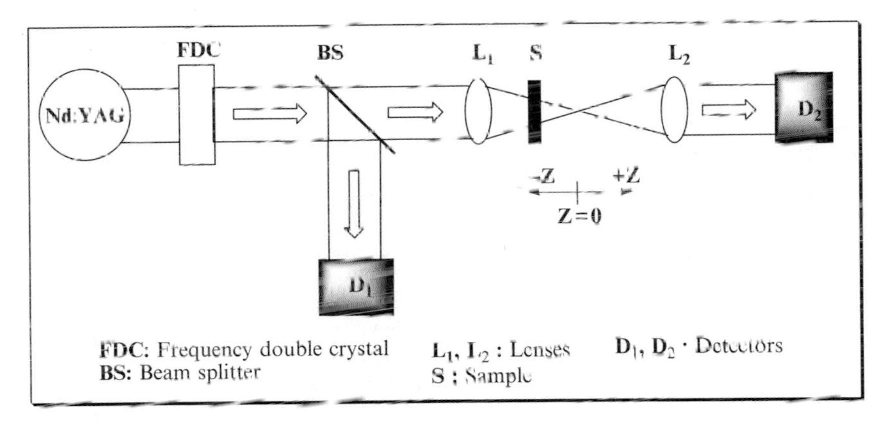

Figure 3.23 Experimental setup for the measurement of optical-limiting properties

Optical limiters are devices that strongly attenuate optical beams at high intensities while exhibiting higher transmittance at low intensities [130]. Such devices are useful for the protection of both the human eye and optical sensors from intense laser beams. Examination of the nonlinear optical transmittance characteristics of **28** at 532 nm (Figure 3.23) showed that it is an excellent optical limiter with performance comparable or superior to those of the state-of-the-art reverse saturable absorption dyes such as fullerene C_{60} and metal phthalocyanine complexes (Figure 3.24) [131]. The optical-limiting threshold for **28** was 0.12 J/cm^2 at 82% linear transmittance. It was conceived that this class of polymeric materials would be promising candidates for use in practical optical-limiting devices. At present, the predominant materials practically employed for optical limiting are small molecules and much less work has been done on the optical-limiting properties of metal-containing polyyne polymers. To our knowledge, related studies for metal-aryleneethynylene copolymers of the type *trans*-[-Pt$(PBu_3)_2$-C$\equiv$CArC$\equiv$C-]$_n$ remain very sparse.

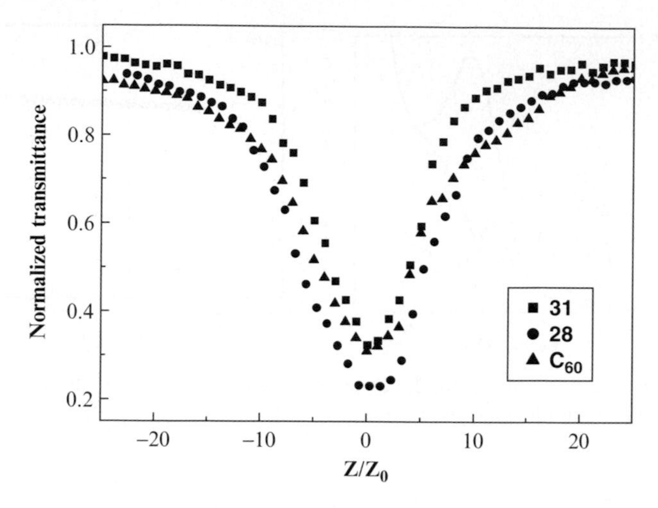

Figure 3.24 The open aperture Z-scan results for **28** and C_{60} solutions at the same linear transmittance of 82%

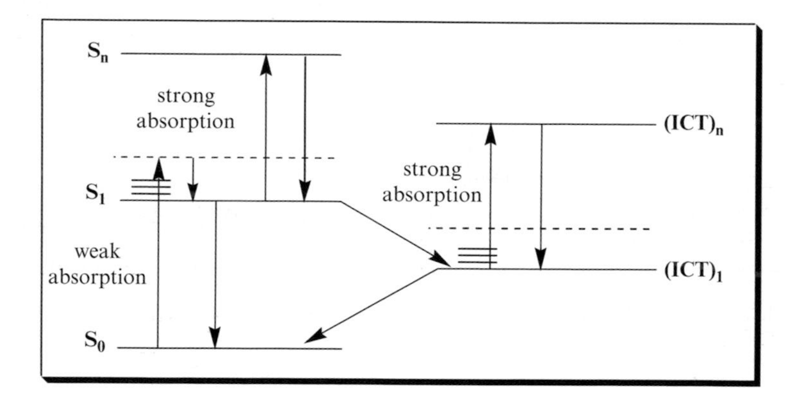

Figure 3.25 The proposed optical-limiting mechanism for the platinum polyyne **28**

From Figure 3.19b, there is no obvious triplet emission in the PL spectrum of **28** at room temperature and its optical-limiting effect may therefore not be induced by the triplet absorption. Thiophene is well known as an electron-rich five-membered aromatic system [132]. Dithienyl unit can act as a donor (D) in the main chain, and the platinum ion as a strong acceptor (A), leading to a donor–acceptor (D–A) structure along the polymeric backbone in **28**. Because of the presence of strong D–A components, when polymer **28** was exposed to the strong incident irradiance of the laser beam, there were intramolecular charge-transfer (ICT) states between the metal ion and the ligand. In this case, the proposed optical-limiting mechanism for **28** is outlined in Figure 3.25. The molecules leapt to the S_1 state from the S_0 state upon absorption of optical energy. Following this, they are capable of absorbing optical energy in order to get promoted to the higher singlet state (S_n), thereby inducing the optical-limiting effect or undergoing the nonradiative decay process to form the $(ICT)_1$ state. When the molecules in the $(ICT)_1$ state gained certain population, absorption of optical energy took place, followed by electronic excitation to the higher $(ICT)_n$ state. The strong energy absorption, together with the $(ICT)_1 \rightarrow (ICT)_n$ process, was responsible for the optical-limiting behavior of **28**. The same phenomenon applied to complex **31**. As shown in Figure 3.24, polymer **28** exhibited better optical-limiting response than **31**, revealing that the nonlinear effect was caused not only by the single repeating segment but also the entity that was delocalized over more than one repeating unit.

3 Further Considerations and Perspectives

This chapter has focused on the chemistry and properties of a novel family of synthetic metal-containing polymers that carry main group elements in the main chain in various capacities. These metal-based macromolecular materials have been recognized for their exciting photophysical properties, structural variability, and potentially important commercial applications. We envisage that the electronic interactions along the π-conjugated backbone will be strongly influenced by the nature of the heteroatom fragment. For instance, conjugation should be disrupted by incorporation of group 13 boron or enhanced by group 15 elements such as phosphorus. With this in mind, it would be interesting to prepare metal polyynes bridged by these main group units [9]. We believe that incorporation of different metal centers to these alkynyl species will likely lead to new metallated materials. In-depth exploration of these newly synthesized compounds as semiconducting and light-emitting materials in the fabrication of LEDs and photocells will be of great interest. The solid-state chemistry of these compounds (e.g., within the realm of ceramization) will also be studied, particularly at the nanostructural level. Success in this work could open new avenues for production of nonoxide ceramics and mixed-metal carbides from their molecular precursors. In the long run, the possible use of materials such as metal-containing organosilicon materials as molecular precursors to the preparation of various nanoparticles or other related materials would be interesting areas of research [11]. In the short term, development of a completely new class of inorganic materials is therefore likely and will be of interest to a wide cross-section of the research community; applications in the area of novel functional materials that combine the essential features of organic polymers, metals, and ceramics may arise in the future.

Acknowledgments Financial support from the Research Grants Council of the Hong Kong SAR (Grant no. HKBU 2054/02P) and the Hong Kong Baptist University is gratefully acknowledged. The author is also indebted to his postgraduate students, postdoctoral associates, and collaborators whose names appear in the references.

References

1. Manners I. (2004) Synthetic metal-containing polymers. Wiley, Weinheim, Germany.
2. Nguyen P, Gómez-Elipe P, Manners I. (1999) Chem. Rev. 99:1515.
3. Manners I. (1996) Angew Chem Int Ed Engl. 35:1602.
4. Abd-El-Aziz AS. (2002) Macromol Rapid Commun. 23:995.
5. Yam VWW. (2004) J Organomet Chem. 689:1393.
6. Long NJ, Williams CK. (2003)Angew Chem Int Ed. 42:2586.
7. Fujikura Y, Sonogashira K, Hagihara N. (1975) Chem Lett. 1067.
8. Sonogashira K, Takahashi S, Hagihara N. (1977) Macromolecules. 10:879.
9. Takahashi S, Kariya M, Yatake T, Sonogashira K, Hagihara N. (1978) Macromolecules. 11:1063.
10. Sonogashira K, Kataoka S, Takahashi S, Hagihara N. (1978) J Organomet Chem. 160:319.
11. Davies SJ, Johnson BFG, Khan MS, Lewis J. (1991) J Chem Soc. Chem Commun. 187.
12. Atherton A, Faulkner CW, Ingham SL, et al. (1993) J. Organomet. Chem. 462:265.
13. Johnson BFG, Kakkar AK, Khan MS, Lewis J. (1991) J Organomet Chem. 409:C12.
14. Davies SJ, Johnson BFG, Lewis J, Raithby PR. (1991) J Organomet Chem. 414:C51.
15. Brédas JL, Cornil J, HeegerAJ. (1996) Adv Mater. 8: 447.
16. Sariciftci NS (ed). (1997) Primary photoexcitations in conjugated polymers: molecular exciton vs semiconductor band model. World Scientific, Singapore.
17. Skotheim TA, Reynolds JR, Elsenbaumer RL (eds). (1998) Handbook of conducting polymers, 2nd ed. Marcel Dekker, New York.
18. Friend RH, Gymer RW, Holmes AB, Jet al. (1999) Nature, 397:121.
19. Wilson JS, Dhoot AS, Seeley AJAB, Khan MS, Köhler A, Friend RH, (2001) Nature. 413:828.
20. Köhler A, Wilson JS, Friend RH. (2002) Adv. Mater. 14:701.
21. Adachi C, Baldo MA, Thompson ME, Forrest SR. (2001) J Appl Phys. 90:5048.
22. Lupton JM, Pogantsch A, Piok Y, List EJW, Patil S, Scherf U. (2002) Phys Rev Lett. 89:167401.
23. Romanovskii YV, Gerhard A, Schweitzer B, Scherf U, Personov RI, Bässler H. (2000) Phys Rev Lett. 84:1027.
24. Gong X, Ostrowski JC, Bazan GC, et al. (2003) Adv. Mater. 15:45.
25. Dhoot AS, Greenham NC. (2002)Adv Mater. 14:1834.
26. Brédas JL, Chance RR (eds). (1990) Conjugated polymeric materials: opportunities in electronics, optoelectronics and molecular electronics. Kluwer Academic Publishers, Dordrecht.
27. Ho PKH, Kim JS, Burroughes JH, et al. (2000) Nature, 404:481.
28. Cao Y, Parker ID, Yu G, Zhang C, Heeger AJ. (1999) Nature. 397:414.
29. Cleave V, Yahioglu G, Le Barny P, Friend RH, Tessler N. (1999) Adv Mater. 11:285.
30. Baldo MA, O'Brien DF, You Y, et al. (1998) Nature. 395:151.
31. Baldo MA, Thompson ME, Forrest SR. (2000) Nature. 403:750.
32. Martin RE, Diederich F. (1999) Angew Chem Int Ed. 38:1350.
33. Stang PJ, Diederich F. (eds). (1995) Modern Acetylene Chemistry. Wiley, Weinheim, Germany.
34. Diederich F. (2001) Chem Commun. 219.
35. Bunz UHF. (2000) Chem Rev. 100:1605.
36. Wong W-Y. (2005) J Inorg Organomet Polym Mater. 15:197.
37. Beljonne D, Wittmann HF, Köhler A, et al. (1996) J Chem Phys. 105:3868.
38. Wong W-Y, Chan S-M, Choi K-H, Cheah K-W, Chan W-K. (2000) Macromol Rapid Commun. 21:453.
39. Lewis J, Raithby PR, Wong W-Y. (1998) J Organomet Chem. 556:219.
40. Wong W-Y, Lu G-L, Choi K-H, Shi J-X, (2002) Macromolecules, 35:3506.
41. Wong W-Y, Wong W-K, Raithby P-R. (1998) J Chem Soc Dalton Trans. 2761.

42. Wong W-Y, Choi K-H, Lu G-L, Shi J-X. (2001) Macromol Rapid Commun. 22:461.
43. Khan MS, Al-Mandhary MRA, Al-Suti MK, et al. (2003) Dalton Trans. 74.
44. Khan MS, Mandhary MRA, Al-Suti MK, et al. (2002) J Chem Soc. Dalton Trans. 2441.
45. Khan MS, Al-Mandhary MRA, Al-Suti MK, et al. (2002) J Chem Soc. Dalton Trans. 1358.
46. Wilson JS, Köhler A, Friend RH, et al. (2000) J. Chem. Phys. 113:7627.
47. Wong W-Y, Poon S-Y, Wong C-K. (2005) Polymeric materials: science and engineering. 93:791.
48. Chawdhury N, Köhler A, Friend RH, Younus M, Long NJ, Raithby PR, Lewis J. (1998) Macromolecules, 31:722.
49. Ho C-L, Wong W-Y. (2006) J Organomet Chem. 691:395.
50. Wong W-Y, Liu L, Shi J-X. (2003) Angew Chem Int Ed. 42:4064.
51. Wong W-Y, Liu L, Poon S-Y, Choi K-H, Cheah K-W, Shi, J-X (2004) Macromolecules, 37:4496.
52. Wilson JS, Chawdhury N, Al-Mandhary MRA, et al. (2001) J Am Chem Soc. 123:9412.
53. Gong X, Robinson MR, Ostrowski JC, Moses D, Bazan GC, Heeger AJ. (2002) Adv Mater. 14:581.
54. Baldo MA, Thompson ME, Forrest SR. (1999) Pure Appl Chem. 71:2095.
55. Corriu RJP. (2000) Angew Chem Int Ed. 39:1376.
56. Abd-Aziz AS, Carraher CE Jr, Pittman CU, Jr. Sheats JE, Zeldin M. (eds). (2005) Macromolecules containing metal and metal-like elements, Vol. 4, Group IVA Polymers, Wiley, New Jersey,.
57. Zeigler JM, Gordon FW (eds) (1990) Silicon based polymer science: A comprehensive resource, advances in chemistry series 224, ACS, Washington, DC.
58. Chicart P, Corriu RJP, Moreau JJE, Garnier F, Yassar A. (1992) In: Laine RM (Ed). Inorganic and Organometallic Polymers with Special Properties. Kluwer, p 179.
59. Birot M, Pillot JP, Dunoguès J. (1995) Chem Rev. 95:1443.
60. Corriu RJP, Douglas WE, Yang Z-X (1990) J Polym Sci. Polym Lett Ed. 28:431.
61. Tamao K, Uchida M, Izumizawa T, Furukawa K, Yamaguchi, S.(1996) J Am Chem Soc. 118:11974.
62. Loy DA, Shea KJ. (1995) Chem Rev. 95:1431.
63. Matsuo T, Uchida K, Sekiguchi A, (1999) Chem Commun. 1799.
64. Pschirer NG, Fu W, Adams RD, Bunz UHF. (2000) Chem Commun. 87.
65. Corriu R, Douglas W, Yang ZX. (1993) J Organomet Chem. 455:69.
66. Douglas WE, Guy DMH, Kar AK, Wang C, (1998) Chem Commun. 2125.
67. Ishikawa M, Ohshita J. (1997) in: Nalwa HS (ed). Silicon and germanium containing conductive polymers in handbook of organic conductive molecules and polymers, Vol. 2. Wiley, New York, p 685,.
68. Adachi A, Manhart SA, Okita K, Kido J, Ohshita J, Kunai A. (1997) Synth Met. 91:333.
69. Manhart SA, Adachi A, Sakamaki K, et al. (1999) J Organomet Chem. 592:52.
70. Itoh M, Mitsuzaka M, Iwata K, Inoue K. (1997) Macromolecules. 30:694.
71. Ohshita J, Shinpo A, Kunai A. (1999) Macromolecules. 38:5998.
72. Ishikawa M, Nate K. (1988) Inorganic and organometallic polymers, M. Zeldin, K. J. Wynne, H. R. Allcock (Eds.), (1988) ACS Symp. Ser. 360, ACS, Washington, DC.
73. Ohshita J, Ishikawa M. (1996) In: Salamone JC (ed). Polymeric materials encyclopedia. CRC Press, New York, p. 892.
74. Corriu R, Gerbier P, Guerin C, Henner B, Jean A, Mutin A. (1992) Organometallics. 11:2507.
75. Garten F, Hilberer A, Cacialli F, et al. (1997) Adv Mater. 9:127.
76. Silcoff ER, Sheradsky T, (1998) Macromolecules. 31:9116.
77. Boyer-Elma K, Carrè FH, Corriu RJP, Douglas WE. (1995) J Chem Soc Chem Commun. 725.
78. Kim ZD, Park JS, Kim HK, Lee TB, No KT. (1998) Macromolecules. 31:7267.
79. Kim HK, Ryu MK, Lee SM. (1997) Macromolecules. 30:1236.
80. Kim HK, Ryu MK, Kim KD, Lee SM, Cho SW, Park JW. (1998) Macromolecules 31:1114.
81. Gleiter R, Schäfer W, Sakurai H. (1985) J Am Chem Soc. 107:3046.
82. Sakurai H, Sugiyama H, Kira M. (1990) J Phys Chem. 94:1837.

83. Yao J, Son DY. (1999) Organometallics. 18:1736.
84. van Walree CA, Roest MR, Schuddeboom W, et al. (1996) J. Am. Chem. Soc. 118:8395.
85. Wong W-Y, Wong C-K, Lu G-L, Cheah K-W Shi J-X, Lin S. (2002) J Chem Soc Dalton Trans. 4587.
86. Wong W-Y, Wong C-K, Lu G-L. (2003) J Organomet Chem. 671:27.
87. Johnson BFG, Kakkar AK, Khan MS, et al. (1991) J Mater Chem. 1:485.
88. Lewis J, Khan, MS Kakkar AK, Johnson BFG, et al. (1992) J Organomet Chem. 425:165.
89. Wong W-Y, Lee AW-M, Wong C-K, et al. (2002) New J. Chem. 26:354.
90. Wong W-Y, Wong C-K, Lu G-L, Lee AW-M, Cheah K-W, Shi J-X. (2003) Macromolecules. 36:983.
91. Wong W-Y, Poon S-Y, Lee AW-M, Shi J-X, Cheah K-W. (2004) Chem Commun. 2420.
92. Cummings SD, Eisenberg R. (1996) J Am Chem Soc. 118:1949.
93. Wittmann HF, Friend RH, Khan MS, Lewis J. (1994) J Chem Phys. 101:2693.
94. Demas NJ, Crosby GA. (1970) J Am Chem Soc. 92:7262.
95. Turro NJ. (1991) Modern molecular photochemistry. University Science Books, Mill Valley, CA.
96. Wong W-Y, Wong C-K, Poon S-Y, Lee AW-M, Mo T, Wei X. (2005) Macromol Rapid Commun. 26:376.
97. Yamaguchi S, Tamao K. (1998) J Chem Soc. Dalton Trans. 3693.
98. Yamaguchi S, Iimura K, Tamao K. (1998) Chem. Lett. 89.
99. Yamaguchi S, Endo T, Uchida M, Izumizawa T, Furukawa K, Tamao K. (2000) Chem Eur J. 6:1683.
100. Chen H-Y, Lam WY, Luo JD, et al. (2002) Appl Phys Lett. 81:574.
101. Yamaguchi S, Endo T, Uchida M, Izumizawa T, Furukawa K, Tamao K. (2001) Chem Lett. 98.
102. Kakkar AK, Khan MS, Long NJ, et al. (1994) J Mater Chem. 4:1227.
103. Zhan X, Liu Y, Wu X, Wang S, Zhu, D.(2002) Macromolecules 35:2529.
104. Kraft A, Grimsdale AC, Holmes AB. (1998) Angew Chem Int Ed. 37:402.
105. Lee AWM, Yeung ABW, Yuen MSW, Zhang H, Zhao X, Wong W-Y. (2000) Chem Commun. 75.
106. Leuninger J, Uebe J, Salbeck J, Gherghel L, Wang CS, Müllen K. (1999) Synth. Met. 100:79.
107. Shahram MA, Yaghoub S, Mehdi H, Leila AF. (2005) Eur Polym J. 41:491.
108. Kim JP, Lee WY, Kang JW, Kwon SK, Kim JJ, Lee JS. (2001) Macromolecules. 34:7817.
109. Hwang SW, Chen Y. (2002) Macromolecules. 35:5438.
110. Hale WF, Farnham AG, Johnson RN, Clendinning RA. (1967) J Polym Sci Part A: Polym Chem. 5:2399.
111. Johnson RN, Farnham AG, Clendinning RA, Hale WF, Merriman CN. (1967) J. Polym. Sci. Part A: Polym. Chem. 5:2375.
112. Bottino FA, Pasquale GD, Scalia L, Pollicino A. (2001) Polymer. 42:3323
113. Jenks WS, Lee W, Shutters D. (1994) J Phys Chem. 98:2282.
114. Lee,W, Jenks WS. (2001) J Org Chem. 66:474.
115. Poon S-Y, Wong W-Y, Cheah K-W, Shi J-X, (2006) Chem Eur J. 12:2550.
116. Liu L, Poon S-Y, Wong W-Y. (2005) J Organomet Chem. 690:5036.
117. Hertel S, Setayesh S, Nothofer HG, Scherf U, Mullen K, Bässler H. (2001) Adv Mater. 13:65.
118. Monkman AP, Burrows HD, Hartwell LJ, Horsburgh LE, Hamblett I, Navaratnam S. (2001) Phys Rev Lett. 86:1358.
119. Zhang H, Lee AW-M, Wong W-Y, Yuen MS-M. (2000) J Chem Soc Dalton Trans. 3675.
120. Chawdhury N, Köhler A, Friend RH, et al. (1999) J Chem Phys. 110:4963.
121. Lewis J, Long NJ, Raithby PR, Shields GP, Wong W-Y, Younus M. (1997) J Chem. Soc. Dalton Trans. 4283.
122. Cornil J, Beljonne D, dos Santos DA, Shuai Z, Brédas JL. (1996) Synth Met. 78:209.
123. Beljonne D, Cornil J, Brédas JL, Friend RH. (1996) Synth Met. 76:61.
124. Köhler A, Wittmann HF, Friend RH, Khan MS, Lewis J. (1994) Synth Met. 67:245.

125. Younus M, Köhler A, Cron S, et al. (1998) Angew Chem Int Ed. 37:3036.
126. Chawdhury N, Younus M, Raithby PR, Lewis J, Friend RH. (1998) Opt Mater. 9:498.
127. Wong KM-C, Zhu S, Hung L-L, Zhu N, Yam VW-W, Kwok H-S. (2005) Chem Commun. 2906.
128. Lu W, Mi B-S, Chan MCW, et al. (2004) J Am Chem Soc. 126:4958.
129. Lin Y-Y, Chan S-C, Chan MCW, et al. (2003) Chem Eur J. 9:1264.
130. Tutt LW, Kost A. (1992) Nature. 356:225.
131. Zhou G-J, Wong W-Y, Cui D, Ye C, (2005) Chem Mater. 17:5209.
132. Wong W-Y (2005) Comment Inorg. Chem. 26:39.

The Spectroscopy and Photophysical Behavior of Diphosphine- and Diisocyanide-Containing Coordination and Organometallic Oligomers and Polymers: Focus on Palladium and Platinum, Copper, Silver, and Gold

Pierre D. Harvey

1 Introduction

The field of organometallic and coordination oligomers and polymers has rapidly flourished in recent years [1–11]. The quest for finding new materials is driven by applications in photonics such as light-emitting-diodes, nonlinear optical and photoconducting materials and photovoltaic cells, electronic devices including materials based on semi-conductivity properties, liquid crystals useful in digital display, nanomaterials, and heterogeneous catalysis. Although the field is "dominated" by the use of N-containing and organometallic assembling ligands, relatively little is known about oligomers and polymers built on diphosphines and diisocyanides [2]. This chapter deals more particularly with recent advances in spectroscopy and photophysical behavior of coordination and organometallic polymers based on the diphosphine and diisocyanide bridging ligands involving direct P-M and NC-M bonds in the backbone, mainly for cases where the metals are palladium, platinum, cooper, silver, and gold. The photophysical parameters and phenomena addressed in this chapter are triplet energy extracted from the emission maxima, excited state lifetimes extracted from luminescence kinetics, emission quantum yields, emission quenching, photoinduced electron and energy transfers, and exciton phenomena. In addition,

A.S. Abd-El-Aziz et al. (eds.), *Inorganic and Organometallic Macromolecules: Design and Applications.*

photovoltaic cell performance for one case is also addressed. Although this field appears to be in its infancy when compared with the well established organometallic and N-donor ligand-containing polymers, very clear advances are being reported at an ever-increasing pace. This chapter stresses the comparison between the polymers, oligomers, and the mononuclear related complexes in an attempt to show how the phophysical parameters and phenomena are influenced by the presence of interacting short-, medium-, and long-chains.

2 Palladium and Platinum

2.1 Valence +2

Palladium(II)- and platinum(II)-containing oligomers and polymers built with diphosphine and diisocyanide assembling ligands—which allow for the presence of metal in the backbone—are not luminescent based on the author's personal experience (e.g., **1** and **2**) [12], but also according to literature (e.g., polymers **3–5**) [13,14]. This contrasts drastically with the well-known palladium(II)- and platinum(II)-containing oligomers and polymers built with ethynyl assembling fragments (also called polyynes), with the general formula as shown for polymer **6** [15,16].

The nonluminescent versus luminescent (and strongly luminescent) photophysical property differences between polymers **1–5** and **6** stem from the nature of the ligands and the strong Pt-CC- bond. The bis(ethynyl)aryl spacers are already, in most cases, strongly luminescent and intense fluorescence can be readily detected with a simple ultraviolet light (UV) that is used for thin layer chromatography (TLC). This is not true for most common diphosphines and diisocyanides. Exceptions do exist, but literature shows that such ligands have not yet been employed to synthesize coordination and organometallic polymers. In addition, photo-induced M-P and M-CN bond scission are also possible, which contribute to the waste of absorbed light energy (energy wasting process); luminescence is therefore not

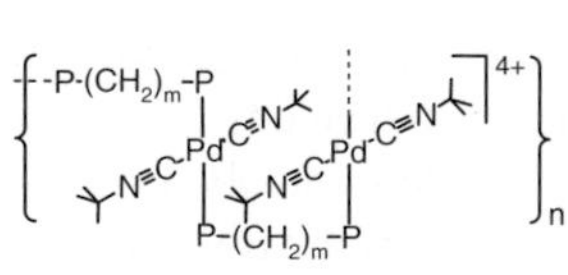

1 (m = 5; P = PPh_2)
2 (m = 6; P = PPh_2)

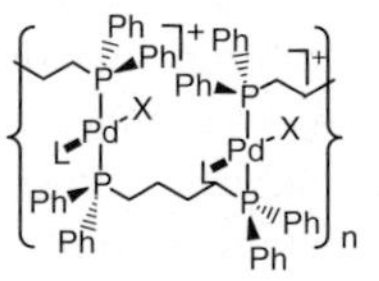

3 (L = MeCN; X = Me)
4 (L = Me(CO); X = Me)

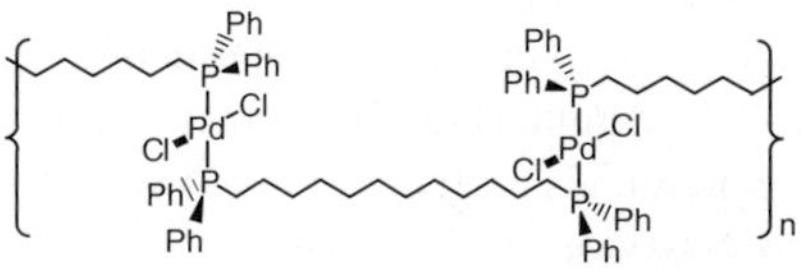

5

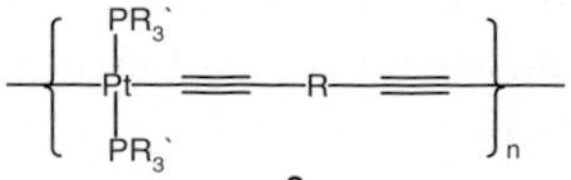

R = aryl
R` = Phoralkyl

6

detected at room temperature in fluid solution, and in general, the luminescence is very weak in both the solid and frozen (77 K) state. On the other hand, the M-CC- bond is very strong and does not undergo photo-induced scission. This results from the stronger conjugation between the –C≡C– fragment and the M *d* orbitals, in comparison with P *d*- and CN-R π-orbitals, and the extra electrostatic attraction between the mono-negatively charged –C≡C– ligand and the positively charged metal (2+). Both phosphine and isocyanide are neutral. Rigidity also contributes to enhance the radiative processes by reducing the number of low-frequency vibrational channels for nonradiative excited state relaxation.

2.2 *Valence +1*

With the lower valence species—such as +1, +0.5, and 0—medium to weak luminescence arising from oligomers and polymers (along with their "related monomeric" models) was detected on many occasions. The simplest examples are the d^9-d^9 Pd-Pd bonded complexes, $Pd_2(dmb)_2X_2$ (dmb = 1,8-diisocyano-*p*-menthane; X = Cl, Br), which were originally synthesized in 1992 [17], and investigated for their photophysical properties the following year [18]. But it was only recently that it was unambiguously demonstrated that these simple complexes exist as polymers in the solid state and a stressed Pd-Pd-bonded complex in solution (Figure 4.1). In the solid state, the $Pd_2(dmb)_2Cl_2$ (**7**) is amorphous and exhibits a glass transition at about 5 °C (from differential scanning calorimetry [DSC]) and can make stand alone films [19]. In solution, the polymer becomes a binuclear complex as is demonstrated by the estimation of the molecular dimension from the measurements of spin-lattice relaxation time (T_1) and nuclear Overhauser enhancement (NOE) constants using closely related binuclear palladium complexes. The technique was recently established [20,21], and used successfully on several occasions thereafter [12,22,23].

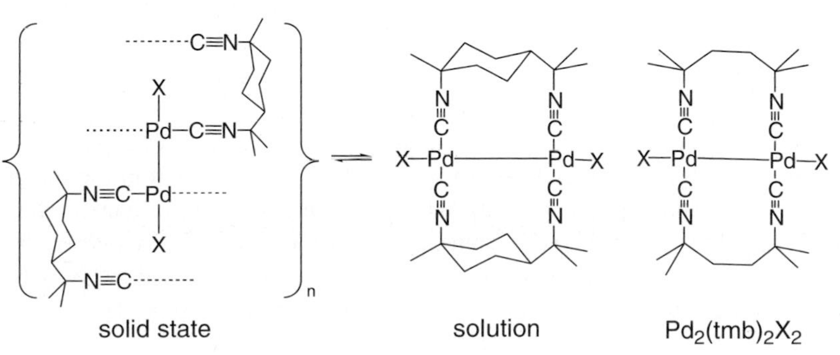

solid state — solution — $Pd_2(tmb)_2X_2$

7 (X = Cl), **8** (X = Br) — **9** (X = Cl), **10** (X = Br)

Figure 4.1 *(Left)* Equilibrium taking place when the solid $\{Pd_2(dmb)_2X_2\}_n$ polymers (X = Cl, Br) are dissolved in solution to generate stressed Pd-Pd-bonded d^9-d^9 complexes. *(Right)* structure of the $Pd_2(tmb)_2X_2$ complexes in solution

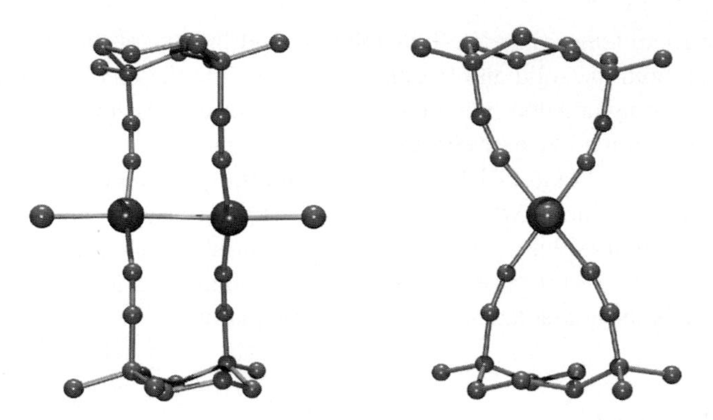

Figure 4.2 Computer modeling showing the binuclear species, $Pd_2(dmb)_2X_2$, resulting from the dissolution of the corresponding organometallic polymers, stressing the ring deformation in the $Pd_2(dmb)$ fragment

The driving force for such phenomena is the the ease of dissolving small molecules—with better solute-solvent interactions—rather than large polymers, particularly when the polymers do not incorporate soluble side-chains. The process is obviously aided by the rather facile lability of the isocyanide ligands, which is completely consistent with the low valence of the palladium (I). The stressed Pd-Pd bond can be seen in Figure 4.2 (computer modeling) where the natural bite distance in dmb is 4.4 Å whereas the normal Pd-Pd bond distance is about 2.6 Å [17,18]. This stress is responsible for the ring-opening-polymerization when the solute-solvent interactions decrease during precipitation.

The emission spectra of $Pd_2(dmb)_2Br_2$ in ethanol at 77 K is shown in Figure 4.3 as an example, and was assigned to a triplet state originating from a $d\sigma$-$d\sigma^*$ excitation localized mainly in the Pd-Pd bond based upon **extended Hückel molecular orbital** (EHMO) calculations and polarization of the emitted light. The emission lifetimes (τ_e) and emisson maxima (λ_{emi}) for 4 related compounds are as follows: (1) 71 ns and 625 nm for $Pd_2(dmb)_2Cl_2$ (**7**), (2) 177 ns and 650 nm for $Pd_2(dmb)_2Br_2$(**8**), (3) 81 ns and 625 nm for $Pd_2(tmb)_2Cl_2$ (**9**), and (4) 125 ns and 650 nm for $Pd_2(tmb)_2Br_2$ (**10**) where tmb = $Me_2(CN)CCH_2CH_2C(NC)Me_2$. The shot emission lifetime was explained by both photo-induced homolytic Pd-Pd and Pd-X bond cleavages based on ps flash photolysis measurements. The goal of replacing dmb by the more flexible tmb ligand was to reduce the ring stress within the material. The emission quantum (Φ_e) in all cases could not be measured because their intensity was too weak ($<10^{-4}$). In the solid state, the emissions were barely perceptible at about the same wavelength as mentioned above, but they were too weak to be analyzed accurately, suggesting efficient intra-polymer emission quenching. The nature of this quenching has not yet been elucidated.

The replacement of the halides by triphenylphosphine and diphosphines such as $Ph_2P(CH_2)_mPPh_2$ (m = 4; bis(diphenylphosphino)butane; m = 5, bis(diphenylphosphino)pentane; m = 6, bis(diphenylphosphino)hexane) and Ph_2P-C≡C-PPh_2 (bis(diphenylphosphino)acetylene) new model complex (**11**), and polymers (**12–15**) are obtained, respectively [19].

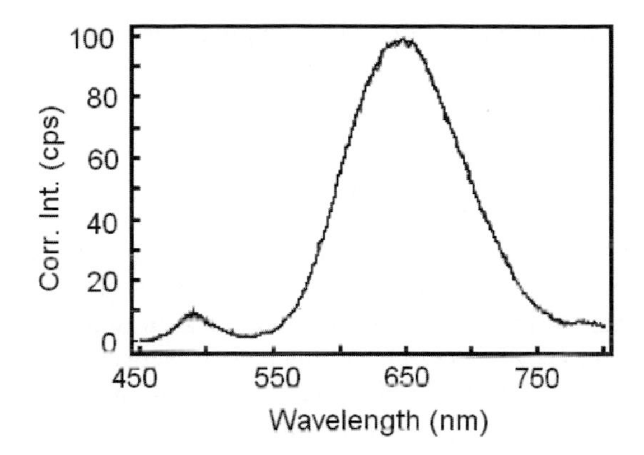

Figure 4.3 Emission spectrum of $Pd_2(dmb)_2Br_2$ (**8**) in ethanol at 77 K. Excitation is at 420 nm in the *dσ-dσ*[*] band located in the Pd-Pd chromophore. The band located at about 440 nm is due to an impurity. Modified from reference 18

11 **12** (m = 4), **13** (m = 5), **14** (m = 6) **15**

Table 4.1 Photophysical parameters of **11**–**15** in butyronitrile at 77 K [19]

Compound or polymer	λ_{abs}(*dσ-dσ*[*])/nm	λ_{emi}/nm	τ_e/ns	Φ_e
11	475	627	2.75	0.14
12	480	632	1.87	0.026
13	484	634	2.70	0.071
14	485	636	2.24	0.046
15	488	638	2.30	0.15

These species are moderately luminescent at 77 K in butyronitrile, but are not luminescent at room temperature, both in the solid state and in solution. Their luminescence also originates from a *dσ-dσ*[*] triplet excited state based on DFT computations. Details of the photophysical parameters are presented in Table 4.1.

The position of the dσ-dσ[*] absorption and emission bands were found to be phosphine-dependent. The trend going from **11** to **14** is explained by the increase in electronic density on the P-atom by inductive effect destabilizing the highest occupied molecular orbital (HOMO) when Ph, $(CH_2)_4$, $(CH_2)_5$, and $(CH_2)_6$ are compared. The stronger red-shift in absorption and emission between **12–14** and **15** is explained by the change in C hybridation state going from sp^2 to sp, stabilizing the lowest unoccupied molecular orbital (LUMO) via coupling of the empty *d* orbitals on P and the empty π-orbital on the ethynyl moiety. The emission lifetimes are particularly short (ns time scale) which was consistent with the data found for **7**, and do not follow the quantum yield data, which are larger than expected for such short lifetimes. At this time, this phenomenon has yet to be explained. It is interesting to note that both the lifetimes and quantum yields decrease from **11** to **12–14**, stressing the presence of a slight emission quenching when flexible ligands are attached to the chromophore. This behavior is known as the loose bolt effect. The full-width-at-half-maximum (FWHM) of the emission band remains about constant all along the series (4,900–5,200 cm^{-1}) indicating absence of significant electronic interactions between the units.

Other examples of d^9-d^9 Pd-Pd-bond containing polymers were reported (19). These polymers, **16–18**, are structurally related to **12–15** but differ in the relative position of the phosphine and isocyanide ligands (equatorial versus axial, where axial is along the M-M bond). As a consequence, the electronic density is not distributed the same along the Pd-Pd bond (i.e., the chromophore), and the absorption and emission dσ-dσ* bands are significantly more blue-shifted with respect to **12–15** (Table 4.2) which indicates a stabilization of the HOMO by having less electronic density on the Pd-Pd axis. This observation is entirely consistent the weaker σ-donor ability of the isocyanide ligand in comparison with the phosphines in general.

Polymer **18** was synthesized in order to verify whether the molecular weight is influenced by side chain groups (it is not), but a notable difference in λ_{abs}(dσ-dσ*) and λ_{emi} was observed (**17** and **18**) despite the fact that the FWHM (2,300–2,500 cm^{-1}) were relatively constant all along this series. And similarly, the FWHM for the emission band varied only from 5,300 to 5,600 cm^{-1} going from **17** to **18**. The loose bolt effect does not apply here because the photophysical parameters increase for

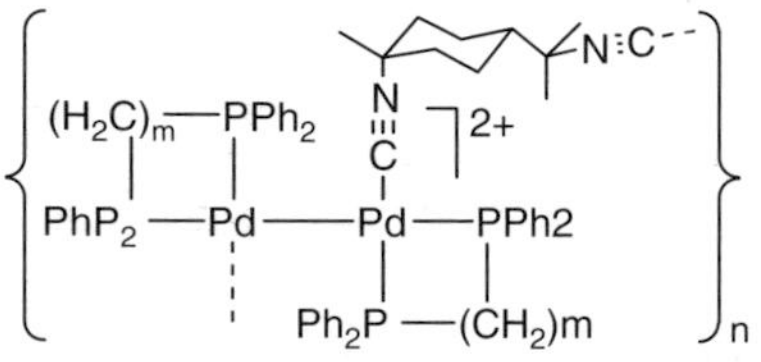

16 (m = 2), **17** (m = 3), 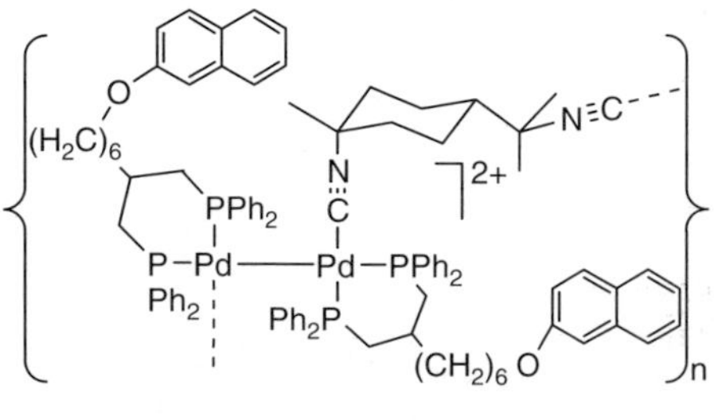 **18**

Table 4.2 Photophysical parameters of **16–18** in butyronitrile at 77 K [19]

Compound or polymer	λ_{abs}(*d*σ-*d*σ*)/nm	λ_{emi}/nm	τ_e/ns	Φ_e
16	424	509	1.94	0.13
17	406	508	1.50	0.12
18	390	500	1.98	0.17

18 whereas the opposite was anticipated. All in all, there is no clear evidence that the different unit electronically interacts with each other.

The model compounds **19** and **20**, along with polymers **21** and **22**, are interesting because the effect of substituting all 4 equatorial positions with P-donors and the 2 axial ones with CNR ligands can be compared [24]. Table 4.3 summarizes some photophysical data for **19–22**.

From this table one can now state that $\lambda_{abs}(d\sigma\text{-}d\sigma^*)$ for the Pd-containing materials vary as **19** and **21** (equatorial 4 P; axial 2 CNR) <**16–18** (equatorial 2 P and 2 CNR; axial 2 P) < **11–15** (equatorial 4 CNR; axial 2 P), consistent with the electronic density argument. The emission lifetime for Pd-containing materials are still in the ns regime, but the Pt-materials exhibit longer lifetimes going all the way up to the μs. Such lifetimes are normal for such experimental conditions and metal, and may reflect the stronger Pt-Pt and Pt-L bonds where excited state deactivation via a nonradiative photo induced homolytic bond cleavage and reformation does not occur efficiently.

The investigations of the model compounds **23** and **24** and the polymers **25** and **26** provided the first clear evidence for inter-unit interactions [24]. Whereas all previous model compounds and polymers discussed in this chapter showed very little clear-cut evidence for inter-unit interactions based on the comparison between band shift and FWHM within the model compounds and the polymers, where (at best) several ~20-nm shifts (in one case), were observed, **23–26** show otherwise. Modest red-shifts in $\lambda_{abs}(d\sigma\text{-}d\sigma^*)$ were observed for **23** and **25** and **24** and **26**, but the red-shifts were far more spectacular for the λ_{emi} data (85 and 42 nm) (Table 4.4). This is clear evidence for electronic conjugations along the backbone, which places the triplet state—a state from which luminescence arises—lower in energy.

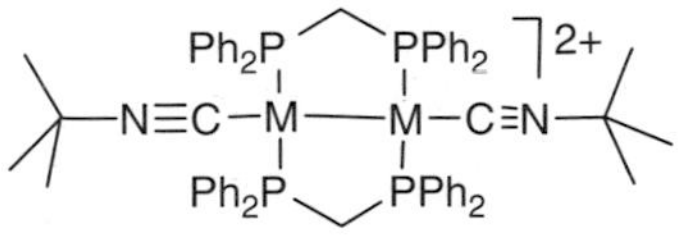

19 (M = Pd), **20** (M = Pt) 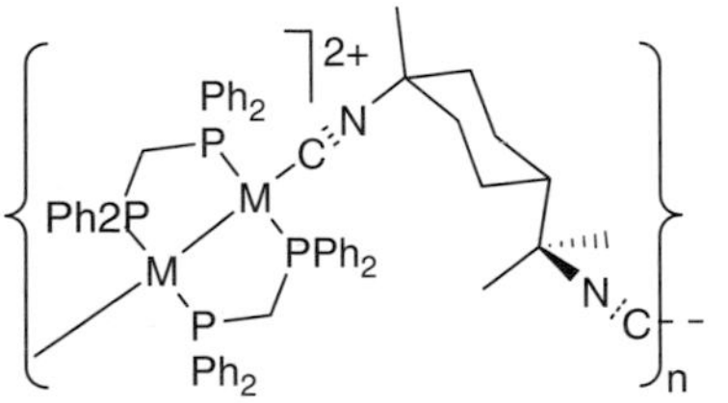 **21** (M = Pd), **22** (M = 23)

Table 4.3 Photophysical parameters of **19–22** in butyronitrile at 77 K [24]

Compound or polymer	$\lambda_{abs}(d\sigma\text{-}d\sigma^*)$/nm	FWHM/cm^{-1}	λ_{emi}/nm	τ_e
19 (M = Pd)	366	3,100	730	too weak
21 (M = Pd)	366	2,830	750	6.2 ns
20 (M = Pt)	328	2,150	675	3.2 μs
22 (M = Pt)	332	2,440	676	4.1 μs

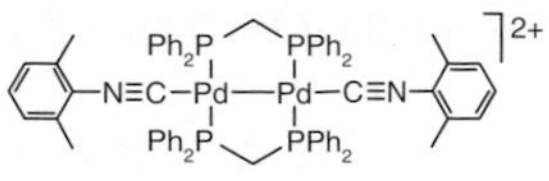

23 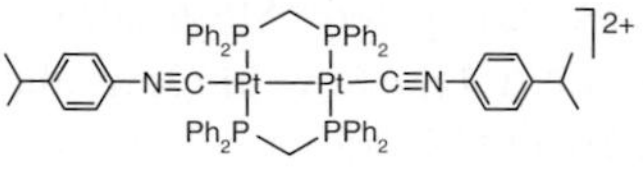 **24** 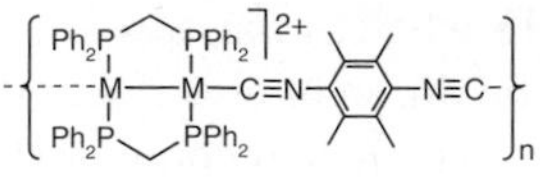 **25** (M = Pd), **26** (M = Pt)

Table 4.4 Photophysical parameters of **23–25** in butyronitrile at 77 K [24]

Compound or polymer	$\lambda_{abs}(d\sigma\text{-}d\sigma^*)$/nm	FWHM/cm^{-1}	λ_{emi}/nm	τ_e
23	364	–	635	4.5 ns
25 (M = Pd)	370	2,640	720	3.7 ns
24	340	1,680	673	3.2 μs
26 (M = Pt)	344	1,160	715	3.1 μs

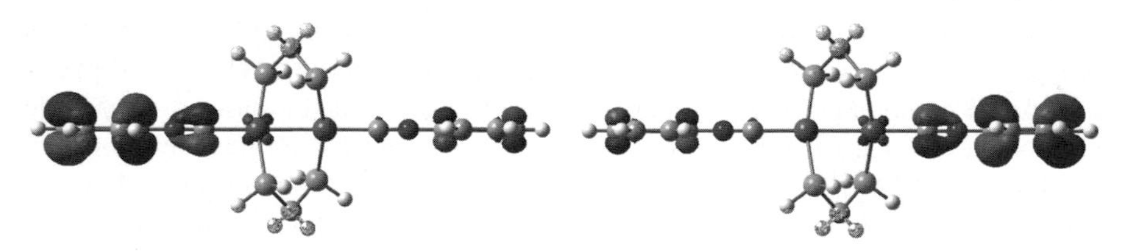

Figure 4.4 MO drawings for the frontier orbitals (HOMO and HOMO-1: degenerated) for $Pd_2(H_2PCH_2PH_2)_2(CNMe)_2^{2+}$.[24]

The presence of conjugation was also indirectly noted by the important gain in thermal stability when the alkyl- and the aryl-isocyanide series (aryl far more stable than alkyl) were compared.

Evidence for conjugation is easily depicted in the pictures of the frontier orbitals (HOMO and HOMO-1 extracted from DFT computations; Figure 4.4), where atomic contributions from the Pd *d* orbitals placed parallel with are depicted. In many ways (i.e., structures, presence of thermal stabilities, and luminescence), the arylisocyanide-containing polymers have much in common with the corresponding aryl polyynes. The major difference is that the former exhibit a lesser degree of emission intensity, particularly at room temperature.

2.3 Valence +0.5

Polymers of clusters are a very rare type of coordination and organometallic polymers. The mixed-valent model clusters **27** and **28** [i.e., Pd(0.5) and Pt(0.5)] and polymers **29–32** were prepared and thoroughly investigated from a spectroscopic and photophysic standpoint [25,26]. From crystallography, the M-M bond distances were obtained and ranged from 2.52(1) to 2.60(2) Å for **27** and **29**, and 2.65(2) Å for **28**. Evidence of M-M bonding was obtained via Raman and UV-vis spectroscopy. Strong low-frequency Raman-active modes associated with $\nu(Pd_2)$ at 165 and 86 cm^{-1} for **27**, and to $\nu(Pt_2)$ at 162, 84 cm^{-1} for **28**, were detected. Similarly, a strong and narrow absorption associated with a $d\sigma$-$d\sigma^*$ absorption—which also undergo a large decrease in FWHM upon cooling (due to the disappearance of vibrationnaly excited modes known as "hot vibronic bands")—were observed at 534, 488, 412 and 394, 394, and 394 nm for **27, 29, 28**, and **30–32**, respectively.

The photophysical data are compared in Table 4.5. The most striking information is the very low emission lifetimes in all cases (lower than all those previously discussed), and the rather weak intensity for all emission bands despite the fact that emissions are triplet $d\sigma$-$d\sigma^*$ excited states. The weakness in intensity may result from facile photo-induced M-M and M-L bond cleavages, which make them particularly subject to oxidation. They are indeed air sensitive in solution, but also decompose easily into oxidized species when submitted to mild redox agents [25,27]. No emission was observed at room temperature, both in fluid solution and in the solid state.

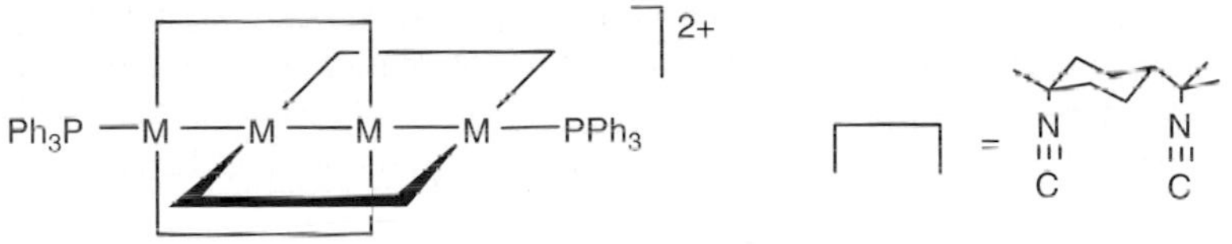

27 (M = Pd), **28** (M = Pt)

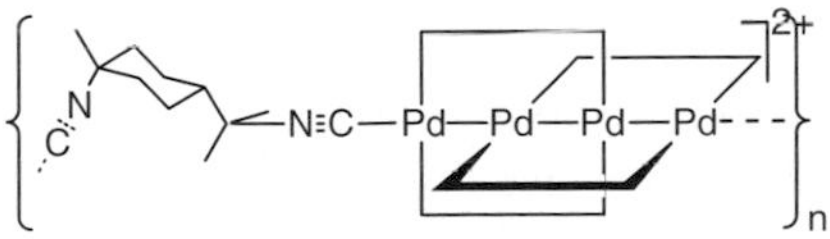

29

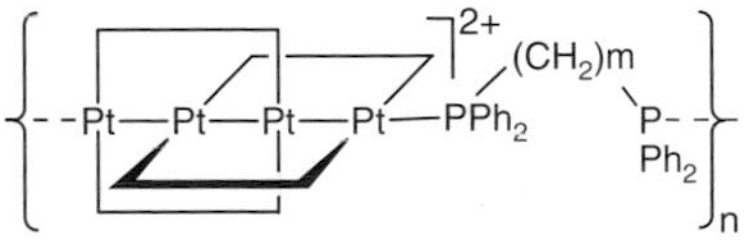

30 (m = 4), **31** (m = 5), **32** (m = 6)

Table 4.5 Photophysical parameters of **27–32** in EtOH at 77 K [25,26]

Compound or polymer	$\lambda_{abs}(d\sigma$-$d\sigma^*)$/nm	λ_{emi}/nm	τ_e/ns
27	534	635	0.67
29	488	720	1.14
28	412	750	2.71
30	394	736	4.78
31	394	750	5.15
32	394	755	5.17

2.4 Zero Valence

To our knowledge, no investigation on organometallic and coordination polymers of palladium and platinum has so far been reported, presumably because these materials exhibit extreme air-sensitivity in solution in most cases and the particularly enhanced lability of the ligands around the metal atom prevent proper and adequate investigation of the structure. However, strong luminescene is documented for such phosphine- and isocyanide-containing mono- and polynuclear complexes [28–33]. There is still room for investigation, notably with the strongly luminescent d^{10}-d^{10} $M_2(dba)_3$ precursors (M = Pd, Pt; dba = dibenzylideneacetone), which are robust in solution while in the presence of air [28,29,31].

3 Copper and Silver

These metals are very different from palladium and platinum. Silver presents almost one exclusive oxidation state the d^{10} metal Ag(I), whereas copper exhibits two—Cu(I) and Cu(II). Mixed-valent species are extremely rare for these metals. Conversely, the Cu(I)- and Ag(I)-containing species, have a tendency, when bearing phosphine and isocyanide ligands, to luminesce in the solid state, in solution at room temperature, and at lower temperatures [34–36]. This makes them interesting targets for new photonic materials. Because of the very strong similarity in photophysical behavior, these two metals are addressed simultaneously in this chapter.

3.1 Homoligand-Containing Polymers

Very little research has been performed on this category of polymers. The earliest example is the $\{Ag(dmb)_2^+\}_n$ polymer (**33**; dmb = 1,8-diisocyano-p-menthane) [37]. The X-ray results revealed a one-dimensional polymeric structure with Ag atoms tetra-coordinated by isocyanide fragments. Each metal was doubly bridged by dmb ligands and formed a zig-zag at the 104° Ag_3 angle; the Ag-Ag separation was 5.0 Å. Overall, these polymers form rigid sticks as shown in Figure 4.5. The rigidity can be seen in the rapid conformation change, from deformed to linear, when a bent structure is set as a starting point in a computer-modeling file.

The counter ion can be simple anions such as PF_6^-, BF_4^-, and NO_3^- and find a vacancy in the lattice, thereby playing no role in the photonic behavior of the materials. The polymers are luminescent in both solid state and solution, particularly at 77 K (Figure 4.6).

The unresolved emission band takes its origin from a metal-to-ligand-charge-transfer (MLCT) state according to EHMO calculations [38]. This is not surprising because the metal center is electron-rich (d^{10}) and the isocyanide fragment exhibits π-acid empty orbitals. The absorption spectrum peaks placed at 204 nm

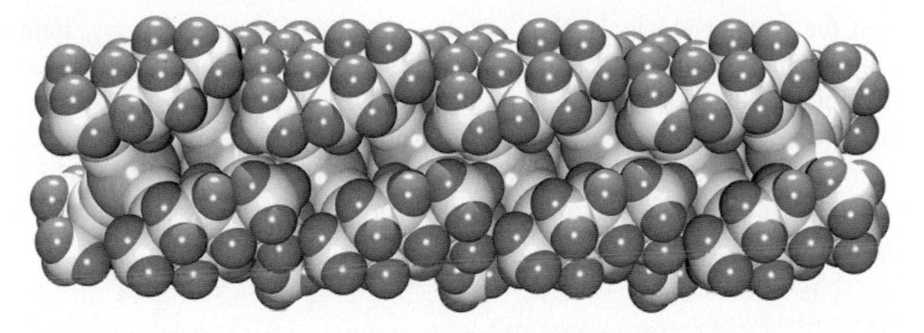

Figure 4.5 Space filling model of rigid-stick polymer **33** showing the zig-zag arrangement of the silver metals

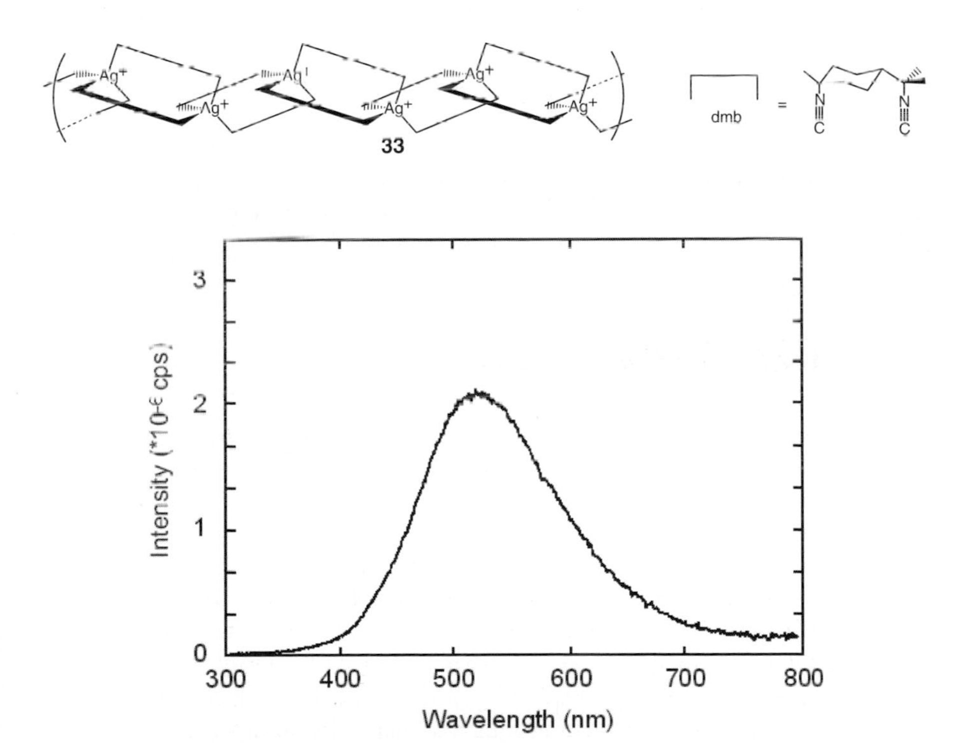

Figure 4.6 Emission spectrum of **33** (as a PF_6^- salt) in the solid state at 77 K stressing the particularly large width

(UV-excitation), making this singlet MLCT one of the most high energy states known. It has received worldwide attention [39,40]. The emission lifetime is found in the μs regime, which is normal for such a metal and ligand. However, anomalies have been noted by time-resolved emission and polarized emission spectroscopy. These anomalies are discussed below.

When the counter-ion is $TCNQ^-$ (tetracyanoquinodimethane anion), luminescence is no longer detected [41]. This likely results from a photo-induced electron transfer from the high-energy electron-rich MLCT triplet state (one can see that the unresolved emission band starts at about 400 nm at 77 K, designating the location of the 0-0 band), to the $TCNQ^-$ acceptor (a single TCNQ may exits under three forms: neutral, mono- and di-anion). The reaction is shown in Eqs. 4.1 and 4.2:

$$\{Ag(dmb)_2^{\ +}\}_n^{\ *} + TCNQ^- \rightarrow [\{Ag(dmb)_2^{\ +}\}_n]^+ + TCNQ^{2-} \qquad (4.1)$$

$$[\{Ag(dmb)_2^{\ +}\}_n]+ + TCNQ^{2-} \rightarrow \{Ag(dmb)_2^{\ +}\}_n + TCNQ^- \qquad (4.2)$$

But efficient back electron transfer occurs, which leads to nonet photochemistry (in the absence of oxygen). It is important to note that Ag(II) is an unlikely oxidation state, but the hole created by this electron transfer must be delocalized on the polycation polymer, $[\{Ag(dmb)_2^{\ +}\}_n]^+$. The supporting argument for this is that the oxidation waves measured by cyclic voltammetry (unpublished results) are extremely large, which is consistent with the existence of oligomers of different length in solution for which a delocalized hole would result in different oxidation potential. Further research should prove extremely useful in the design of photonic materials based on this polymer.

3.2 *Mixed Ligand-Containing Polymers*

This category of polymers is interesting because of the possible structural diversity for such materials and the "mixed-ligand" nature of the excited state. As for the structures, both the diphosphines and diisocyanides can act as assembling ligands for the metal. The mixed-nature of the excited state in the "poly-chromophore" is easy to predict from EHMO calculations on the luminescent mixed-ligand di-copper and di-silver models such as **34–37** [42,43], which can serve as useful polymer models for polymer (**38**) [44]. In complex **34**, the atomic contributions for the HOMO are mainly the P-lone pairs and Cu *d*-orbitals, whereas for the LUMO, these were the CO_2

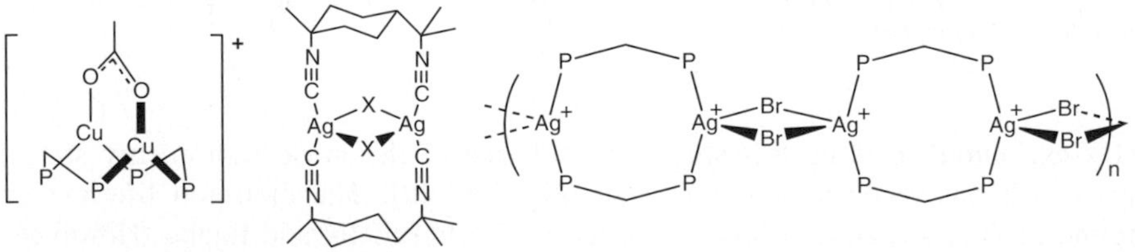

34 **35** (X = Cl), **36** (X = Br), **37** (X = I) **38** (P = PMe_2)

π^*-system, making the lowest energy excited state a mixed metal-ligand charge transfer excited state (L/ML′CT) [42]. Similar conclusions were obtained for the complexes **35–37** with L = X [43]. As far as interunit interactions are concerned, a comparison of the spectroscopic signature and photophysical behavior between the model compounds and polymers is needed to address them.

Two series of model compounds (**39, 40, 43, 44**) and coordination polymers (**41, 42, 45, 46**) were recently investigated using dmb as the assembling ligand, and the diphosphine as a simple chelating ligand [45]. Noting the d^{10} configuration of the metal, a L/ML′CT (L = P, M = Cu or Ag, L′= CNR) absorption and emission were anticipated. Figure 4.7 shows an example of absorption and emission spectra of one of these materials (**42**).

For all cases, the emission bands are unresolved and very broad. This is often encountered for such tetra-coordinated materials [33,42,43], but no anomaly is detected in the decay traces. The lifetime decays are all mono-exponential, suggesting the absence of excitonic phenomena, as discussed later in this chapter. However, the emission position red-shifts by about 22 nm going from **40** to **42** and from **43** to **45** (Table 4.6), suggesting interactions between the polymeric units. However, taking these uncertainties into account, the emission lifetimes are mainly constant. The large blue-shift from **39** to **41** is not explained at this time.

Another family of coordination polymers of the same type, **47–52**, where a diphosphine was used as an assembling unit and *tert*-butyl isocyanide as the monodentate ligand (to be placed at the other vacant site of the tetrahedral about the

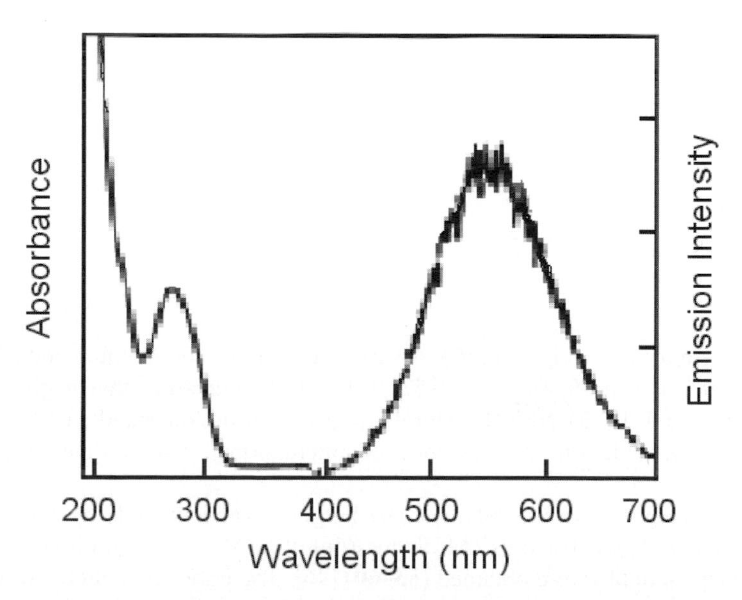

Figure 4.7 UV-vis spectrum *(left)* in MeCN and solid state emission spectrum *(right)* for **42** at 298 K. (Modified from reference [45])

Table 4.6 Solid-state emission data for **39–46** at 298 K [45]

Model compound or polymer	λ_{emi}(L/ML′CT)/nm	τ_e/μs
39	540	42 ± 4
40	515	21 ± 4
43	478	18 ± 3
41	480	38 ± 5
42	548	27 ± 2
45	500	22 ± 4
46	480	48 ± 4

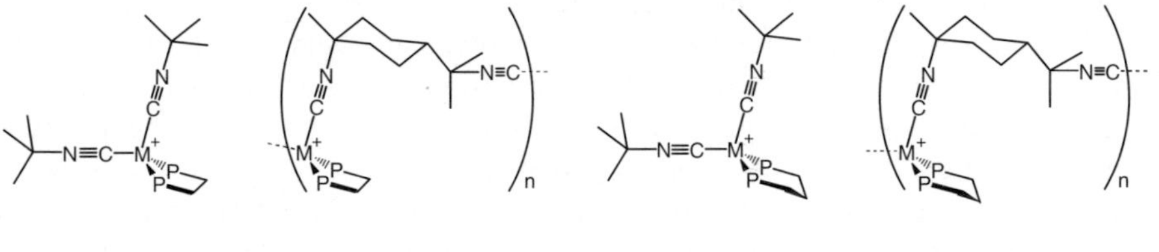

39 (M=Cu), 40 (M=Ag) 41 (M=Cu), 42 (M=Ag) 43 (M=Cu), 44 (M=Ag) 45 (M=Cu), 46 (M=Ag)
P = PPh_2 (counter-anion = BF4)

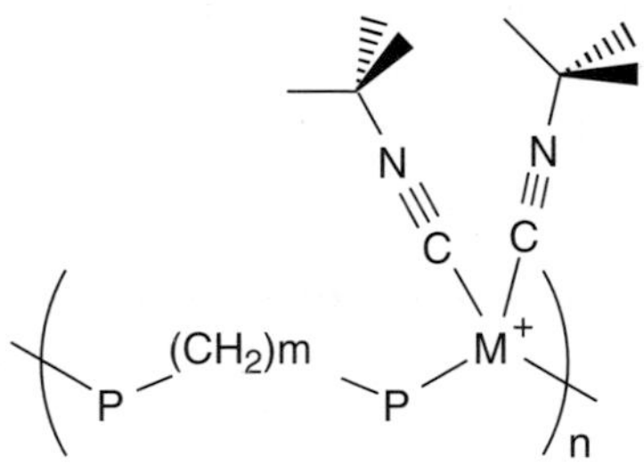

M	dppb (m = 4)	dpppen (m = 5)	dpph (m = 6)
Cu	**47**	**48**	**49**
Ag	**50**	**51**	**52**

metal), were also investigated [46]. A similarity in emission maxima and lifetimes was noted between **39–46** and **47–52** (Table 4.7), indicating that a given chromophore (here $P_2M(CNR)_2^+$) exhibits a given emission band (with varying maxima depending on the presence of interactions), and a given range for lifetime.

By replacing the two monodentate *tert*-butyl isocyande ligands by dmb, weakly soluble materials identified as $\{M(diphos)(dmb)^+\}_n$ (M = Cu, Ag; diphos = dppm, dppb, dpppen, dpph) were obtained (**53–60**) [46]. The polymers tend to swell when dissolving, consistent with their polymeric nature. One of them made crystal of suitable quality for X-ray analysis (**54**). However, the strucutre for the dppb-,

Table 4.7 Solid-state emission data for **47–52** at 298 K [46]

Model compound or polymer	λ_{emi}(L/ML′CT)/nm	τ_e/μs
47	475	22 ± 5
50	481	14 ± 3
48	483	31 ± 5
51	491	55 ± 4
49	485	12 ± 3
52	486	56 ± 4

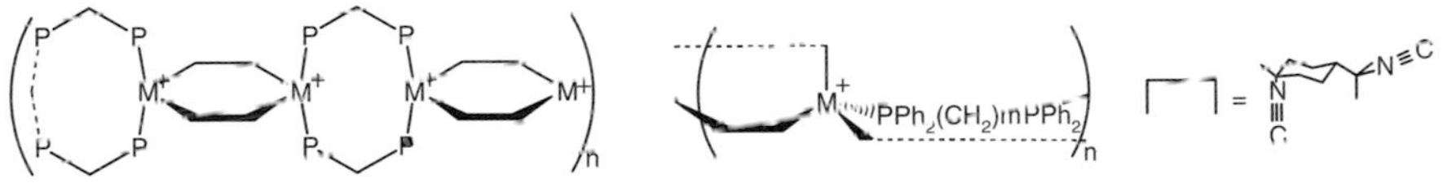

M	(P − PPh$_2$)	dppb (m = 4)	dpppen (m = 5)	dpph (m = 6)
Cu	53	55	57	59
Ag	54	56	58	60

Table 4.8 Solid-state emission data for **53–60** at 298 K [46]

Model compound or polymer	λ_{emi}(L/ML′CT)/nm	τ_e/μs
53 $\{Cu(dppm)(dmb)^+\}_n$	495	42 ± 7
54 $\{Ag(dppm)(dmb)^+\}_n$	499	27 ± 3
55 $\{Cu(dppb)(dmb)^+\}_n$	490	15 ± 5
56 $\{Ag(dppb)(dmb)^+\}_n$	487	26 ± 5
57 $\{Cu(dpppen)(dmb)^+\}_n$	492	24 ± 5
58 $\{Ag(dpppen)(dmb)^+\}_n$	483	48 ± 4
59 $\{Cu(dpph)(dmb)^+\}_n$	485	18 ± 3
60 $\{Ag(dpph)(dmb)^+\}_n$	487	6 ± 1

dpppen-, and dpph-containing polymers cannot be addressed at this time. Polymer **54** formed a one-dimensional stick baring a strong similarity to polymer **33**, but the chromophore was obviously different.

Table 4.8 gathers the photophysical properties of polymers **53–60** and again similar conclusions can be drawn about the nature of the chromophore.

3.3 Electron Transfer and Photovoltaic Cells

The photo-induced electron transfer in polymers **33** and **61** was also investigated in some detail and a design of photovoltaic cells was made based on the research [47,48].

Both **33** and **61** form semi-conducting materials when the counter-anion is exchanged with $TCNQ^-$ followed by a doping with neutral $TCNQ^0$. The materials exhibit formula of the type $\{[M(dmb)_2]TCNQ \cdot xTCNQ^0\}_n$ where M = Cu, Ag,

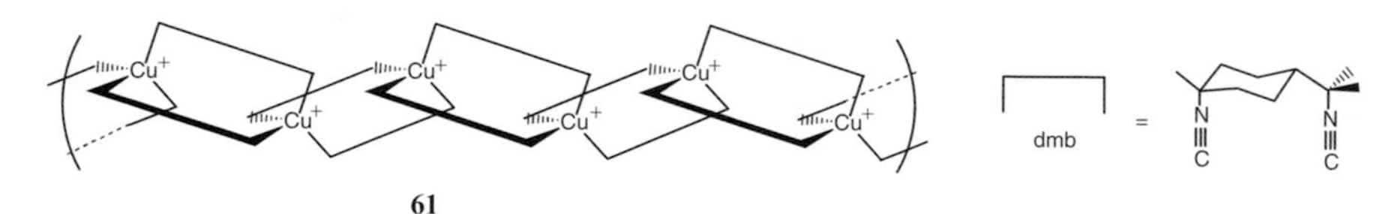

and x = 1 and 1.5. The Ag-containing materials exhibited better conductivity, which was consistent with the better crystalline nature of the materials as found by X-ray powder diffraction methods. However, the Cu-containing materials demonstrated photoconductivity (with a rather slow kinetic) where the Ag materials did not—likely because Ag(II) is thermodynamically difficult to stabilize for a long period of time). This highly reversible and reproducible slow kinetic is shown in Figure 4.8 for a pressed pellet of the pure Cu-containing material using the 4-point probe.

The Cu-containing materials are semi-crystalline (X-ray powder diffraction pattern) making it more difficult to extract crystals that are suitable for X-ray structure determination. The crystalline part of these materials can be analyzed in the X-ray powder diffraction patterns, and it can be noted that both Cu- and Ag-containing materials are isostructural. Luckily, a crystal structure was obtained for the semi-conducting $\{[Ag(dmb)_2]TCNQ \cdot 1.5TCNQ^0\}_n$ materials (Figure 4.9), which establishes the structure for the Cu-analogs.

Polymer **61** (as PF_6^-, BF_4^-, ClO_4^- salts) is also strongly luminescent at room temperature and at 77 K (λ_{emi} = 517 nm, solid state). From quenching experiments, the emission of a pellet-containing polymer **61** is quenched, or almost totally quenched, when in the presence of both solid $TCNQ^0$ and $TCNQ^-$ (as with the Ag-analog). The photo-induced reactions (Eqs. 4.3–4.6) are electron transfers from the electron rich d^{10} copper center to the electron acceptors $TCNQ^{n-}$:

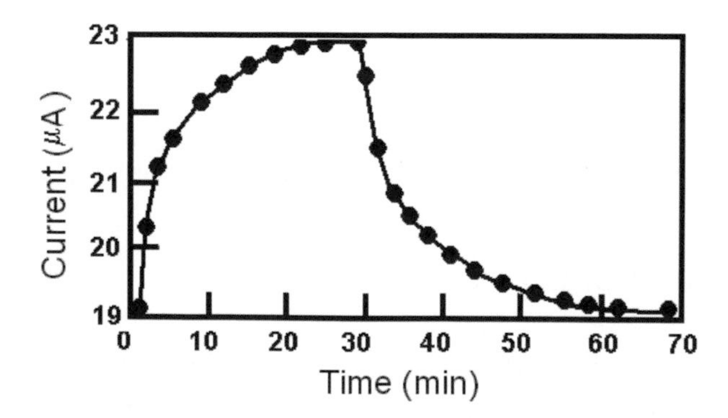

Figure 4.8 Evolution of the photo-current when $\{[Cu(dmb)_2]TCNQ \cdot TCNQ^0\}_n$ is pressed as a pellet and broadband irradiated with visible light. The light was water-jacketed to prevent heat from interfering with the measurements

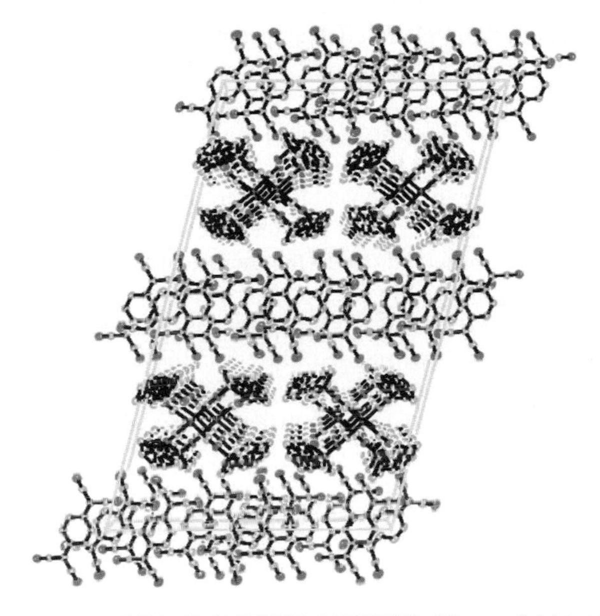

Figure 4.9 X-ray structure of $\{[Ag(dmb)_2]TCNQ \cdot 1.5TCNQ^0\}_n$ The parallel $\{Ag(dmb)_2^+\}_n$ polymer chains are seen along their z-axis where the Ag zig-zag are clearly seen. The $\{TCNQ_{2.5}^-\}_n$ bi-dimensional network lies between the polymer rigid rods. The *box* indicate the unit cells

$$\{Cu(dmb)_2^+\}_n^* + TCNQ^- \rightarrow [\{Cu(dmb)_2^+\}_n]+ + TCNQ^{2-} \quad (4.3)$$

$$[\{Cu(dmb)_2^+\}_n]+ + TCNQ^{2-} \rightarrow \{Cu(dmb)_2^+\}_n + TCNQ^- \quad (4.4)$$

$$\{Cu(dmb)_2^+\}_n^* + TCNQ^0 \rightarrow [\{Cu(dmb)_2^+\}_n]+ + TCNQ^- \quad (4.5)$$

$$[\{Cu(dmb)_2^+\}_n]+ + TCNQ^- \rightarrow \{Cu(dmb)_2^+\}_n + TCNQ^0 \quad (4.6)$$

It is assumed that these photo-induced electron transfers for $TCNQ^{n-}$ acceptors (n = 0, 1) should occur for any n value between 0 and 1, which explains the lack of luminescence in the TCNQ-containing materials. As the number of electrons in the conducting band increases, the conductivity of the materials increases as well, explaining the trend depicted in Figure 4.8. The slow kinetic is first explained by the large distance that the electron has to travel (through an insulating organic layer of dmb) when moving from the chromophore, $Cu(CN)_4^+$ to the $\{TCNQ_{2.5}^-\}_n$ bi-dimensional network. Second, based on crystal field theory the photo-product—a d^9 Cu(II)-containing center—prefers octahedral geometry over the imposed rigidity of a tetrahedral polymer backbone. This makes the activation energy barrier for electron transfer higher than that of a structurally flexible mononuclear complex.

Taking advantage of the photo-conducting ability of the $\{[Cu(dmb)_2]TCNQ\cdot XTCNQ^0\}_n$ (X = 1, 1.5), photovoltaic cells were built using the design presented in Figure 4.10.

The photo-response, or amount of electricity produced at a given wavelength (photo-current), is presented in Figure 4.10 *(right)*. The signal exhibits a long tail extending from the UV- down to the visible region. This trace is the same as the tail observed in the absorption spectrum of polymer **61**, which exhibited a maximum at 253 nm—which was not observed due to the the cut-off absorption of the glass-tin oxide layers. The quantum yield (i.e., number of photons sent versus number of electrons measured) range from 10^{-4} to 10^{-3}, which is modest, but again explainable by the slow kinetic previously discussed. Based on these results, it does not seem that the use of such materials should ultimately lead to industrial applications. The ligand, which is useful to keep the cooperative chromophores together, is too bulky in these cases.

3.4 Exciton Phenomena

The "exciton phenomena" is a delocalization of energy through a material. This energy can be electricity (i.e., electrons or holes) or simply energy (i.e., involving electrons and holes). While some circumstantial evidence from cyclic voltammetry was observed for a hole being delocalized along the backbone of the $\{Ag(dmb)_2^+\}_n$ polymer as discussed above, there was, until recently, no evidence for a delocalized energy transfer (exciton) along the backbone of a single polymer chain. Before the findings on the subject are presented, a description of the phenomenon is necessary. Figure 4.11 shows a schematic representation of a one-demsional coordination or organometallic polymer denoted by $-M_n-M_n-M_n-M_n-$, etc, where M_n represents a mono- (n = 0) or polynuclear center (n > 0) somewhat similar to that described above. The incident irradiation is absorbed by a single chromophore, M_n, along the backbone ($h\nu_1$). The stored energy is then reversibly transmitted via an energy transfer process to the neighboring chromophore, with no thermodynamic gain or

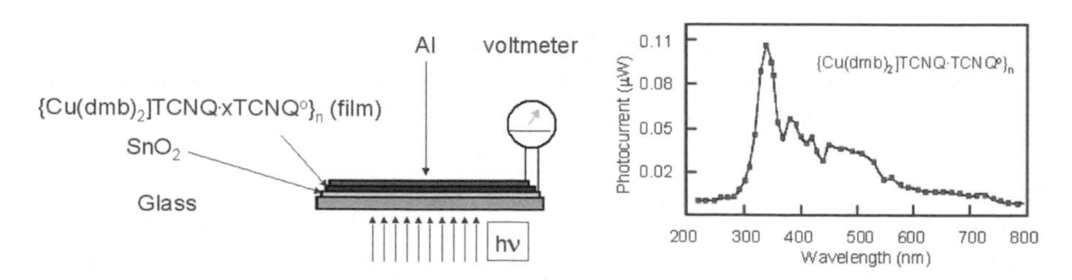

Figure 4.10 Structure of the photovoltaic cell built with the $\{[Cu(dmb)_2]TCNQ \bullet xTCNQ^0\}_n$ materials *(left)*. Photo-current as a function of irradiation wavelength *(right)*. The cut-off below 300 nm results from the absorption of the tin oxide layer

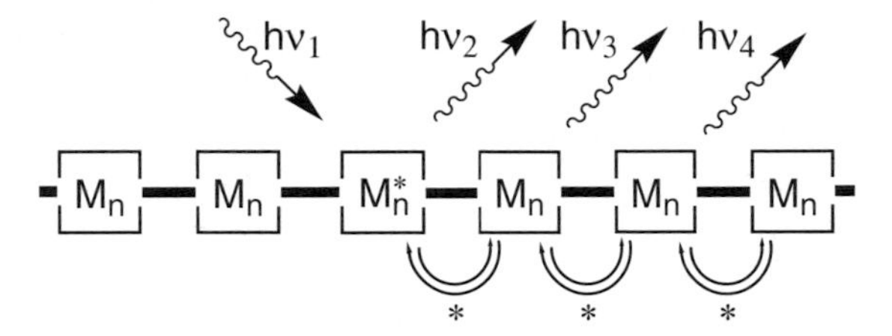

Figure 4.11 An exciton phenomenon where the incident irradiation is absorbed by a single chromophore along the backbone ($h\nu_1$), where the energy is reversibly transmitted to the neighboring chromophore which can re emit the light ($h\nu_2$, $h\nu_3$, $h\nu_4$,...) at any given moment

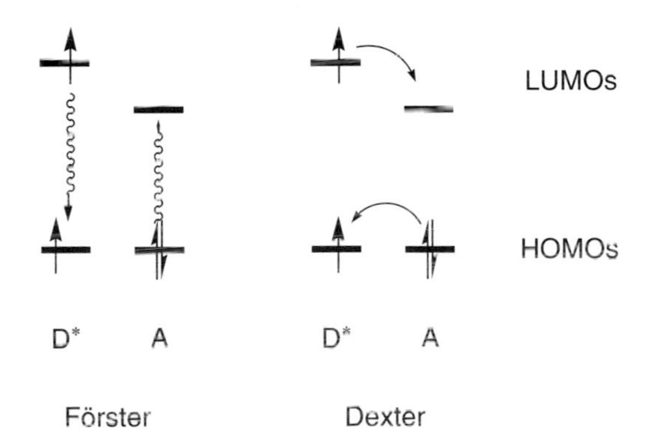

Figure 4.12 Scheme showing the Dexter *(left)* and Förster *(right)* mechanims between a donor and an acceptor. For reversible exciton phenomena, the molecular enegy levels (HOMO and LUMO) are of the same energy

loss (i.e., $\Delta G = 0$), which can either retain or re-emit the light ($h\nu_2$, $h\nu_3$, $h\nu_4$, ...) at any given moment

The energy transfer phenomena have been well known for several decades. Two mechanisms are known: (1) Förster (the most common) and (2) Dexter [49–51]. The processes are shown in Figure 4.12.

The Förster mechanism consists of interactions via the electric field of the excited molecule or metal complex, either in the singlet or triplet excited states for diamagnetic molecules, The interactions are known as donors or accecptors. Currently, the energy gap between the HOMO and LUMO designate the donor (larger gap) and acceptor (smaller gap). At any given time, the excited electron on the LUMO can be relaxed to the HOMO, thereby inducing an excitation of one electron on the acceptor HOMO up to the LUMO, again via electric field interactions. The Dexter mechanism is far different and consists of a concerted double electron transfer—HOMO-HOMO and LUMO-LUMO—between the donor an acceptor. When the chromophores are identical, this process can go back and forth endlessly, inducing formation of a delocalized state. The choice of mechanism, Förster or Dexter, depends entirely on the distance of the process [52,53]. The transfer rate is fast in the singlet state. The rate of the triplet state is significantly slower because only the Dexter process can occur due to the diradical nature of the excited species. Figure 4.13 shows the exciton phenomenon in a polymer.

The energy levels in a polymeric material are almost identical taking into account that a chromophore placed at the end of the polymer rod cannot have the same chemical environment as one placed in the middle. Nonetheless, assuming that they identical, the energy delocalization proceeds via a reversible migration of energy. For coordination and organometallic polymers, the presence of heavy atoms promotes spin-orbit coupling so that the triplet-excited states are heavily populated. As a result, the mechanism is Dexter and therefore strongly dependent on distance. Efficient exciton processes should occur only for closely separated chromophores. The rate of transfer according to Dexter is shown in Eq. 4.7 [51]:

$$k_{ET}(\text{Dexter}) = (2\pi/h)\, KJ\, \exp(-2d/L) \qquad (4.7)$$

where d is the distance between the two chromophores, J is the integral overlap, K is an experimental constant, and L is the average Bohr radius. The spectral signature for the identification of an exciton phenomenon are: (1) difference in emission positions between the model compounds and the corresponding polymers (polymers exhibit broader and red-shifted emissions), (2) presence of an infinite number of emission bands in the time-resolved emission spectra, (3) presence of nonexponential decay traces for the emissions, and (4) presence of depolarized emissions.

The first report on coordination of exciton processes with organometallic polymers was for polymers **33** and **61** [38]. The model compounds are the corresponding $M(CN\text{-}t\text{-}Bu)_4^+$ complexes (M = Cu (**62**), Ag (**63**)), assuming that the steric and electronic environment are comparable to the corresponding polymers **33** and **61** as illustrated in Figure 4.14. The emission maxima for **62** and **63** are 490 and 474 nm, respectively (solid state at 77 K), comparatively at 517 and 500 nm for **61** and **33**, respectively. The 26–27 nm red shifts indicate the presence of interactions in the triplet excited states.

Time-resolved emission spectroscopy demonstrated the presence of an infinite number of emission bands with no evidence of an isosbestic point, as exemplified by Figure 4.15. At early photophysical event (delay time = 0.1 ms), the emission band was blue-shifted at 511 nm, a value moving in the direction of model compound

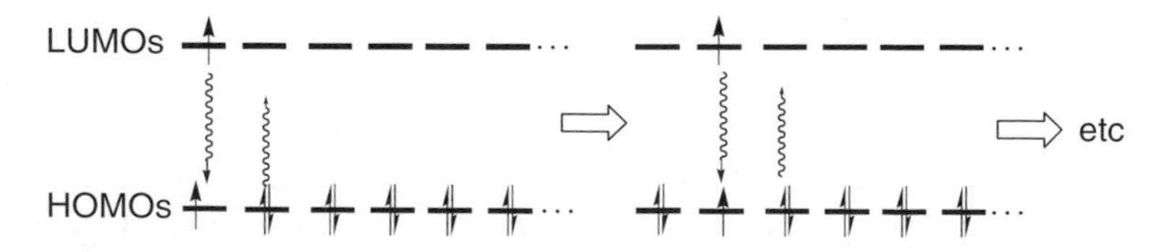

Figure 4.13 Scheme showing how the exciton propagates along the chromophores placed in the backbone of a polymer. Only a Förster mechanism is represented and only a limited number of energy levels (i.e., for 6 chromophores) are shown

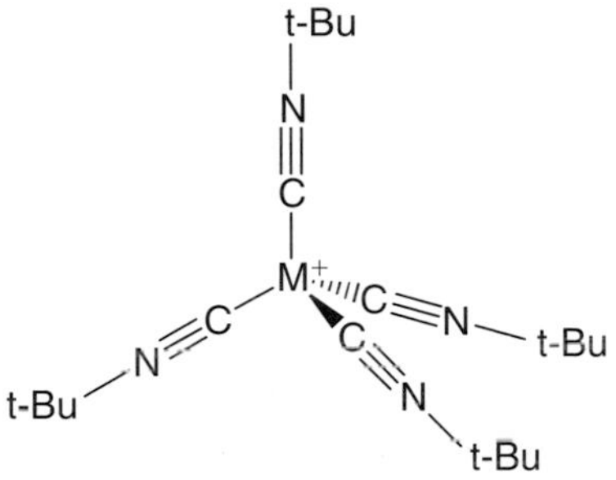

62 (M = Cu), **63** (M = Ag)

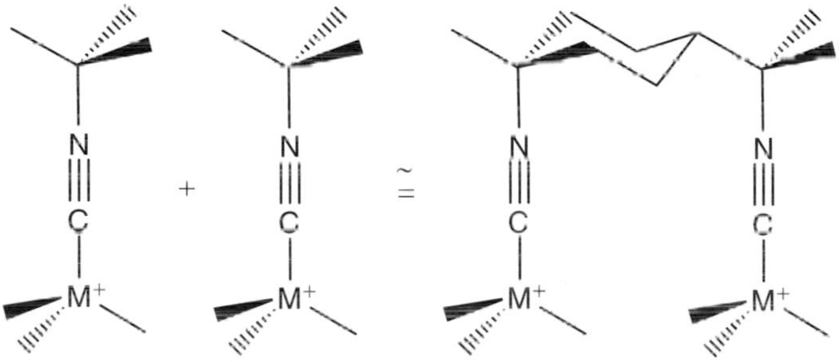

Figure 4.14 Drawing showing the similar chemical environments of model compounds **62** and **63** and organometallic polymers **33** and **61**

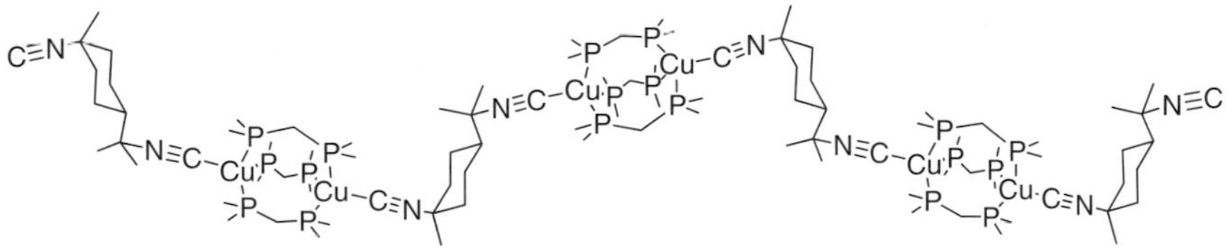

$\{Cu_2(dmpm)_3(dmb)^{2+}_{1.33}\}3$, (**64**)

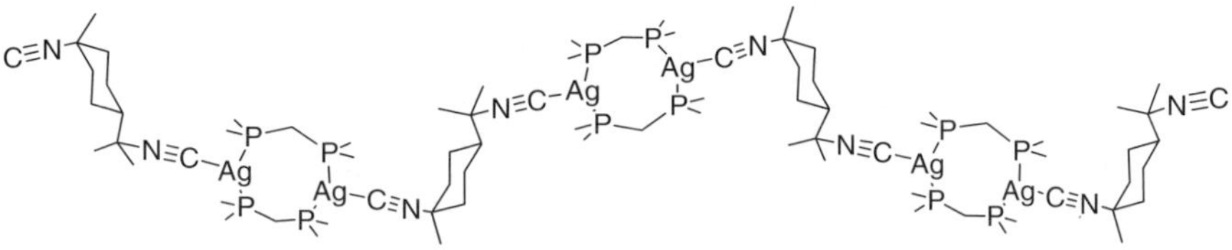

$\{Ag_2(dmpm)_2(dmb)^{2+}_{1.33}\}3$, (**65**)

62. As the delay increased, the overall emission intensity decreased, consistent with a relaxation process of the excited state. However, the measured band maxima had a substantial red-shift (up to 548 nm), larger than 517 nm, the value measured by conventional steady state methods. Overall, the steady state method resulted in a weighted average of all the components. In addition, whereas the emissions of the model compounds were polarized, those for **33** and **61** were not. The most striking piece of evidence came from the emission decays of the polymers. The model compounds exhibited linear plots of ln of the emission intensity after laser pulse (from which emission lifetimes are extracted) yet the corresponding plots for the polymers were nonexponential (Figure 4.16).

It was also noted for polymer **33** that the ln plots of the emission intensity matched the delay time after the excitation pulse in crystalline form, as fine powder,

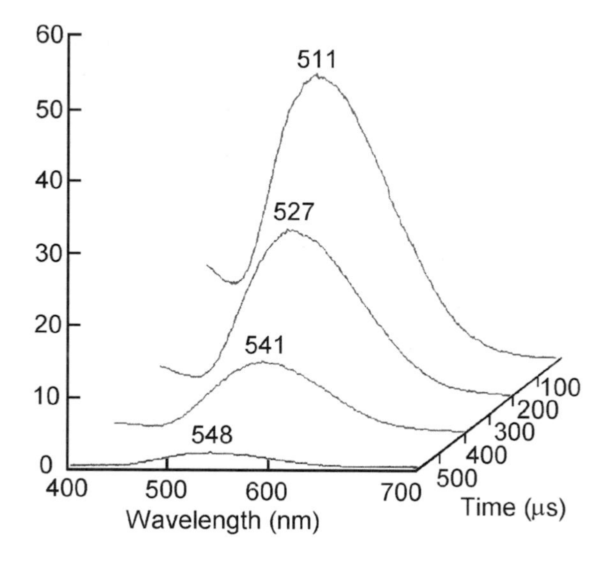

Figure 4.15 Time resolved emisson spectra of **61** in the solid state at 293 K. The delay times after excitation pulse are 0.1 *(purple)*, 2.0 *(green)*, 4.0 *(orange)*, and 6.0 ms *(blue)*

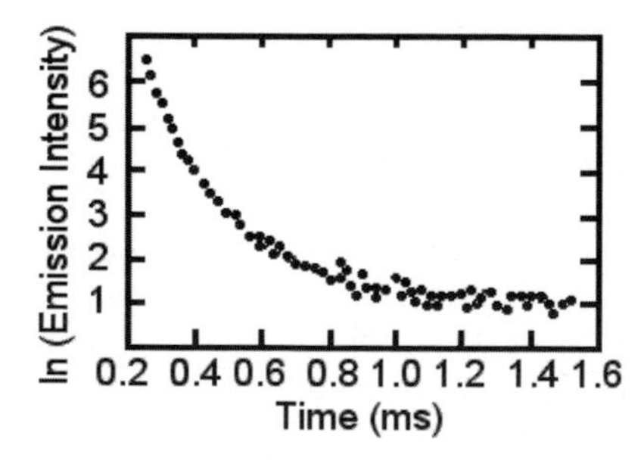

Figure 4.16 Decay trace of the emission intensity of polymer **33** (as a BF_4^- salt). Modified from reference [38]

and in solution at 77 K. This means that the morphology of the material or the presence of neighboring chains did not interfere with the process. The exciton phenomenon occurred strictly within the backbone of the polymer chain. In the solid state, the distance separating the inter-chain chromophores (>8 Å) is too large to assume that any process will ocurr with efficiency. In fact, the intermolecular N-N distances (8.25 and 8.67 Å) [38] indicate that the chains are relatively isolated in the solid state, whereas the intramolecularN-N separations are on the order of 4.5 Å (dmb in its *U*-conformation). If this process occurs within the backbone of the chain, then it should be length-dependent. The PI recently reported two oligomers, **64** and **65**, which both averaged 3 units based on ^{1}H nuclear magnetic resonance (NMR), infrared (IR), chemical analyses, and thermal gravimetric analyses (TGA) [54].

The model compounds, **66–68**, were characterized by X-ray crystallography and exhibited rich luminescence as illustrated by the data presented in Table 4.9.

Despite the fact that no X-ray data is available for **64** and **65**, it is possible to anticipate what the approximate inter-chromophore distances in the solid state are by examining the X ray data for the building blocks **66–68**. For these species, the closest intermolecular P-P distances are 6.683 Å for **66**, and 6.328, 6.585, and 6.982 Å for **2** [54]. These distances are greater than the intramolecular N-N separation in the linking dmb ligand (Z-conformation) in the recently reported computed model compound $Cu_6(dmpm)_9(dmb)_2(CN\text{-}t\text{-}Bu)_4^{6+}$ (~5.8 Å; PC-Model; Figure 4.17) [54] and in the polymer $\{[Pd_4(dmb)_4(dmb)]^{2+}\}_n$ (5.549 Å; X-ray) [25]. If one accepts that the rate for exciton hopping (or energy transfer; k_{ET}) varies as $k_{ET} \propto 1/r^6$ (r = interchromophore distance; for a Förster mechanism) [49,50], then the contribution of the intermolecular process is minor.

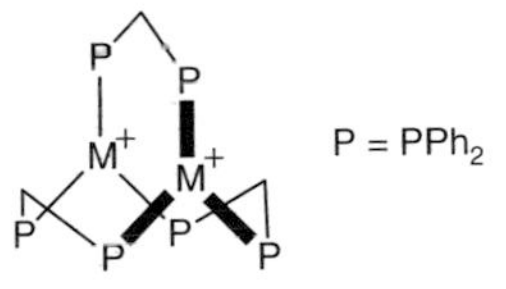

P = PPh_2

66 (M = Cu), **67** (M = Ag)

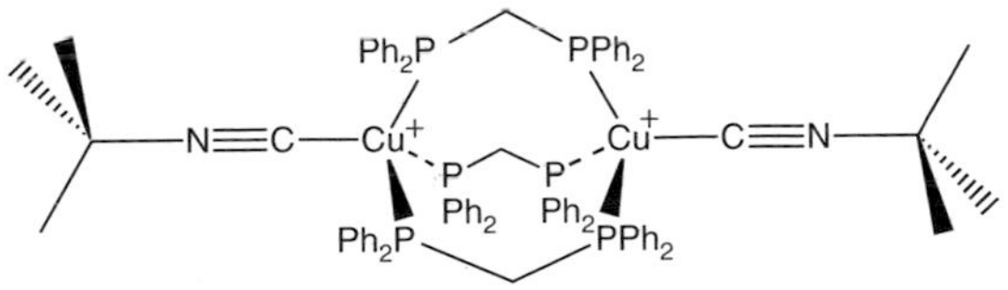

68

Table 4.9 Spectroscopic and photophysical data for **64–68** [54]

	Solid state (298 K)		MeCN solution (298 K)			PrCN solution (77 K) or polymer	
Model compound	λ_{emi}/nm	τ_e/µs	λ_{emi}/nm	τ_e/µs	Φ_e	λ_{emi}/nm	τ_e/µs
66	Not measured		500	16.5	0.020	505	840
68	476	291	501	3.8	0.0043	502	712
64	482	257	550	<2	0.0013	474	523
67	445	41	466	3.0	0.0031	482	6300
65	447	31	520	~2	0.0014	500	542

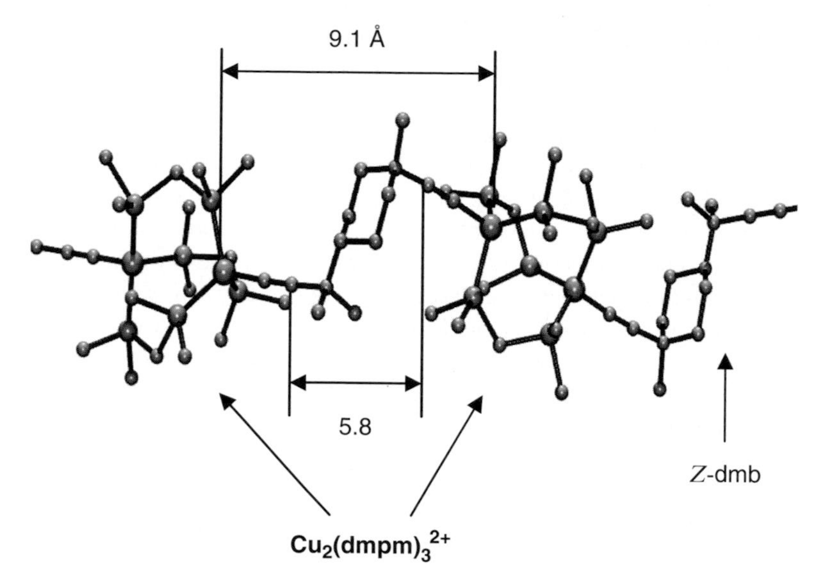

Figure 4.17 Representation of a fragment of the $\{[Cu_2(dmpm)_3(dmb)_{1.33}](BF_4)_2\}_3$ oligomer (**64**), where the computed intrachain interchromophore distances are indicated (5.8 and 9.1 Å for the N-N and Cu-Cu separations, respectively) [54]

Figure 4.18 shows the time resolved emission spectra (20–2,000 μs) for **64** and **65** in the solid state. **64** and **65** exhibit blue emission maxima at 482 and 447 nm, respectively, when submitted to continuous wave excitation [55]. At the early event after the light pulse, the recorded emission band is blue-shifted with respect to the emission band measured in continuous wave mode. As the delay time increases, the observed emission band red-shifts constantly and the intensity decreases. All in all, the emission bands measured with continuous wave light are composed of a number of blue- and red-shifted components. The maximum band shifts are ~10 and ~34 nm for **64** and **65**, respectively [55], which are smaller than those observed for the longer $\{M(dmb)_2^+\}_n$ polymers (**33** and **61**, up to 40 nm) [38]. The 298 K decay traces are found to be linear for **66, 67**, and **68** with emission lifetimes of 251 (472), 41 (445), and 291 μs (476 nm), respectively [54]. These results indicate for the first time that, qualitatively, the amplitude of red-shift increases as the polymer length increases. The shorter lifetimes normally encountered for the Ag species with respect to the Cu homologs results from the larger spin-orbit coupling of the heavier element, commonly called "heavy atom effect" [56].

As stated, the decay traces are nonexponential for **64, 65, 33**, and **61**. A typical example is shown in Figure 4.19, where a straight line is observed for the model compound **3** and a curve is measured for the polymer **33** in the ln plot of the emission intensity versus delay time after the excitation pulse.

In these cases, the data were analyzed using the ESM, and the results were plotted as population distribution (or relative amplitude) versus lifetimes in Figures 20

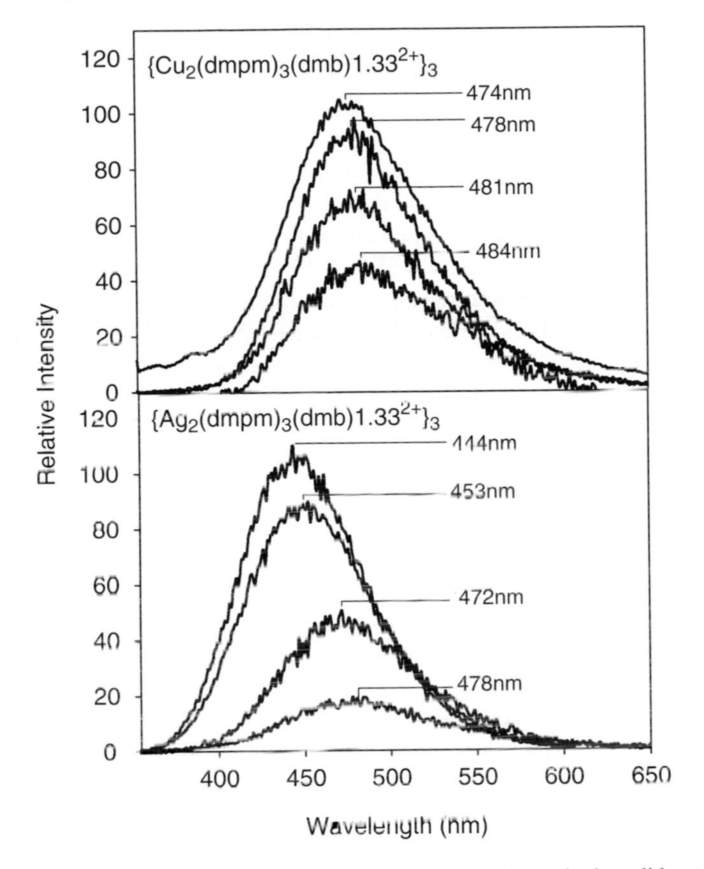

Figure 4.18 Time-resolved emission spectra for **64** *(up)* and **65** *(down)* in the solid state at 298 K. The measurements were made in the following time frames: for **64**: 474 nm, 20–70; 478, 500–600; 481, 1,000–1,300; 484; 2,000–2,500 μs; for **65**: 444 nm, 20–70; 453, 300–400; 472, 500–600; 478, 1,000–1,300 μs [55]

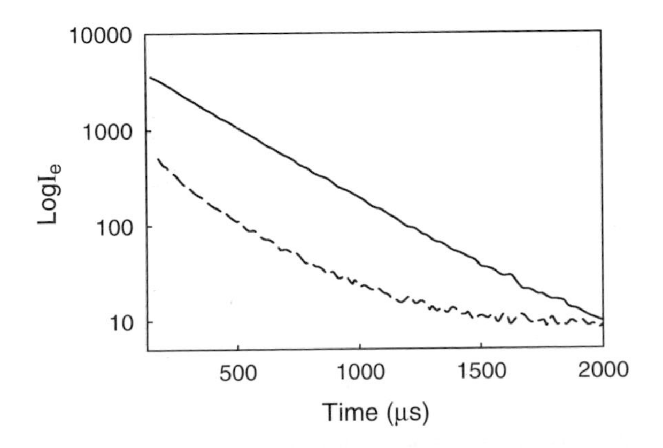

Figure 4.19 Solid-state decay traces for the emission of **68** *(—)* versus **61** *(----)* at 298 K [55]

and 21. The maximum of probability represents an average lifetime or the most probable lifetime, and these are 257 (**64**) and 31 μs (**65**). The width of the distribution is related to the curvature of the decay traces (log plot); as the width increases, the curvature increases. The data are not dependent on the excitation intensity (using neutral density filters), indicating that local heating has little or no effect on the results.

However, one interesting question does arise. The number of units is 1, 3, and ~ 45 (as evaluated here by the measurement of the intrinsic viscosity) for **66, 64**, and **61**, and 1, 3, and "very large" for **67, 65**, and **33**, respectively. The width of the population distribution plots in Figures 20 and 21 does not follow the trend proportionally, particularly for **64** and **61** as a large change in the number of units should be accompanied by a large change in width. This observation cannot be explained straightforwardly. First, one has to consider that the exciton process is reversible (Figure 4.22). So a small-chain can exhibit the same spectroscopic feature as a long-chain depending on the efficiency of exciton hopping.

Secondly, it is not necessarily true that the extent of exciton migration is as extensive for one chromophore as another, simply because the atomic distribution of the HOMO and LUMO is not identical for all chromophores. Therefore, the probability of energy transfer cannot be the same. In addition, the distances between the chromophores measured as MC≡N···N≡CM or M···M or P···P are different as well.

One interesting and important remark concerning this point is the comparison of the photophysical properties of the Cu(I) and Ag(I) species described above with Pd-Pd and Pt-Pt bond-containing oligomers and polymers are also built with dmb and diphosphine bridging ligands, such as $\{Pd_2(dmb)_2(diphos)^{2+}\}_n$, $\{Pd_4(dmb)_4(dmb)^{2+}\}_n$, and $\{Pt_4(dmb)_4(diphos)^{2+}\}_n$ [19,25,26]. Whereas the former polymers exhibited

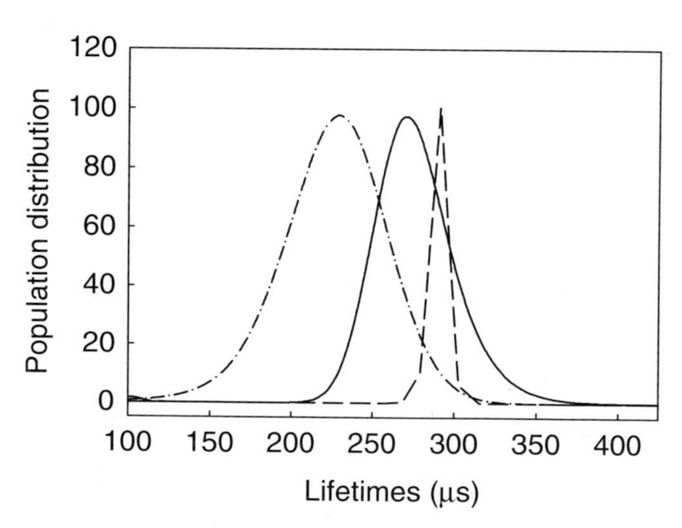

Figure 4.20 Comparison of the distribution of lifetimes as a function of lifetime fitting the emission decay traces for **66** (...), **64** (—), and **61** (...) in the solid state at 298 K [55]

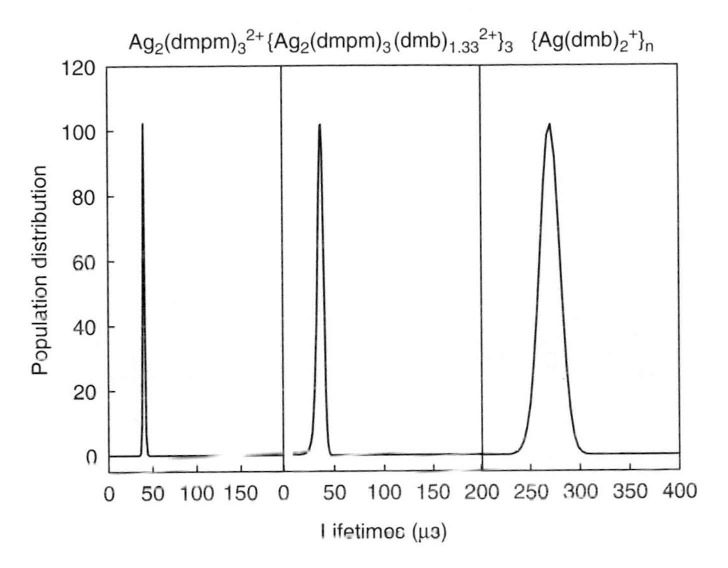

Figure 4.21 Comparison of the distribution of lifetimes as a function of lifetimes fitting emission decay traces for **67**, **65**, and **33** in the solid state at 298 K [55]

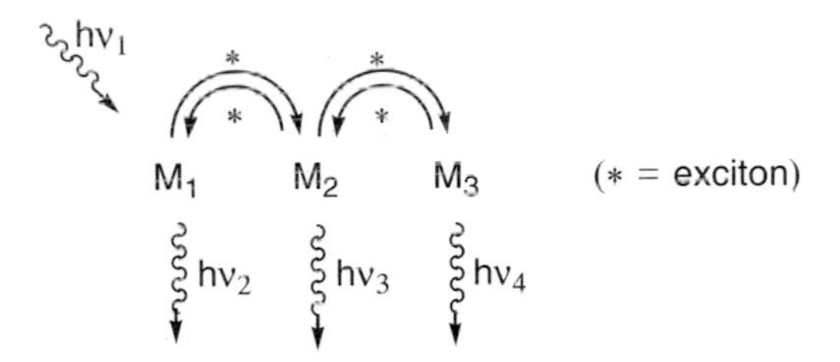

Figure 4.22 Scheme showing energy hopping over a short oligomer

nonexponetial decay traces and depolarized broad emissions, the latter ones exhibited monoexponetial luminescence decays, indicating the presence of noninteracting chromopohores within the chain as well as polarized emissions, strongly suggesting that the exciton process is either very weak or absent in these cases. A close examination of the atomic contributions of the HOMO and LUMO revealed that a large contribution of the C and N *p* orbitals (π and π^* orbitals) are computed for the Cu(I) and Ag(I) chromophores (Table 4.10) [38], whereas these are minor or nonexistent for the Pd and Pt species as exemplified in Figure 4.23 [25,26]. In other words, the electronic density is more spread out in the former.

Knowing that dmb exists in a U- and a Z-conformation (Figure 4.24), the distance between the last atoms involved in the chromophores is a function of both the conformation and the atomic contribution of this chromophore, and then the exciton process will be equally dependent.

Table 4.10 Selected atomic contribution for the HOMO and LUMO in $Ag(CNH)_4^+$ [38]

Atomic contributions	
LUMO	HOMO
60% C(*p*)	55% Ag(d_{xz})
26% N(*p*)	10% C(*p*)
	25% C(*s*)

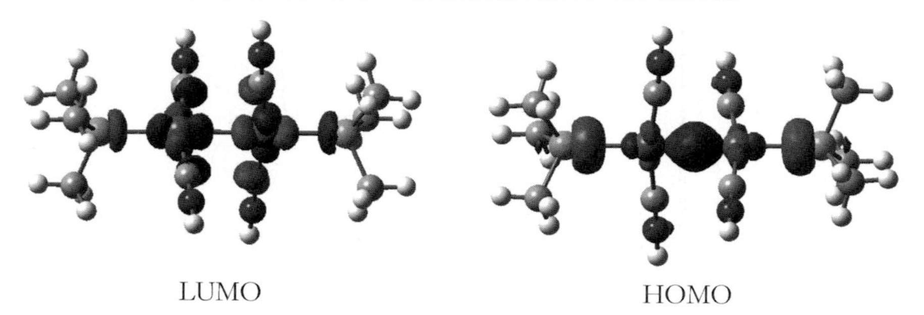

Figure 4.23 Pictures of the HOMO and LUMO of $Pd_2(CNH)_6(PMe_3)_2^{2+}$ giving rise to singlet and triplet $d\sigma$-$d\sigma^*$ excited states. Note the modest atomic contribution of the equatorially placed isocyanide ligands with respect to the axially coordinated phosphines

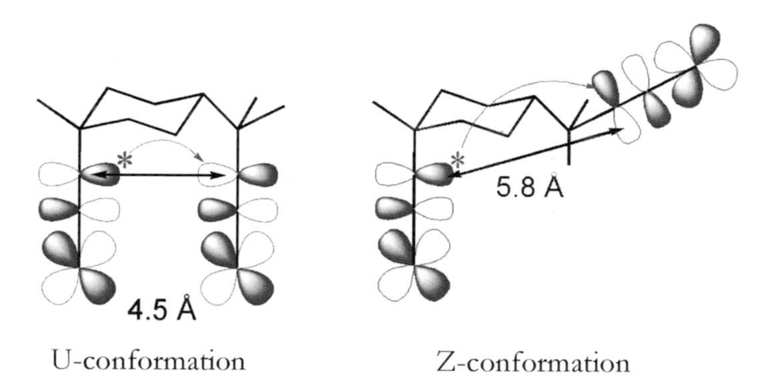

Figure 4.24 Drawings showing the dmb ligand in its U- and Z-conformations stressing the distances between the N-atoms, the closest atom between the two chromophores

In Figure 4.24, the distances between the two N and M atoms are about the same depending whether the N≡C—M axes are parallel or not. This means if the atomic contributions on N and C are important for Cu- and Ag-containing chromophores, but less or absent for Pd- and Pt-containing species, the exciton phenomenon will be as important for the U-conformation for both types of chromophores. On the other hand in the Z-conformation, the N···N separation is about 5.8 Å. If the atomic contribution on the N atom is important as indicated for Cu- and Ag-containing chromophores (Table 4.10), then energy migration is still possible. On the other hand, if this contribution is weak or nil, then transfer is very inefficient. In the cases of Pd- and Pt-containing species discussed earlier in this chapter, the contribution is negligible (Figure 4.23). So, the closest separation is that for M···M, which is in the order of 11.4 Å. At such a distance, through space energy transfer and exciton processes are extremely inefficient, particularly in the triplet states where only the Dexter mechanism operates (Eq. 4.7). In support for this affirmation, Raman spectroscopy was extensively used to demonstrate evidence for M···M interactions in dimers and polymers [57]. Ultimately, it was demonstrated that exceeding 5 Å, M···M interactions are so weak that they are barely detected [58].

4 Gold

Gold(I) and gold(III) have the tendency to be involved in M···M interactions (aurophilic interactions) and in the solid-state form polymeric structures for which properties in solution, for example, are different from those in the solid state. For example, calix[4]arene-containing ligands have the tendency to form oligomers and polymers [59], and in a recent work on mono- (**69**) and tetra- functionalized calix[4]arene (**70**) by isocyanidegold(I) chloride. Their emission spectra are slightly different from each other in the solid state as presented in Figure 4.25, suggesting

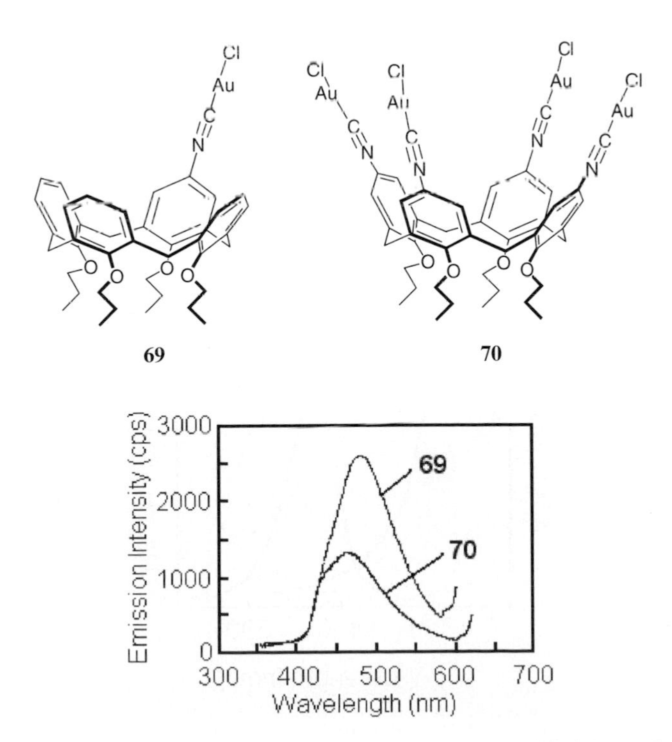

Figure 4.25 Solid-state emission spectra of **69** and **70** at 298 K. Modified from reference [60]

the presence of either intra- or intermolecular interaction [60]. Because the luminescence spectra are identical in solution, the difference in the solid-state spectra can only result from the presence of intermolecular Au···Au interactions. The matrix assisted laser desorption/ionization time-of-light (MALDI-TOF) spectrum did indeed confirmed the presence of at least a dimer in the solid state [61].

A related binuclear gold(I) compound, **71**, crystallizes to form an Au···Au polymer (Figure 4.26) [61]. A strong Raman scattering at 50 cm^{-1} is consistent with

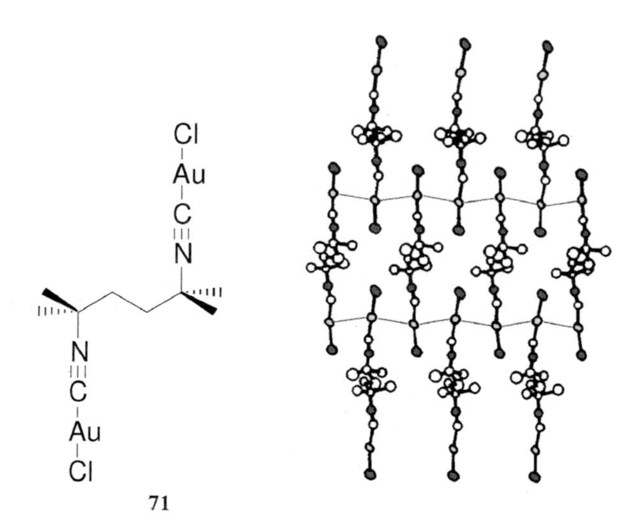

Figure 4.26 Drawing of **71** *(left)* and X-ray structure of it polymeric form in the solid state *(right*: r(Au···Au) = 3.3063 (3) Å) The line in *red* represents the Au···Au interactions (Au = *yellow*, Cl = *green*, N = *blue*, the other are C atoms). The H-atoms are not shown for clarity. Modified from reference [62]

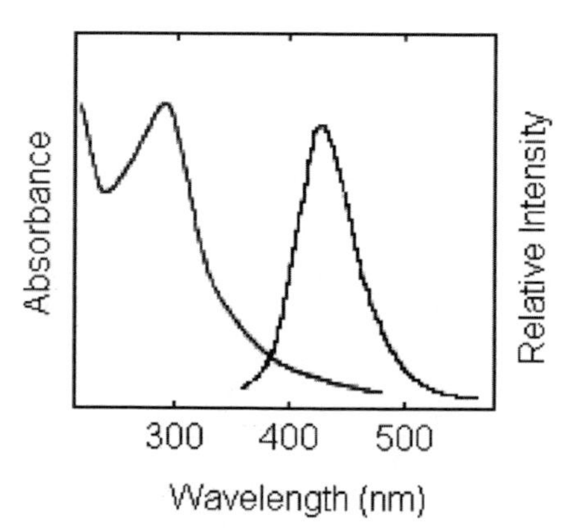

Figure 4.27 Solid-state absorption *(left)* and emission spectra *(right)* of **71** at room temperature. Modified from reference [62]

the presence of weak d^{10} Au···Au interactions. The solid-state absorption spectrum exhibit an absorption band at about 290 nm as shown in Figure 4.27, which is absent for the spectra measured for solution, assignable to a $d\sigma^*$-$p\sigma$ along the metallic chain. The solid-state emission spectrum exhibits a strong structureless emission at about 435 nm (bright yellow) with an excited lifetime of 0.70 µs, assignable to a $d\sigma^*$-$p\sigma$ phosphorescence. In solution this emission disappears and only a very small emission centered at 403 nm is perceptible.

Polymers **72–74** emit strongly at 495 nm with emission lifetimes of 8.5, 8.3, and 5.7 µs, respectively, assignable to $d \rightarrow p$ triplet excited states [63]. A spectacular emission quenching in the solid state at room temperature has been reported for **75** [63]. This quenching was not investigated in detail but an intra- or inter-chain triplet-triplet through space energy transfer, or photo-induced electron transfer from the rich d^{10} metal to the aromatic is suspected. The former explanation appears more probable because of the lower triplet energy of the naphthyl group, and would also suggest that the aryl must be placed close to the Au^+ atom.

On the other hand, model compound **75** and polymer **76** emit strongly at 462 and 450 nm, respectively, in EtOH solution at room temperature [64]. The small blue-shift of the emission band relative to dimeric species suggest that no Au···Au interaction occurs. The absorption spectra of both species superpose each other, with no band attributable to a $d\sigma^*$-$p\sigma$ witnessing the Au···Au interactions.

Puddephatt et al. also investigated an exhaustive list of rigid-rod gold-containing model compounds and polymers built upon diphosphine and diisocyanide assembling ligands [65]. Eight were model compounds (**78–81**) and organometallic polymers (**81–85**) containing a bis(ethynylgold(I)) fragment, all of which were found to be luminescent in the solid state at room temperature. The wavelength

Atomic Contributions	
LUMO	HOMO
60 % C(p)	55 % Ag(d_{xz})
26 % N(p)	10 % C(p)
	25 % C(s)

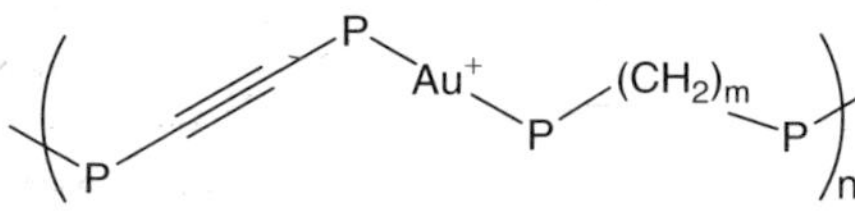

P = PPh2; **72** (m = 3), **73** (m = 4), **74** (m = 5)

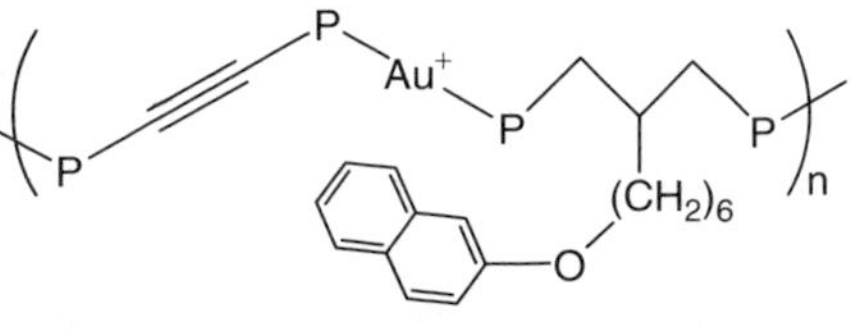

75

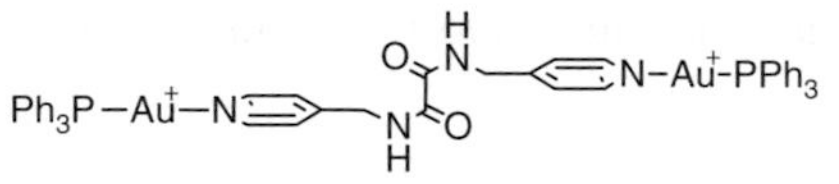

76 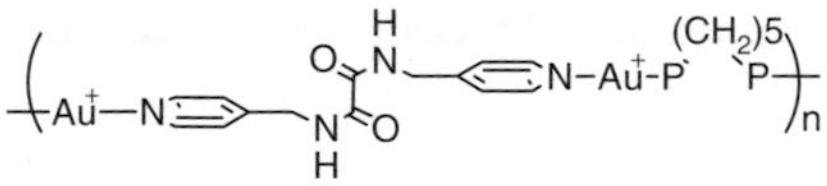 **77**

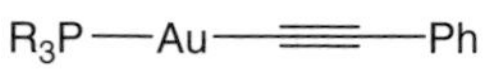

78, λ_{emi} = 455 nm, R = H
79, λ_{emi} = 455 nm, R = Me

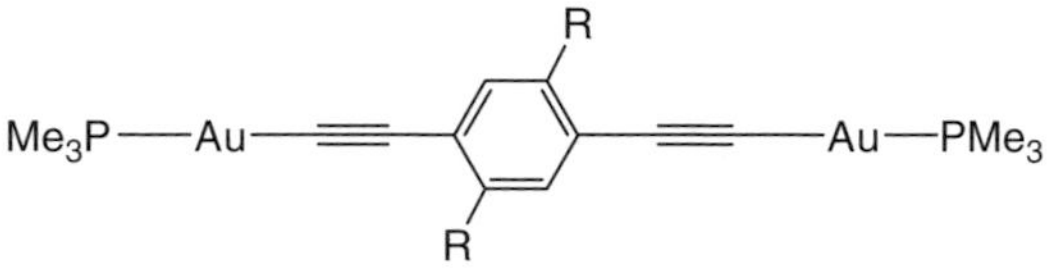

80, λ_{emi} = 540 nm, R = H
81, λ_{emi} = 540 nm, R = Me

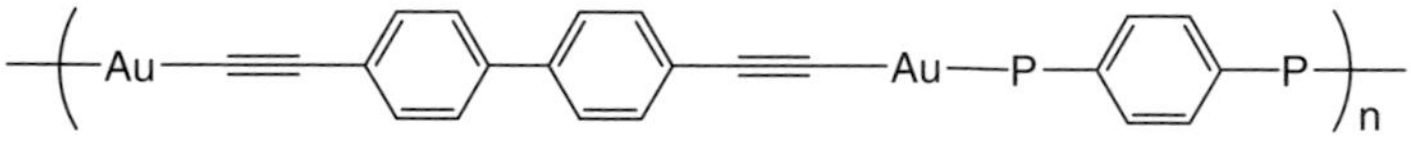

P = PPh_2

82, λ_{emi} = 600 nm

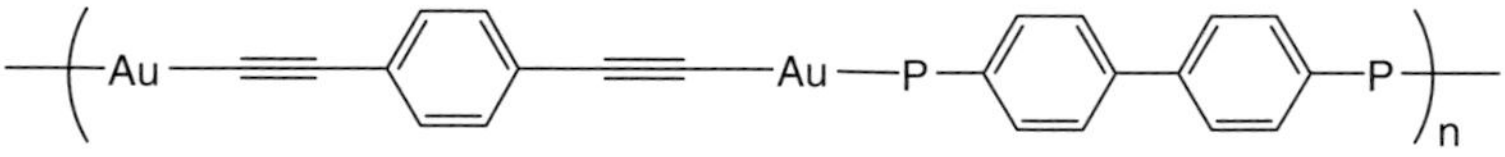

P = $P(iPr)_2$

83, λ_{emi} = 600 nm

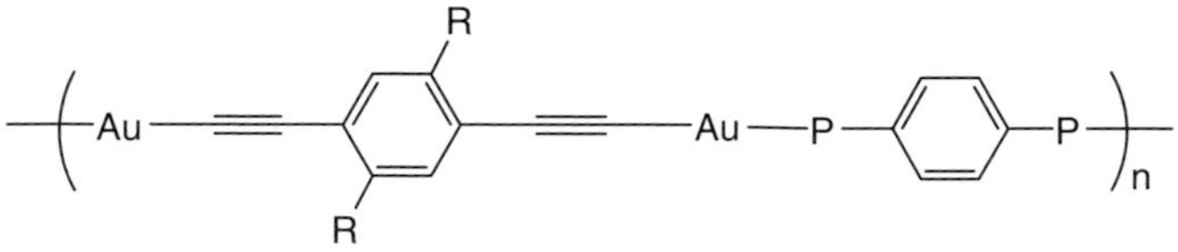

P = PPh2

84, λ_{emi} = 600 nm, R = H
85, λ_{emi} = 600 nm, R = Me

of maximum emission is indicated below with the molecule number. The emission bands are located between 455 and 600 nm, and, surprisingly, the polymers turn out to be strongly red-shifted. The important feature is that the solid-state emission maxima are red-shifted in comparison with the solution data (**78**, 424 nm; **79**, 410 nm; **80**, 415 nm; **81**, 415 nm, in CH_2Cl_2). These shifts result from the presence of Au···Au interactions. Possibly, the lowest energy excited state for the species in solution when no Au···Au interactions occur, is an intra-ligand $\pi\pi^*$ mixed with some Au *d* orbitals. The assignment for the nature of the excited state was aided by EHMO calculations.

A list of 12 model compounds (**86–91**) and organometallic polymers (**92–96**) where a diisocyanide is employed are shown below [65]. Again, the luminescence band associated with the polymers was also red-shifted in comparison with the model compounds with the exception of **89**. In this case, the presence of the NO_2 moiety may play a role in the nature of the excited state. Similarly, the emission band in CH_2Cl_2 solution (**86**, 429 nm; **87**, 432 nm; **89**, 503 nm) was also blue-shifted with respect to the solid state, and similar conclusions about the presence of Au···Au interactions occurring in the solid state were drawn.

The presence of luminescence is not only reserved for linear model compounds and one dimensional organometallic and coordination polymers. Two-dimensional polymers using tri-coordinated 16-electron gold(I) center were also found to emit in the solid state (very broad emission centred at about λ_{emi} = 420 nm) [66]. In this case, no Au···Au interactions were possible because of obvious steric hindrance from the PPh_2 fragments (inter-planar contacts). The assignment for the emitting state must be primarily $d \rightarrow p$ mixed with some π^* phenyl contributions, as described above for other d^{10} systems (i.e., see for example model compound **34**) [42].

It is interesting to note that the one-dimensional polymer **98**, also exhibited the same maximum at that found for **97** (also very broad emission centered at about λ_{emi} = 420 nm) [66]. Steric must certainly play a major role in the "choice" of nature of excited state in these unsaturated gold species.

5 Conclusion

There is a welt of photophysics in diphosphine- and diisocyanide-containing coordination and organometallic oligomers and polymers. This chapter is limited to palladium, platinum, copper, silver, and gold, but other metal-containing oligomers and polymers do exist. Some of the results on photo-induced electron transfer are very promising in that photonic devices such a photovoltaic cells and polymer light emitting diodes are possible. The discovery that energy transfer excitons occur in the backbone of the polymer is unprecedented, and to this day has major relevance in electron transfer excitons as well. Thus, even though the coordination and organometallic polymers are not conjugated, electron exciton along the backbone is possible. This statement is supported by the electrochemical findings on the $\{M(dmb)_2^+\}_n$ polymers (M = Cu, Ag). The presence of Au···Au interactions in gold-containing polymers is also an interesting feature because the color of the emission is regulated by simple structure modification (i.e., steric hidrance). Overall, the potential for using such species to design new photonic materials is enormous and needs to be explored and exploited to its full capacity. Based on the growing interest in such species, future developments are anticipated.

Acknowledgment The Natural Sciences and Engineering Research Council (NSERC) is acknowledged for funding over the years that led to the works described in this chapter. The authors would

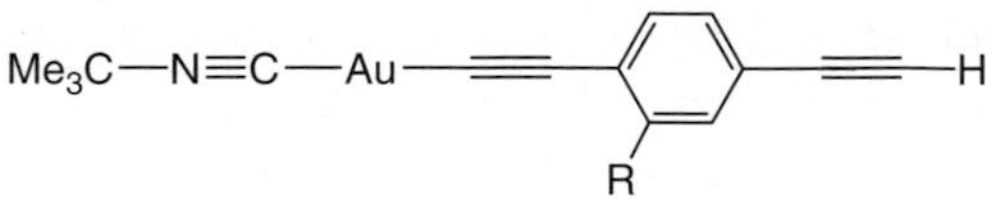

$Me_3C—N≡C—Au—≡—C_6H_3(R)—≡—H$

86, λ_{emi} = 461 nm, R = H
87, λ_{emi} = 460 nm, R = Me

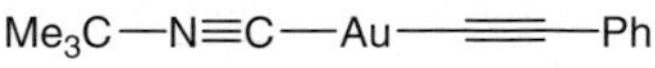

$Me_3C—N≡C—Au—≡—Ph$

88, λ_{emi} = 470 nm

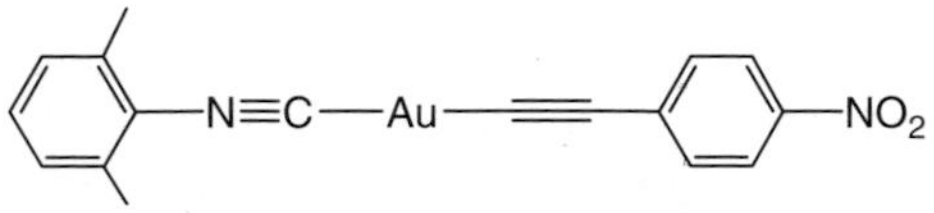

—N≡C—Au—≡— —NO_2

89, λ_{emi} = 633 nm

R
Ph—≡—Au—C≡N— —N≡C—Au—≡—Ph
R

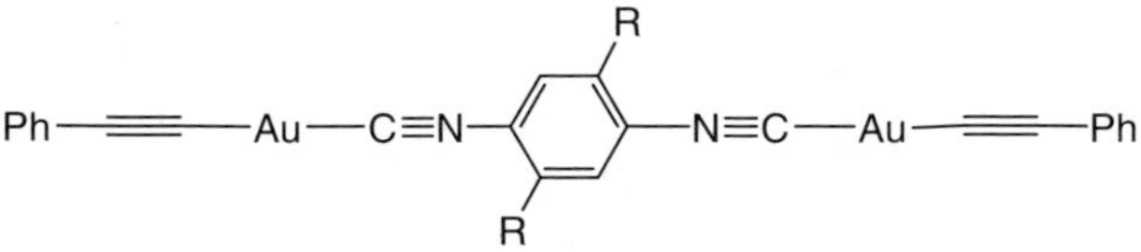

90, λ_{emi} = 550 nm, R = H
91, λ_{emi} = 501 nm, R = Me

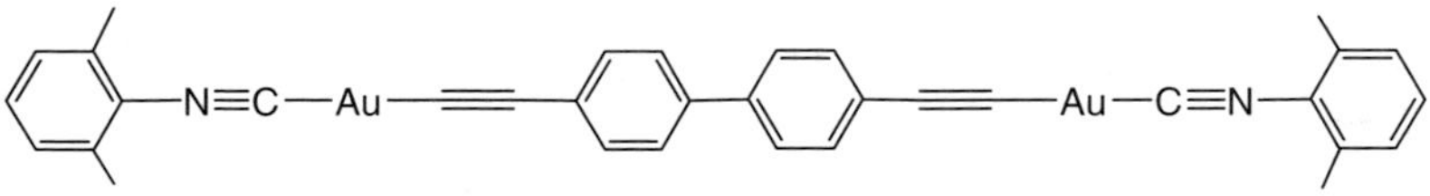

—N≡C—Au—≡— —≡—Au—C≡N—

92, λ_{emi} = 550 nm

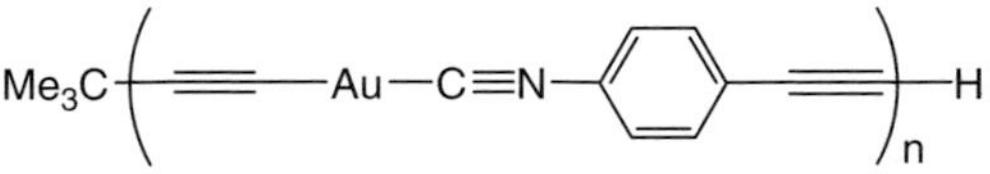

Me_3C —≡—Au—C≡N— —≡— H n

93, λ_{emi} = 585 nm

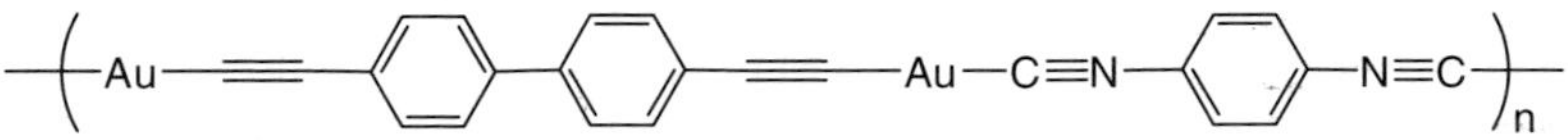

Au—≡— —≡—Au—C≡N— —N≡C n

94, λ_{emi} = 585 nm

R R`
Au—≡— —≡—Au—C≡N— —N≡C n
R `R

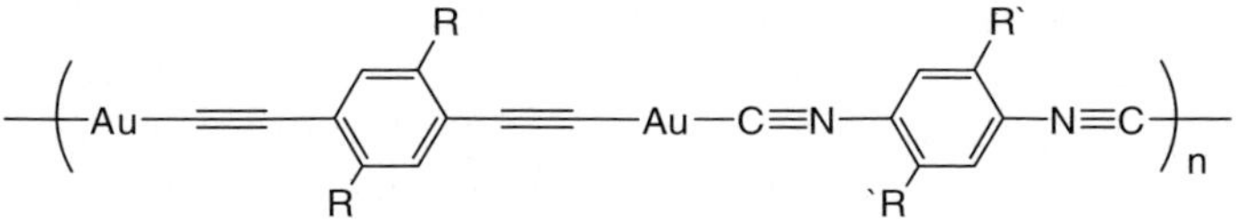

95, R = Me, R`= H; **96**, R = H, R`= CMe_3; **97**; R = Me, R`= CMe_3 : λ_{emi} = 585 nm

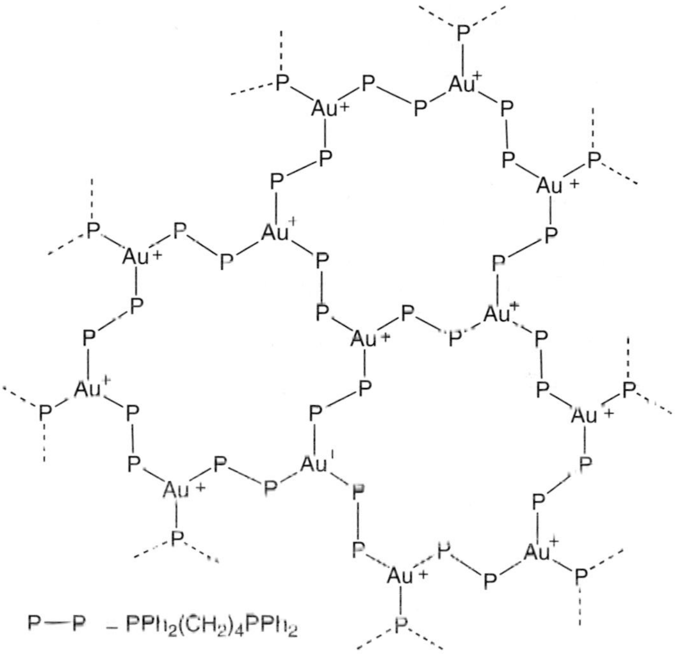

P—P = $PPh_2(CH_2)_4PPh_2$

98

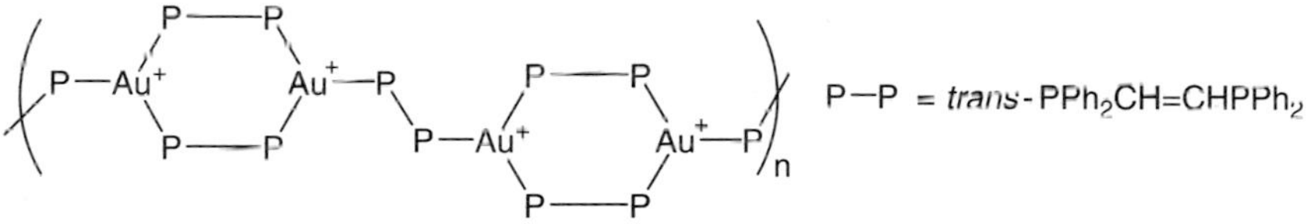

P—P = *trans*-$PPh_2CH{=}CHPPh_2$

99

like to thank the undergraduate and graduate students and postdoctoral fellows that for their research assistance. Their names are listed with the references.

References

1. Abd-El-Aziz AS, Carraher Jr. CE, Pittman Jr. CU, Sheats JE, Zeldin M, (eds). (2005) Macromolecule containing metal and metal-like elements, Vol. 5. Wiley Interscience, John Wiley and Sons Inc, New York.
2. Abd-El-Aziz AS, Manners I. (eds). (2006) Frontiers in transition metal-containing polymers. John Wiley and Sons Inc, New York.
3. Manners I. (2001) Science 294:1664.
4. Manners I. (2004) Synthetic metal-containing polymers; Wiley-VCH: Weinheim.

5. Newkome GR, He E, Moorefield CN. (1999) Chem Rev 99:1689.
6. Astruc D, Chardac F. (2001) Chem Rev 10:2991.
7. Abd-El-Aziz AS. (2003) Ed Macromol Symp. 196:1.
8. Abd-El-Aziz AS, Harvey PD. (2004) Eds Macromol Symp. 209:1.
9. Kingsborough RP, Swager TM. (1999) Prog Inorg Chem. 48:123.
10. Wolf MO. (2001) Adv Mater 13:545.
11. Archer RD. (2001) Inorganic and organometallic polymers. John Wiley & Sons Inc, New York.
12. Samar D, Fortin J-F, Decken A, Fortin D, Harvey PD.J. (2006) Inorg Organomet Poly Mat., 2006: in press.
13. Harvey PD. (2005) In: Abd-El-Aziz AS, Carraher Jr, CE, Pittman Jr, CU, Sheats JE, Zeldin M, (eds). Macromolecule containing metal and metal-like elements, Vol. 5. Wiley Interscience, John Wiley and Sons Inc, New York, p, 83.
14. Harvey PD. (2006) In: Abd-El-Aziz A, Manners I. (eds). Frontiers in transition metal-containing polymers. Wiley Interscience, John Wiley and Sons Inc, New York.
15. Köhler A, Wittmann HF, Friend RH, Khan MS, Lewis J. (1996) Synth Met. 7:147.
16. Sonogashira K, Asamai K., Takeuchi NJ. (1985) Mat Sci Lett. 4:737.
17. Perreault D, Drouin M, Michel A, Harvey PD. (1992) Inorg Chem. 31:2740.
18. Harvey PD, Murtaza Z. (1993) Inorg Chem. 32:4721.
19. Sicard S, Bérubé J.-F, Samar D, et al. (2004) Inorg Chem. 43:5321.
20. Turcotte M, Harvey PD. (2002) Inorg Chem. 41:1739.
21. Harvey PD. (2004) Macromol Symp. 209:81.
22. Plourde F, Gilbert K, Gagnon J, Harvey PD. (2003) Organometallics 22:2862.
23. Mongrain P, Harvey PD. (2006) Macromo Rapid Commun. 2006.
24. Bérubé JF, Gagnon K, Fortin D, Decken A, Harvey PD. (2006) Inorg Chem, 45:2812.
25. Zhang T, Drouin M, Harvey PD. (1999) Inorg Chem 38:1305.
26. Zhang T, Drouin M, Harvey PD. (1999) Inorg Chem 38:957.
27. Zhang T, Drouin M, Harvey PD. (1999) Inorg Chem. 38:4928.
28. Gan L, Harvey PD. (1991) Inorg Chem. 30:3239.
29. Harvey PD, Gray HB. (1990) Polyhedron 9:1949.
30. Harvey PD, Dallinger RF, Woodruff WH, Gray HB. (1989) Inorg Chem. 28:3057.
31. Harvey PD, Adar F, Gray HBJ. (1989) Am Chem Soc., 111:1312.
32. Harvey PD, Gray HBJ. (1988) Am Chem Soc. 110:2145.
33. Harvey PD, Schaefer WP, Gray HB. (1988) Inorg Chem. 27:1101.
34. Harvey PD. (1995) Inorg Chem. 34:2019.
35. Harvey PD. (2001) Coord Chem Rev. 17:219.
36. Harvey PD. (2004) Macromol Symp. 209:67.
37. Perreault D, Drouin M, Michel A, Harvey PD. (1992) Inorg Chem. 31:3688.
38. Fortin D, Drouin M, Turcotte M, Harvey PDJ. (1997) Am Chem Soc. 119:531.
39. Yan B (2004) Faguang Xuebao 25:47.
40. Zaoui B, Fortin D, Harvey PDJ. (1996) Soc Alg Chim. 6:231.
41. Fortin D, Drouin M, Harvey PD.J. (1998) Am Chem Soc. 120:5351.
42. Harvey PD, Drouin M, Zhang T. (1997) Inorg Chem. 36:4998.
43. Piché D, Harvey PD. (1994) Can J Chem. 72:705.
44. Perreault D, Drouin M, Michel A, et al. (1992) Inorg Chem. 31:695.
45. Fournier É, Sicard S, Decken A, Harvey PD. (2004) Inorg Chem. 43:1491.
46. Fournier É, Lebrun F, Drouin M, Decken A, Harvey PD. (2004) Inorg Chem. 43:3127.
47. Fortin D, Harvey PD. (1998) Coord Chem Rev. 171:351.
48. Fortin D, Drouin M, Harvey PD. (2000) Inorg Chem. 39:2758.
49. Förster T. (1946) Naturwissenschaften 33:166.
50. Förster T. (1948) Ann Phys. 2:55.
51. Dexter DLJ. (1953) Chem Phys. 21:836.
52. Faure S, Stern C, Espinosa E, Guilard R, Harvey PD. (2005) Chem Eur J. 11:3469.
53. Faure S, Stern C, Guilard R, Harvey PD.J. (2004) Am Chem Soc. 126:1253.
54. Fournier É, Decken A, Harvey PD. (2004) Eur J Inorg Chem. 4420.

55. Harvey PD, Fournier É. (2006) ACS Symposium Series 928:472.
56. Turro NJ. (1978) Modern photochemistry. Benjamen Cummings, Menlo Park.
57. Harvey PD. (1996) Coord Chem Review. 153:175.
58. Harvey PD, Truong KD, Aye KT, Bandrauk AD. (1994) Inorg Chem. 33:2347.
59. Harvey PDJ. (2004) Inorg Organomet Polym. 14:211.
60. Gagnon J, Drouin M, Harvey PD. (2001) Inorg Chem 40:6052.
61. Gagnon J. (2001) PhD Dissertation, Université de Sherbrooke, Canada.
62. Perreault D, Drouin M, Michel A, Harvey PD. (1991) Inorg Chem. 30:2.
63. Harvey PD. (2003) Macromol Symp. 196:173.
64. Tzeng BC, Yeh HT, Kuo JH, Lee GH, Peng S-M. (2006) Inorg Chem. 45:591.
65. Irwin MJ, Vittal JJ, Puddephatt RJ. (1997) Organometallics. 16:3541.
66. Brandys M-C, Puddephatt RJJ. (2001) Am Chem Soc. 123:4839.

Metal Binding Studies of Ferrocene Peptides in Solution

Francis E. Appoh and Heinz-Bernhard Kraatz

1 Introduction

The electrochemical detection of cations binding to a ligand that is equipped with a redox active group, such as Fc or cobaltocene, is of great interest. The specificity of ion detection is dictated by ligand-ion interactions and the efficiency of the electronic communication between the ligand-ion complex and the redox active group, either through-space or through-bond interactions [1].

Various Fc-substituted macrocycles (Schemes 5.1 and 5.2) such as crown ethers **1** [2], cryptand 2 [3,4], closed aza-oxa crown **3** [2,5], cylam **4** [3], open aza-oxa **5** [3], and calixarenes **6, 7** [6] have been shown by electrochemical methods to sense the coordination of alkaline-, alkaline earth-, transition- metals, and lanthanide ions to the macrocycle.

Peptides are known to show some specificity toward metal binding, as was shown by titration experiments using ^{1}H-NMR [7,8], potentiometry [9], luminescence measurements [10], infrared (IR), circular dichroism (CD), and ultraviolet (UV_ spectroscopies [11]. Poly-L-aspartic acid binds to metals such as Eu^{3+}, Ce^{3+}, La^{3+}, Cu^{2+}, and Pb^{2+}, and acts as a corrosion inhibitor for steel and iron. This property has been ascribed to the carboxylate side chain of aspartic acid [12,13].

Lanthanide, alkaline earth metals, and alkaline metals coordination to peptides and amino acids generally involves the carboxylate terminal. Various coordination modes are identified in the literature. Monodentate modes are generally associated with alkaline metal-carboxylate interactions. Bidentate carboxylate coordination networks in the typical syn-syn bridging, chelate bidentate and tridentate modes

A.S. Abd-El-Aziz et al. (eds.), *Inorganic and Organometallic Macromolecules: Design and Applications.*

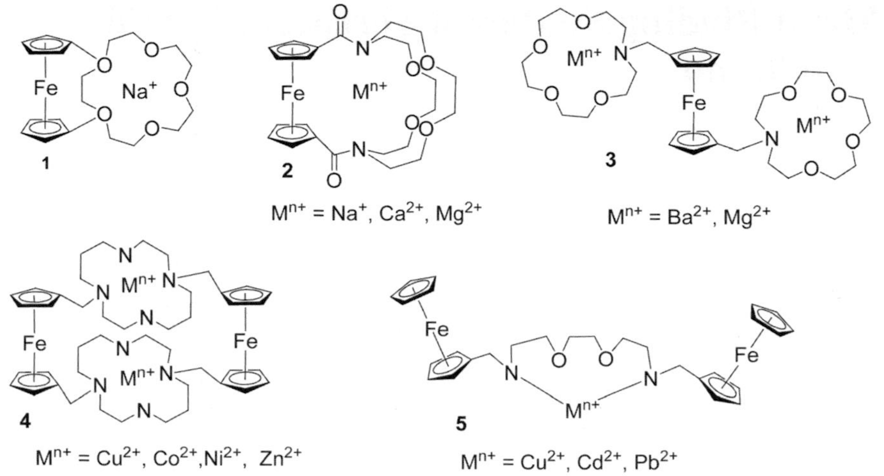

Scheme 5.1 Fc functionalized macrocylics for metal ion detection. crown ethers **1** [2], cryptands **2** [3,4], closed aza-oxa crown **2** [2], cylams **4** [3], open aza-oxa **5** [3]

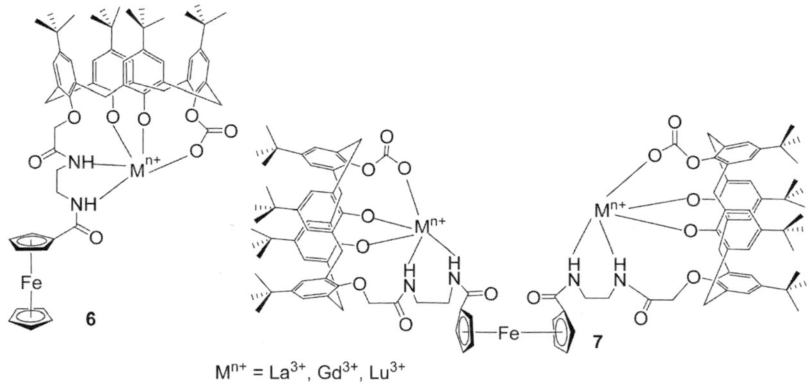

Scheme 5.2 Fc-calixarenes **6** and **7** for lanthanide ion detection [6]

have been observed in the crystal structure of N-methylglycine Eu(III) complexes and glycine-Nd^{3+} complexes, as well as in lanthanide glutamate and aspartate complexes [14-19]. IR spectroscopy is a diagonistic tool which provides structural insight into the coordination mode of the carboxylate group [20, 21]. The acid group in the IR spectra is a single band in the range of 1700 to 1730 cm^{-1} resulting from antisymmetric C=O stretching mode [22]. In peptide-metal complexes this singlet disappears and is replaced by a doublet corresponding to the symmetric $\nu_s(COO^-)$ and asymmetric $\nu_{as}(COO^-)$ stretching modes. The difference between the symmetric and asymmetric stretching vibration of the carboxylate group ($\Delta\nu$) in complexes with alkaline metals, alkaline earth metals, and transitional metals can be used qualitatively to determine the coordination mode in the complex. Tacket has proposed an empirical relationship for $\Delta\nu$ bridging carboxylates showing a $\Delta\nu = 140–170\,cm^{-1}$,

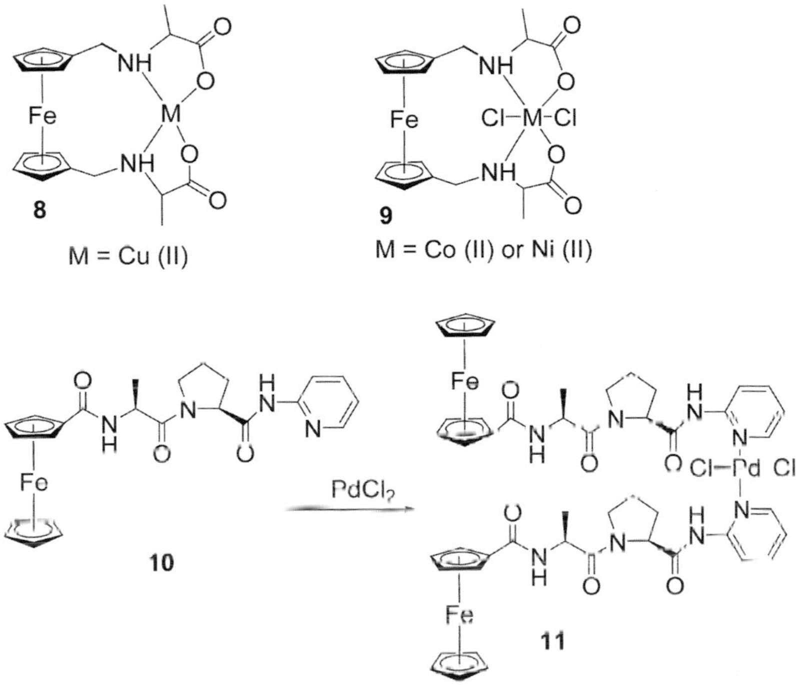

Scheme 5.3 Examples of Fc amino acid and peptide conjugates for metal ion binding [24,25]

whereas for bidentate carboxylates $\Delta\nu = 40–80\,cm^{-1}$, and for unidentate carboxylates $\Delta\nu = 200–300\,cm^{-1}$ [21]. Gericke and Huhnerfuss proposed a similar empirical relationship from interactions of saturated fatty acids with metals and reported $\Delta\nu$ values of $\Delta\nu = 158\,cm^{-1}$ for ionic complexes, of $\Delta\nu = 150\text{-}200\,cm^{-1}$ for bridging bidentate, and $\Delta\nu = 80–110\,cm^{-1}$ for bidentate chelating [23].

Fc-functionalized amino acids and peptides have received some attention as metal binding ligands (Scheme 5.3). Chohan has shown from spectroscopic evidence that 1.1′-dimethyl-Fc derivatives of amino acids formed square planar metal complexes with Cu(II) **8**, whereas with Co(II) and Ni(II) octahedral metal complexes of **9** were formed. [24]. Electrochemical detection of metal ion interaction was demonstrated by Hirao, who showed that the electrode potential of Fc-peptide **10** shifted cathodically by 20 mV when complex **11** was formed [25].

Cheng and co-workers recently reported Fc-cyclopeptides **12–15** (Scheme 5.4) and showed that they acted as redox-switchable cation receptors [26]. These Fc-cyclopeptides exhibited strong anodic shifts of their electrode potentials in the presence of alkaline earth metals and lanthanides. The extent of the anodic shift can be correlated with the charge density of the metal ion with a bias toward binding of lanthanides, alkaline earth metals, and the least sensitivity to alkaline metals [26]. Chowdhury et al, reported the syntheis and interaction of cyclic Fc-Histidine conjagates **16** with metal ions (Scheme 5.5). Electrochemical measurements showed that the compound exhibited cathodic shifts in the order $Na^+ > Li^+ > K^+ > Cs^+$ which in the order of their ionic sizes and suggest that the observed shift relates to the cavity of the compound [27].

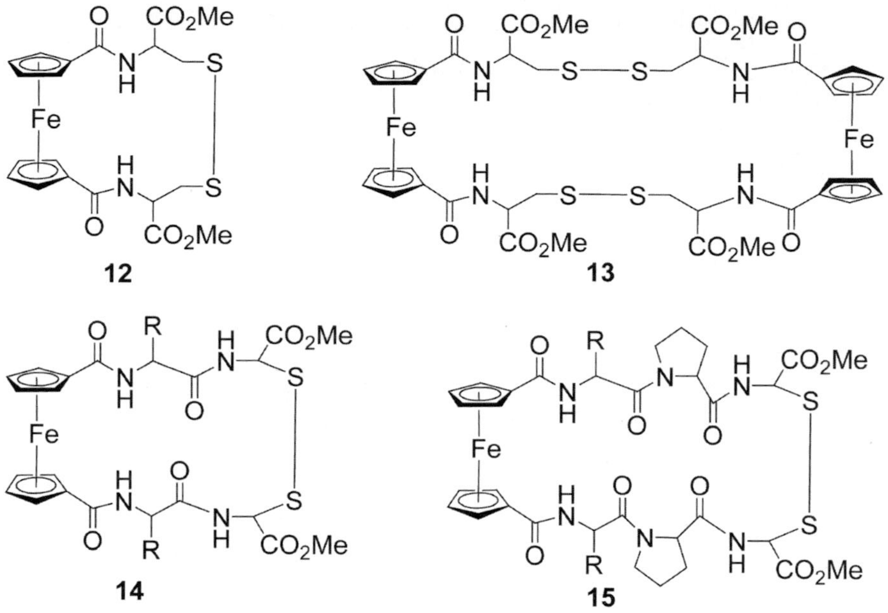

Scheme 5.4 Fc-cyclopeptides for metal ion detection [26]

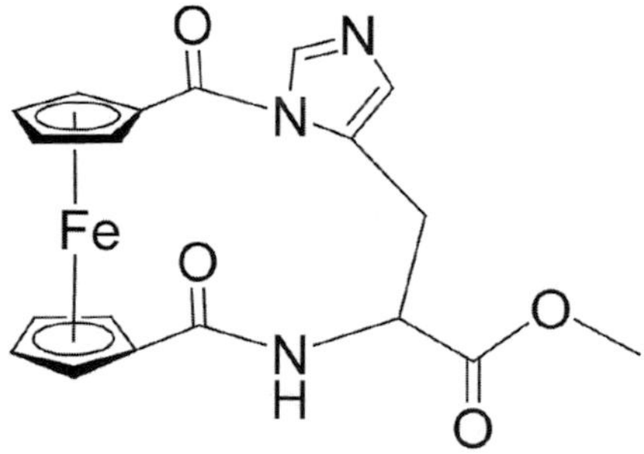

Scheme 5.5 Cyclic Fc-Histidine conjagates **16**

To date, no systematic studies of Fc-peptide-metal interactions have been carried out. A more systematic study involving linear Fc-peptide-metal interaction is therefore required to answer questions such as: Is it possible to use electrochemical methods to observe metal binding to Fc-peptide conjugates? Do these systems display specificity towards certain metals? How are the metals coordinated?

This chapter presents the synthesis and characterization of mono- and disubstitute Fc-peptides. Their characterization and their solution properties are discussed. Quantification of the metal binding to Fc-peptide/acids is reported. The coordination environment of the Fc-peptide binding site is examined by spectroscopic and electrochemical methods.

2 Synthesis and Characterizations of Ferrocene Peptides

The synthesis of monosubstituted Fc-peptide esters, Fc-GlyOEt **17**, Fc-AspOMe **18**, Fc-GluOEt **19** (Scheme 5.6) and their disubstituted analogs Fc-[GlyOEt]$_2$ **23**, Fc–[AspOBz]$_2$ **24**, and Fc–[GluOEt]$_2$ **25** (Scheme 5.7) was achieved with moderate to good yield using the carbodiimide-HOBt method.

Ester deprotection by base hydrolysis yielded the corresponding acids Fc-GlyOH **20**, Fc-AspOH **21**, Fc-GluOH **22**, Fc–[GlyOH]$_2$ **26**, Fc–[AspOH]$_2$ **27**, and Fc–[GluOH]$_2$ **28**.

Scheme 5.8 shows the pathway for the synthesis of Fc-GluOEt **19**, Fc–[AspOBz]$_2$ **24**, and Fc–[GluOEt]$_2$ **25**. The deprotection of Fc-GluOEt **19** and Fc–[GluOEt]$_2$ **25** with NaOH in methanol yielded the desired Fc-GluOH **22** and Fc–[GluOH]$_2$ **28**. The synthesis of Fc–[AspOH]$_2$ **27** from its corresponding ester Fc–[AspOBz]$_2$ **24**

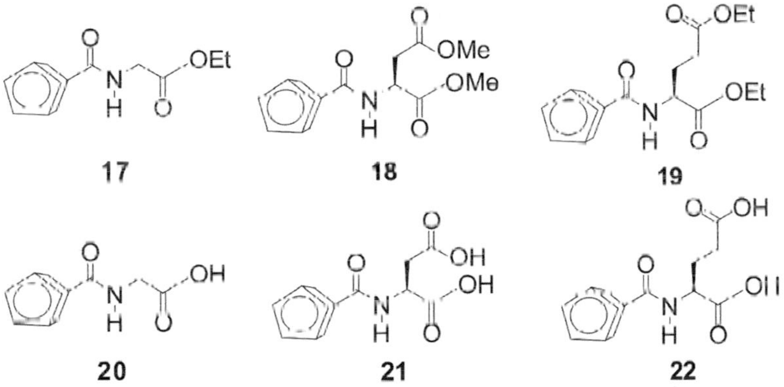

Scheme 5.6 Monosubstituted Fc-peptides esters Fc-GlyOEt **17**, Fc-AspOMe **18**, Fc-GluOEt **19**, and acids Fc-GlyOH **20**, Fc-AspOH **21**, and Fc-GluOH **22**

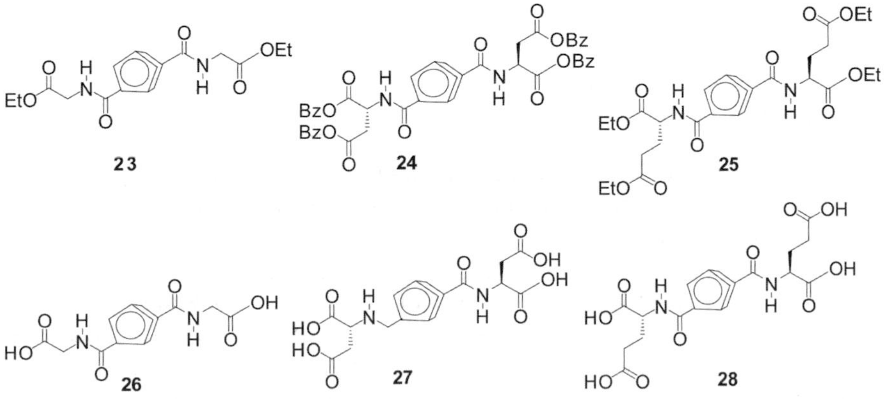

Scheme 5.7 Disubstituted Fc-peptides esters Fc–[GluOEt]$_2$ **23**, Fc–[AspOBz]$_2$ **24**, and acids Fc–[GlyOH]$_2$ **25**, Fc–[AspOH]$_2$ **26**, and Fc–[GluOH]$_2$ **27**

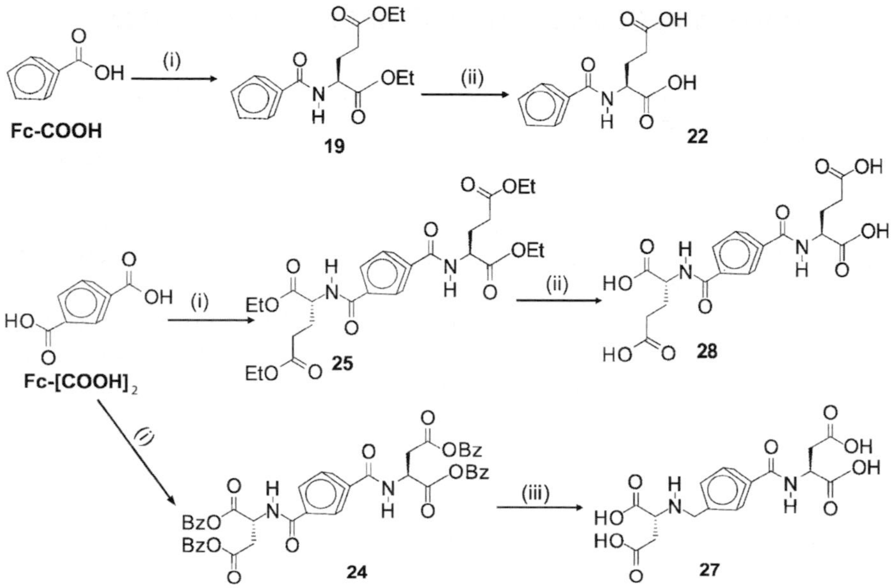

Scheme 5.8 Synthesis of compounds **19**, **24**, and **25** starting from monosubstituted and the disubstituted Fc carboxylic acid, respectively. *(i)* Dry $CHCl_2$/HOBt/EDC at 0 °C and H–Glu(OEt)–OEt or H–Asp(OBz)–OBz/Et_3N/CH_2Cl_2. Deprotection of **19** and **25** with *(ii)* NaOH aq/MeOH yielded **22** and **28** or by *(iii)* hydrogenolysis of **24** with H_2 using Pd/C catalyst in wet methanol to obtain **27**

was achieved by hydrogenolysis of the benzyl protecting group in the presence of a Pd/C catalyst at 60 psi in wet methanol.

Figure 5.1 shows representative ^{1}H-NMR spectra of compounds **19** and **24**. The position of the amide NH in Fc–[GluOEt]$_2$ **25** is further downfield compared with Fc-GluOEt **19**. Such a downfield shift of amide NH is generally the result of H-bonding. The use of different solvents has an effect on the position of the chemical shift, especially for species that have the tendency to form H-bonds with the solvent. It was observed that the amide NH of the monosubstituted **17–19** was shifted further downfield in $CDCl_3$ as compared with DMSO-d6 (Tables 5.1 and 5.2). Kraatz previously reported the effect of protic solvents on the NH proton chemical shifts in a series of Fc-peptides that were attributed to strong H-bonds between the solvent and the amide moiety, which also affected the electronic structure of the Fc group [28].

Some selected ^{1}H-NMR chemical shifts in $CDCl_3$ for Fc-GlyOEt **17**, Fc-AspOMe **18**, Fc-GluOEt **19**, and their disubstituted analogs Fc–[GlyOEt]$_2$ **23**, Fc–[AspOBz]$_2$ **24**, and Fc-[GluOEt]$_2$ **25** are presented in Figure 5.1. Table 5.2 also presents some selected ^{1}H-NMR chemical shifts in DMSO-d6 for Fc-GlyOEt **17**, Fc-AspOMe **18**, Fc-GluOEt **19**; the disubstituted analogs Fc–[GlyOEt]$_2$ **23**, Fc–[AspOBz]$_2$ **24**, Fc–[GluOEt]$_2$ **25**; and corresponding acids Fc-GlyOH **20**, Fc-AspOH **21**, Fc-GluOH **22**, Fc–[GlyOH]$_2$ **26**, Fc–[AspOH]$_2$ **27**, and Fc–[GluOH]$_2$ **28**. The Fc moiety of the monosubstituted systems showed a 2:2:5 peak ratio for the 2 ortho, the 2 meta protons, and the 5 protons for the unsubstituted Cp ring. The Asp and Glu conjugates

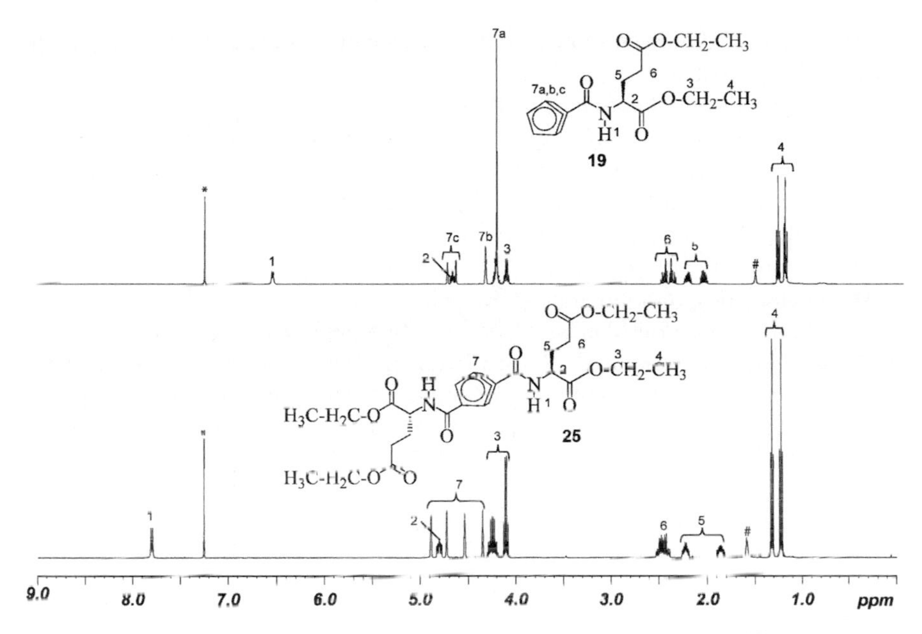

Figure 5.1 Representative ^{1}H-NMR spectra for compound for Fc-GluOEt **19** *(top)* and Fc–[GluOEt]$_2$ **25** *(bottom)* (5 mM) with some prominent labeled peaks taken in $CDCl_3$. Labeled peaks amide NH (1), α H (2), O-CH_2CH_3 (3) and (4), CH_2 of Glu (5) and (6). * and # are the residual $CHCl_3$ and H_2O, respectively

Table 5.1 ^{1}H-NMR chemical shifts of Fc-peptides esters in $CDCl_3$. δ in ppm versus TMS

Compounds	NH	Cp	ester
Fc-GlyOEt **17**	6.15	4.25, 4.81, 4.36	4.13, 1.05
Fc-AspOMe **18**	6.75	4.73, 4.65, 4.34, 4.24	3.62
Fc-GluOEt **19**	6.55	4.75, 4.66, 4.35, 4.25	4.05, 1.22, 1.17
Fc–[GlyOEt]$_2$ **23**	7.94	4.84, 4.48	4.10, 1.19
Fc–[AspOBz]$_2$ **24**	7.56	4.80, 4.73, 4.46, 4.36	7.35, 5.10
Fc–[GluOEt]$_2$ **25**	7.84	4.99, 4.75, 4.56, 4.37	4.07, 1.21, 1.17

Table 5.2 ^{1}H NMR chemical shifts of Fc-peptides **17–28** in DMSO-d6. δ in ppm versus TMS

Compounds	NH	Cp	OH
Fc-GlyOEt **17**	8.31	4.80, 4.39, 4.24	–
Fc-AspOMe **18**	8.23	4.80, 4.76, 4.38, 4.20	–
Fc-GluOEt **19**	8.01	4.89, 4.82, 4.38, 4.21	–
Fc-GlyOH **20**	8.14	4.79, 4.35, 4.24	12.51
Fc-AspOH **21**	7.98	4.80, 4.79, 4.35, 4.21	13.12
Fc-GluOH **22**	7.84	4.87, 4.82, 4.36 ,4.21	12.40
Fc–[GlyOEt]$_2$ **23**	8.30	4.85, 4.44	–
Fc–[AspOBz]$_2$ **24**	8.30	4.80, 4.78, 4.29	–
Fc–[GluOEt]$_2$ **25**	8.30	4.83, 4.82, 4.42, 4.39	–
Fc–[GlyOH]$_2$ **26**	8.22	4.80, 4.38	12.40
Fc–[AspOH]$_2$ **27**	8.38, 8.01	4.83, 4.38	12.38
Fc–[GluOH]$_2$ **28**	8.12, 7.92	4.90, 4.87, 4.42, 4.37	12.35

18, **19**, **24**, and **25**, however, showed some differentiation of the 2 ortho protons of the substituted Cp moiety.

For the disubstituted compounds, two singlets in a 1:1 ratio were observed for the Fc moiety in compounds **23**, **26**, **27**, whereas for **24**, **25**, and **28** a 4 proton in a 1:1:1:1 ratio was observed. The successful deprotection of the esters to the free acids is indicated by both the appearance of a resonance at δ 12.3–13.2 as a result of the acid OH (Table 5.2) and the disappearance of the resonances as a result of the ester protecting groups (Table 5.1).

IR spectroscopy in solid state (KBr pellets) has provided valuable information about the structural identity of Fc-modified peptides. The salient IR peaks of interest for compounds **17–25** are listed in Table 5.3. The position of the Amide A band between 3,400 and 3,300 cm^{-1} is quite informative. Amide NH stretches below 3,400 cm^{-1} are diagnostic of the H-bonded amide NH [29,30]. The peak positions of all compounds except that for 83 are indicative of H-bonded structures.

In addition all the esters showed additional non-H-bonded peaks. It was surprising that compound **23** appeared to be non-H-bonded whereas the disubstituted analog Gly **26** was H-bonded. Intramolecular H-bonding involving the two podant amino acid groups is a common structural motif in 1,n′-bis-amino acid substituted ferrocenes [31,32]. This was comfirmed by the crystal structure of Fc-$[GlyOH]_2$ **26** which showed strong intramolecular H-bonding between the two amide NH and the opposite carbonyl on the two podant amino acid substituents (Figure 5.6).

Table 5.3 Selected IR and UV data for Fc-peptides

	IR/cm^{-1}					UV/vis
Compound	Amide A	Fc	C=O	Amide I	Amide II	λ_{max} (ε)
Fc-GlyOEt	3,475, 3,253	3,079	1,747	1,637	1,547	444
Fc-AspOMe	3,450, 3,300	3,081	1,734	1,627	1,537	446
Fc-GluOEt	3,447,3,288	3,095	1,731	1,639	1,560	442
Fc-GlyOH	3,422	3,099	1,746	1,604	1,548	444
Fc-AspOH	3,344	3,089	1,752	1,600	1,539	442
Fc-GluOH	3,322	3,095	1,752	1,616	1,562	446
Fc–$[GlyOEt]_2$	3,340, 3,269	3,090	1,743	1,632	1,540	450
Fc–$[AspOBz]_2$	3,425, 3,365	3,089	1,735	1,653	1,532	449
Fc–$[GluOEt]_2$	3,442, 3,306	3,094	1,733	1,634	1,541	440
Fc–$[GlyOH]_2$	3,385	3,096	1,715	1,631	1,548	460
Fc–$[AspOH]_2$	3,356	3,118	1,731	1,693	1,576	446
Fc–$[GluOH]_2$	3,342	3,099	1,733	1,619	1,540	446

IR was recorded in KBr for both free acids and esters. UV recorded in methanol in 1 to 10 mM concentration range. λ_{max} in nm and (ε) extinction coefficient in $Lmol^{-1}cm^{-1}$

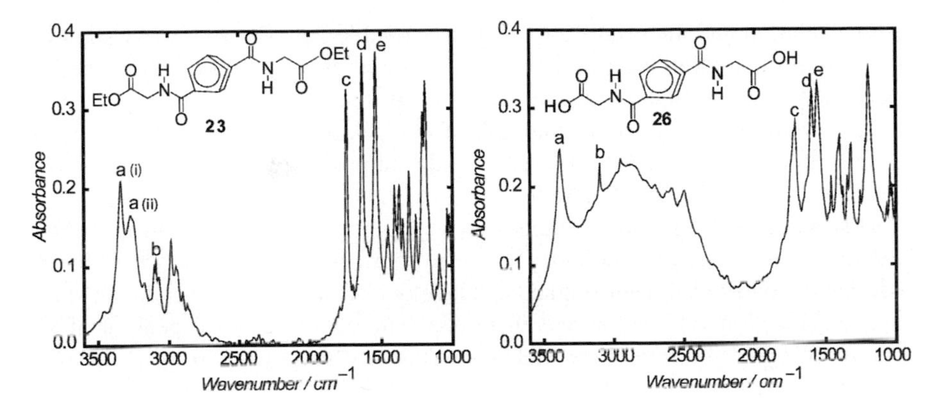

Figure 5.2 Partial IR spectra for compounds Fc-[GlyOEt]$_2$ **23** and Fc-[GlyOH]$_2$ **26**. Amide A, non-H-bonded *(i)*, H-bonded *(ii) (a)*, Cp-H *(b)*, C=O from ester or acid *(c)*, Amide I *(d)*, Amide II *(e)*. Compound **89** shows a broad peak in the region of 2500 to 3500 cm^{-1}, which is characteristic of associated COOH groups

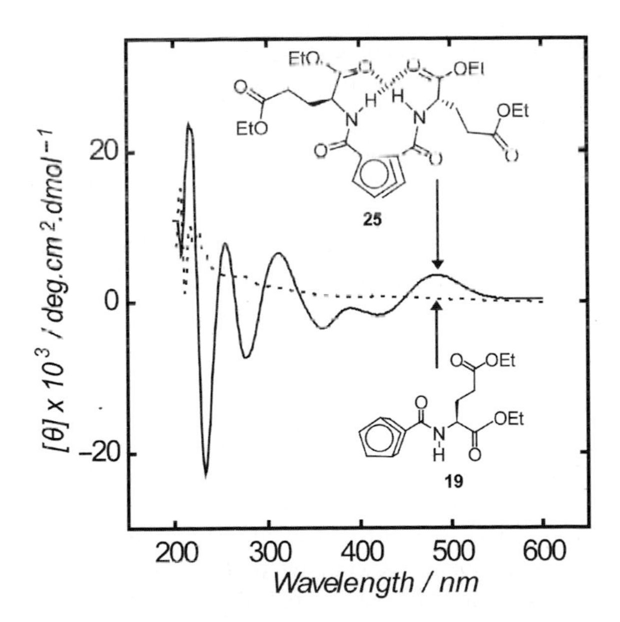

Figure 5.3 CD spectra of Fc-GluOEt **19** *(broken line)* and Fc-[GluOEt]$_2$ **25** *(solid line)* in MeOH (1.0×10^4 M)

Thus, these systems exist in conformations that facilitate intramolecular H-bonding between the two podant amino acids. The presence of a broad peak in the 3,600 to 2,600 cm^{-1} region of the deprotected acids shows the presence of OH stretching vibrations which are characteristic of acids in their associated form (Figure 5.2, **26**) [33,34]. The UV-vis spectra for all Fc-peptide were recorded in

MeOH solution and the results are tabulated in Table 5.3. The Fc moiety shows a broad weak absorption band in the visible region at about 450 nm with extinction coefficients lower than 400 $L mol^{-1} cm^{-1}$ that is attributed to the d–d transition.

CD spectra of the 1,n′-compounds **24**, **25**, **27**, **28**, with exception of the glycine analogs **23** and **26**, in MeOH showed an induced positive Cotton effect in the Fc region at $\lambda = 440–480$ nm (Figure 5.3).

This indicates that the two L-amino acids on both rings force the Fc into a P-helical structure. This is usually attributed to the formation of intramolecular H-bonds between the two substituents on the Fc. The monosubstituted analogs do not show such organization in solution and therefore exhibit only a weak peak at about 405 nm. It was demonstrated by Metzler-Nolte that this weak peak is an intrinsic property of the molecule and is not caused by intermolecular H-bonding [35].

2.1 X-ray Crystal Structures

Single crystals of compounds **82**, **86**, and **89** were obtained by the diffusion of hexane into dichloromethane or methanol solutions of the peptides giving yellow to orange crystals. Fc-GluOEt **82** crystallizes in a trigonal space group $P3_1$ [36]. The asymmetric unit contained two independent molecules, in which the most notable difference is in the conformations of the diethyl ester groups (Figure 5.4). The C-C bond distance in the substituted Cp rings 1.419(6) Å and 1.428(4) Å were longer than in the unsubstituted 1.402(6) Å and 1.403(4) Å. Selected Fe-C distances included Fe(1)-C(11) = 2.018(3) Å , Fe(1)-C(14) = 2.053(3) Å , Fe(1)-C(17) = 2.035(3) Å , and Fe(2)-C(31) = 2.017(3) Å, Fe(2)-C(34) = 2.054(3) Å , Fe(2)-C(37) = 2.048(3) Å. The bond angles between *N*(2)-C(21)-C(31)-(35) were 177.9(3)° whereas those for O(6)-C(21)-C(31)-C(32) were178.3(3)°, showing the coplanarity of the amide moiety with the Cp ring.

The nonbonding distance between the carbonyl group and an amide NH of an adjacent molecule are d(O(1)···N(2) = 2.822(3) Å and *d*(O(6)···N(1) = 2.861(3) Å. There is no intramolecular H-bonding but intermolecular H-bonding is possible (Scheme 5.9), leading to extended H-bonded networks in the solid-state however linking the amide NH(1) and O(6) in a head to head fashion (Scheme 5.9). Structural arrangements found for compounds **82**, has been reported for crystal structures Fc-Glu$(OBz)_2$ and Fc-AlaOMe **92** [37,38].

The X-ray crystal structure of **21** shown in Figure 5.5 was obtained from a solution of CH_2Cl_2 layered with Et_2O [28]. The geometric features of **21** shows two Cp rings of the Fc group are parallel with a small bent angle CpFeCp of 1.1° and adopt an eclipsed conformation. Compound **21** crystallizes with a molecule of water, which occupies a position that enables H-bonding with the Fc-amide NH and the Asp C=Ogroups. The Fc-C=O group is involved in hydrogen bonding interactions with the acid groups of the Asp acid side chain as well as the α-carboxyl group of two adjacent molecules, forming a bifurcated H-bond with O-H···O=C(Fc) contacts of 2.678(2) and 2.605(2) Å. The H-bonding distance to the α-carboxy group

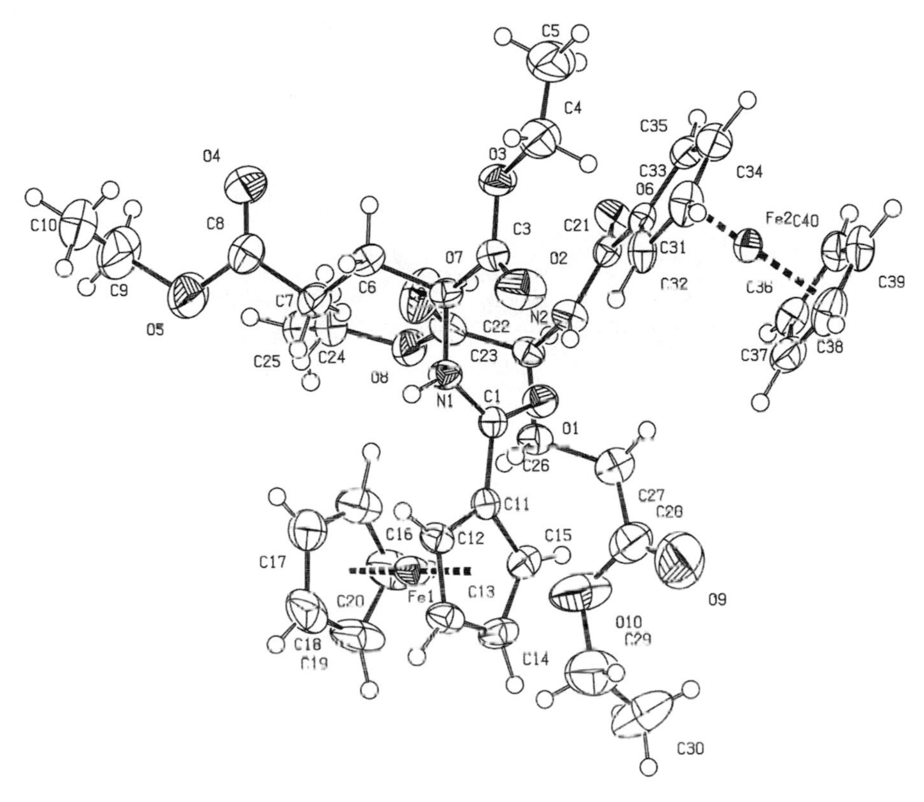

Figure 5.4 ORTEP plot of Fc-GluOEt **82** showing the asymmetric unit

Scheme 5.9 Head-to-head intermolecular H-bonding patern in Fc-GluOEt **19**

is shorter, indicating a stronger H-bond, which is in line with the lower p*K*a for this group (p*K*a = 1.88) compared with the side chain carboxyl group (p*K*a = 3.65). The compound shows extensive hydrogen bonding network involving the water mole-

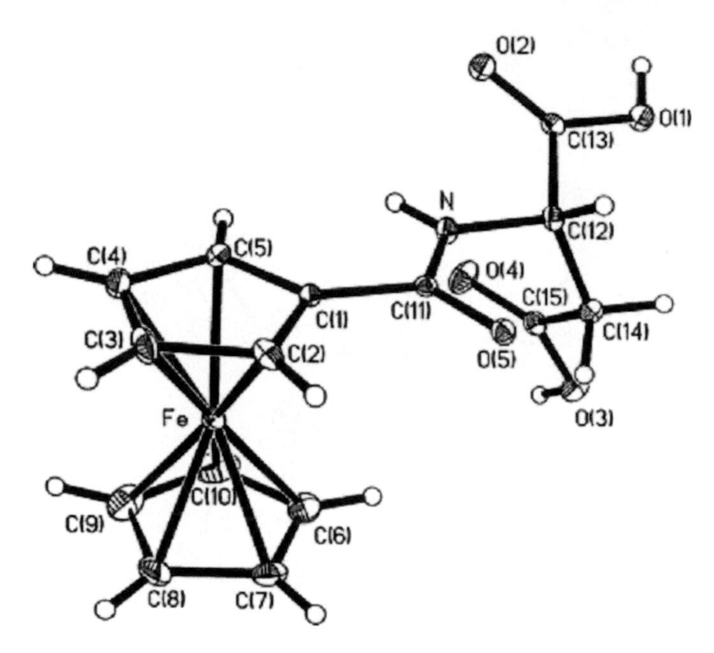

Figure 5.5 ORTEP diagram for Fc-Asp(OH)-OH showing a single molecule of **21**

cule and three Fc-aspartic acid molecules. The water molecule is nestled in a pocket created by the Fc-aspartic acid moiety, which enables it to establish a series of H-bonding contacts [28].

The structure of the disubstituted Fc glycine-ethylester **23** and its corresponding acid **26** are shown in Figures 5.6 and 5.7. The crystal structure of compound **23** (Figure 5.6) shows that the substituents on the Fc are in the 1,3′-conformation which allows it to engage in intermolecular H-bonding between N(1) and (O1) of neighbouring molecules [39]. No intramolecular H-bonds are formed in the molecular structure. Compound **86** exhibits features common to many monosubstituted Fc-amides [40, 41]. The coplanarity of the 2 Cp rings with a small Cp–Fe–Cp bent angle of 1.8°. In addition, both amide groups exhibit only a slight twist with regard to their respective Cp planes (Cp1-amide: 14.1° and Cp2-amide 15.2°) allowing the interaction between the π-systems of the Cp ring and the amide group. The Cp–C(O) distances of 1.482(5) Å and 1.487(5) Å for C(1)–C(11) and C(6)–C16), respectively, are within the range of other Fc-amino acids and peptides, and simple Fc-amides [42, 43]. Similarly, the amide C=O (O(1)–C(1) = 1.235(5) Å, O(4)–C(6) = 1.230(4) Å) and amide C–N (C(1)–N(1) = 1.326(5) Å, C(6)–N(2) = 1.344(5) Å) bond distances are normal compared to related Fc-amino acids and peptides.

The molecule interacts with its adjacent neighbours through H-bonding, resulting in the formation of a one-dimensional H-bonded chain, in which O(1) interacts with

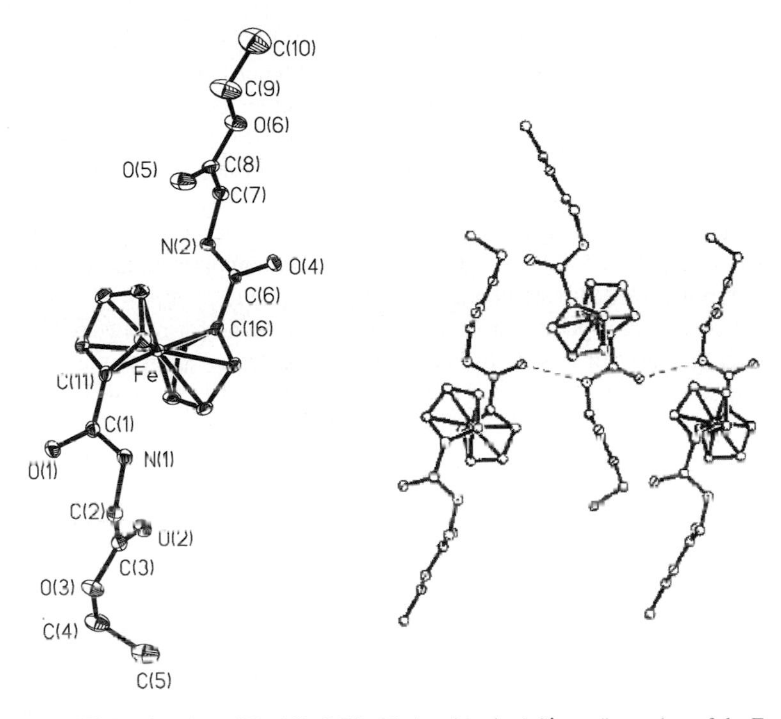

Figure 5.6 ORTEP drawing of Fc-[GlyOEt]$_2$ **23** showing the 1,3′- conformation of the Fc group. Ellipsoids are drawn at the 30% probability level. All hydrogen atoms are omitted for clarity *(left)*. Formation of a one-dimensional H-bonded polymeric chain involving H-bonding of N(1) and O(1) of adjacent molecules (O(1)···N(1) = 2.839(5) Å *(right)*

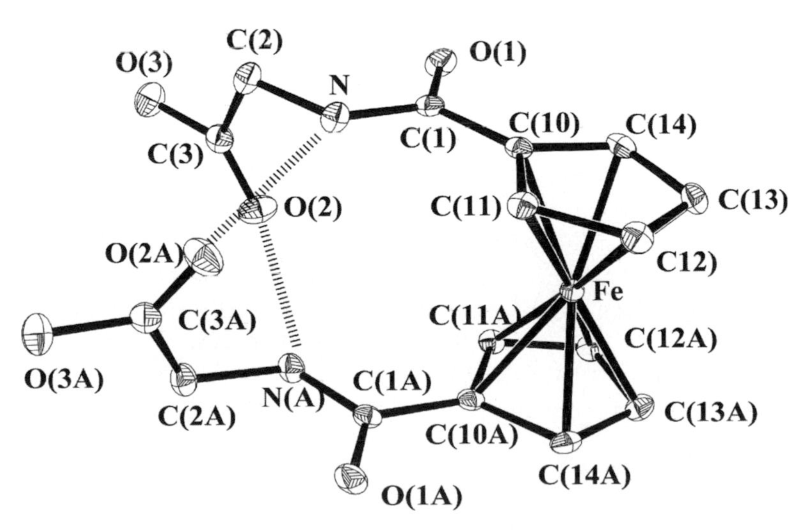

Figure 5.7 ORTEP drawing of Fc-[GlyOH]$_2$ **26** showing the 1,2′-conformation of the two podents peptide substituents. The intramolecular H-bonding distances between N and O(2A) is 2.875(3) Å [39]

the N(1) of the Gly group of an adjacent molecule (O(1)···N(1) = 2.839(5) Å (Figure 5.6). Interestingly, the amide group on the other Cp ring is not involved in H-bonding and is well-separated from other molecules (~ 4 Å). This H-bonding pattern is reminiscent of that reported by Hirao and coworkers for the monosubstituted Fc-Gly-Gly-OEt and Fc-Ala-Pro-OEt with an alternating up-down orientation of the molecules [40,44].

The resulting dihedral angles of the podant glycine ethylester which were involved in H-bonding are $\Phi_1 = 65.4(4)°$ and $\Psi_1 = -158.4(3)°$. In comparison, those of the other substituent not involved in H-bonding are $\Phi_1° = -69.9(5)°$, $\Psi_1° = 162.3(3)°$. The individual chains are separated from each other by 4 Å and form a layered structure with a layer thickness of 11.4 Å. In sharp contrast, the deprotection of the ester group in Fc-[GlyOEt]$_2$ **23** to the free acid Fc-[GlyOH]$_2$ **26** causes a change in the conformation from a 1,3′ into a 1,2′-conformation (Scheme 5.10) [39].

The crystal structure of Fc-[GlyOH]$_2$ **26** is shown in Figure 5.7 and its packing diagram is shown in Figure 5.8. The crystal structure of this centro-symmetric molecule, exhibits a strong intramolecular H-bonding between the two podant substitutents which force the Fc framework into a 1,2′-conformation (d(N···O(2A) and O(2)···N(A) = 2.875(3) Å. This provides a rigid framework, in which the amide twist is reduced considerably (2.9°). The resulting dihedral angles Φ_1 and Ψ_1 are significantly different from the ethyl ester ($\Phi_1 = 79.0(2)°$ and $\Psi_1 = -178.60(18)°$).

The intramolecular H-bonding interaction involving the two amide NH and the two acid C=O on opposite Cp rings forces them into a 1,2′-conformation. This allows the formation of a two-dimensional H-bonded network, in which there are strong interactions between the Fc-C=O and the OH groups of adjacent molecules

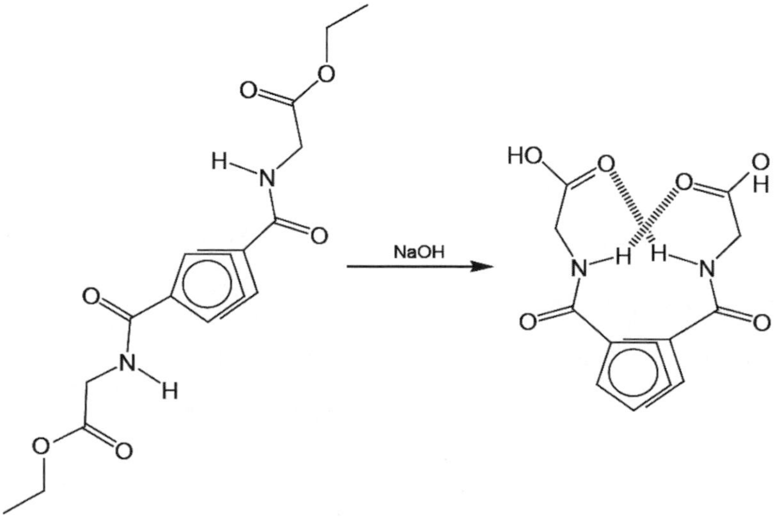

Scheme 5.10 Structural transformation by hydrogen bonding. Conformational change from 1,3′-conformation in Fc-[GlyOEt]$_2$ **23** into the 1,2′-conformation in the free acid Fc-[GlyOH]$_2$ **26**

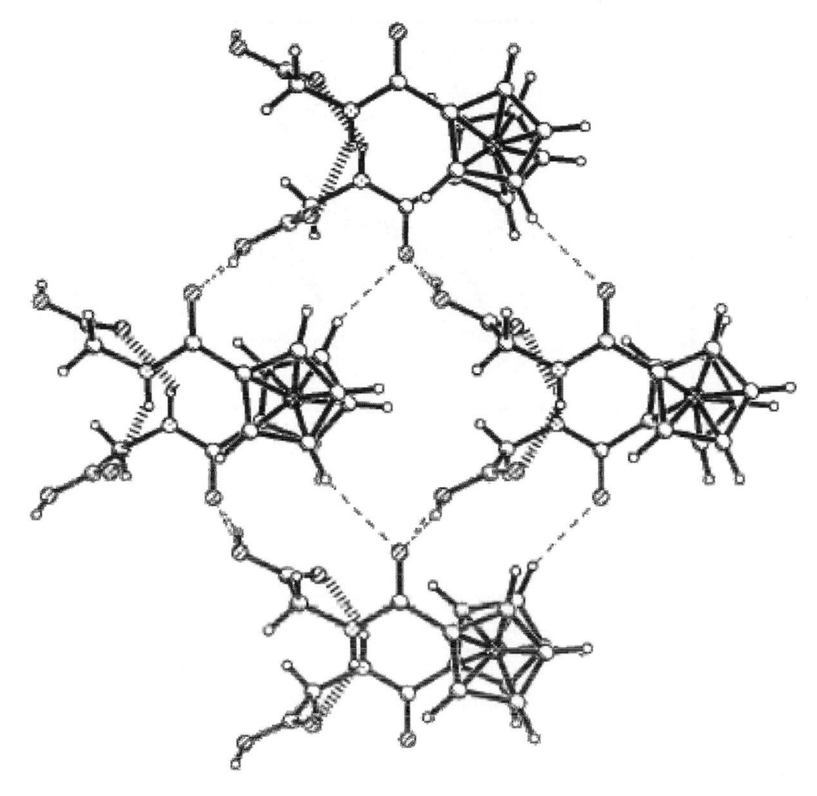

Figure 5.8 Two-dimensional network formed by the interaction of adjacent molecules of Fc-[GlyOH]$_2$ **26**. H-bonding interactions between adjacent molecules involving the interaction of the Fc-C−O group with the acid OH (d(O(1)ċċċO(3*) = 2.620(3) Å. Individual molecules display intramolecular H-bonding pattern with d(N–O(2)) and d(N(A)–O(2)) = 2.875(3) Å. Additional weak O···H–C interactions between O(1) and H(12) on the Cp ring, which may support the structural motif are shown

(d(O(1)···O(3*) = 2.620(3) Å). In addition, there are weak OċċċH–C interactions between O(1) and H(12) on the Cp ring, which may support the structural motif. Weak C–H···O and C–H···N H-bonding interactions are well documented and known to be important factors in stabilizing and sometimes even controlling the structure in the solid state (Figure 5.8) [45].

2.2 Electrochemical Characterization of Ferrocene Peptides

In this section, the effect of solvent and the nature of the peptide substituent on the electrochemical properties of the Fc redox probe are discussed. The redox properties of Fc-peptides esters **17–19** and **23–25** were investigated in methanol, acetonitrile and dichloromethane, while the acid analog **20–22** and **26–28** were studied in

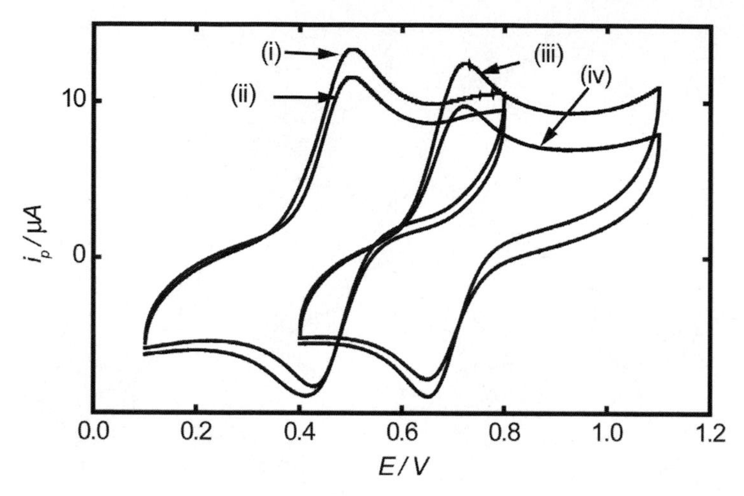

Figure 5.9 Cyclic voltamograms of Fc-peptides; Fc-AspOMe **18** *(i)*, Fc-AspOH **21** *(ii)*, and Fc-$[AspOBz]_2$ **24** *(iii)*, and Fc-$[AspOH]_2$ **27** *(iv)*. 2.0 M $NaClO_4$/MeOH glassy carbon working electrode, Pt counter electrode, and Ag/AgCl reference electrodes at a scan rate of 100 mVs^{-1}

methanol and water. The choice of solvent is related to the solubility of the compounds. Figure 5.9 shows CVs for the monosubstituted and disubstituted Asp ester **18**, **24** and their corresponding acids **21** and **27** as representative examples.

In all systems, the Fc-peptides displayed a single oxidation step Fc → $Fc^+ + e^-$ which is quasi-reversible under all conditions. The scan rate (ν is related to the peak current i_p by the Randle-Seveik relationship at 25 °C as shown in Eq. 5.1[46].

$$i_p = 2.59 \times 10^5\ n^{3/2}\ AD_o^{1/2} C\ \upsilon^{1/2} \tag{5.1}$$

where i_p is the peak current of the CV peak, A is the area of the electrode (cm^2), C represents the bulk concentration of the electrochemically active species (mol/l), and n is the number of electrons transferred per active species ($n = 1$ for the Fc-peptides). D_o is the diffusion coefficient. Thus for systems where diffusion is the sole mode of transport of redox ions, the cathodic and anodic peak currents vary linearly with the square root of the scan rate. Figure 5.10, shows a representative CV for Fc–$[GlyOEt]_2$ **25** at various scan rates.

Table 5.4, summarizes the $E_{1/2}$ and ΔE_p for compounds **17–28**. Fc-amino acid conjugates exhibit a single redox peak cathodically shifted by about 30 mV compared to the Fc-COOH. In contrast, the disubstituted conjugates shift to more anodic potentials of the same magnitude from Fc-$[COOH]_2$. This indicates that the Fc-amino acid conjugation has a profound effect on the electronic properties of the Fc moiety. Kraatz and coworkers have shown that the electronic difference between a Fc-CO–NH_2 versus Fc-COOH originates from the differences in the interaction of the π-obitals of the substituent with those of the Cp ring [40]. The LUMO

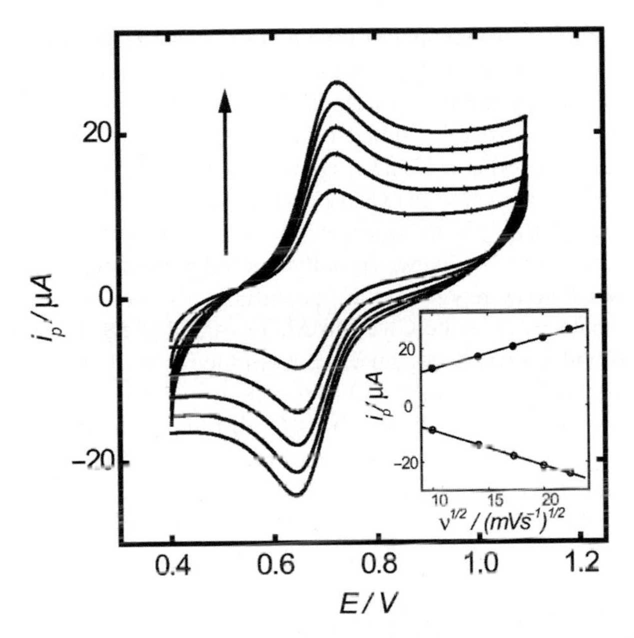

Figure 5.10 CV for Fc-$[GlyOEt]_2$ **25** at scan rate 100, 200, 300, 400, and 500 mV/s (direction of the arrow indicating increasing scan rate). Inset Plot of oxidation and reduction peak currents versus the square root of the scan rate in 2.0 M $NaClO_4$ /MeOH carbon working, Pt counter electrode, and Ag/AgCl reference electrode

Table 5.4 Halfwave potentials ($E_{1/2}$) and peak separation (ΔE_p) for Fc-peptides at 25 °C, glass carbon working electrode, Pt counter electrode, and Ag/AgCl reference electrode in selected solvent systems

	2 M $NaClO_4$/MeOH		0.2 M TBAP/CH_3CN		0.2 M TBAP/CH_2Cl_2		2 M $NaClO_4$/aq	
Compound	$E_{1/2}$	ΔE_p	$E_{1/2}$	ΔE_p	$E_{1/2}$	ΔE_p	$E_{1/2}$	ΔE_p
Fc-COOH	491	85	–	–	–	–	–	–
Fc–$[COOH]_2$	654	104	–	–	–	–	–	–
Fc-GlyOEt **17**	461	75	608	89	626	165	–	–
Fc-AspOMe **18**	458	91	601	85	640	113	–	–
Fc-GluOEt **19**	465	73	608	100	636	145	–	–
Fc-GlyOH **20**	459	75	–	–	–	–	373	73
Fc-AspOH **21**	464	75	–	–	–	–	384	69
Fc-GluOH **22**	460	76	–	–	–	–	379	66
Fc–$[GlyOEt]_2$ **23**	681	76	810	98	828	146	–	–
Fc$[AspOBz]_2$**24**	688	75	818	105	840	160	–	–
Fc–$[GluOEt]_2$ **25**	680	85	811	92	847	200	–	–
Fc–$[GlyOH]_2$ **26**	677	70	–	–	–	–	612	72
Fc–$[AspOH]_2$ **27**	686	85	–	–	–	–	612	66
Fc–$[GluOH]_2$ **28**	674	88	–	–	–	–	633	62

Standard seviations on $E_{1/2}$ = 5-8 mV and ΔE = 5–10 mV. All values in mV

Standard seviations on $E_{1/2}$ = 5-8 mV and ΔE = 5–10 mV. All values in mV

of the Fc-amide is lower than that of the Fc-acid, which makes the amide slightly more electron withdrawing, and thus makes it more difficult to oxidize.

There is also the general trend of the disubstituted Fc conjugates **23–28** having electrode potentials anodically shifted by about 200 to 240 mV compared to the monosubstituted counterparts **17–22** (Figure 5.8). Similar differences in electrode potentials have been observed between Fc–$(Pro)_n$OBz **29–32** and its disubstituted analogs Fc–$[(Pro)_nOBz]_2$ **33–36** (where $n = 1$ to 4) (Scheme 5.11) [47,48].

For Fc-amino acid conjugates, virtually no differences were observed in the redox potentials. Sheely reported redox potentials of closely related monosubstituted Fc conjugates such as Fc-GlyOMe **37**, Fc-AlaOMe **38**, Fc-LeuOEt **39**, and Fc-PheOEt **40** with electrode potential values in the range of 598 to 608 (± 5) mV in 0.2 M TBAP/CH_3CN (Scheme 5.12) [49].

It is known that increasing the number of substitutents on the Fc moiety has a significant effect on the electrode potential of the Fc redox probe [50]. For example the redox potental of $[Fc]^{+/-}$ **41**, $[Me_5Fc]^{+/-}$ **42**, and $[Me_{10}Fc]^{+/-}$ **43**, showed cathodic shifts of up to 500 mV in going from compound **41** to compound **43**, which was attributed to the electron donating properties of the methyl substituents on the Cp ring (Scheme 5.13) [50].

Increasing the chainlength of the Fc peptides, however, has an influence on the electrode potential. In a series of Fc-oligoproline conjugates Fc–$(Pro)_n$OBz **29–32**, Fc–$[(Pro)_nOBz]_2$ **33–36**, and $[Fc–(Pro)_nCSA]_2$ **44–47** (where $n = 1$ to 4) (Scheme 5.11)

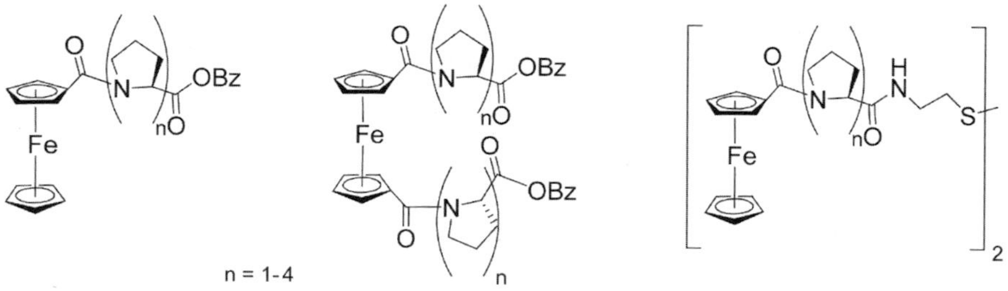

Scheme 5.11 Fc-proline analogs **29–32, 33–36** and **44–47**

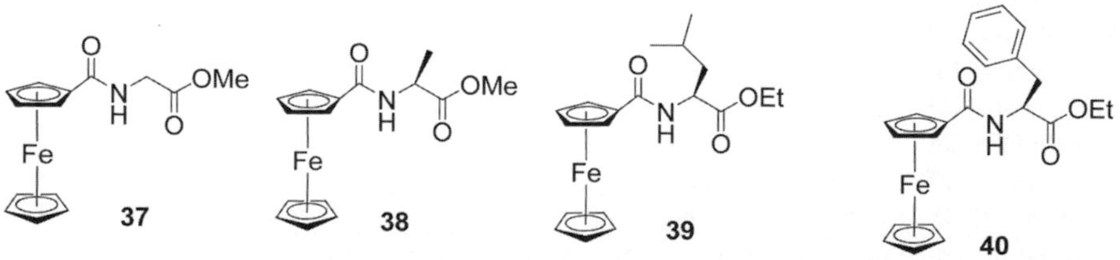

Scheme 5.12 Fc-amino acid analogs **37–40**

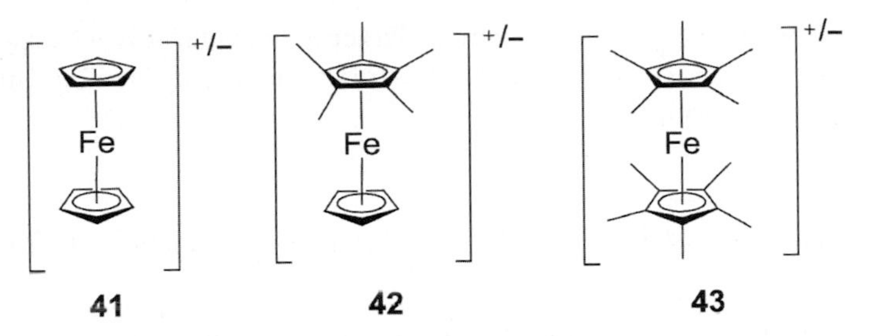

Scheme 5.13 Structures of redox probes $[Fc]^{+/-}$ **41**, $[Me_5Fc]^{+/-}$ **42**, $[Me_{10}Fc]^{+/-}$ **43**

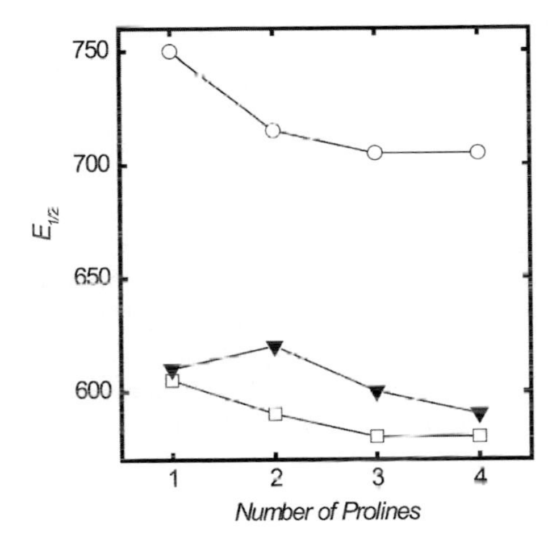

Figure 5.11 Electrode potentials of Fc–$(Pro)_n$OBz **29–32** —□—, Fc–$[(Pro)_n OBz]_2$ **33–36** —○—, and $[Fc–(Pro)_n CSA]_2$ **44–47** —▼— (where n = 1 to 4) [47,48]

cathodic shifts of their electrode potentials were observed with increasing chain length (Figure 5.11) [47,48].

Significant solvent influences on electrode potential are observed (Table 5.4). In solvents that are also capable of forming H-bonds such as H_2O and MeOH, electrode potentials were observed at more cathodic potentials compared with those measured in solvents that are unable to establish H-bonds with the Fc-peptides. The observed cathodic shift order has been linked to the ability of solvents to H-bond with the amide linkage of these compounds [28]. This observation is consistent with the NMR results described above where changes in solvent from the nonpolar and non-H-bonding $CDCl_3$ to the polar and H-bond acceptor DMSO affected the chemical shift of the ortho H's, which is ascribed to the interaction of the solvent with the amide NH.

ΔE_p (peak separations) values shown in Table 5.4 are greater than 59/n mV (wheren n is number of electrons) and therefore generally denote quasi-reversible behaviour. However ΔE_p values were much larger in nonpolar solvent, especially in CH_2Cl_2, than in polar solvents, which may be attributed to incomplete iR compensation in nonpolar solvent [50].

2.3 *pH Effects on Electrode Potentials of Fc-Peptide Acids*

The effect of pH on the redox properties of the Fc-peptide acids Fc-GlyOH **20**, Fc-AspOH **21**, Fc-GluOH **22** and Fc–$[GlyOH]_2$ **26**, Fc–$[AspOH]_2$ **27**, and Fc–$[GluOH]_2$ **28** were studied in phosphate buffer (containing 5% methanol to ensure solubility) to provide an approximate p*K*-of the Fc-peptide acids and to determine the effect of substituents on the electrochemistry of the Fc moiety.

Figure 5.12 shows the CV of compound **26** at pH 2.0 and pH 8.0. The redox process remained fully reversible over the entire pH region, with $\Delta E = 50$–69 mV and ratio of peak currents close to unity. This indicates that the redox process is under thermodynamic control with respect to proton transfer. The faradaic response however changes drastically from the acidic to basic condition. A plausible explanation for this effect is that H^+ promotes the voltammetric response by reducing electrostatic repulsion between the negatively charged Fc-peptide and the negatively charged electrode, thus resulting in an increase in the rate of electron transfer [51,52].

Figures 5.13 and 5.14 show the plots of electrode potentials of the monosubstituted Fc-peptide acids Fc-GlyOH **20**, Fc-AspOH **21**, Fc-GluOH **22** and their disubstituted analogs Fc–$[GlyOH]_2$ **26**, Fc–$[AspOH]_2$ **27**, and Fc–$[GluOH]_2$ **28** as well as Fc-COOH and Fc–$[COOH]_2$.

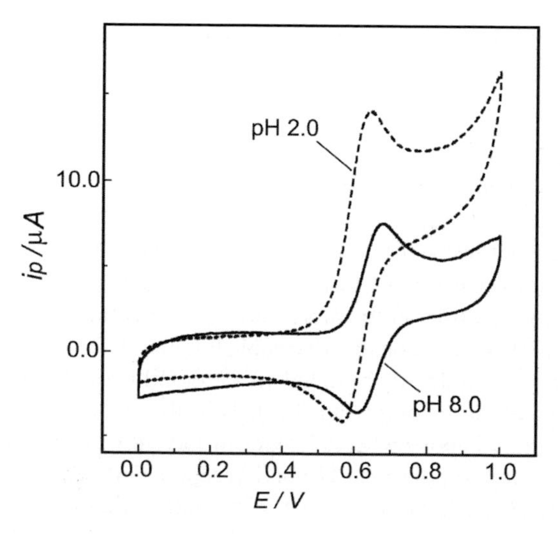

Figure 5.12 CV of Fc-$[GlyOH]_2$ **26** at pH 2.0 and pH 8.0 (T = 25 °C ± 1, glassy carbon working electrode, Ag/AgCl reference electrode and Pt counter electrode 0.5 mM in 2.0 M phosphate buffers pH 2.0 to 8.0 (containing 5% methanol to ensure solubility)

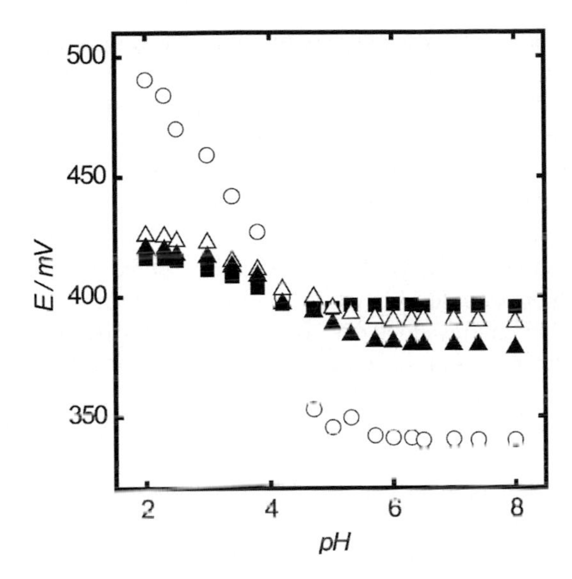

Figure 5.13 Effect of pH on the electrode potentials of monosubstituted compounds Fc-COOH (o), Fc-GlyOH **20** (■), Fc-AspOH **21** (Δ), Fc-GluOH **22** (▲). (T = 25 °C ± 1, glassy carbon working electrode, Ag/AgCl reference electrode and Pt counter electrode 0.5 mM in 2.0 M phosphate buffers pH 2.0 to 8.0 (containing 5% methanol to ensure solubility)

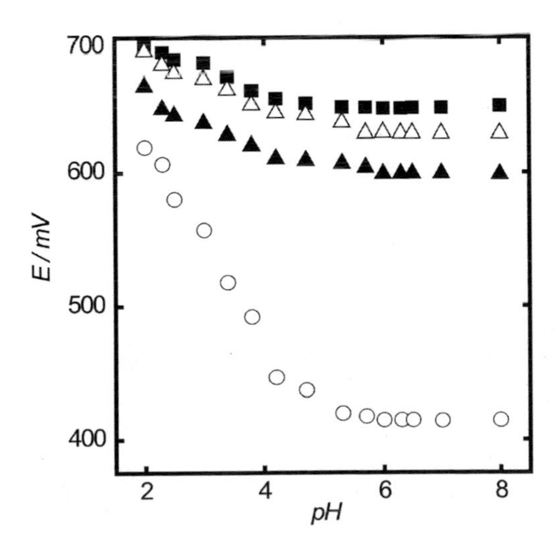

Figure 5.14 Effect of pH on the electrode potentials of disubstituted Fc compoGuds; $Fc(COOH)_2$ (o), Fc–$[GlyOH]_2$ **26** (■), Fc–$[AspOH]_2$ **27** (Δ), Fc–$[GluOH]_2$ **28** (▲). (T = 25 °C ± 1, glassy carbon working electrode, Ag/AgCl reference electrode and Pt counter electrode 0.5 mM in 2.0 M phosphate buffers pH 2.0 to 8.0 (containing 5% methanol to ensure solubility)

Two conclusions can be reached. Firstly, the half-wave potential ($E_{1/2}$) is dependent on the proton concentration of the solution, reaching a limit at pH 5.0 to 6.0 below which the $E_{1/2}$ is independent. The region of dependence is associated with the deprotonation of the carboxylate groups of the compounds. The redox signals shifts cathodically with increasing pH and demonstrates the sensing of carboxylate protonation and deprotonation. Secondly, the slope for the acids **16** and **17** compared with that observed for the Fc-peptides Fc-GlyOH **83**, Fc-AspOH **84**, Fc-GluOH **85** and Fc–[GlyOH]$_2$ **89**, Fc–[AspOH]$_2$ **90**, Fc–[GluOH]$_2$ **91** is steeper.

For the protonation/deprotonation equilibria of compound **16** and **17**, (Scheme 5. 14) the limiting slope $\delta E_{1/2}/\delta$pH of –50 and –56 mV per pH unit respectively is obtained which is close to the –57 mV per pH unit value expected for a one electron-one proton equilibrium calculated from the Nernst equation (Eq. 5.2). For the monosubstituted Fc-peptide conjugates Fc-GlyOH **20**, Fc-AspOH **21**, Fc-GluOH **22**, a slope of $\delta E_{1/2}/\delta$pH 10–14 mV/pH is observed. Similarly, for the 1,n′-disubstituted analogs Fc–[GlyOH]$_2$ **26**, Fc–[AspOH]$_2$ **227**, Fc–[GluOH]$_2$ **28** a slope $\delta E_{1/2}/\delta$pH, of –18 to –23 mV/pH was observed.

The shifts for the disubstituted are about two times the value for the monosubstituted analogs. These show that the pH-potential relationship of Fc modified compounds is largely governed by its substituent. The $E_{1/2}$-pH dependence is accounted for by the presence of ionizable functional groups on the Fc-peptide that undergoes a change in p*K*a upon electron transfer [53,54]. With increasing pH

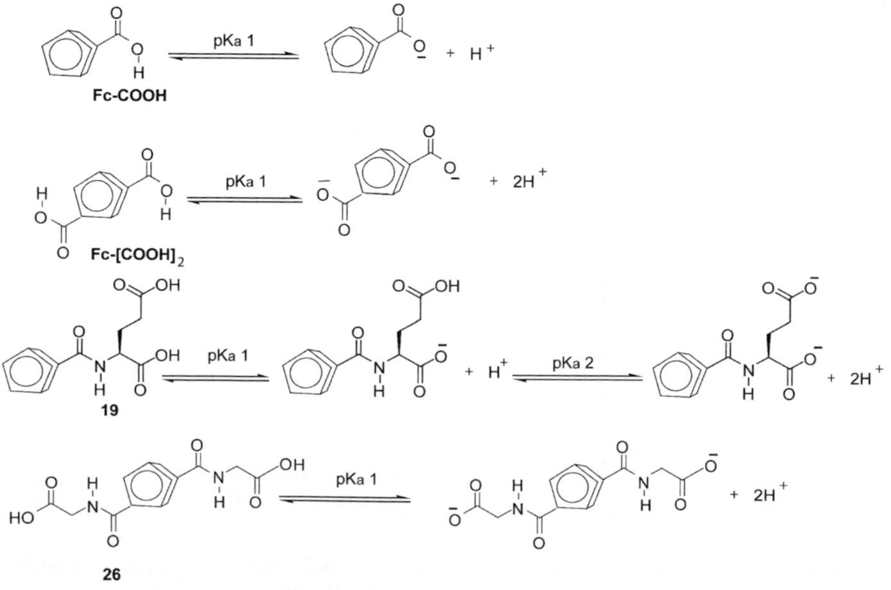

Scheme 5.14 Scheme showing the protonation/deprotonation equilibra for Fc-COOH, Fc–[COOH]$_2$, Fc-GluOH **19**, and Fc–[GlyOH]$_2$ **26**

values the carboxylates stabilize the Fc^+ state compared with the Fc, and as a consequence the redox potential shifts cathodically [55]. The magnitude of the cathodic shift per pH unit may be related to the proximity of the acidic substitute to the Fc group as well as conformational or polarity changes caused by the ionization of the amino acids. As shown above the electronic difference between Fc-CO–NH_2 versus Fc-COOH, which originates from the differences in the interaction of the π-obitals of the substituent with those of the Cp ring Fc [40], has considerable influence on the electronic properties and may explain the significant differences in their pH dependent electrochemical behavior.

The measured $E_{1/2}$ is predicted by the Nernst equation to have a pH dependence according to Eq. 5.2 [56].

$$E_f = E_o - (0.059/n)\log\{(K_{ox} + [H^+])/(K_{red} + [H^+])\} \tag{5.2}$$

where n is the number of electrons, $[H^+]$ is the proton ion concentration, E_o and E_f are the initial and final halfwave electrode potential respectively, and K_{ox} and K_{red} are the apparent equilibrium constants for the deprotonation in the oxidized and reduced states, respectively. From Eq. 5.2, one can deduce that

$$\Delta E_{1/2} = -59/n\,(pK_{ox} - pK_{red}). \tag{5.3}$$

Thus for $n = 1$, we would expect a limiting slope $\delta E_{1/2}/\delta pH = -59$ mV per pH unit. Martinez-Máñez has shown that if the interaction between redox active centre and the charged substrate is electrostatic, the $\Delta E_{1/2}$ vs pH response is a function of the charge of the oxidized framework, the substrate, the number of electrons, number of electroactive groups, the distance between the redox-active center and the charge substrates and the relative permittivity [57]. The method however gives approximate pKa's of redox-active pH-responsive molecules and has been applied to Fc-COOH, Fc-$[COOH]_2$ and compounds such as Fc-boronic acid **48**, open Fc-aza macrocycles **49** and **50** and to Fc-$(CH_2)_n$–COOH (where n = 0-2) **51–53** (Scheme 5.15) [56,58–60]. For compounds **51–53**, it was reported that $\delta E_{1/2}/\delta pH$ decreased with an increasing aliphatic chain length [60].

Least-square curve fitting of the experimental $E_{1/2}$-pH curves using Equation 2.3 allows for the determination of the pKa's for protonation and deprotonation of the oxidized and reduced species. For example, the curve fitting applied to compound **26** is shown in Figure 5.15. However, the CV titration does not allow us to distinguish between a stepwise deprotonation involving the mono-deprotonated species Fc–($GlyO^-$)(GlyOH) and fully deprotonated Fc–$[GlyO]_2^{2-}$. A pKa of 3.8 ± 0.2 for Fc–$[GlyOH]_2$ and 3.3 ± 0.2 for $[Fc–(Gly--OH)]^+$ was obtained [39].

To test the validity of the method, an NMR titration experiment was carried out by following the changes in the proton shift for the methylene protons of Fc–GlyOH as a function of pH (Figure 5.15). The δ of CH_2 experiences a pH dependence over the range of pH 2.0 to 8.0 and allows us to extract information regarding the acidity of the carboxyl group. Importantly, electrochemical and NMR pH titrations give identical pKa values for the free acid **26** of 3.8 ± 0.2 (3.8 ± 0.15 for NMR). Again,

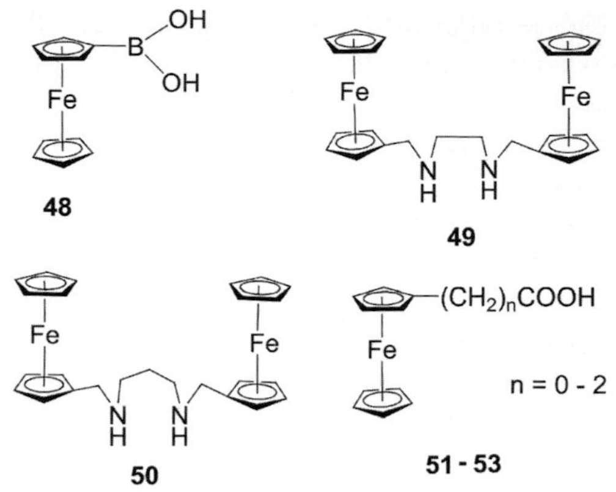

Scheme 5.15 Structures of compounds **100–115**

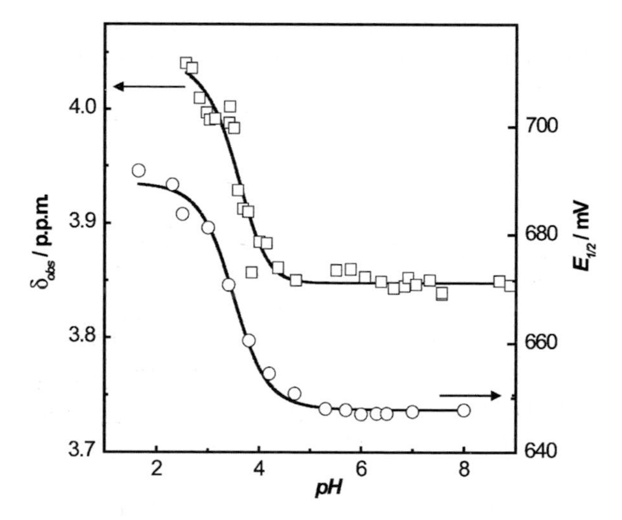

Figure 5.15 Dependence of $E_{1/2}$ (o), and δ_{CH2} (□) on pH for Fc-[GlyOH]$_2$ **26** (T = 25 °C ± 1, glassy carbon working electrode, Ag/AgCl reference electrode and Pt counter electrode 0.5 mM in 2.0 M phosphate buffers pH 2.0 to 8.0 (containing 5% methanol to ensure solubility) (T = 25 °C ± 1)

the NMR titration experiments do not allow us to distinguish between two separate deprotonation events. Thus the p*K*a of both protons is assumed to be identical or close to identical [39].

The absolute value of the p*K*a is in good agreement with values reported before for the p*K*a's of pyruvic acid (3.2–3.9) and other N-substituted-Gly (3.75–4.07) [61-63]. The deprotonation/protonation equilibrium for compounds **19** and **26** is shown in Scheme 5.14. Unlike in Fc-GluOH **19**, the two carboxylic acid groups (α p*K*a1 and γ p*K*a2) of Fc–[GlyOH]$_2$ **26** are completely distinguishable. p*K*a values were determined from a least-square fit of the experimental $E_{1/2}$ data. Data for

Fc-peptides are presented in Table 5.5. The measured p*K*a's are close to those reported for n-terminal functionalized amino acid also shown in Table 5.5 [61].

2.4 Ferrocene Peptides as Ligands for Metal Ion Interaction

Metal ion interaction with Fc-peptides was studied by two methods: (i) electrochemical titrations of the Fc-peptides with metal ions in aqueous solution having the pH adjusted to 7.0, and by (ii) synthesis of metal complexes of the Fc-peptides from aqueous solution at pH 7.0 and the spectroscopic examination of the products.

The direct electrochemical measurement of interactions of a Fc-peptide host (FcL) with metal ions M forming the redox active complex FcL–M, can be represented by Scheme 5.16.

The four square reactions constitute a closed loop which gives a total Gibbs free energy change of zero (Eq. 5.4),

Table 5.5 Approximate p*K*a determination from voltammetric measurements.

	p*K*a1		p*K*a2	
Compound	Fc^+	Fc	Fc^+	Fc
Fc-COOH	3.3 (0.2)	5.4 (0.4)	–	–
Fc–[COOH]$_2$	3.9 (0.4)	5.9 (0.3)	–	–
Fc-GlyOH **20**	3.3 (0.1)	3.6 (0.1)	–	–
Fc-AspOH **21**	2.0 (0.1)	2.1 (0.1)	3.8 (0.1)	4.3 (0.1)
Fc-GluOH **22**	1.9 (0.1)	2.0 (0.2)	3.9 (0.1)	4.5 (0.1)
Fc–[GlyOH]$_2$ **26**	3.3 (0.2)	3.8 (0.2)	–	–
Fc–[AspOH]$_2$ **27**	1.4 (0.1)	2.1 (0.1)	3.8 (0.2)	4.1 (0.1)
Fc–[GluOH]$_2$ **28**	2.0 (0.1)	2.3 (0.1)	4.0 (0.2)	4.2 (0.1)
Fc-COOH *	3.34 (0.04)	6.44 (0.01)	–	–
Fc–[COOH]$_2$ *	3.48 (0.07)	6.74 (0.01)	3.00 (0.10)	5.76 (0.01)

Standard deviations in brackets. (*literature values) [57]

$$
\begin{array}{ccc}
\text{FcL} & \xrightleftharpoons[e^-]{E_{free}} & \text{Fc}^{(+)}\text{L} \\
{+M}\ \Big\updownarrow\ K_{red} & & {+M}\ \Big\updownarrow\ K_{ox} \\
\text{FcL-M} & \xrightleftharpoons[e^-]{E_{comp}} & \text{Fc}^{(+)}\text{L-M}
\end{array}
$$

Scheme 5.16 Square scheme of the complexation of metal ion by Fc-peptide. The free FcL forms a complex with M in their reduced and oxidized forms. $E_{1/2}{}^{f}$ and $E_{1/2}{}^{c}$ are the half-wave potentials and K_{ox} and K_{red} are the stability constants

$$\Delta\Sigma G = \Delta G_{FcL} + \Delta G_{ox} + \Delta G_{FcL\text{-}M} + \Delta G_{red} = 0 \tag{5.4}$$

$$nF(E - E_{free}) - RT\ln(K_{ox}) + nF(E_{comp} - E) + RT\ln(K_{red}) = 0 \tag{5.5}$$

Rearranging Eqs. 5.4 and 5.5 which links the change in half-wave potential $\Delta E_{1/2}$ between the free ligand ($E_{1/2}^{F}$) and the complexed state ($E_{1/2}^{C}$), is indicative of the binding strength shown in the following Eq. 5.6 [1]:

$$\Delta E_{1/2} = E_{1/2}^{C} - E_{1/2}^{F} = (RT/nF)\ln(K_{ox}/K_{red}) \tag{5.6}$$

where K_{red} and K_{ox} are the stability constants in the reduced and oxidized states, respectively. The magnitude of K_{ox}/K_{red} is a measure of the binding strength. The measurement of the electrochemical response of the compounds in the presence of either La^{3+}, Tb^{3+}, Zn^{2+}, Mg^{2+} or Ca^{2+} as guest cation species was investigated by CV by addition of aliquots of aqueous solutions of metal salt to 5.0 mL, 0.5 mM solution of the free Fc-peptide acid in 0.2 M $NaClO_4$ at pH 6.8 to 7.0.

Figure 5.16 shows a typical CV and DPV obtained for a solution of Fc-GluOH **22** before and after the addition of 2 equivalents of Tb^{3+} at pH 6.8. Table 5.6 summarizes the changes in electrode potentials obtained from CV measurements for the Fc-peptides Fc-GlyOH **20**, Fc-AspOH **21**, Fc-GluOH **22** and Fc–$[GlyOH]_2$ **26**, Fc–$[AspOH]_2$ **27**, Fc–$[GluOH]_2$ **28** in the absence and in the presence of metal ions. In all cases, addition of metal ions to the solution causes anodic shifts. For the monosubstituted acidic systems, Fc-AspOH **21** and Fc-GluOH **22**, the addition of La^{3+} and Tb^{3+} caused large shifts in the $E_{1/2}$ (Table 5.6), whereas shifts caused by the addition of Mg^{2+} and Ca^{2+} were much smaller. Unexpectedly, addition of metal ions to the disubstituted Fc derivatives **27** and **28** caused significantly smaller changes in $E_{1/2}$. The $E_{1/2}$ for Fc–$[GlyOH]_2$ **26** for the addition of Mg^{2+} and Ca^{2+} was surprisingly larger. The anodic shift is rationalized by a through-space coulombic interaction of the metal ion and the positively charged ferrocenium [64]. Similarly, Beer and coworkers observed anodic shifts of 15 to 55 mV and 80 to 205 mV for the interaction of Fc-Calixarances **6** and **7** with La^{3+}, Gd^{3+}, and Lu^{3+} [6]. The interaction of the Fc-amino acids with La^{3+} and Tb^{3+}, causes anodic shifts that appear to be related to an increase of the ionic radii and the charge/size ratio [1,26,65]. Huang has shown that Fc-cyclopeptides **14** showed strong anodic shifts in their electrode potentials in the presence of alkaline earth metals and lanthanides [26]. For La^{3+} and Yb^{3+}, anodic shifts of 132 mV and 151 mV were observed. The extent of anodic shift correlated with the charge density of the metal.

A reduction in peak current after addition of the metal is also observed. This results from changes in the diffusion coefficient of the complex compared to the free ligand, which in turn is caused by the increase in molecular weight as expected from the Randles-Seveik equation. This may therefore not be a direct measure of the binding affinity for the metal ion by the ligand even though it has been used as a means of quantification by other researchers in which the metal ion is usually the redox probe [66–68]. We therefore chose to express the binding activity by

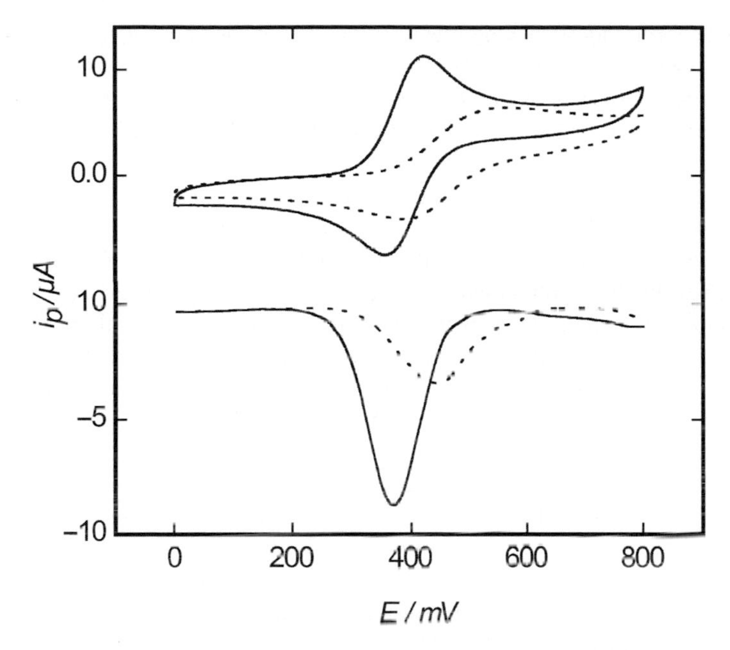

Figure 5.16 CV *(top)* and DPV *(bottom)* plots for Fc-GluOH **22** before *(solid lines)* and after addition of 2 mol equivalents of Tb^{3+} at pH 6.8, 0.2 M $NaClO_4$ glassy carbon working electrode, Pt counter electrode, and Ag/AgCl reference electrode. For the CV studies the scan rate was 100 mV/s whiles for the DPV experiments a scan rate of 20 mV/s and pulse amplitude of 50 mV was used

Table 5.6 Electrochemical data for Fc-modified compounds **20–22** and **26–28** from CV measurements

M^{n+}	20	21	22	26	27	28
Free	376(69)	388(76)	392(87)	632 (80)	636(83)	605(80)
Mg^{2+}	424(83)	394(80)	400(93)	744(180)	646(90)	611(88)
Ca^{2+}	422(80)	393 (88)	401(130)	720(144)	652(81)	616(80)
Zn^{2+}	412(96)	403 (93)	421(120)	646(114)	652(82)	625(89)
La^{3+}	400(90)	424(110)	459(120)	716(180)	656(83)	639(82)
Tb^{3+}	398(92)	452(120)	471(125)	669(184)	678(78)	655(88)

$E_{1/2}$ values are reported in mV and numbers in parentheses are the peak separation ($\Delta E = E_{pc} - E_{pa}$) values in mV

the changes in the electrode potential of the Fc group. This effect may be brought about by through-bond, through-space or conformational interaction which is detected by the redox probe [64] and is related to the relative ratio of K_{ox}/K_{red} from Eq. 5.6. [1,65]. The metal complexation results are summarized in Table 5.6.

In general, binding to La^{3+} and Tb^{3+} is stronger for compounds **21** and **22** whereas, the Gly-substituted compounds, **20** and **26**, exhibit preference for Mg^{2+} over the other ions. Hall and Chu reported the results of a CV study investigating the co-ordination of alkaline earth and lanthanide metal cations to a series Fc-cryptants **2** [4]. It was noted that there is a correlation between the $\Delta E_{1/2}$ and the charge/radius ratio of the metal ion. Similar trends were observed for **12–15** complexes with alkaline metals, alkaline earth metals and lanthanides [26]. Thus, on the basis of the ratio of charge to ionic radius which increases in the order $Tb^{3+} > La^{3+} > Zn^{2+} > Mg^{2+} > Ca^{2+}$, one would expect lanthanides to show much larger shifts compared to the alkaline earth metals, as was observed for Fc-AspOH **21** and Fc-GluOH **22** [64].

The preference of the Fc-GlyOH **20** and Fc–[GlyOH]$_2$ **26** to bind with Mg^{2+} and Ca^{2+} is rationalized by the peptides flexibility and its preference for smaller cations and lower coordination numbers. From potentiometric titrations, Cefola calculated the stability constant for the amino acid-metal complexes Gly–Mg^{2+}, Gly–Ca^{2+}, and Gly–La^{3+} complexes as 3.44, 1.38, and 2.24, and for Asp-Mg^{2+}, Asp–Ca^{2+}, and Asp-La^{3+} as 2.43, 1.60, and 3.42, respectively [69]. This means that while Gly was more selective toward Mg^{2+}, Asp was selective towards La^{3+}. Therefore, it is not surprising for Fc-GlyOH **20** and Fc-[GlyOH]$_2$ **26** to show a higher affinity toward Mg^{2+} and Ca^{2+} compared with the Fc-AspOH **21**, Fc–[AspOH]$_2$ **27**, and Fc-GluOH **22** and Fc–[GluOH]$_2$ **28** (Table 5.7).

The addition of the lanthanides to Fc-GluOH **22** resulted in a maximum anodic shift $\Delta E_{1/2}$ of 70 to 80 mV after the addition of two equivalents of Ln^{3+} (Figure 5.17). The alkaline earth metals resulted in a negligible increase with metal additions. The disubstituted conjugates Fc–[AspOH]$_2$ **27** and Fc–[GluOH]$_2$ **28** exibited maximum anodic shifts after four equivalents of metal ions were added. For Fc–[GlyOH]$_2$ **26** the maximum shift is reached after the addition of two equivalents of metal ions (Table 5.5). Maximum equivalence are observed at 2:1 molar ratio for the monosubstituted Fc-peptide and at 4:1 molar ratio for disubstituted Fc-peptide. We speculate reasonable structures of the Fc-peptide complexes to be as shown in Scheme 5.17.

To further investigate the metal coordination properties of Fc-peptides, metal complexes of Fc-–GlyOH **20** and Fc-GluOH **22** were synthesized in aqueous

Table 5.7 Electrochemical response of Fc-modified **20–22** and **26–28** to metal additions from CV measurements

	Mg^{2+}		Ca^{2+}		Zn^{2+}		Tb^{3+}		La^{3+}	
	0.86[a], 2.3[b]		0.99[a], 2.0[b]		0.74[a], 2.7[b]		0.92[a], 3.3[b]		1.06[a], 2.8[b]	
	$\Delta E_{1/2}$	C	$\Delta E_{1/2}$	C	$\Delta E_{1/2}$	C	$\Delta E_{1/2}$	C	$\Delta E_{1/2}$	C
20	48(2)	6.3	46(2)	5.9	36(2)	4.0	22(2)	2.3	24(2)	2.5
21	10(4)	1.5	8(4)	1.4	18(4)	2.0	66(2)	12.7	39(2)	4.5
22	12(4)	1.6	16(4)	1.9	29(4)	3.1	80(2)	21.7	68(2)	13.7
26	115(4)	83.3	88(4)	29.5	14(4)	1.7	37(2)	2.6	80(2)	21.7
27	11(4)	1.5	24(4)	2.5	13(4)	1.6	42(4)	5.0	20(4)	2.2
28	6(4)	1.3	11(4)	1.5	20(4)	2.2	50(4)	6.8	34(4)	3.7

The values in parentheses represent the number of metal equivalents required to cause maximum change in $E_{1/2}$

[a] Ionic radii of cation in Å

[b] Charge/size ratio, $\Delta E_{1/2} = E_{1/2}^{Complex} - E_{1/2}^{Free}$ values in brackets represent the number of metal equivalents required to cause maximum change in $E_{1/2}$, (where $C = K_{ox}/K_{red}$)

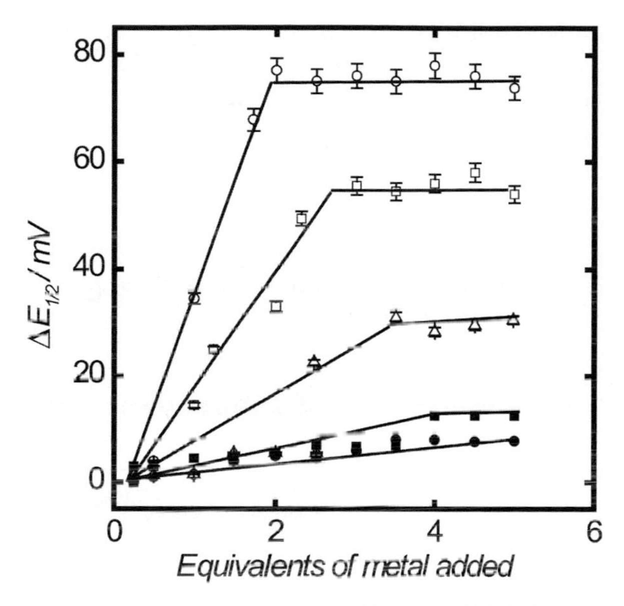

Figure 5.17 Graph of metal ion titration with FcGluOH **22** at pH 6.8. Tb^{3+}, La^{3+}, Ca^{2+}, Mg^{2+}, Zn^{2+} Δ. (Lines are only to guide the eye)

Scheme 5.17 Proposed schematic representations of Fc-GlyOH **20** (**a**), Fc-AspOH **21** (**b**), Fc-GluOH **22** (**c**), Fc–[GlyOH]$_2$ **26** (**d**), Fc–[AspOH]$_2$ **27** (**e**), and Fc–[GluOH]$_2$ **28**-metal complexes from electrochemical deductions (**f**)

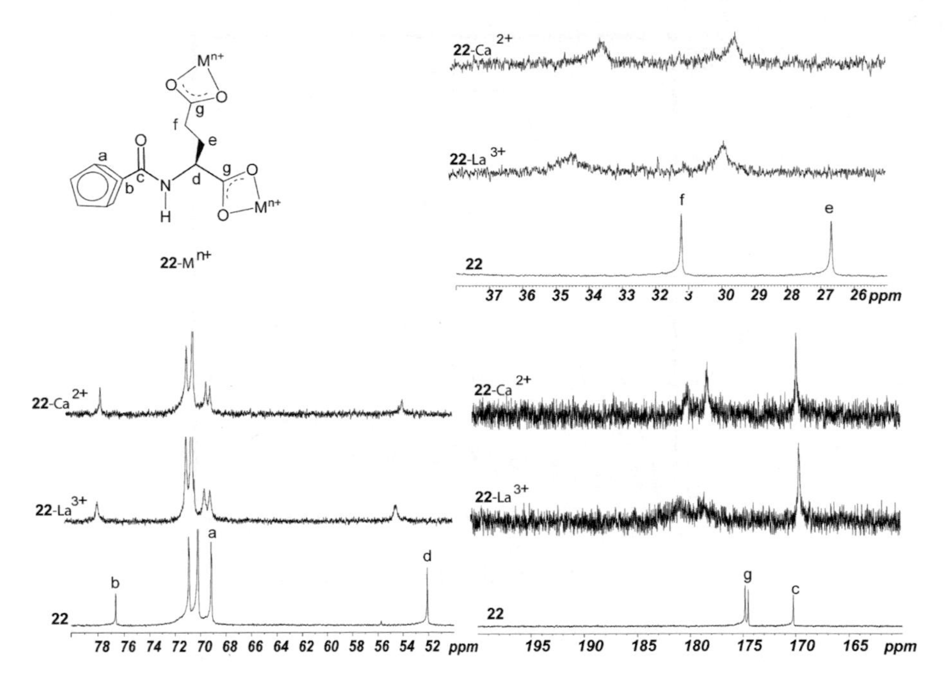

Figure 5.18 Portions of ^{13}C-NMR spectra of Fc-GluOH **22** and its Ca^{2+} and La^{3+} complexes in DMSO-d6. ortho C of substituted Cp (**a**) and C ^{ipso} of Fc (**b**); C=ONH (**c**), C^{α} (**d**), C $^{\beta}$ (**e**), C^{γ} (**f**) , COOH of Glu (**g**)

solution at pH 7.0 and analyzed by spectroscopic methods. Figure 5.18 shows portions of a stack plot of the ^{13}C- NMR spectra of Fc-GluOH **22** complexed with Mg^{2+} and La^{2+}. The formation of the complexes results in changes in the chemical shifts of various carbons. The study shows that the carboxylate carbon is the most affected while the amide C=O is virtually unaffected by the complexation. It is also fascinating to observe that the ortho C of the substituted Cp ring that are magnetically equivalent in the free Fc-GluOH **22**, is differentiated into 2 signals upon complexation. This suggests that the coordination influences the spatial environment of the Fc. Table 5.8 lists the magnitude of $\Delta\delta$ experienced by the various C atoms of Fc-GlyOH **20** and Fc-GluOH **22** as a result of complex formation.

For the complexes of **20** and **22** with metal ions, the carboxylate carbon experiences downfield shifts from the uncoordinated state of $\Delta\delta = 3.3\text{-}7.0$. The $\Delta\delta$ for the aliphatic C centre were $\Delta\delta$ 2.7-3.4 for γ-CH_2, 2.3- 2.4 for the β CH_2, and 2.0-3.4 for $\Delta\alpha$-CH (Figure 5.17; Table 5.8). The fact that the β CH_2 for Glu shows values lower than α-CH and γ-CH_2 demonstrates the proximity of the latter two aliphatic C to the site of coordination. Interaction of Asp, Glu, Phe, and glutathione with Pd^{2+} [70] and Ln^{3+} [8,10] have been reported to exhibit similar magnitudes of $\Delta\delta$ for the carboxylate and aliphatic carbons.

Table 5.8 Metal complexation induced Δδ of identified peaks of the Fc-GlyOH **20**-metal and Fc-GluOH **22**-metal complex

Compd	C=ONH	C=OOH	αC	β-CH_2	γC (CH_2)	C^{ipso} (C_P)
20–Mg^{2+}	0.5	3.3	2.0	–	–	0.8
20–Ca^{2+}	0.5	4.1	2.7	–	–	0.9
20–La^{3+}	0.7	3.6	2.8	–	–	0.7
22–Mg^{2+}	0.6	3.7, 4.0	3.0	2.3	2.7	0.6
22–Ca^{2+}	0.5	4.7, 4.8	3.4	2.0	2.9	0.8
22–La^{3+}	0.5	6.7, 7.0	3.2	2.4	3.4	0.8

Values are in ppm

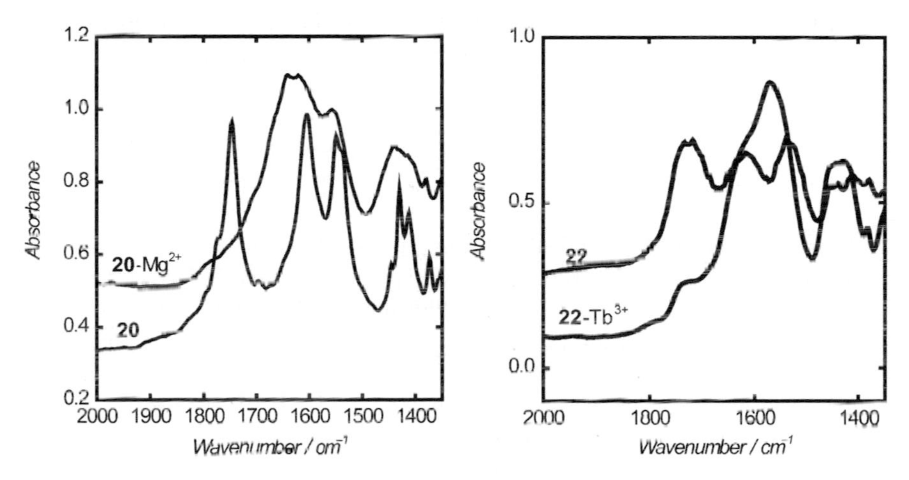

Figure 5.19 Partial IR of 20 and 20-Mg^{2+} complex *(left)*, 22 and 22-Tb^{3+} complex *(right)* as KBr pellets in the carboxylate and Amide I and II region

The downfield shifts of the carbonyl carbon chemical shifts are consistent with complexation with the metals because it causes a decrease in the electron density around the corresponding carbon atom [71]. The amide carbon and carbons of the Cp ring showed shifts of less than 1.0 ppm indicating that they do not participate directly in binding of the metal ions. The fact that the C^{ipso} of the substituted Cp exhibits remains virtually unchanged in the presence of the metal ion excludes through bond interactions. This means that metal ion binding in these systems may be purely electrostatic through- space interactions.

IR spectroscopy provides useful information about the coordination mode of the carboxylate. Typical IR spectra are shown in Figure 5.19 for Fc-GlyOH **20** and Fc-GluOH **22** and their metal ions complexes. After complexation, the characteristic C=O stretches of the free acid at 1756 cm^{-1} for Fc-GlyOH **20** and 1733, 1717 cm^{-1} for Fc-GluOH **22** are absent.

The carboxylate groups in the metal complexes give rise to bands at 1,550 to 1,575 cm^{-1} which are assigned to the antisymmetric stretching vibration $\nu_{as}(COO^-)$.

The bands in the 1419 to 1435 cm^{-1} region are the symmetric stretching vibration $\nu_s(COO^-)$ [20,21,72]. These values are consistent with established values of Mg^{2+}, Ca^{2+}, and Ln^{3+} complexes of Glu. Literature values for free and complexed Glu are 1,700 cm^{-1} for ν(COOH), 1,540 to 1,680 cm^{-1} for $\nu_s(COO^-)$ and 1,410 to 1,420 cm^{-1} for $\nu_{as}(COO^-)$ after complexation [73–75]. Table 5.9 summarizes the stretching vibrations for the free acids 83 and 85 and their metal complexes. The difference between the symmetric and asymmetric stretching vibration of the carboxylate group (Δν) is a diagnostic tool that provides structural insight into the coordination mode of the carboxylate group [20,21]. In systems with $\Delta\nu = 105–110$ cm^{-1} a bidentate chelating coordination is suggested, while for systems with $\Delta\nu = 135–155$ cm^{-1} a bidentate bridging mode is most likely.

The results indicate that Fc-GlyOH **20** and FcGluOH **22** coordinated differently to different metal ions. For example FcGlyOH **20** coordinates to Mg^{2+} by a bidentate chelating fashion whereas with Ca^{2+} it is mainly a bidentate bridging mode and a mixture of the two modes with La^{3+} and Tb^{3+} ions. Schemes 5.15 and 5.16 show proposed structural representations of the complexes and reported crystal structures of Gly and Glu metal complexes. Except for the bidentate bridging Fc-GlyOH–Ca coordination mode, deductions from the IR confirms earlier proposed structures made on the basis of the electrochemically determined stoichiometry of Fc-peptide-metal complexes in Scheme 5.17. Bidentate carboxylate coordination networks in the typical syn-syn bridging bidentate, chelate bidentate and a mixture of the two modes shown in the proposed structure (Scheme 5.18), were observed before in the crystal structure of N-methylglycine Eu(III), glycine-Nd^{3+}, as well as in other lanthanide complexes of glutamate and aspartate (Scheme 5.19) [14–17,19,76–78].

ESI-MS is a useful tool in the identification of metal complexation and allows deductions about the complexation behaviour of ligands [79,80]. In order to confirm the stoichiometry of the metal complexes in solution, ESI-MS experiments were carried out.

The ESI-MS spectrum of a solution of ligand Fc-GluOH **22** with Tb^{3+} is shown in Figure 5.20. The spectrum shows peaks related to the ligand as well as the

Table 5.9 Selected IR stretching frequencies of the carboxylate group (KBr disc)

Compound	νCOOH	ν_{as} C=O	ν_{as} C=O	$\Delta(\nu_{as\ C=O}-\nu_{asC=O})$
Fc-GlyOH **20**	1756	–	–	–
Fc-GluOH **22**	1733, 1717	–	–	–
Fc-GlyOH **20**-Mg^{2+}	–	1556	1,438	118
Fc-GlyOH **20**-Ca^{2+}	–	1579	1,444; 1,422	135 and 157
Fc-GlyOH **20**-La^{3+}	–	1570	1,455; 1,415	125 and 155
Fc-GlyOH **20**-Tb^{3+}	–	1572	1,443; 1,422	129 and 150
Fc-GluOH **22**-Mg^{2+}	–	1538	1,440; 1,425	98 and 113
Fc-GluOH **22**-Ca^{2+}	–	1550	1,445; 1,419	105 and 131
Fc-GluOH **22**-La^{3+}	–	1550	1,440; 1,418	110 and 132
Fc-GluOH **22**-Tb^{3+}	–	1571	1,435; 1,418	105 and 153

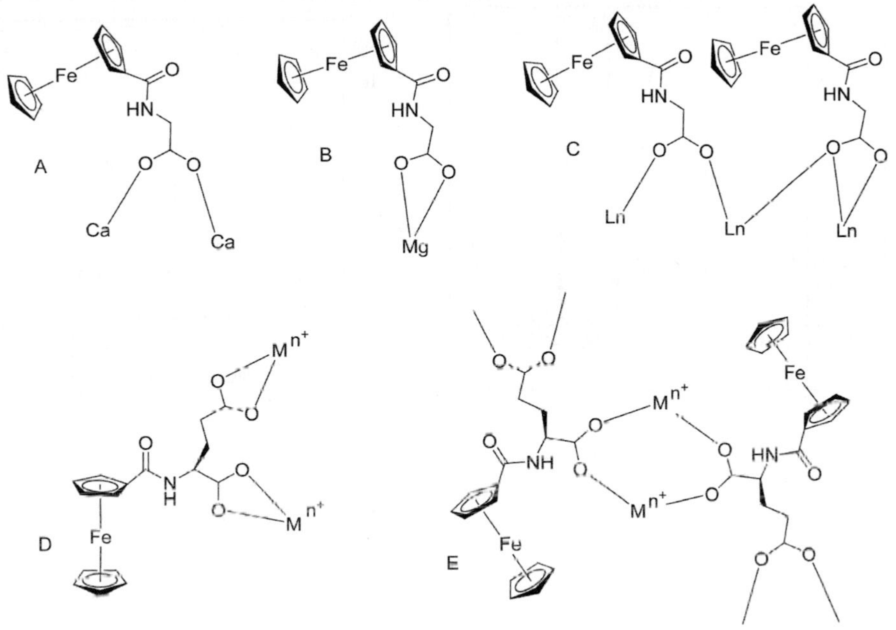

Scheme 5.18 Proposed structure of metal complexes bidentate bridging Fc-GlyOH–Ca (**a**), bidentate Fc-GlyOH–Mg (**b**), mixed bidentate and bidentate bridging Fc-GlyOH–Ln (**c**), bidentate Fc-GluOH–metal complexes (**d**), and bidentate bridging Fc-GlyOH–Ln (**e**)

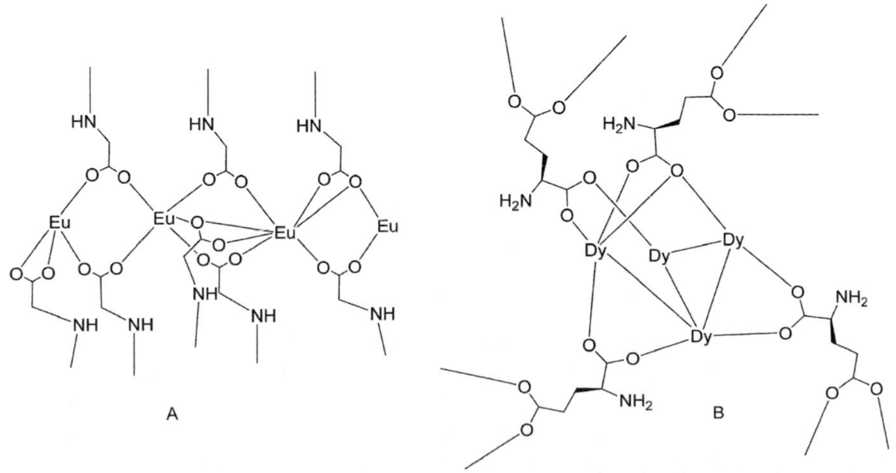

Scheme 5.19 Reported crystal structures of Gly and Glu metal complexes: Mixed bidentate, and bidentate bridging CH_3–Gly-Eu^{3+} complex (**a**) [14] bidentate bridging Glu-Dy (**b**) [78]

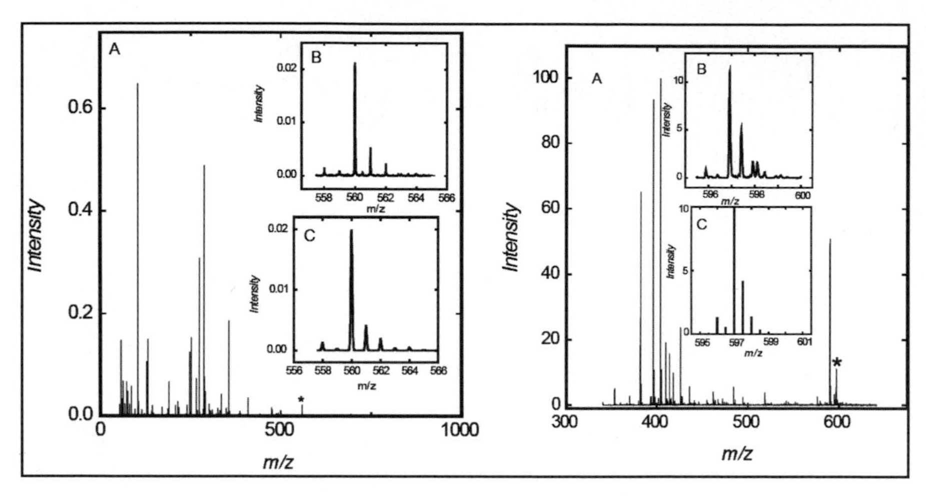

Figure 5.20 ESI-MS of $[M_2L_2(H_2O)_9]^{2+}$ of 4:1 Fc-GluOH **22**-Ca^{2+} *(left)*, $[M_2L_2(H_2O)_9]^{2+}$ of 4:1 Fc-GluOH **22**-Tb^{3+} in 1:1 methanol/H_2O at pH 6.5 *(right)*. **a** Raw data, **b** Expanded region of molecular ion and, **c** simulated isotopic ratio

complex. The ligand is indicated by peaks *m/z* at 382.0342 and 404.0163 representing $[L+Na]^+$ and $L–H+2Na]^+$ respectively.

The spectrum also shows a prominent peak at *m/z* 597.0024, which is assigned to $[Tb_2L_2(H_2O)_9]^{2+}$(L = 7) on the basis of modeling the correct isotopomer. The isotopic peaks are separated by 0.5 *m/z* units, which clearly reveal the +2 charge state of this ion. Minor peaks are the result of other complexes such as $[M_2L_2(H_2O)_7(CH_3OH)]^{2+}$ which is observed at *m/z* 595.0052 which shows the coordination of a solvent molecule to the metal. A similar experiment for Fc-GlyOH **20** with Ca^{2+} and Tb^{3+}, showed ligand peaks at *m/z* 287.0234, 310.0156 representing $[L+H]^+$ and $[L+Na]^+$, respectively. Peaks relating to m/z 410.0261, 637.1237, were identified for $[ML(H_2O)_3(CH_3OH)]^+$ and $[ML(H_2O)_9(CH_3OH)]^+$ for the Ca^{2+} and Tb^{3+} complexes, respectively [64].

3 Conclusions

In this chapter, the synthesis and characterization of Fc-functionalized amino acids were reported. The monosubstituted Fc-peptides, generally are involved in intermolecular H-bonding while the corresponding disubstituted systems are generally stabilized by intramolecular H-bonding involving the "Herrick" pattern. The crystal structure of Fc–$[GlyOEt]_2$, however, exhibited an unusual 1,3′-conformation allowing only intermolecular H-bonding between adjacent molecules. Deprotecting the ester moiety of Fc–$[GlyOEt]_2$ to the free acids Fc–$[GlyOH]_2$ results in a structural change from the 1,3′-conformer to a 1,2′-conformer which is stabilized by two strong cross-strand intramolecular H-bonds.

Electrochemical characterization of Fc-GlyOEt, Fc-AspOMe, Fc-GluOEt, Fc–$[GlyOEt]_2$, Fc–$[AspOBz]_2$, Fc–$[GluOEt]_2$ and their free acid analogues shows that (i) they exhibit quasireversible redox behavior in a variety of solvents, (ii) the amino acid had no significant effect on the halfwave potential, and (iii) the type of solvent had significant influence on the halfwave potential. Expectedly the redox potentials of the monosubstituted Fc-peptide are cathodically shifted compared to their disubstituted analogs.

Acidic Fc-peptides display a protonation-deprotonation equilibria, which was probed by electrochemical and NMR titrations. The p*Ka*'s values obtained through electrochemical titration compare well with those obtained with NMR titrations. Using CV and DPV, it was shown that the Fc group provides an electrochemical response to metal binding to the carboxylate group. Acidic Fc-peptide selectively bind to some metal ions. It has been established that Fc-GlyOH shows selectivity towards Ca^{2+} and Mg^{2+}, whereas Fc-AspOH and Fc-GluOH are selective towards lanthanide metal ions. This preference is rationalized in terms of their charge to size ratios. Structural evidence was obtained from IR, NMR spectroscopy and mass spectrometry. Bidentate and bidentate bridging carboxylate–metal complexation modes were predominant with each carboxylate forming a metal binding center. The amide group is not involved in metal complexation which reflects the findings of most published crystal structures of Gly, Asp, Glu with oxophilic metals such as the alkaline earth metals and the lanthanides [14-17,19,76–78,81]. These findings are invaluable for metal complexation studies of surface-bound peptides, since it is expected that the coordination behavior is similar on the surface and in solution.

References

1. Beer PD, Gale PA, Chen GZ. (1999) J Chem Soc Dalton Trans. 1897.
2. Beer PD, Chen Z, Pilgrim J. (1995) J Chem Soc Faraday Trans. 4331.
3. Plenio H, Aberle C, Shihadeh Ya, et al. (2001) Eur J Chem. 7:2848.
4. Hall CD, Chu SYF. (1995) J Organomet Chem. 498:221.
5. Beer PD, Chen Z, Pilgrim J. (1995) J Chem Soc Faraday Trans. 995:4331.
6. Brindley GD, Fox OD, Beer PD. (2000) J Chem Soc, Dalton Trans. 4354.
7. Legendziewicz J, Kozioski H, Jezowska-Trzebiatowska BHuskowska E. (1979) Inorg Nuc Chem Lett. 15:349.
8. Podanyi B, Reid RS. (1988) J Am Chem Soc. 110:3805.
9. Pessoa JC, Gajda T, Gillard Rd, et al. (1998) J Chem Soc, Dalton Trans. 3587.
10. Legendziewicz J, Huskowska E, Kozioski HJezowska-Trzebiatowska B. (1981) Inorg Nuc Chem Lett. 17:57.
11. Brittain HG. (1979) Inorg Chem. 18:1740.
12. Silverman DC, Kalota DJ, Stover FS. (1995) Corros Sci. 51:818.
13. Qutierrez E, Miller TC, Gonzalez-Redondo JR, Holcombe JA. (1999) Environ Sci Technol. 33:1664.
14. Gawryszewska PP, Jerzykiewicz L, Sobota P, Legendziewicz J. (2000) J Alloys Compd. 300-301:275.

15. Glowiak T, Dao-Cong NLegendziewicz J. (1986) Acta Crystallogr Sect C. C42:1491.
16. Legendziewicz J, Huskowska E, Waskowska A, Argay G. (1984) Inorg Chim Acta. 92:151.
17. Legendziewicz J, Huskowska E, Argay G, Waskowska A. (1989) J Less-Common Metals. 146:33.
18. Bao-Qing M, Zhang D-S, Gao S, Jin T-Z, Yan C-HG.-X.X. (2000) Angew Chem Int Ed Engl. 39:3644.
19. Wang R, Liu H, Carducci MD, Jin T, Zheng C, Zheng Z. (2001) Inorg Chem. 40:2743.
20. Deacon GB, Phillips RJ. (1980) Coord Chem Rev. 33:227.
21. Tackett JE. (1989) Applied Spectroscopy. 43:483.
22. Nakamoto K. (1986) Wiley, New York.
23. Gericke A, Huhnerfuss H. (1994) Thin Solid Films. 245:74.
24. Chohan ZH. (2000) Metal Based Drugs. 7:177.
25. Moriuchi T, Yoshida KHirao T. (2001) J Organomet Chem. 637-639:75.
26. Huang H, Mu L, He JCheng J-P. (2003) J Org Chem. 68:7605.
27. Chowdhury S, Schatte G, Kraatz H-B. (2006) Eur J Inorg Chem. 988.
28. Baker Mv, Kraatz H-BQuail Jw. (2001) New J Chem. 25:427.
29. Tsang Ky, Diaz H, Graciani NKelly Jw. (1994) J Am Chem Soc. 116:3988.
30. Gellman SH, Dado GP, Liang G-P, Adams BR. (1991) J Am Chem Soc. 113:1164.
31. Nomoto A, Moriuchi T, Yamazaki S, Ogawa A, Hirao T. (1998) Chem Commun. 163.
32. Moriuchi T, Nomoto A, Yamauchi O, Ogawa A, Hirao T. (2001) J Am Chem. Soc 123:68.
33. Pavia D, Lampman GM, Kriz GS. (1996) Introduction to spectroscopy, Harcourt College Publishers, USA, p 27.
34. Wu N, Fu L, Su M, Aslam M, Wong KC, Dravid Vp. (2004) Nano Lett. 4:383.
35. De Hatten X, Weyhermueller T, Metzler-Nolte N. (2004) J Organomet Chem. 689:4856.
36. Appoh FE, Sutherland TC, Kraatz H-B. (2003) Acta Crystallographica. E59:m1174.
37. Kraatz H-B, Lusztyk J, Enright GD. (1997) Inorg Chem. 36:2400.
38. Bauer W, Polborn K, Beck W. (1999) J Organomet Chem. 579:269.
39. Appoh Fe, Sutherland TC, Kraatz H-B. (2004) J Organomet Chem. 689:4669.
40. Lin L, Berces A, Kraatz H-B. (1998) J Organomet Chem. 556:11.
41. Carr JD, Coles SJ, Hassan WW, et al. (1999) J Chem Soc, Dalton Trans 57.
42. Gallagher JF, Kenny PTM, Sheehy MJ. (1999) Inorg Chem Commun. 2:200.
43. Kraatz H-B, Lusztyk J, Enright GD. (1997) Inorg Chem. 36:2400.
44. Moriuchi T, Nomoto A, Yoshida K, Hirao T. (1999) J Organomet Chem. 589:50.
45. Desiraju GR, Steiner T. (1999) "The weak hydrogen bond", Oxford University Press, Oxford.
46. Bard AJ, Faulkner LR. (1980) Electrochemical methods: Fundamentals and applications, Wiley, New York.
47. Kraatz H-B, Leek Dm, Houmam A, Enright GD, Lusztyk J, Wayner DM. (1999) J Organomet Chem. 589:38.
48. Xu Y, Saweczko P, Kraatz H-B. (2001) J Organomet Chem. 637-639:335.
49. Sheely MJ, Gallagher JF, Yamashita M, et al. (2004) J Organomet Chem. 689:1511.
50. Noviandri I, Brown KN, Fleming DS, et al. (1999) J Phys Chem B. 103:6713.
51. Armstrong FA, Hill HAO, Oliver BN, Whitford D. (1985) J Am Chem Soc. 107:1473.
52. Büchi FN, Bond AM, Codd R, Freeman HC . (1992) Inorg Chem. 31:5007.
53. Battistuzzi G, Borsari M, Loschi LSola M. (1997) J Bioinorg Chem. 2:350.
54. Reid LS, Taniguchi VT, Gray HB, Mauk AG. (1982) J Am Chem Soc. 104:7516.
55. Beulen MWJ, Van Veggel FCJM, Reinhoudt DN. (1999) Chem Commun. 503.
56. Moore ANJ, Wayner DDM. (1999) Can J Chem. 77:681.
57. Martinez-Máñez R, Soto J, Tendero MJL. (1997) J Chem Soc, Faraday Trans. 93:2175.
58. Benito A, Martinez-Máñez R, Soto J, Tendero MJL. (1997) J Chem Soc, Faraday Trans. 933:2175.
59. Tendero MJL, Benito A, Martinez-Máñez R, Soto J. (1996) J Chem Soc, Dalton Trans. 4121.
60. De Santis G, Fabbrizzi L, Lichelli M, Pallvicini P. (1994) Inorganica Chimica Acta. 225:239.

61. Kresge AJ. (1991) J Pure Applied Chem. 63:213.
62. Curley K, Pratt RF. (1997) J Org Chem. 62:4470.
63. Bryson A, Davies NR, Serjeant EP. (1963) J Am Chem Soc. 85:1933.
64. Appoh FE, Sutherland TC, Kraatz H-B. (2004) J Organomet Chem. 690:1209.
65. Medina JC, Goodnow TT, Rojas MT, et al. (1992) J Am Chem Soc. 114:10583.
66. Lu X, Zhu K, Zhang M, Liu HKang J. (2002) JBiochem Biophys Methods. 52:189.
67. Mannan RJ, Quayum ME, Sarker GC. (1997) J Bangladesh Acad Sci. 21:35.
68. Kramer C, Dunsch L. (1998) Electrochimica Acta. 44:819.
69. Cefola M, Tompa AS, Celiano AV, Gentile PS. (1962) Inorg Chem. 1:290.
70. Yamauchi O, Odani A. (1981) J Am Chem Soc. 103:391.
71. Carubelli CR, Massabni AMG, Leite SRDA. (1997) J Brazil Chem Soc. 8:597.
72. Hou H, Li G, Li L, Zhu Y, Meng X, Fan Y. (2003) Inorg Chem. 42:428.
73. Devereux M, Jackman M, McCann M, Casey M. (1998) Polyhedron. 17:153.
74. Malorova IV, Formina TA, Dobrynima NA, Sevryugina YY. (1992) Khimiya. 33:565.
75. Yu H-G, Dong J-X, Liu Y, Qu S-S. (2003) J Chem and Engineneering Data. 48:1279.
76. Csoregh I, Huskowska E, Legendziewicz J. (1992) Acta Crystallogr Soct B. C48:1030.
77. Torres J, Kremer C, Kremer E, et al. (2002) J Chem Soc, Dalton Trans. 4035.
78. Ma B-Q, Zhang D-S, Gao S, Jin T-Z, Yan C-H, Xu G-X. (2000) Angew Chem Int Ed. 39:3644.
79. Burford N, Eelman MD, Groom K. (2005) J Inorg Biochem. 99:1992.
80. Burford N, Eelman MD, Leblanc WG. (2004) Canadian J Chem. 82.1254.
81. Dudev T, Lim C. (2004) J Phy Chem B. 108:4546.

Metal Ion Binding to Ferrocene Peptide Dendrimer Films

Francis E. Appoh and Heinz-Bernhard Kraatz

1 Introduction

Self-assembled monolayers (SAMs) of peptides molecules linked to a gold surface via gold-thiol linkages have received considerable attention [1–4]. Recently, the focus has shifted from understanding the formation and structure of such SAMs toward their application. Electrochemical sensing of metal ions using SAMs is such an application [5]. The ability to control the density and conformation of functional groups on surfaces have allowed the design of systems that exhibit some selectivity toward binding of certain metal ions, which is useful for chemical sensing [6,7], as well as for environmental cleanups [8].

The detection of electrochemically active metal ions at SAM interfaces is typically done by voltammetric methods [9]. The ability to alter amino acid sequences within oligopeptides provides considerable control over the affinities of the peptide ligands for specific metal ions. For example, using an L-cysteine modified gold electrode, Chen et al. showed that Cu^{2+} can be detected at a limit below 5 ppb [10]. Films of poly(L-aspartate) conjugates of 3-mercaptopropionic acid exhibit a detection limit of 0.2 ppb for Cu^{2+} [11]. This approach was extended to the Cu-binding oligopeptide sequence HS-CA-Gly-Gly-His (Figure 6.1a, [12], which exhibits a sub-ppt detection limit for Cu^{2+}. *Ab initio* calculations suggest that Cu binding involves the formation of a wrap-around complex as shown in Figure 6.1, involving coordination to the imidazole of His as well as coordination to three additional backbone Ns [13]. The detection of metal ions that are not redox active is possible using external redox probes such as $[Fe(CN)_6]^{3-/4-}$ or $[Ru(NH_3)_6]^{3+/2+}$. The film controls

A.S. Abd-El-Aziz et al. (eds.), *Inorganic and Organometallic Macromolecules: Design and Applications.*

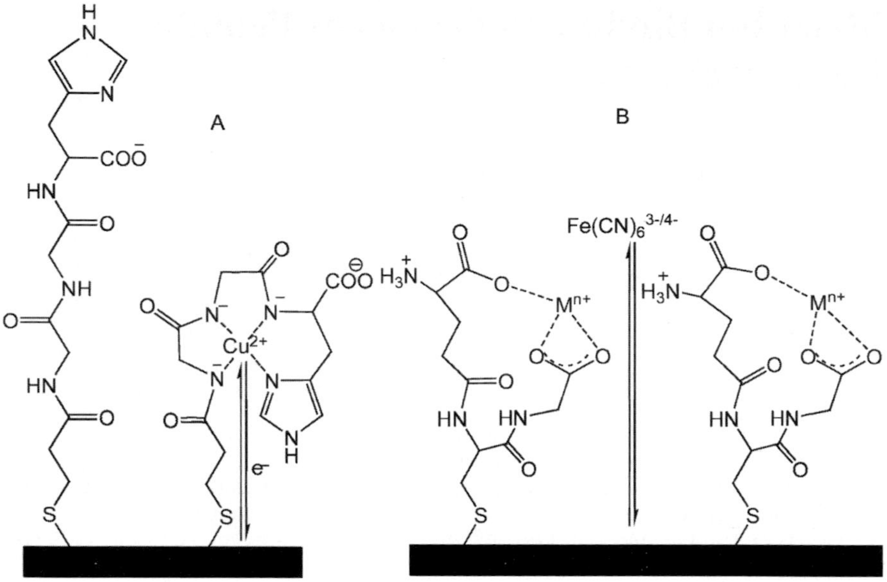

Figure 6.1 Metal ion detection at monolayer interface. **a** the direct electrochemical method with HS-CA-Gly-Gly-His detects Cu2+ in the sub-ppt limit [12]. **b** Ion gated channel method based on Fe(CN)63–/4– at a glutathione monolayer interface with a detection limit of about 1 mM La3+ [16–18]

the ability of the redox probe to approach the surface, a process that has been compared with gated ion channels. Complex formation between the peptide and the nonredox active metal ion controls the permeability of the film toward the redox probe as indicated in Figure 6.1b [14,15]. For example, using glutathione-modified Au surfaces, it was possible to detect nonredox active metal ions such as La^{3+}, Eu^{3+}, Lu^{3+}, Ca^{2+}, Mg^{2+}, and K^{+} in the presence of the redox probe $[Fe(CN)_6]^{3-/4-}$. The detection limit towards lanthanides (La^{3+} 1 μM) was about three orders of magnitude greater than that for alkaline earth metals [16–18].

Films that contain redox active groups as well as metal binding sites have shown to be particularly useful. There are several examples of metal ion detection at electropolymerized films possessing a metal binding site. Examples are the detection of Cu^{2+} and Ni^{2+} using films prepared from thioazo-phenol [19] and the detection of alkaline metals using films of tetrathiofulvalene (TTF)–dithia-crown ethers [20]. Majoral and coworkers demonstrated that electrodeposited films of a third generation TTF containing dendrimer (Scheme 6.1) give an anodic electrochemical response towards binding of Ba^{2+} [21].

Another interesting example is the detection of alkaline and alkaline earth metals at Fc-crown polypyrrole films prepared by electropolymerizing a substituted pyrrol (Scheme 6.2) [22,23]. Upon binding of Ca^{2+} and Ba^{2+} to these Fc-crown polypyrrole films, the signal at 565 mV disappeared and a new signal at 680 mV appeared as a result of the formation of the 1:1 metal ion complex.

In our work, we were interested in exploiting the metal ion recognition of peptides, while maintaining the ability to electrochemically detect nonredox active

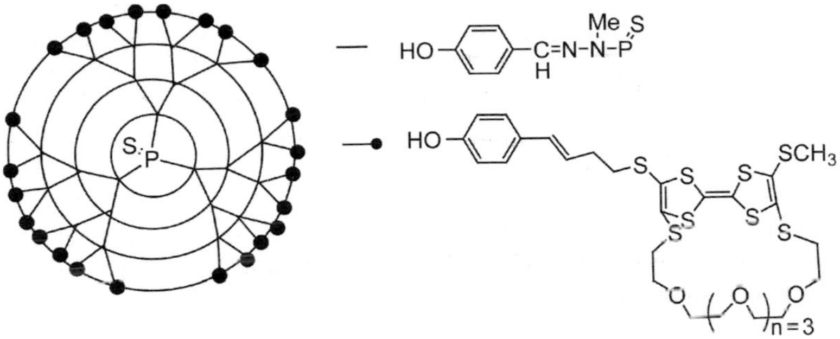

Scheme 6.1 Third generation TTF crown of metal ion detection

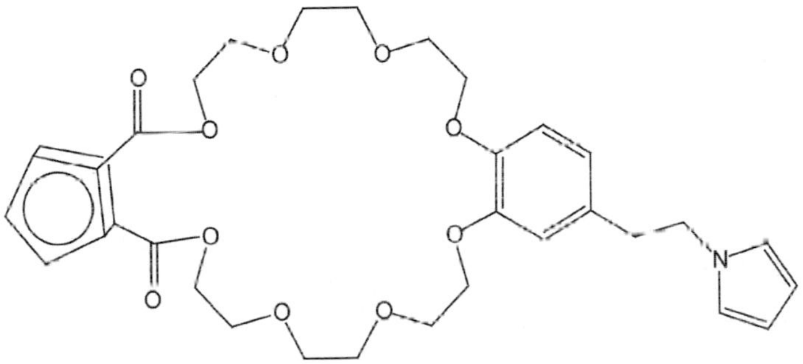

Scheme 6.2 Structure of Fc-functionalized pyrrole crown ether used for the formation of Fc-crown polypyrrole films. This polymer is active as sensor for Ca^{2+} and Ba^{2+}

metal ions. For this purpose, we began to speculate that Fc-peptide conjugates might enable us to achieve this goal. This approach would allow us to monitor the electrochemical characteristics of the Fc group as a function of metal ion concentration, and provide an alternative to an ion-gated approach that relies on solution-based redox probes. Prior to our work, Fc-peptide functionalized surfaces had not been exploited as metal ion sensors. In fact, the Fc group may interfere with metal ion binding and this posed obvious questions about the ability of metal ions to interact with Fc-peptide modified surfaces as well the magnitude of the Fc-based electrochemical response. Thisa chapter outlines the use of Fc-glutamic acid cystamine conjugates for thin film formation and metal binding.

2 Synthesis of Ferrocene Peptide Cystamines

The synthesis of the Fc cystamine conjugates $[Fc\text{-}G1(OH)CSA]_2$ **2** and $[Fc\text{-}G2(OH)CSA]_2$ **4** is shown in Schemes 6.3 and 6.4. Carbodiimide coupling of FcCOOH and $[Glu\text{-}CSA]_2$ results in the formation of the two esters $[Fc\text{-}G1(OBz)CSA]_2$ **1**,

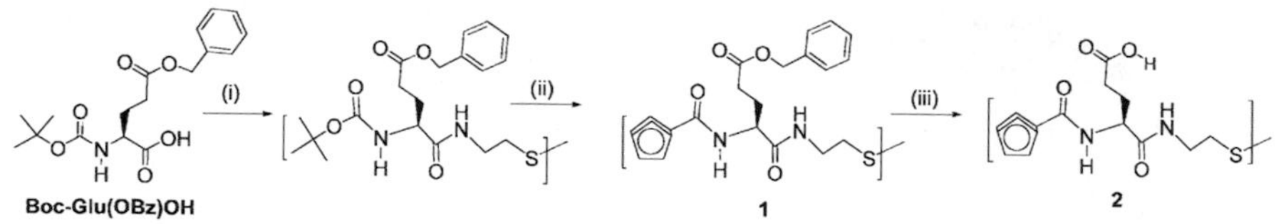

Scheme 6.3 Synthesis of $[Fc\text{-}G1(OBz)CSA]_2$ **1** and $[Fc\text{-}G1(OH)CSA]_2$ **2**. (i) cystamine · HCl/Et_3N, HOBt/EDC in CH_2Cl_2; (ii) in situ TFA deprotection of Boc group followed by coupling with Fc-COOH in presence of EDC/HOBt in CH_2Cl_2; (iii) base hydrolysis of $[Fc\text{-}G1(OBz)CSA]_2$ **1** with NaOH in methanol

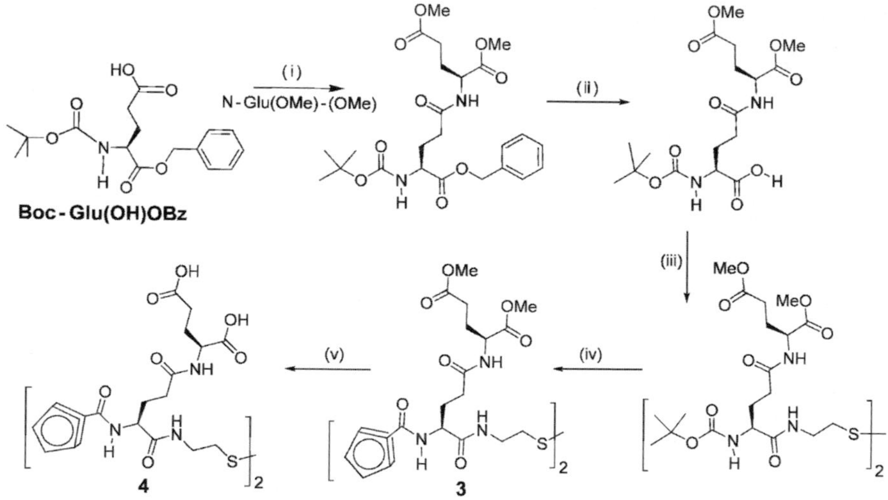

Scheme 6.4 Synthetic scheme for formation of $[Fc\text{-}G2(OMe)CSA]_2$ **3** and $[Fc\text{-}G2(OH)CSA]_2$ **4**. (i) Et3N, EDC/HOBt, CH2C12; (ii) hydrogenolysis in methanol; (iii) coupling of cystamine and Boc-Glu(Glu)-OH with EDC/HOBt in CH2C12; (iv) Boc deprotection and coupling with Fc-COOH; (v) base hydrolysis

$[Fc\text{-}G2(OMe)CSA]_2$ **3**, which are subjected to base hydrolysis to yield the corresponding acids $[Fc\text{-}G1(OH)CSA]_2$ **2** and $[Fc\text{-}G2(OH)CSA]_2$ **4**.

The Fc-conjugates were obtained as orange solids after purification by column chromatography and were characterized spectroscopically. Representative ^{1}H-NMR spectra of compounds **1** and **3** are shown in Figure 6.2.

For compounds **1** and **3** three classes of amide NH appear downfield including the Fc-CO-NH doublet NH, the cystamine triplet NH, and the side-chain glutamate amide doublet NH. The downfield position of these amide protons is indicative of their involvement in H-bonding (Table 6.1) [24–26]. The α-proton appears as a multiplet in the δ 4.5–4.8 region. The number increases with increasing number of generations. The methylene protons of cystamine are differentiated into two multiplets at δ 3.5–3.6 and a triplet at δ 2.8–2.6 whereas those due to the side-chain glutamate occur in the δ 2.0–2.6 range. The complete deprotection of the esters **1** and **3** by

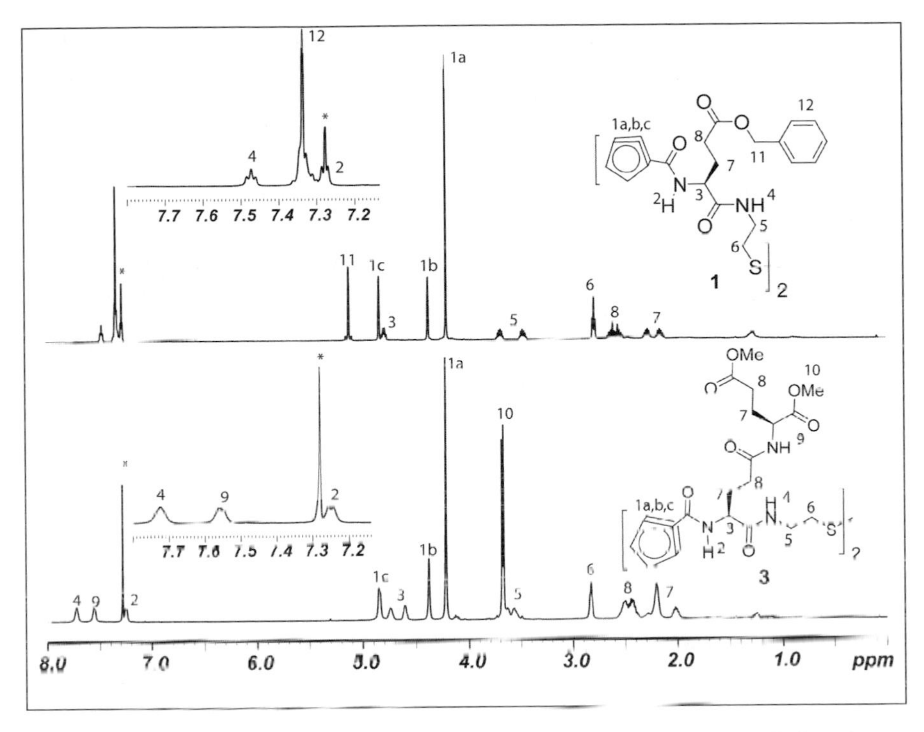

Figure 6.2 ^{1}H-NMR spectra of [Fc-G1(OBz)CSA]$_2$ **1** and [Fc-G2(OMe)CSA]$_2$ **3**. Cp: H^{1a}, H^{1b}, H^{1c}; Amide NH: Fc-CO-NH doublet NH^2, the cystamine triplet NH^4, Glu amide doublet NH^9 in **3**. The α-proton H^3 appears as a multiplet in the δ4.5–4.8 region. The CH_2 of cystamine are differentiated into two multiplets H^5 and a triplet H^6 (δ 2.8–3.6) while those due to the side chain Glu, H^7 and H^8 occur in the δ 2.0–2.6 region. Protecting groups for **1**: the OCH_2 (H^{11}) and Ph (H^{12}); for **3**: OMe (H^{10}) * Residual solvent

Table 6.1 Selected ^{1}H-NMR (δ in ppm) signals of Fc-peptide cystamine dendrimers

Compounds	NH^a	NH peptide	Cp	CH_2 (CSA)
1	7.47	7.28	4.21 4.38, 4.84	2.78 3.47, 3.68
2	8.39	7.50	4.12 4.33, 4.82	2.82 3.43, 3.86
3	7.72	7.24 7.56	4.22 4.38, 4.85	2.83 3.43, 3.86
4	8.18	7.84 8.11	4.20 4.33, 4.89	2.79 3.42, 3.86

Spectra for ester 1 and 3 were recorded in $CDCl_3$ whereas those of the acids compounds 2 and 4 were recorded in DMSO-d6

base hydrolysis can be seen from the loss of the signals resulting from benzyl and methyl protons and the appearance of the acid protons in ^{1}H-NMR at about δ 12.0 in dimethylsulfoxide (DMSO)-d6.^{13}C-NMR spectroscopy shows the loss of resonances resulting from the methyl and benzyl carbons at δ52 and, δ 136–129.

IR spectroscopy data on the Fc dendrimer conjugates is summarized in Table 6.2.

Figure 6.3 shows IR plots of [Fc-G2(OMe)CSA]$_2$ **3** and [Fc-G2(OH)CSA]$_2$ **4**. As expected the deprotected free acids exhibited a characteristic strong broad band in the region of 3,500 to 2,500 cm^{-1}, because of OH stretching vibration of the free acids which is absent in the protected dendrimers [27,28]. The Amide A peaks are observed at about 3,300 cm^{-1} which are indicative of hydrogen bonded NH protons [24,29].

The lower wavenumber region for the compounds shows peaks for the stretching vibrations of ester carbonyl C=O ranging from 1,730 to 1,722 cm^{-1} from 1,647 to 1623 cm^{-1} for the Amide I band originating predominantly from the C=O stretching vibrations, and from 1,535 to 1,525 cm^{-1} for the Amide II bending vibration. The C-H stretching of the Fc group occurs in the range of 3,095 to 3,080 cm^{-1} [30–32]. The UV-vis spectra for all compounds exhibit a single broad absorption with absorption maxima (λ_{max} in the range 450 to 442 ± 4 nm) that is independent of the peptide substituent.

Table 6.2 IR and UV-vis data for Fc-peptide cystamine dendrimers

	IR					UV-vis
Compound	AmideA	Cp	C=O	Amide I	Amide II	λ_{max} (ε)
1	3,300	3,087	1,735	1,627	1531	442(240)
2	3,299	3,091	1,739	1,629	1525	446(220)
3	3,297	3,086	1,733	1,629	1529	446(255)
4	3,313	3,086	1,722	1,630	1530	450(320)

IR (KBr, cm^{-1}); UV-vis (MeOH, c = 1.0–6.0 mM) λ_{max} in nm, ε in $cm^{-1}M^{-1}$

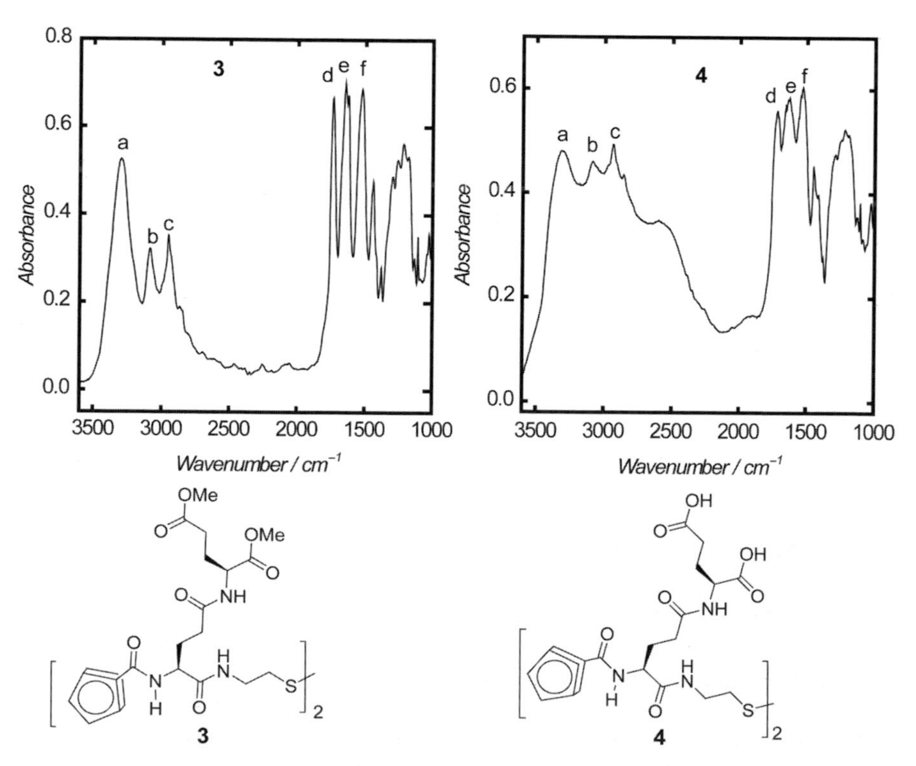

Figure 6.3 Partial IR spectra of $[Fc\text{-}G2(OMe)CSA]_2$ **3** and $[Fc\text{-}G2(OH)CSA]_2$ **4** as KBr pellets. Amide A *(a)*, Cp C-H *(b)*, C-H's of peptide *(c)*, C=O carbonyl acid/ester *(d)*, Amide I *(e)*, Amide II *(f)*

2.1 *Characterization of Surface Immobilized Films*

Films of Fc-peptide conjugates **1–4** adsorbed on Au surfaces were prepared via immersion of Au electrodes in ethanolic solution of compounds **1–4**. Cyclic voltammetric (CV) studies of the resulting films were carried out using a 2.0 M $NaClO_4$ as a supporting electrolyte at pH 7.0. A representative CV of film **2** at various scan rate and a plot of oxidation and reduction peak currents i_p versus scan rate ν are shown in Figure 6.4.

The CVs show the characteristic reversible one-electron oxidation-reduction redox couple associated with the Fc group. The relationship between i_p and ε is linear, indicating the presence of surface-bound Fc groups [33]. The formal potential E^o and the peak separation ΔE for **1–4** are listed in Table 6.3. The formal potential E^o of **3** is shows a slight cathofic shift compared with **1**, which is attributed to a better stabilization of the Fc^+. Abruña and coworkers observed similar relationship and concluded that from a one-dimensional model perspective, there is one metal centre per dendrimer molecule, so that the larger sized dendrimers allow a more stable charge distribution as a result of lower charge density [34]. Earlier reports have shown that the amino acid sequence influenced the redox potential [35,36]. However, the same can not be said of **2** and **4**, and the difference may result from the ionization of their carboxylate interface at the prevailing pH of the experiment. The E^o for the acid films was also cathodically shifted by about 30 to 40 mV compared with their ester counterpart (Table 6.3).

Ideally, CV of adsorbed redox species should exhibit a sharp, symmetric anodic and cathodic peak with a peak separation $\Delta E = 0$ mV [37]. However, a nonzero peak separation was observed for all films of these compounds, which might result from

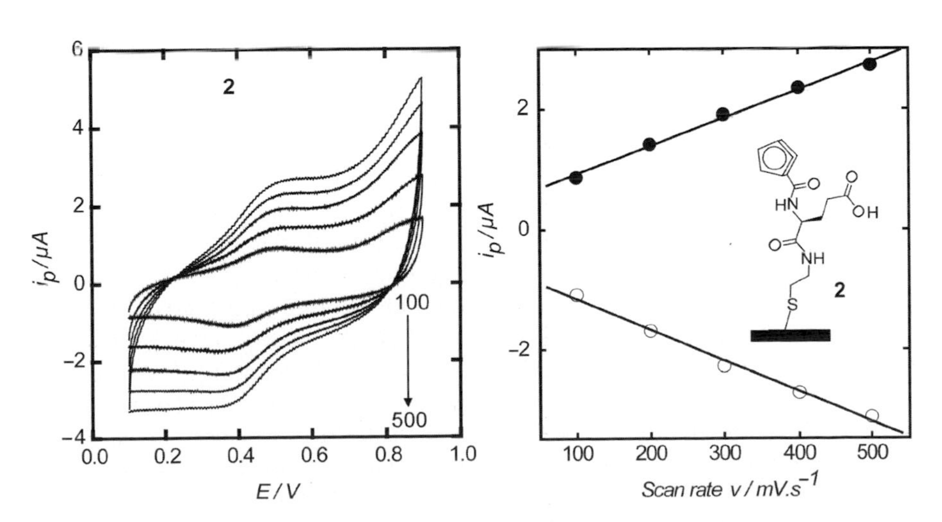

Figure 6.4 CV of compound **2** at scan rates between 100 and 500 mV/s *(left)*. A plot of anodic peak current versus scan rate *(right)*. 2.0 M $NaClO_4$ at pH 7.0, a reference electrode Ag/AgCl, BAS Au electrode (geometric area 0.025 cm^2) working electrode

Table 6.3 Summary of the electrochemical properties of films of compounds **1–4** on gold electrodes (0.025 cm^2)

Films	E^o	ΔE	ΔE_{fwhm}	Γ_{Fc}	Sp.A	Cal.Sp.A.
1	467 ± 20	73 ± 5	163 ± 18	3.12 ± 0.40	530 ± 60	390
2	425 ± 18	47 ± 5	125 ± 25	3.75 ± 0.09	440 ± 70	370
3	455 ± 15	109 ± 12	184 ± 20	3.35 ± 0.20	500 ± 50	480
4	424 ± 15	42 ± 6	130 ± 28	3.03 ± 0.40	550 ± 40	440

Surface concentration? Γ_{Fc} in 10^{-11} mol·cm^{-2}, experimental specific area (Sp.A) and calculated specific areas (Cal.Sp.A.) in Å^2. E^o, ΔE, and ΔE_{fwhm} in mV. Supporting electrolyte 2.0 M $NaClO_4$ at pH 7.0, Ag/AgCl reference, and Pt counter electrode

slow kinetics. The electron transfer rate k_{ET} associated with surface confined redox active species is known to affect the ΔE value [38]. The peak separation ΔE for the acids **2** and **4** was lower than that of corresponding esters **1** and **3**. The relatively small ΔE values (about 40 mV) for the acids suggest a faster kinetics for the acids compared with the corresponding esters.

The peak full width at half maximum, ΔE_{fwhm}, is used to evaluate the organizational behavior of monolayers [38]. The parameter provides a quantitative measure of the relative interactions taking place between the electroactive groups. In systems of minimal interactions the ideal value is 90.3/n, when n is the number of electrons) [38]. A ΔE_{fwhm} of 90 to 100 mV for Fc-containing films indicates a homogenous environment around the redox centers [39]. ΔE_{fwhm} in the range of 125 to 184 mV was observed for films of the Fc-conjugates **1–4**, which suggest the presence of Fc moieties with different environments, brought about by electrostatic repulsion between the redox moiety and disorder within the film [40]. Surface concentrations Γ_{Fc} of 3.0–3.7 × 10^{-11} mol·cm^{-2} were obtained for all films. These values are slightly higher than the 1.0–2.5 × 10^{-11} mol/cm^2 reported for Fc-functionalized poly(propylene imine) dendrimers [41]. The experimentally observed specific areas for these dendritic compounds range from 440 to 820 Å^2, which is close to the theoretical values derived from Spartan modelling. This shows that the Fc-dendrimer films constitute approximately monolayer coverages.

Blocking studies were carried out on modified Au electrodes of the Fc dendrimer esters and acids using CV experiments in order to probe the compactness of films. Figure 6.5 shows the results of the blocking study with $[Fe(CN)_6]^{3-/4-}$ in the absence and in the presence of Fc-peptide films of compound **1** and **2**. In the potential range of –0.2 to 0.8 V, both systems showed some blocking toward $[Fe(CN)_6]^{3-/4-}$, as indicated by the lack of the signal of the $[Fe(CN)_6]^{3-/4-}$ redox probe. Instead, the only redox signal observed is that of the Fc/Fc$^+$ of the Fc-dendrimer film itself.

It was observed that the anodic peak current of Fc was enhanced in the presence of the $[Fe(CN)_6]^{3-/4-}$ redox probe. This effect was rationalized by a possible communication between the $[Fe(CN)_6]^{3-/4-}$ redox probe in close proximity to the Fc/Fc$^+$ couple. Thus, the oxidation Fc→ Fc$^+$ + e^- is followed by a reduction process which involves $[Fe(CN)_6]^{4-}$, and which allows another reduction process to the Fc to continue the CV cycle. Similar observations have been reported for immobilized Fc-peptides and Fc-functionalized poly(propylene imine dendrimers [41,42]. The differential in the

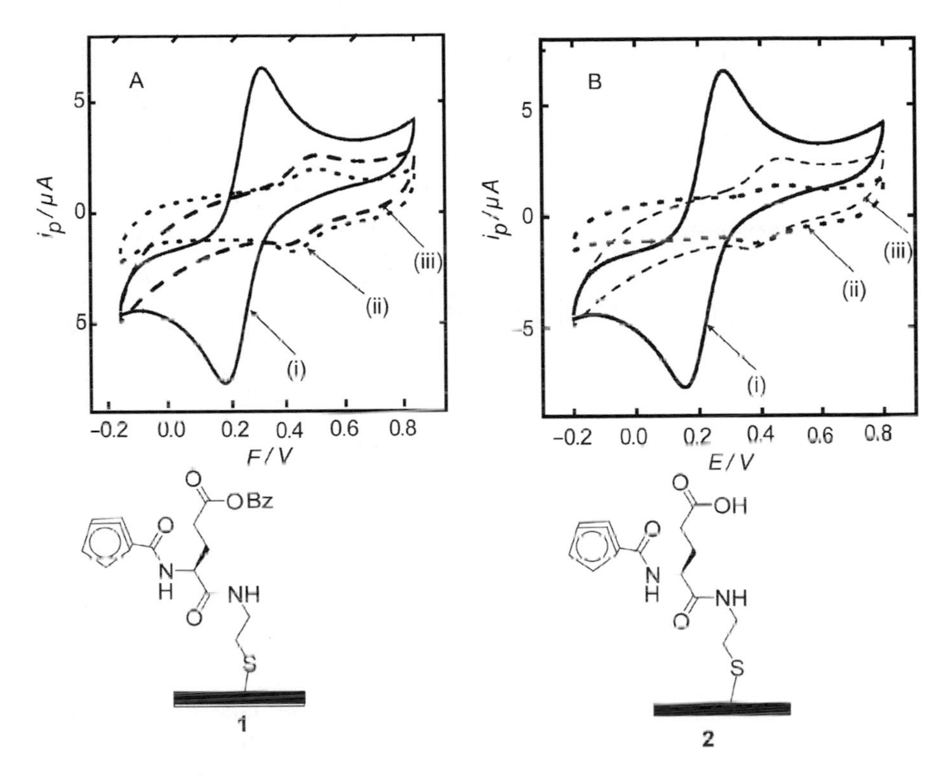

Figure 6.5 Blocking studies with films of compounds **1** and **2** in 2.0 M $NaClO_4$ at pH 7.0 aq at a scan rate of 100 mVs^{-1}. **a** $[Fe(CN)_6]^{3-/4-}$ at bare gold *(i)*, films **1** on Au in $[Fe(CN)_6]^{3-/4-}$ *(ii)*, and films **1** on Au *(iii)*. **b** $[Fe(CN)_6]^{3-/4-}$ at bare gold *(i)*, film **2** on Au in $[Fe(CN)_6]^{3-/4-}$ *(ii)*, and film **2** alone *(iii)*. Using a, Ag/AgCl reference electrode, Au working electrode (geometric area 0.025 cm^2)

anodic and cathodic peak currents has also been associated with differences in the kinetics of ferrocyanide mediated ET rates [41,42]. It is believed that the rate of the Fc-mediated $[Fe(CN)_6]^{3-/4-}$ oxidation is faster than its reduction, therefore resulting in a larger signal [41,42].

2.2 Electrochemical Studies of Metal Interactions at Monolayer Interface

The electrochemical response of **1**, **2**, and **4** films to Ca^{2+} and Tb^{3+} were examined by CV. Figure 6.6 shows a representative CV of films **2** before and after addition of 2.0 μM of Tb^{3+}. Unlike films involving Fc-crownether, metal interactions with films **1**, **2**, and **4** do not lead to the appearance of a new redox signal. Instead, the E^o of the Fc/Fc^+ couple is slightly anodically shifted. There is however a drastic decrease in the i_p and, hence, the surface charge Q_{Fc} with each addition of metal ions which

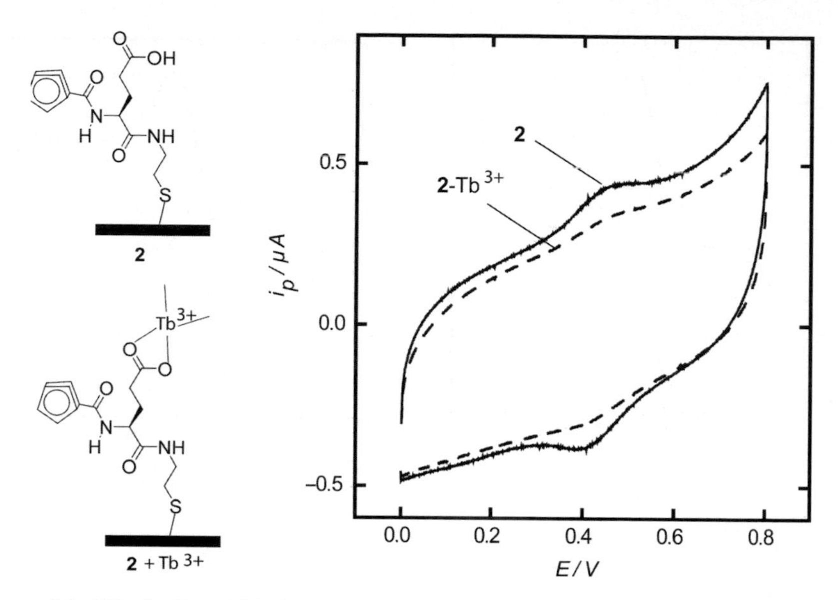

Figure 6.6 CV of a film of **2** before (**a**) and after (**b**) addition of Tb^{3+}. 2.0 M $NaClO_4$ supporting electrolyte at pH 7.0, versus Ag/AgCl, Pt counter electrodes

eventually reaches a steady state value. This Q_{Fc} decrease inthe analyte-binding films has been rationalized by others in terms of an ion-pair effect [43–55]. Based on the ion pair theory [44,56], the counterions in the electrolyte necessary to support the Fc redox process are blocked with each addition of M^{n+} to the monolayer. The formation of layers of metal ion-counter ion $M^{n+}[ClO_4^-]_n$ instead of $Fc^+[ClO_4^-]$ is the result. This blocks the penetration of ClO_4^- counter anions, resulting in the decrease in Q_{Fc} [49,57].

Ulman and coworkers have shown that binding of Cd^{2+}, Ca^{2+}, Pd^{2+}, and Ba^{2+} to ω-COOH SAMs on Au formed dense overlayers at the interface the between the electrolyte and the functionalized monolayer. The investigations revealed that strong chemical interactions between the carboxylate groups, the metals, and the counter ion induced dense counter-ion overlayers [58]. Figure 6.7 shows plots of Q_{Fc} versus molar concentrations of metal ion added for films of **1**, **2**, and **4**, showing a correlating decrease in Q_{Fc} with an increase in the addition of $[M^{n+}]$ [46,48,49]. There was an initial sudden decrease in the films of **1**, which may have resulted from reorganization of the film followed by a more gradual decrease. The rates of change for acidic films were more gradual throughout the titration.

The changes in Q_{Fc} versus $[M^{n+}]$ were fitted to the Langmuir adsorption isotherm [43,46,49]. The Langmuir model is based on the assumptions that adsorption is limited to one monolayer that all surface sites are equivalent and the adsorption to one site is independent of the occupancy condition of the adjacent sites. The complexation equilibrium is given by:

$$\text{Au-(Fc-peptide)} + \text{M}^{n+}\,(\text{aq}) \rightleftarrows \text{Au-(Fc-peptide-m}^{n+})$$

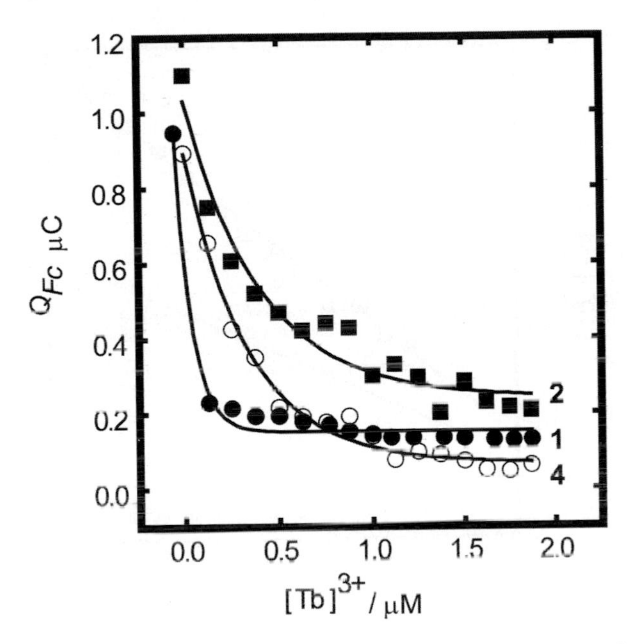

Figure 6.7 Graph of the apparent surface charge Q_{Fc}/μC of films of **1**, **2**, **4**, versus molar concentration of metal ion added ($[M^{n+}]$/μM). 2.0M $NaClO_4$ at pH 7.0 versus Ag/AgCl. (line = guide to the eye)

The conditional equilibrium constant K can be determined using the Langmuir adsorption isotherm equation. [43,46,49]:

$$\theta = K[M^{n+}]/(1+K[M^{n+}]) \tag{6.1}$$

where θ is the ratio of the surface charge Q_{Fc} of metal ion at a concentration $[M^{n+}]$ to its maximum surface charge Q_{max}. Equation 1 thus becomes [43,46,49]:

$$[M^{n+}]/Q = (1/K\,Q_{max}) + ([M^{n+}]/Q_{max}). \tag{6.2}$$

The surface charge of M^{n+} is equal to the surface charge of electroinactive Fc, which can be evaluated from the difference in Fc surface charge measured in the presence and absence of Tb^{3+}.

Linear plots of $[M^{n+}]/Q_{Fc}$ versus $[M^{n+}]$ (Figure 6.8) for the oxidation give the conditional equilibrium constant (Table 6.4). The equilibrium constant for the interaction of Tb^{3+} and Ca^{2+} with **3** shows a higher binding constant compared with that of **4**. This result is similar to observations made by Blankespoor in the study of Cu^{2+} interactions on nitrilotriactetic and imminodiactetic surfaces [59]. As expected, the ester showed very weak binding to the metal ions. These log$K$ values of Tb^{3+} and Ca^{2+} interaction with **2** and **4** are comparable to the literature values of metal-carboxylate interactions for some mono- and dicarboxylate ligands [60–64]. The lowest detection limit for Tb^{3+} and Ca^{2+} interaction at Fc-dendrimers films is 10.0 μM.

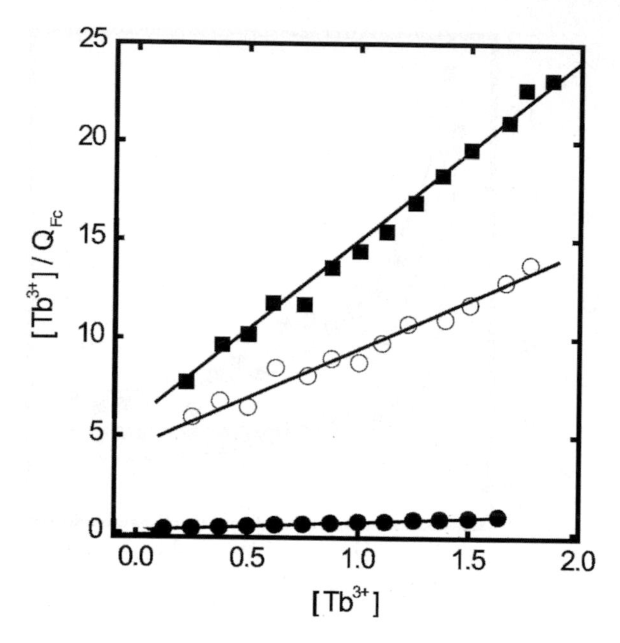

Figure 6.8 Langmuir plots of $[M^{n+}]/Q_{Fc}$ of films of **1** ●, **2** ■, **4** ○ versus $[Tb^{3+}]$ /μM

Table 6.4 Data for interaction of Tb^{3+} and Ca^{2+} with Fc-dendrimer films

	1/Tb^{3+}	**2**/Tb^{3+}	**4**/Tb^{3+}	**2**/Ca^{2+}	**4**/Ca^{2+}
E^o	483 (15)	433 (8)	440 (10)	437 (8)	451 (12)
ΔE^o	21 (5)	7 (4)	15 (6)	11 (6)	18 (8)
K / M	60 (10)	2700 (21)	2100 (15)	1600 (68)	700 (74)
Log$K$	1.76 (0.08)	3.43 (0.02)	2.98 (0.04)	3.20 (0.04)	3.03 (0.05)

The decrease in the charge, Q_{Fc}, which occurs in the presence of additional metal ions leads to the following hypothesis: (i) The facilitation of electron transfer of ClO_4^- across the film interface for ion pair formation of $Fc^+[ClO_4^-]$ is blocked by the formation of $M^{n+}[ClO_4^-]_n$ layers at the interface; (ii) the barrier originates from the compactness of the COO-metal ion complex formation at the interface; (iii) the actual surface concentration of the Fc-dendrimers in the film does not change. The total charge of the electrode depends on the electronic charge of the gold electrode, the charge of the film, and any ions that are adsorbed or tightly ion-paired with the film. Scheme 6.5 illustrates the proposed effect of metal ion complexation on the transfer of anions in solution across the monolayer interface.

The ion-pair theory was invoked by Dong and coworkers to explain the behavior of adsorption of various surfactants at coassembled Fc-teminated alkanethiol-alkylthiolphene thiol SAMs on gold [51]. It was proposed that the adsorption of surfactants at the mixed monolayer created an energy barrier that inhibited the diffusion of counter ions present in solution from diffusing into the adsorbed layer as compensation for the oxidation of the Fc to Fc^+ ions [51]. The creation of a barrier at an interface can be tested with electrochemical impedance spectroscopy (EIS) where an increase

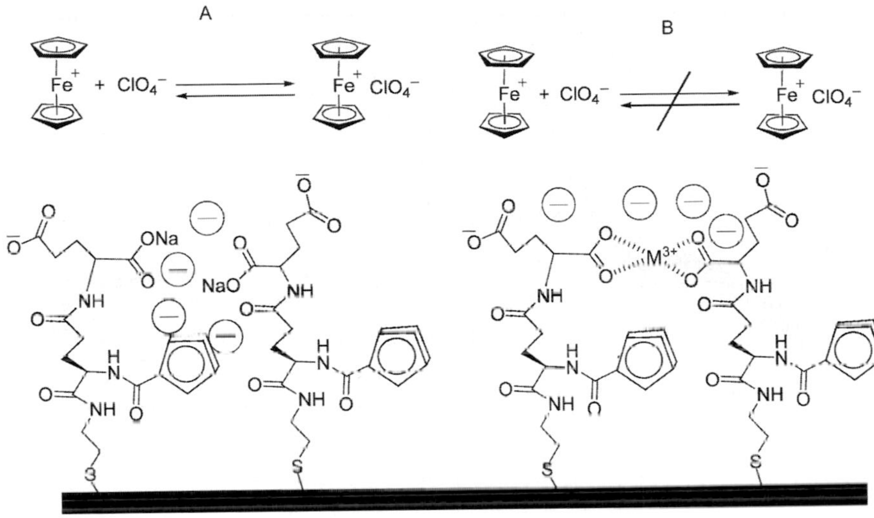

Scheme 6.5 Representation of formation of metal assisted ion pair at the film interface **4**. **a** In the absence of M^{n+} the supporting electrolyte can freely diffuses into the film. **b** Formation of strong M^{n+}-carboxylate layer prevents the effective diffusion of counter ions into the film

in the film thickness will result in the reduction of the total capacitance. Second, the existence of the barrier-effect from the complex formation can be tested by using ethylene diamine tetraacetic acid (EDTA) complexation to regenerate the interface.

Binding of metal ions to the dendrimer immobilized on the electrode is believed to affect the interfacial properties of the monolayer through ionic interactions. These changes are monitored by changes in the capacitance of the film [65,66]. The total capacitance C_{tot} as described by Eq. 6.3 is composed of a series of capacitances ($C_{momolayer}$ and $C_{interface}$ formed by the protonation/deprotonation of the COOH group at the pH 7.0) [40,67–69]. For a capacitance in series:

$$1/C_{tot} = 1/C_{monolayer} + 1/C_{intefacial\ layer} \tag{6.3}$$

$$C_{tot} = \varepsilon\,\varepsilon_o/d \tag{6.4}$$

where ε is the dielectric constant of the monolayer and ε_o is the dielectric constant of the solution and d is the film thickness, $C_{monolayer}$ and $C_{interfacial\ layer}$ are defined by the circuit diagram as shown in Figure 6.9. This means that the capacitance is inversely proportional to the thickness and one would expect a decrease in capacitance with increasing layer formation.

The impedance spectra are acquired at twenty frequencies ranging from 0.1 Hz to 100 kHz at an applied potential equivalent to the E^o of the individual Fc-peptide cystamine film. Figure 6.9a is a typical example of the impedance plot in the capacitance plane for films of **2** in the absence and the presence of Ca^{2+} ions. Fitting of the experimental impedance spectra to an equivalent circuit allows us to evaluate the individual components of the equivalent circuit. A constant phase element

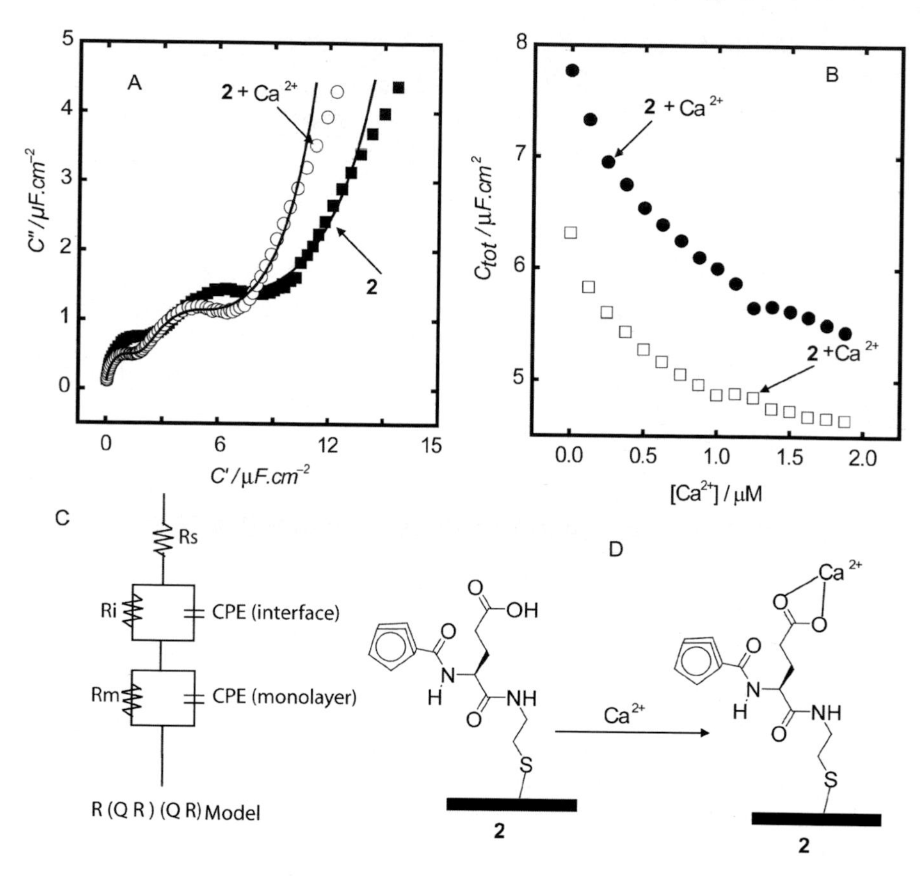

Figure 6.9 a Impedance plot in the capacitance plane **2** and **2** + Ca^{2+} (continuous line is fitting to model R(QR)(QR)). **b** Effect of Ca^{2+}on C_{tot} of films of **2** and **4**. ε **c**Equivalent circuit for fitting. **d** Interaction of Ca^{2+} with a film of **2**

(CPE) is used here is to account for the inhomogeneity of the film because of the surface roughness effects. Therefore the $CPE_{interface}$ or CPE_{mono} are pseudocapacitances that require an additional exponent, n.

Values n from fittings range from 0.85 to 0.95 (Table 6.5), which is sufficiently close to unity for the CPE to be interpreted as capacitance. The titration with metals does not change the equivalent circuits, indicating that conformational integrity on the surface is maintained [65,66,70]. As was observed in metal titrations of other Fc-films [14,15], the capacitance of the systems decreased significantly with the addition of the metal ion (Figure 6.9 b). The most reasonable explanation for the observed drop in capacitance is the stronger neutralization of the COO^- charges on the surface by Tb^{3+} compared with the Na^+ supporting electrolyte [57,70,71]. The differences observed in values for the solution R_s, for the various films are unexpected. Nahir and Bowden from experiments of immobilized *cyt c* on $HS(CH_2)_{13}$-COOH and $HS(CH_2)_5$-COOH have shown that for the shorter films the two parameters R_i and R_s are indistinguishable [71]. Films of the Fc-conjugate are quite thin and the R_s value may have contributions from the R_i and/or R_m, which we cannot separate.

Table 6.5 Data for Impedance titration of Tb^{3+} and Ca^{2+} at film interfaces of **2** and **4**: CPE_{int} (μFcm^{-1}), CPE_{mono} (μFcm^{-1}), R_{dl} ($\mu\Omega cm^2$), R_{ct} ($10^2\Omega cm^2$), R_s (Ωcm^2) from EIS studies taken at E^o of Fc redox probe

	2	2+Tb³⁺	2+Ca²⁺	4	4+Tb³⁺	4+Ca²⁺
R_s	10.9(0.5)	11.3 (0.3)	11.9 (0.4)	5.1 (0.2)	5.1 (0.2)	5.1 (0.5)
CPE_{int} n	43.5 (0.2)	21.8 (0.4)	20.8 (0.1)	14.2 (0.4)	11.9 (0.2)	10.9 (0.2)
	0.85	0.90	0.80	0.80	0.95	0.90
CPE_{mono} n	9.5 (0.2)	8.3 (0.1)	7.4 (0.1)	11.4 (0.2)	11.9 (0.2)	10.9 (0.1)
	0.90	0.95	0.90	0.90	0.85	0.80
R_i	0.6 (0.3)	0.6 (0.2)	0.5 (0.2)	1.5 (0.4)	1.5 (0.3)	1.3 (0.2)
R_m	0.2 (0.1)	0.3 (0.1)	0.3 (0.2)	1.7 (0.3)	2.0 (0.1)	1.9 (0.3)

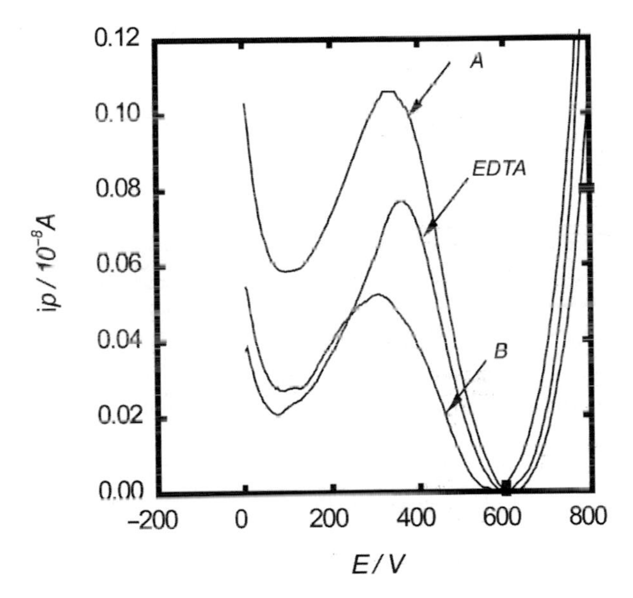

Figure 6.10 Normalized DPV scans showing the regeneration of a **2**+Tb^{3+} film by exposure to a 1.0 mM EDTA solution (EDTA). film of **2** prior to exposure to Tb^{3+}, **2**+Tb^{3+}(**b**). DPV scan rate of 20 mV/s and pulse amplitude of 25 mV at gold electrodes (BAS with geometric area 0.025 cm^2), Reference electrode Ag/AgCl, supporting electrolyte 2.0 M $NaClO_4$ (aq) solution pH 7.0

2.3 Surface Regeneration

EDTA is a strong chelating agent for metal ion such as lanthanides and alkaline earth metals (binding constant EDTA-Ln^{3+} about $\log K = 18.0$) [72]. Fc-peptide acid films were regenerated by stripping the sorbed metal ions using EDTA chelation. This was achieved by soaking the metal-carboxylate bound film in a solution of 1.0 mM EDTA solution for 30 minutes. Figure 6.10 shows DPV scans for films of **3** in the presence and absence of Tb^{3+} and after soaking a Tb^{3+} film in EDTA solution. The results show that treatment with 1.0 mM EDTA solution results in the recovery of the signal, lending further support to the importance of ion-pairing effects that are responsible for the decrease in the peak intensity. Though the films are washed after regeneration experiments, 100% signal recovery was never achieved. This may result from EDTA forming a complex with the peptide on the surface by H-bonding, thus trapping some metal ions in the process. The use of strong acids coupled with mechanical shaking has been reported for stripping lanthanides bound to glycine-functionalized adsorbent materials [73]. The use of strong acids, however, leads to the decomposition of the Fc moiety and cannot be applicable in this experiment.

2.4 Metal Ion -Film Interaction by XPS

X-ray photospectroscopy (XPS) allows an in-depth analysis of the coordination of metal ions to the films of **2** and **4**. All XPS spectra were calibrated with reference to the Au $4f_{7/2}$ peak at 84.00 eV. Studies by Winograd and coworkers describing the interaction of Al^{3+} with mercaptohexadecanoic acid (MPA) films showed that C1*s* and O1*s* core-level peaks shifted in the presences of Al^{3+} [74]. Figure 6.11 shows for example the C1s core peaks for **2** in the absence and presence of Ca^{2+} and Tb^{3+}. The C1s core level of **2** was deconvoluted into 3 peaks 284.68, 285.95, and 288.03 eV corresponding to CH-, C-O/C-N, and C=O moieties respectively [75,76].

Table 6.6 lists the deconvoluted peaks associated with the C1*s*, O1*s*, and N1*s* core elements. The C1*s* peaks for the C-O and C=O experience shifts towards higher binding energies upon addition of Ca^{2+} and Tb^{3+} (Figure 6.11). This is consistent with changes associated with metal ion coordination as observed for Al^{3+} on MPA surfaces [74]. Shifts rangin from 0.35 to 0.40 eV were observed for C=O components which compares well with shifts observed for metal coordination to other surfaces ($\Delta BE = 0.45$ eV for Al^{3+}-MPA; $\Delta BE = 0.5$ eV for Cu^{2+}-ANTA [77]). Shifts associated with the C-O/C-N peak for the Ca^{2+} and Tb^{3+} of 0.71–0.89 eV with **4** were significantly higher than the smaller shifts rangin from 0.15 to 0.20 eV for **2**. Interactions with NH of the peptides are unexpected but a differentiation between the C-N and C-O groups is not possible in this study. The O1*s* spectrum shown in Figure 6.11 is deconvoluted into 2 peaks: the double bonds -C=O at 531.23 and 531.28 eV, and the single bonded C–O-H centered at 532.61and 533.65 eV for **2** and **4** respectively [78].

Table 6.6 XPS analysis of films of **2** and **4** and their interactions with Ca^{2+} and Tb^{3+}

		C1*s*		O1*s*		N1s
Films	CH-	C-O/C-N	C=O	C-O-H	C=O	N
2	284.68	285.95	288.03	532.61	531.23	399.74
2-Ca^{2+}	288.70	286.15	288.38	532.96	531.66	399.79
2-Tb^{3+}	284.68	286.10	288.43	533.43	531.69	399.78
4	284.78	285.49	288.09	532.65	531.28	400.01
4-Ca^{2+}	284.81	286.20	288.43	533.04	531.76	399.96
4-Tb^{3+}	284.78	286.35	288.49	533.04	531.82	399.98

Listed are the C1s, O1s, and N1s levels. All binding energies are in eV

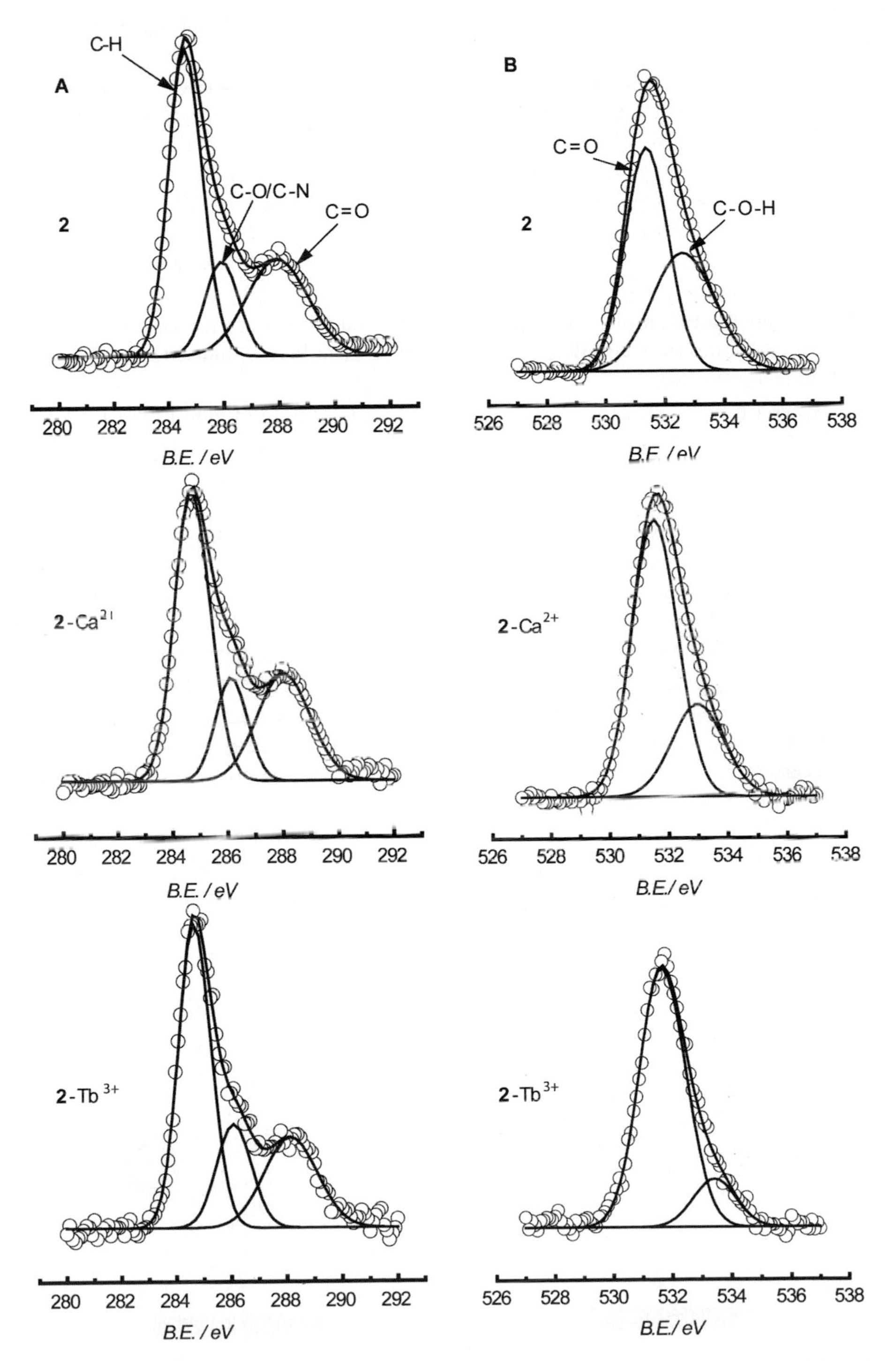

Figure 6.11 Deconvoluted high-resolution spectra of films of **2** and in the presence of Ca^{2+} and Tb^{3+}; showing the C1*s* region (**a**); showing the O1*s* region (**b**)

It has been suggested that the changes observed in these peaks after coordination to metal ions are indications of the involvement of the oxygen in the metal binding process [78]. As can be seen from Table 6.6, O1*s* signal shifts towards higher binding energies, thereby increasing the peak separations between them. Increases in peak separation observed for Al^{3+}-MPA and Cr^{2+}-MUA films have been associated with bidentate coordination [74,79] whereas decreases in peak separation observed for Cu^{2+}-MUA complexes have been linked with monodentate coordination [79].

The bidentate complexation modes of Ca^{2+} and Tb^{3+}, with the acid groups of Fc-peptide acids from the IR and NMR studies, have been alluded to in previous solution studies [80].

Based on our surface spectroscopic results, a bidentate mode of coordination of Ca^{2+} and Tb^{3+} with the carboxylate group of Fc-peptide acid cystamine films is proposed (Scheme 6.6). There is precedence for bidentate metal coordination to surface-bound ligands. In a study of lanthanide selective sorbents, Fryxell and coworkers showed that self-assembly of glycinate monolayers on mesoporous materials bound Eu^{3+} in an 8-coordinated fashion. In this system, the close proximity of the ligands allowed four bidentate ligands to chelate the lanthanide cation [73].

Using scanning force microscopy of elaidic acid monolayers prepared on a terbium-containing subphase, Dérue et al. provided evidence for the formation of a bidentate complex betweeb Tb^{3+} and three elaidic acid molecules [81].

Films of Fc-peptide acids **2** and **4** exhibited a single peak for the N1*s* peak between 399.74 and 400.01 eV resulting from the various amide functional groups (Figure 6.12). Individual amide groups were not distinguishable. The peak position was consistent with the literature as the amide groups gave a peak between 400.0 and 400.5 eV [82]. The peak position and shape did not change in the presence of Ca^{2+} and Tb^{3+}, which suggest that the amide groups are not

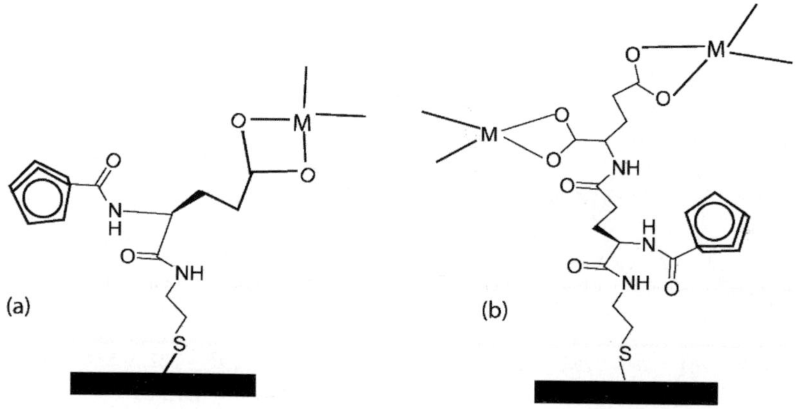

Scheme 6.6 Bidentate complexation mode of Fc-peptide acids linked to gold surfaces and with Ca^{2+} and Tb^{3+}

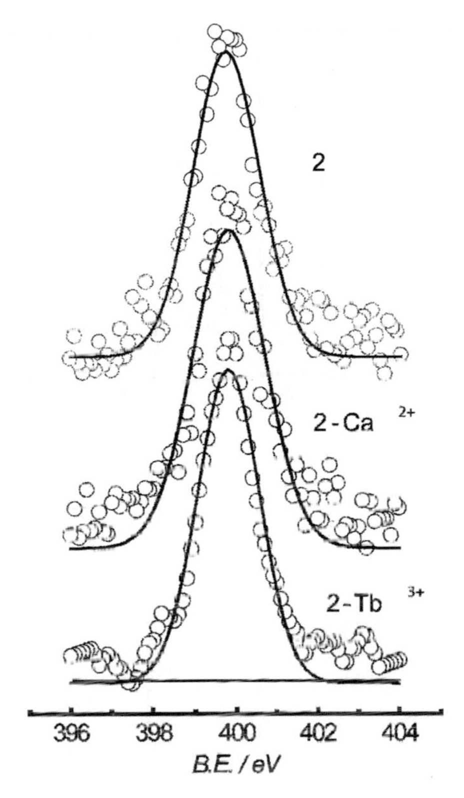

Figure 6.12 N1*s* core-level spectra of films of **2** in the absence and presence of Ca^{2+} and Tb^{3+}

involved in metal binding. This, however, is significantly different from the films of amino acids/peptide, such as Cys, Gly-Cys, and Val-Cys films where terminal NH_3^+ provides an additional binding site. The N1*s* peaks for these films were between 402.2 and 401.4 eV and were shifted upon coordination to Cu^{2+} to 399.4, 399.5, and 399.8 eV [12,83].

The XPS spectrum for films in the presence of Tb^{3+} and Ca^{2+} (Figure 6.13) shows the peaks for Td 3*d* and Ca 2*p*, respectively. The peaks related to the Tb3*d* occur as a doublet at 1243.62 and 1277.43 eV in **2**-Tb film and 1242.41 and 1277.13 eV in **4**-Tb film displaying the spin-orbit splitting of 33.81 and 34.72 eV, respectively, compared with literature value of 34.6 eV which is characteristic of binding energies for Tb^{3+} species [84,85]. The signals related to the 2*p* of Ca^{2+} are centered between 343.32 and 347.13 eV [86,87] and demonstrate their low intensity masked by the presence of the strong interference from the Au4*d* peaks, which occur in the same region. From the XPS section, the attenuation of C=O and O1*s* strongly suggest that both oxygen atoms of the acids functional group reacts with the metal ion to form the metal-carboxylate complexes. Calculations of compositional ratios of Ca^{2+} and Tb^{3+}:S show a 1:1 ratio for films of **2** and 2:1 ratio for **4**.

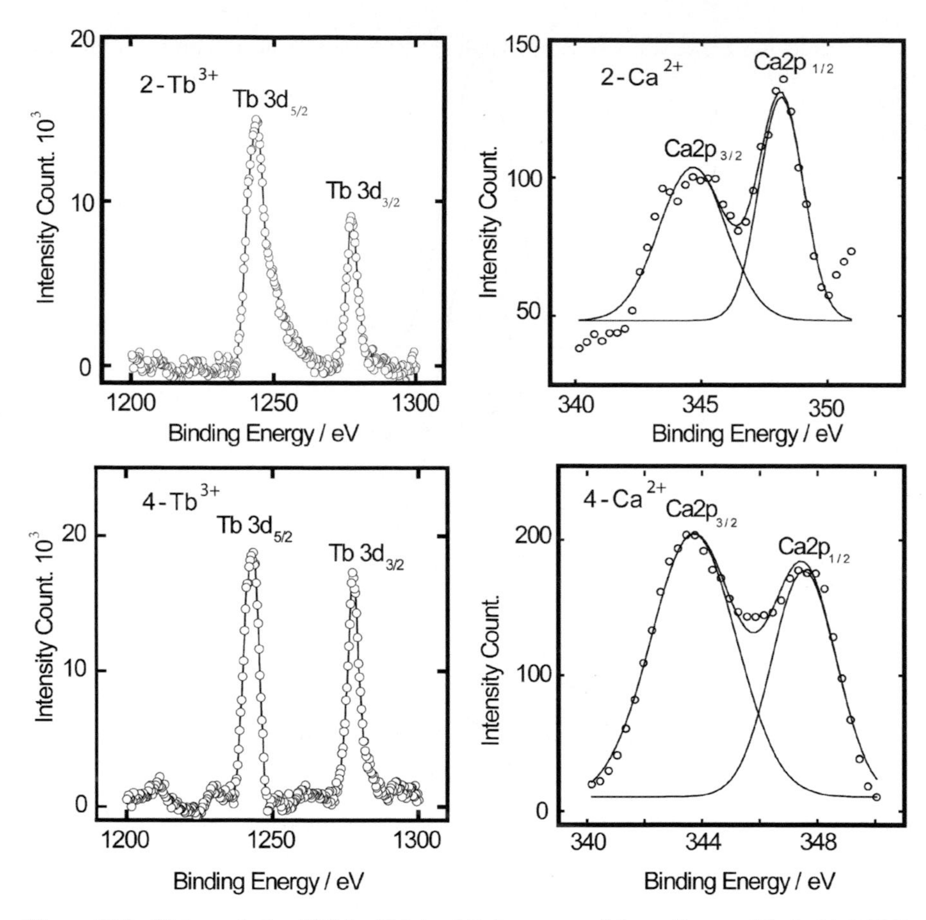

Figure 6.13 High-resolution XPS for Tb3*d* and Ca2*p* spectra of Au surface modified with **2**-Tb^{3+}, **4**-Tb^{3+}, **2**-Ca^{2+}, and **2**-Ca^{2+} films

Angle resolved XPS (ARXPS) measurements were used to measure the thickness of the films of **2** and **4** in the presence and absence of the metal ions from the attenuation of the Au signals. The photoelectron intensity from the thin film-covered substrate varies with the takeoff angle, θ (taken as the angle between the surface plane and entrance to the analyzer), and is given by Eq. 6.5 [88,89]:

$$\ln(I) = -d/(\lambda \sin\theta) + \ln(I_0) \tag{6.5}$$

where I_0 and I are the intensities of the photoelectron from the clean substrate and from the covered substrates respectively, d is the film thickness, and λ is the inelastic mean free paths of photoelectrons. Accordingly, ln (I) should be linearly related to 1/sin θ with a slope of ($-d/\lambda$) according to Eq. 6.5. Figure 6.13 shows the Au_{4f} XP spectra of the for **2**, **2**-Tb^{3+}, **2**-Ca^{2+} films Au measured at various θ angles and the plots ln (I) versus 1/sin θ. The two peaks resulting from $Au_{4f5/2}$ and $Au_{4f7/2}$ were at 87.55 as well as 83.83 eV, respectively. The intensities of these peaks decreased with decreasing θ and plots of ln (*I*) versus 1/sin θ were linear. The slope of $-d/\lambda$ from the ln (*I*) versus 1/sin θ plot for the $Au_{4f7/2}$ at 83.83 eV showed that the slope for films containing metal ions increases considerably (Figure 6.14). Using a λ value of 42 Å, [90,91] the thicknesses are calculated and compared with those from Spartan modeling (Table 6.7).

The thicknesses of the films are increased by 3 Å in the presence of Ca^{2+}, by 7–8 Å in the presence of Tb^{3+}. The differences between the thicknesses of the two metal films may be related to their coordination numbers in aqueous solution Ca^{2+} (C.N. of 6) and Tb^{3+} (C.N. of 8–12). These values are similar to Abbott's XPS measurements of thicknesses of MUA films which showed increases of 3 and 7 Å in the presence of $Ni(ClO_4)_2$ and $Cu(ClO_4)_2$, respectively [92].

Taking the ionic size of ClO_4^- as 2.4 Å, Ca^{2+} as 1.3 Å and Tb^{3+} as 1.2 Å [93], suggests that one layer of $Ca(ClO_4)_2$, and two layers of $Tb(ClO_4)_3$ are formed

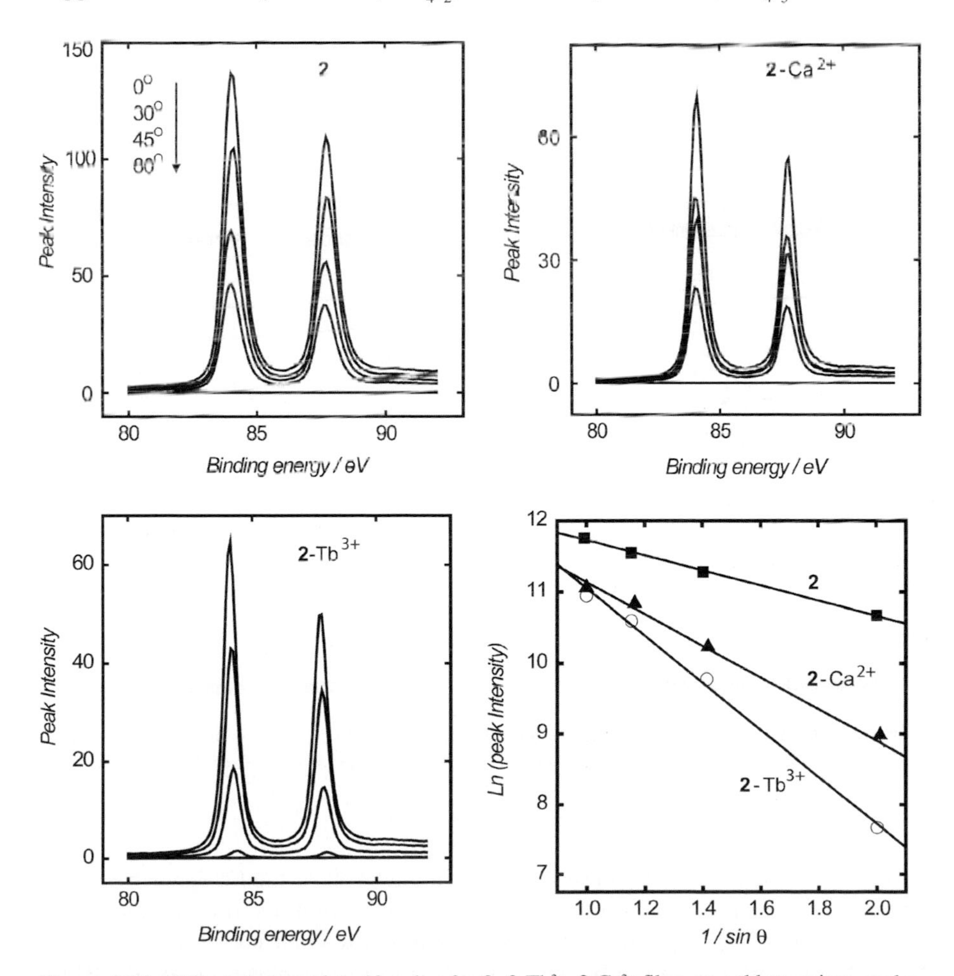

Figure 6.14 XPS stackplots of Au4f region for **2**, **2**-Tb^{3+}, **2**-Ca^{2+} films on gold at various angles. Graphs of relation between 1/sin θ and logarithm of the integrated peak intensities of the Au 4*f* in the XPS of **2**, **2**-Tb^{3+}, **3**-Ca^{2+} films on gold at various angles

Table 6.7 Calculated thickness from XPS and Spartan modeling

	XPS		Spartan	
Film	Slope	λ	*d* Å	*d* Å
2	1.63	42	7	9
2-Tb^{3+}	3.38	42	14	–
2-Ca^{2+}	2.42	42	10	–
4	1.86	42	8	9
4-Tb^{3+}	3.85	42	16	–
4-Ca^{2+}	2.51	42	11	–

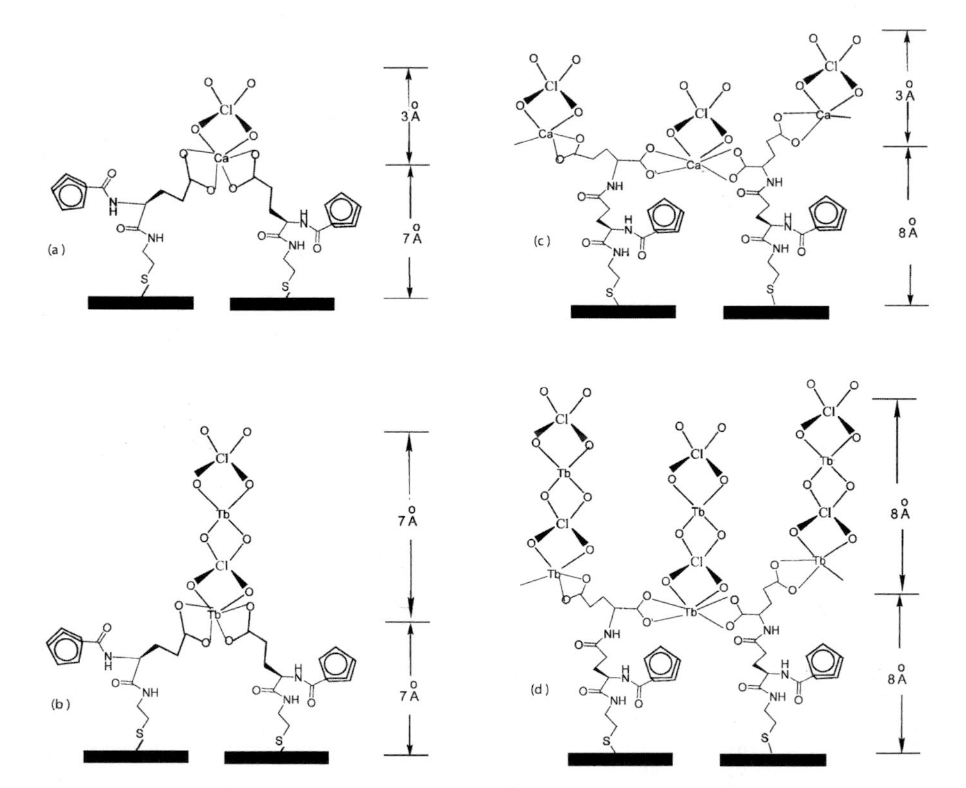

Scheme 6.7 Proposed model for the formation of metal complexes on surfaces for single layer bidendate of **2**-Ca^{2+} (**a**), bilayer bidentate **2**-Tb^{3+} (**b**), single layer bidendate of **4**-Tb^{3+} (**c**), and bilayer, bidentate of 4-Tb^{3+} (**d**)

respectively (Scheme 6.7). These increases in thickness of the Fc-dendrimer films in the presence of metal ions confirm the observation from the titration monitored by electrochemical impedance. The XPS results discussed above are consistent with observations made from IR spectroscopy of metal complexes of Fc-GluOH for bidendate coordination modes [80].

3 Conclusions

This chapter presented the surface characterization of disulfide-linked Fc-peptide $[Fc\text{-}G1(OBz)CSA]_2$ **1**, $[Fc\text{-}G1(OH)CSA]_2$ **2**, $[Fc\text{-}G2OMeCSA]_2$ **3**, and $[Fc\text{-}G2(OMe)CSA]_2$ **4**. These systems have a disulfide group that was used for immobilization on Au. On gold surfaces, the acids, unlike the corresponding esters, formed multilayers held together by intermolecular H-bonding. Monolayer coverages, however, are attainable by sonication in mixtures of solvents. CV measurements in the presence of $[Fe(CN)_6]^{3-/4-}$ reveal that the Fc-peptide-modified electrodes exhibit excellent barrier properties.

The immobilized films of the acids $[Fc\text{-}G1(OH)CSA]_2$ **2** and $[Fc\text{-}G2(OH)CSA]_2$ **4** responded to the presence of Ca^{2+} and Tb^{3+}. The interaction monitored by CV shows an attenuation of the Fc signal, which is the result of changes in ion pairing between the Fc^+ and ClO_4^-. Expectedly, the interactions lead to an increase in the film thickness as reflected in a decrease in the total capacitance C_{tot} from EIS studies and from XPS depth-profiling experiments. Regeneration experiments using EDTA show that biding of the metal ions to the carboxylate groups is virtually reversible. Binding constants are higher for Tb^{3+} than for Ca^{2+} and are comparable with those lanthanide and alkaline earth metal-dicarboxylic acid metal interactions. The detection limit for Tb^{3+} and Ca^{2+} interaction at Fc-dendrimers films is ~ 10.0 μM. XPS analysis of films in the presence of Ca^{2+} and Tb^{3+} indicates that binding takes place at the carboxylate site and that the amide group is not involved in binding. Further compositional analysis of films of Tb^{3+} on $[Fc\text{-}G1(OH)CSA]_2$ and $[Fc\text{-}G2(OH)CSA]_2$ show that a 1:1 complex is formed with $[Fc\text{-}G1(OH)CSA]_2$, whereas $[Fc\text{-}G2(OH)CSA]_2$ can coordinate with two metal ions.

References

1. Scheibler L, Dumy P, Stamou D, Duschl C, Vogel H, Mutter M. (1998) Tetrahedron. 54:3725.
2. Strong AE, Moore BD. (1999) J Mater Chem. 9:1097.
3. Strong AE, Moore BD. (1998) Chem, Commun. 473.
4. Whitesell JK, Chang HK. (1993) Science. 261:73.
5. Flink S, Van Veggel FCJM, Reinhoudt DN. (2000) Adv Mater. 12:1315.
6. Turyan I, Mandler D. (1994) Anal Chem. 66:58.
7. Crego-Calama M, Reinhoudt DN. (2001) Adv Mater. 13:1171.
8. Feng X, Fryxell GE, Wang LQ, Kim AY, Liu J. (1997) Science. 276:923.
9. Steinberg S, Tor Y, Sabatani E, Rubinstein I. (1991) J Am Chem Soc. 113:5176.
10. Liu A-C, Chen D-C, Lin C-C, Chou -HC. (1999) Anal Chem. 71:1549.
11. Yang W, Gooding JJ, Hibbert DB. (2001) Analyst. 126:1573.
12. Yang W, Jaramillo D, Gooding JJ, et al. (2001) Chem Commun. 1982.
13. Kim HT. (2005) Bull Korean Chem Soc. 26:679.
14. Sugawara M, Kojima M, Sazawa H, Umezawa Y. (1987) Anal Chem. 59:2842.
15. Aoki H, Buhlmann P, Umezawa Y. (2000) Electroanalysis. 12:1272.
16. Takehara K, Ide Y, Aihara M. (1992) Bioelectrochem Bioenerget. 29:113.

17. Takehara K, Ide Y, Aihara M, Obunchi E. (1992) Bioelectrochem Bioenerget. 29:103.
18. Takehara K, Aihara M, Ueda N. (1994) Electroanalysis. 6:1083.
19. Wang Z, Cook MJ, Nygård A-M, Russell DA. (2003) Langmuir. 19:3779
20. Liu S-G, Liu H, Bandyopadhyay K, Gao Z, Echegoyen L. (2000) J Org Chem. 65:3292.
21. Le Derf F, Levillian E, Trippé G, et al. (2001) Angew Chem Int Ed. 40:224.
22. Ion A, Ion I, Moutet J-C, et al. (1999) Sensors and Actuators B. 59:118.
23. Ion Ac, Moutet J-C, Pailleret A, et al. (1999) J Electroanal Chem. 464:24.
24. Jiang H, Leger J-M, Dolain C, Guionneau P, Huc I. (2003) Tetrahedron. 59:8365.
25. Boas U, Sontjens SM, Jensen Kj, Christensen J, Meijer EW. (2002) ChemBioChem. 3: 433.
26. Gung BW, Zhu Z. (1996) Tetrahedron. 37:2189.
27. Pavia DI, Lampman GM, Kriz GS. (1996) Introduction to spectroscopy. Harcourt College Publishers, New York, p 27.
28. Wu N, Fu LSM, Aslam M, Wong KC, Dravid VP. (2004) Nano Lett. 4:383.
29. Gung BW, Zhu Z. (1997) J Org Chem. 62:6100.
30. Hou H, Li G, Li L, Zhu Y, Meng X, Fan Y. (2003) Inorg Chem. 42:428.
31. Han SW, Seo H, Chung YK, Kim K. (2000) Langmuir. 16:9493.
32. Leavy MC, Bhattacharyya S, Cleland WE, Hussey CI. (1999) Langmuir. 15:6582.
33. Gomez ME, Kaifer AE. (1992) J Chem Edu. 69:502.
34. Diaz DJ, Bernhard S, Storrier GD, Abruña HD. (2001) J Phys Chem B. 105:8746.
35. Galka MM, Kraatz H-B. (2002) CHEMPHYSCHEM. 356.
36. Kraatz H-B, Leek DM, Houman A, Enright GD. (1999) J Organomet Chem. 589:39.
37. Ravenscoft MS, Finklea HO. (1994) J Phys Chem. 98:3843.
38. Sabapathy RC, Bhattacharyya S, Leavy MC, Cleland WE, Hussey CI. (1998) Langmuir. 14:124.
39. Finklea Ho, Liu L, Pavenscroft MS, Punturi S. (1996) J Phy Chem. 100:18852.
40. Finklea Ho. (1996) In: *Electroananlytical chemistry,* Vol. 19, Bard AJ, Rubinstein I (eds). Electrochemistry of organised monolayer. Marcel Dekker, New York, p 109.
41. Oh S-K, Baker LA, Crooks RM. (2002) Langmuir. 18:6981.
42. Bediako-Amoa I, Sutherland TC, Li C-Z, Silerova RKraatz H-B. (2004) J Phys Chem B. 108:704.
43. Ju H, Leech D. (1999) Phys Chem Chem Phys. 1:1549.
44. Uosaki K, Sato Y, Kita H. (1991) Langmuir. 7:1510.
45. Creager SE, Rowe GK. (1993) Langmuir. 9:2330.
46. Rojas MT, Koniger R, Stoddart JF, Kaifer AE. (1995) J Am Chem Soc. 117:336.
47. Twardowski M, Nuzzo RG. (2003) Langmuir. 19:9781.
48. Wang Y, Kaifer AE. (1998) J Phys Chem, B. 102:9922.
49. Ju H, Leech D. (1998) Langmuir. 14:300.
50. Godinez La, Castro R, Kaifer AE. (1996) Langmuir. 12:5087.
51. Peng Z, Qu X, Dong S. (2004) J Electroanal Chem. 563:291.
52. Yao X, Wang J, Zhou F, Wang J, Tao N. (2004) J Phys Chem B. 108:7206.
53. Andreu R, Calvente JJ, Fawcett WR, Molero M. (1997) J Phys Chem B. 101:2884.
54. Valincius G, Niaura G, Kazakeviciene B, et al. (2004) Langmuir. 20:6631.
55. Stone DL, Smith DK. (2003) Polyhedron. 22:763.
56. Rowe GK, Creager SE. (1991) Langmuir. 7:2307.
57. Jenkins ATA, Le-Meur J-F. (2004) Electrochemistry Comm. 6:373.
58. Li J, Liang KS, Scoles G, Ulman A. (1995) Langmuir. 11:4418.
59. Blankespoor R, Limoges B, Schöllhorn B, Syssa-Magalé J-LYazidi DY. (2005) Langmuir. 21:2362.
60. Duckworth OW, Martin ST. (2001) Geochimica Cosmochimica Acta. 65: 4289.
61. Agarwal RcC, Agarwal SL. (1981) Thermchimica Acta. 44:121.
62. Cefola M, Tompa AS, Celiano AV, Gentle PS. (1962) Inorg Chem. 1: 290.
63. Saladini M, Menabue L, Ferrari E. (2001) Carbohydrate Research. 336:55.
64. Wang Z-M, Van De Burgt LJ, Choppin GR. (2000) Inorganica Chimica Acta. 310:248.

65. Bontidean I, Lloyd JR, Hobman JL, et al. (2000) J Inorg Biochem. 79:225.
66. Bontidean I, Ahlqvist J, Mulchandani A, et al. (2003) Biosensors Bioelectronics. 18:547.
67. Cecchet F, Rudolf P, Rapino S, et al. (2004) J Phy Chem B. 108:15192
68. Smith CP, White HS. (1993) Langmuir. 9:1.
69. Kane V, Mulvaney P. (1998) Langmuir. 14:3303.
70. Ekeroth J, Konradson P, Bjorefors F, Lundstrom I, Liedberg B. (2002) Anal Chem. 74:1979.
71. Nahir TM, Bowden EF. (2002) Langmuir. 18:5283.
72. Wu SL, Horrocks Wd-W. (1997) J Chem Soc Daltons Trans. 1497.
73. Fryxell Ge, Wu H, Lin Y, et al. (2004) J Mater Chem. 14:3356.
74. Fisher GL, Hooper AE, Opila RI, Allara DL, Winograd N. (2000) J Phys Chem B. 104:3267.
75. Wang Y, Cai J, Rauscher H, Behm RJ, Goedel WA. (2005) Chem Eur J. 11:3968.
76. Omoike A, Chorover J. (2004) Biomacromolecules. 5:1219.
77. Madoz-Gúrpide J, Abad JM, Fernández-Recio VM, Vázquez L, Gómez-Moreno C, Fernández VM. (2000) J Am Chem Soc. 122:9810.
78. Yang W, Gooding JJ, Hibbert GD. (2001) J Electroanaly Chem. 516:10.
79. Jung DR, Czanderna AW, Herdt GC. (1996) J Vac Sc Technol A. 14:1779.
80. Appoh FE, Sutherland TC, Kraatz H. (2004) J Organomet Chem. 690:1209.
81. Derue V, Alexandre S, Valleton J-M. (1996) Langmuir. 12:3740.
82. Cecchet F, Rudolf P, Rapino S, et al. (2004) J Phys Chem B. 108:15192.
83. Jung G, Ottnad M, Hartmann HJ, Rupp H, Weser UFres Z. (1973) Anal Chem. 263:282.
84. Sarma DD, Rao CN. (1980) J Electron Spectrosc Rel Phenom. 20:25.
85. Silver RF, Zaniquelli EE, Serra OA, Torriani IL, De Castro SGC. (1998) Thin Solid Films. 324:245.
86. Taylor ET, Kinner AE. (1999) Surface Science Spectra. 6:193.
87. Tarasevich BJ, Chusuei CC, Allara DL. (2003) J Phys Chem B. 107:10367.
88. Laibinis PL, Bain CD, Whitesides GM. (1991) J PhysChem. 95:7017.
89. Yanagida M, Kanai T, Zhang X-Q, Kondo T, Uosaki K. (1998) Bull Chem Soc Jpn. 71:2555.
90. Wen X, Linton RW, Formagglo F, Toniolo C, Samulski ET. (2004) J Phys Chem A. 108:9673.
91. Bain CD, Whitesides GM. (1989) J Phys Chem. 93:1670.
92. Yang K-L, Cadwell K, Abbott NL. (2003) Adv Mater. 15:1819.
93. Douglas B, Mcdaniel DH, Alexander JJ. (1983) Concepts and models of inorganic chemistry. John Wiley & Sons, Canada, Toronto, p 219.

Iron-Containing Polymers with Azo Dyes in their Side Chains or Backbones

Alaa S. Abd-El-Aziz and Patrick O. Shipman

1 Introduction

Interest in organometallic macromolecules has grown exponentially ever since Arimoto and Haven first polymerized vinylferrocene in 1955 [1]. Organometallic polymers are known to possess unique optical, magnetic, and thermal properties which allow for potential applications as chemical sensors, electrocatalysts, modified electrodes, and photo-active molecular devices [2–7]. Organoiron polymers are one of the most prevalent classes of organometallic polymers, with many reports on their synthesis and properties published over the past 50 years [8–11]. Of the many varieties of organoiron species, ferrocene and cationic cyclopentadienyliron complexes are most commonly incorporated into polymers.

The *cis-trans* photoisomerization of azobenzene chromophores is a well-studied phenomena that allows for these materials to be used as electrooptic modulators, photorefactive switches, reversible storage systems, and in nanoscale applications [12–18]. The incorporation of these chromophores into polymers is especially interesting for producing new properties. Manners and coworkers have reported liquid crystalline poly(ferocenylsilanes) containing azo dyes in their side chains [19]. Our group has prepared a number of different organoiron polymers containing azo chromophores in either the main or side chains [20–23]. This chapter reviews the work that has been performed in this area by our research group.

A.S. Abd-El-Aziz et al. (eds.), *Inorganic and Organometallic Macromolecules: Design and Applications.*

2 Synthesis of Organoiron Polymers Containing Azo Dyes in Their Side Chains

2.1 Polyethers and Polythioethers

Over the past 10 years our group has reported the preparation of many different classes of organometallic macromolecules. Using the excellent electron-withdrawing nature of the cationic cyclopentadienyliron moiety, the nucleophilic aromatic substitution of cyclopentadienyliron-coordinated chloro-arenes can be accomplished in mild conditions at room temperature. Our group has prepared the easily functionalized organometallic complex **3** from the reaction η^6-dichlorobenzene-η^5-cyclopentadienyliron (**1**) with **2** (Scheme 7.1). The carboxylic acid function on this complex can be reacted with a number of different compounds to give monomers that can be polymerized to give polymers with a variety of different side chains. Compound **3** has been utilized in the synthesis of linear polymers containing both cationic and neutral organoiron moieties, and in the preparation of star-shaped molecules.

Recently, we reported the synthesis of azo dye-functionalized organoiron complexes from the reaction of the bimetallic organoiron complex **3** with azo dyes **4** (Scheme 7.2). These monomers displayed excellent solubility in polar organic solvents and were isolated as vibrantly colored powders ranging from orange to purple in good yields.

Polymerization of **5** gave organoiron polymers containing azo dyes in the side chains. The polymers (**7, 8, 9**) were prepared by nucleophilic aromatic substitution reactions between **5** and various *O*- and *S*- containing dinulceophiles (Scheme 7.3).

The cationic organoiron polymers were isolated in high yields and displayed good solubility in polar organic solvents. Exposure of these polymers to ultraviolet (UV) light caused the photolytic cleavage of the cationic cyclopentadienyliron moieties and gave organic polymers containing azo dyes in their side chains (Scheme 7.4).

Interactions between the cationic organoiron moieties and the gel permeation column (GPC) prevented direct determination of the molecular weights. However, the molecular weights could be estimated from the molecular weights of the organic polymer analogs. The weight averaged molecular weights (M_w) of polymers **7, 8**, and **9** ranged from 11,800 to 31,600 with PDIs between 1.16 and 2.24. Thermal gravimetric analysis (TGA) of polymers **7, 8**, and **9** showed two weight losses, the first between 200 °C and 250 °C corresponding to both the cleavage of

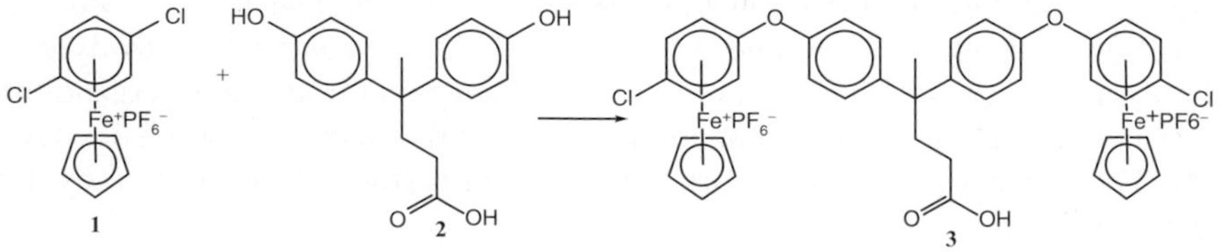

Scheme 7.1

R	R'	R''
H	H	CH_2CH_3
$C(O)CH_3$	H	CH_2CH_3
NO_2	H	CH_2CH_3
CH_3	CH_3	CH_2CH_2CN
$C(O)CH_3$	CH_3	CH_2CH_2CN
NO_2	H	CH_2CH_2CN
Cl	H	CH_2CH_2CN
Cl	CH_3	CH_2CH_2CN

Scheme 7.2

the cationic iron moieties and the degradation of the azo dye. The second weight loss starting between 400 °C and 570 °C was was the result of degradation of the polymer backbone. Differential scanning calorimetry (DSC) indicated that polymers **7**, **8**, and **9** possessed glass transition temperatures (T_gs) ranging from 100 to 180 °C depending on the spacer used.

Studies on the UV-vis properties of polymers showed polymers **7**, **8**, and **9** had λ_{max}s between 417 nm and 490 nm in dimethylformamide (DMF) depending on the substituents present on the azo dye. These absorption are attributed to the $n \rightarrow \pi^*$ transitions of the azo dyes. The polymers containing the most electron withdrawing groups were red-shifted compared to the polymers with less electron-withdrawing groups. Upon addition of a hydrochloric acid solution the polymers underwent a bathochromic shift in their λ_{max} to between 515 nm and 530 nm as a result of the formation of the azonium ion (Figure 7.1). Also, at low to moderate (10–90%) concentrations of acid both the azonium and azo forms of the dye coexist; only at high concentrations of acid does the azo dye completely convert to its azonium species.

The substituents on the azo dye greatly affected the λ_{max} observed for the $n \rightarrow \pi^*$ transition; the more electro-withdrawing substitutent the higher the observed λ_{max}. Figure 7.2 for example shows the UV-vis spectra of **8** where, peak **a** has the shortes wavelength resulting from the presence of the nitro group. Peak **b** shows that the acetyl substituent gives a longer λ_{max} in comparison to the polymer of peak **a**, however, it was shorter than that of peak **c** (nitro group).

R	R'	R"
H	H	CH2CH$_3$
C(O)CH$_3$	H	CH2CH$_3$
NO$_2$	H	CH2CH$_3$
CH$_3$	CH$_3$	CH2CH$_2$CN
C(O)CH$_3$	CH$_3$	CH2CH$_2$CN
NO$_2$	H	CH2CH$_2$CN
Cl	H	CH2CH$_2$CN
Cl	CH$_3$	CH2CH$_2$CN

# R	R'	R"	Spacer
Polymer 7			
H	H	CH_2CH_3	6a
C(O)CH$_3$	H	CH_2CH_3	6a
NO$_2$	H	CH_2CH_3	6a
CH$_3$	CH$_3$	CH_2CH_2CN	6a
C(O)CH$_3$	CH$_3$	CH_2CH_2CN	6a
NO$_2$	H	CH_2CH_2CN	6a
Cl	H	CH_2CH_2CN	6a
Cl	CH$_3$	CH_2CH_2CN	6a
Polymer 8			
H	H	CH_2CH_3	6b
C(O)CH$_3$	H	CH_2CH_3	6b
NO$_2$	H	CH_2CH_3	6b
Polymer 9			
H	H	CH_2CH_3	6c
C(O)CH$_3$	H	CH_2CH_3	6c
NO$_2$	H	CH_2CH_3	6c

Scheme 7.3

2.2 *Polynorbornenes*

Polynorbornene organoiron polymers containing pendent azo dyes were prepared by our groups. The nucleophilic aromatic substitution of the organoiron complex **14** with azo dyes (**13**) gave the azo dye-functionalized organoiron complexes (**15**). These complexes were isolated as brightly colored powders in high yields and showed excellent solubility in polar organic solvents. The reaction of 5-norbornene-2-carboxylic acid (**16**) with the benzylic alcohol function of these organoiron complexes (**15**) gave the organoiron monomers **17** (Scheme 7.5).

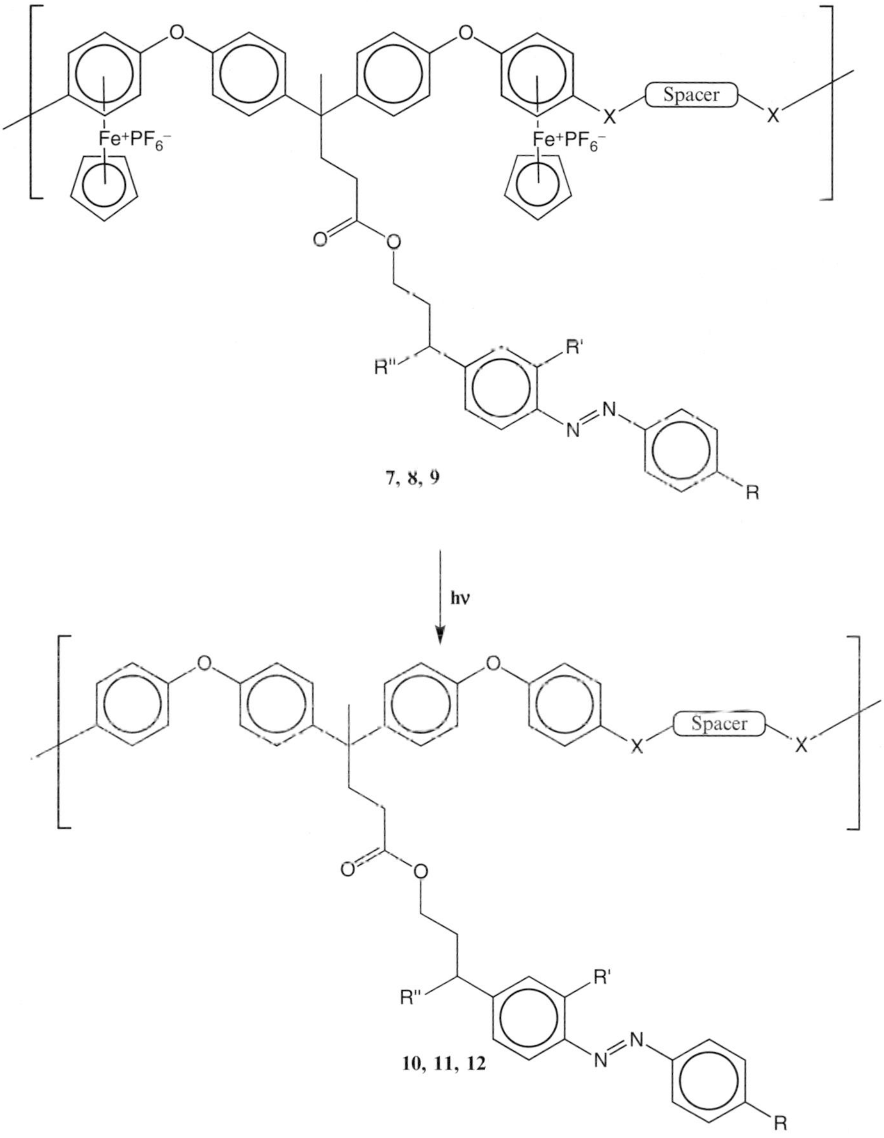

7, 8, 9

hν

10, 11, 12

Scheme 7.4

Ring-opening metathesis polymerization (ROMP) of **18** with Grubbs' catalysts gave the azo dye-functionalized polynorbornenes containing pendent organoiron moieties. Photolytic demetalation of **19** gave their organic analogs (**20**) (Scheme 7.6). The organic polymers possessed M_w ranging from 11,400 to 21,800 with PDIs between 1.10 and 1.11. These molecular weights correspond to M_w ranging between 16,000 and 31,600 for organometallic polymers.

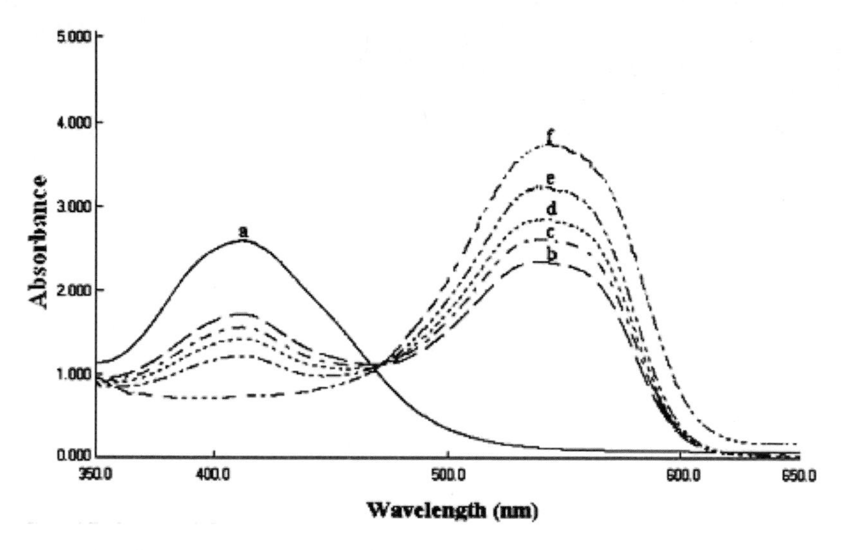

Figure 7.1 UV-Vis Spectrum of **7** ($R = CH_3$, $R' = CH_3$, $R''' = CH_2CH_2CN$) DMF (**a**), 8:2 *(v/v)* DMF/10%HCl (**b**), 8:2 *(v/v)* DMF/20%HCl (**c**), 8:2 *(v/v)* DMF/40%HCl (**d**), 8:2 *(v/v)* DMF/60%HCl (**e**), 8:2 *(v/v)* DMF/100%HCl (**f**)

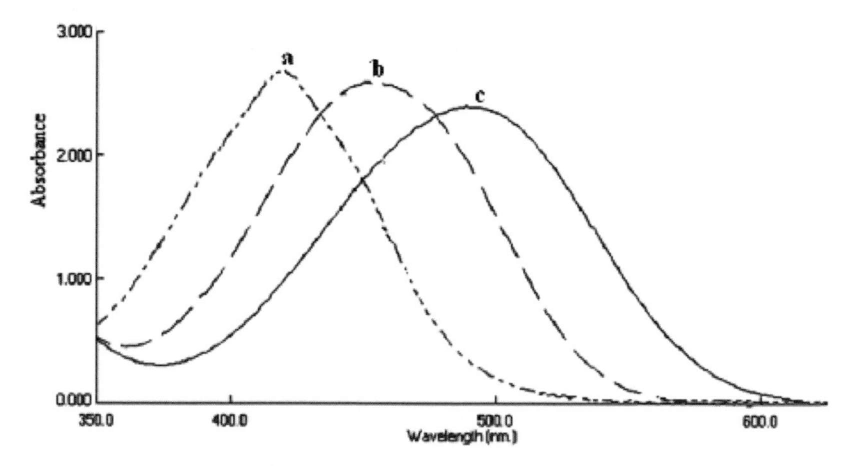

Figure 7.2 The effect on λ_{max} resulting from substituients on the azo dye in **8**

A second method for the synthesis of organometallic polynorbornenes containing a bound azo dye involved the reaction of η^6-chlorotoluene-η^5-cyclopentadienyliron complex **21** with bifunctional azo dyes to yield the azo dye-containing organoiron compounds, which were then further reacted with 5-norbornene-2-carboxylic acid (**16**) to form the azo dye functionalized norbornene monomers, **24,** shown in Scheme 7.7.

ROMP of the azo dye functionalized norbornene monomers **24a,b** with Grubbs' catalyst gave the organometallic polymers. The photolytic cleavage of the cyclopentadienyliron groups yielded their organic analogues (Scheme 7.8). The M_w of the

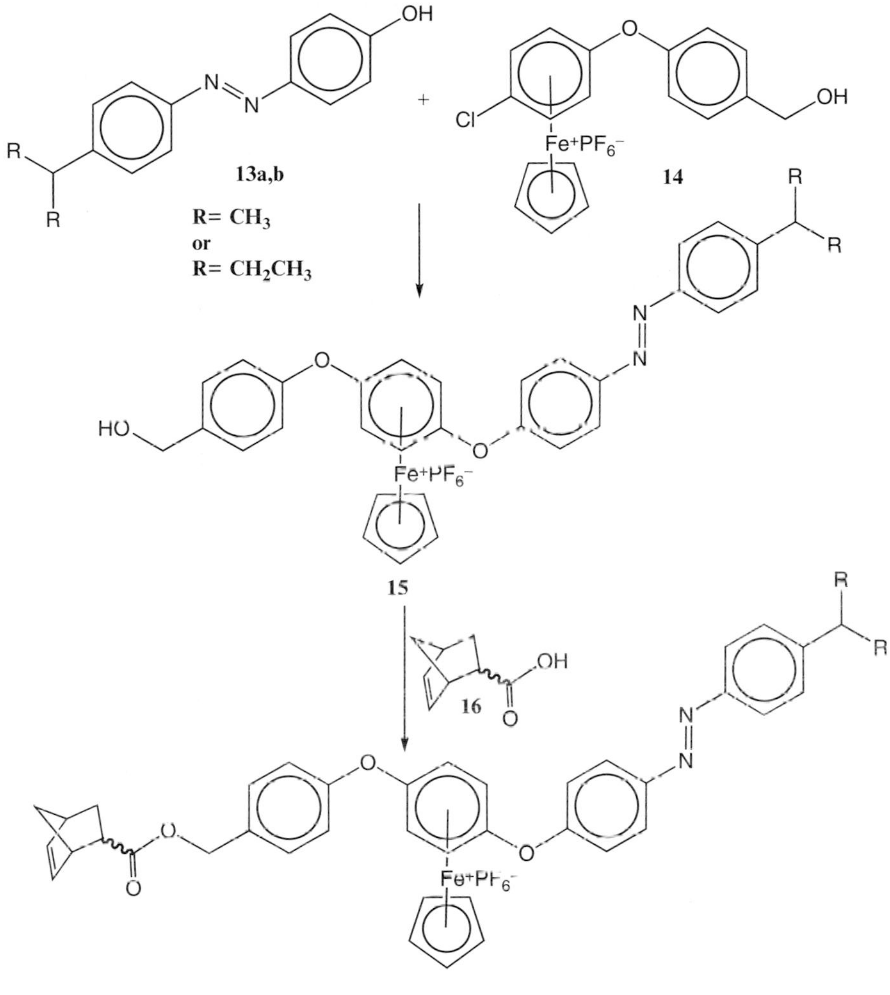

13a,b

R= CH_3
or
R= CH_2CH_3

14

15

16

17

Scheme 7.5

organic polymers ranged between 9,200 and 11,200 with PDIs between 1.12 and 1.14, the M_w of the organometallic polymers ranged between 14,300 and 16,800.

Electrochemical studies on the organometallic polymers (**25**) showed that the cationic iron centres underwent reduction between –1.2V and –1.4V (figure 7.3). UV-Vis investigations showed that the polymers displayed λ_{ma} between 420nm and 430nm in DMF solutions; however, the addition of a HCl solution caused bathochromic shifts (figure 7.3).

TGA analysis showed that polymers **25** were thermally stable with the first weight loss starting between 225 and 241 °C as a result of the decomplexation of the cyclopentadienyliron moieties as well as partial decomposition of the polymer side chains. The second weight-loss beginning between 400 and 450 °C was attributed to the degradation of the polymer backbone. DSC analysis showed that the organoiron

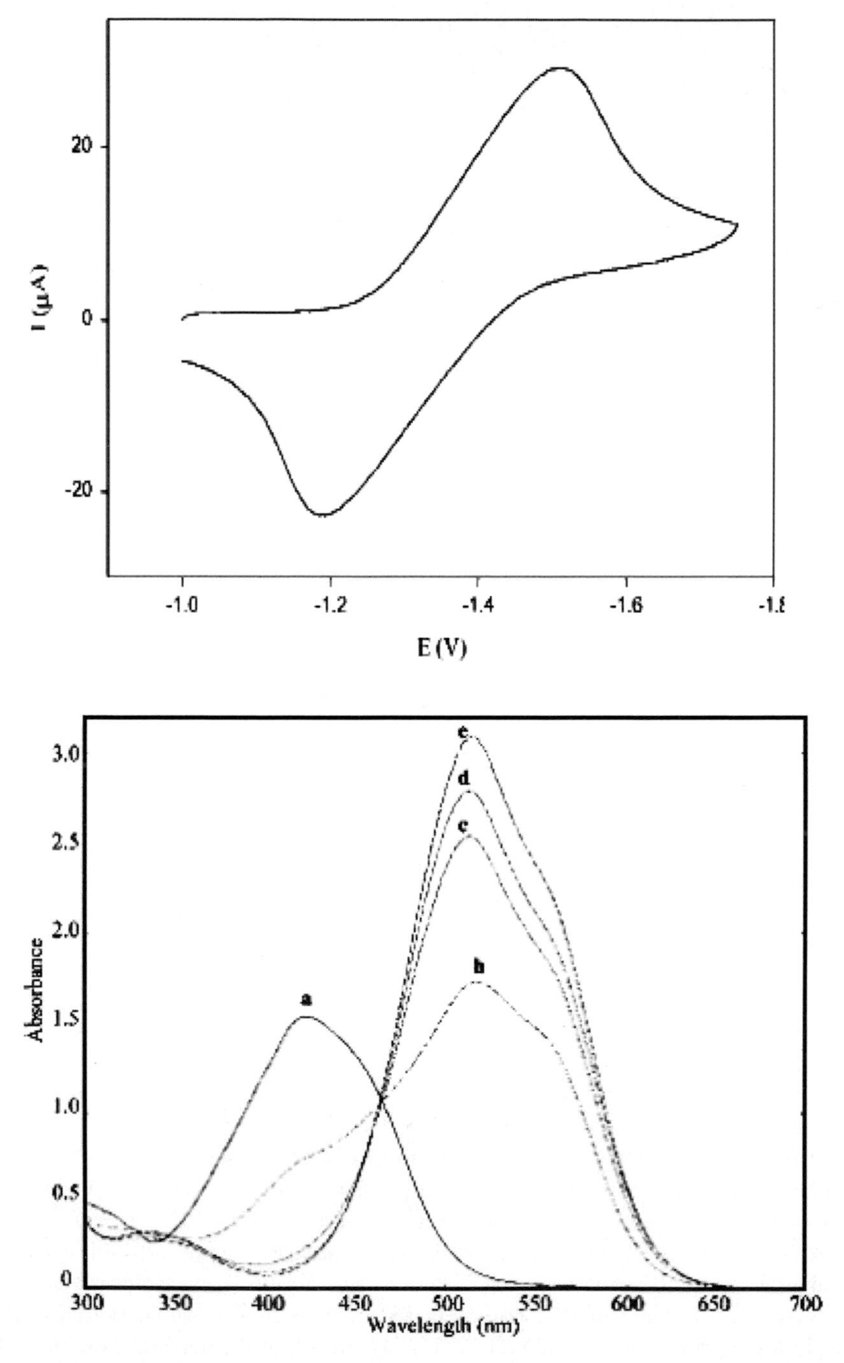

Figure 7.3 Cyclic Voltammogram of polymer **25** ($R = CH_3$) obtained in propylene carbonate at –40 °C using a scan rate of 0.2 V/s *(top)*. UV-vis spectra of polymer 25($R = CH_3$) obtained in DMF (**a**), DMF/10% HCl (**b**), DMF/30% HCl (**c**), DMF/50% HCl (**d**), and DMF/100% HCl (**e**) *(bottom)*

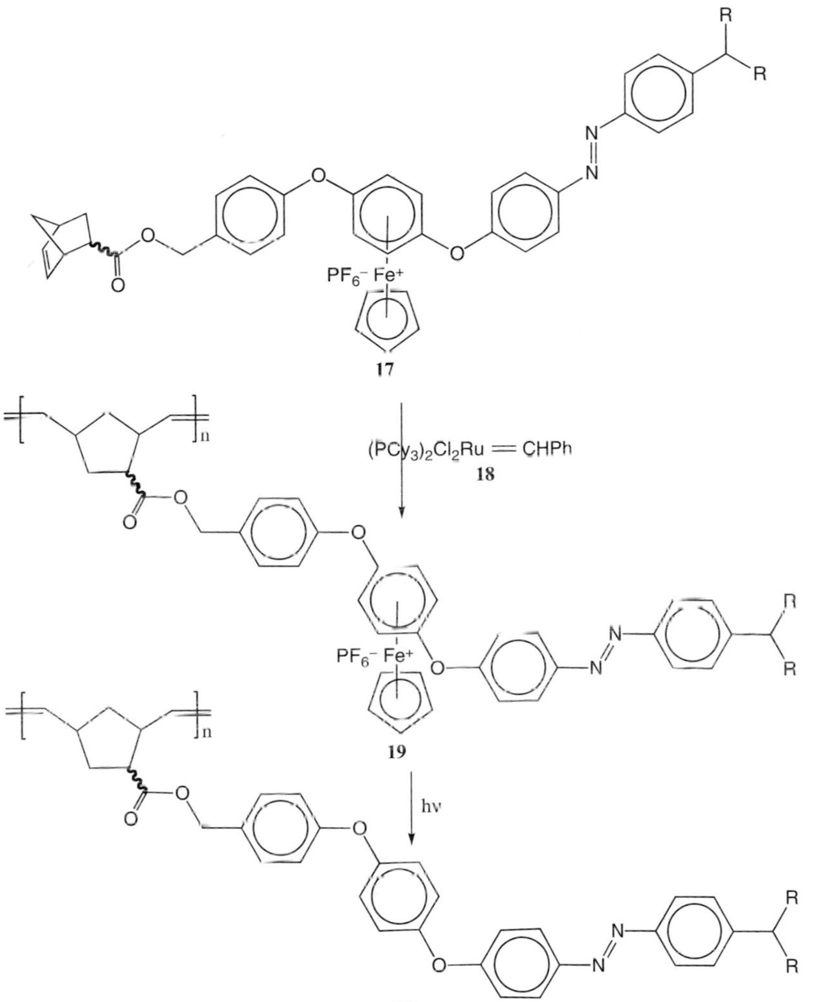

Scheme 7.6

polymers possessed T_gs higher than those of the organic polymers. Polymers **19** also possessed T_gs 30 °C higher than those for polymers **25** resulting from the differences in the lengths of the side chains and the placement of the cationic iron centers within them.

A similar polymer containing a hetaryl-azo dye was prepared by reacting the benzothiazole azo dye **28** with the organoiron complex **27** to give complex **29** (scheme 7.9). The subsequent reaction of **29** with 4-hydroxybenzyl alcohol gave benzothiazole-containing organoiron complex **31**.

Reaction of complex **31** with 5-norbornene-2-carboxylic acid (**16**) yielded the organoiron monomer **32**. ROMP of monomer **32** using Grubbs' catalyst (**18**) gave

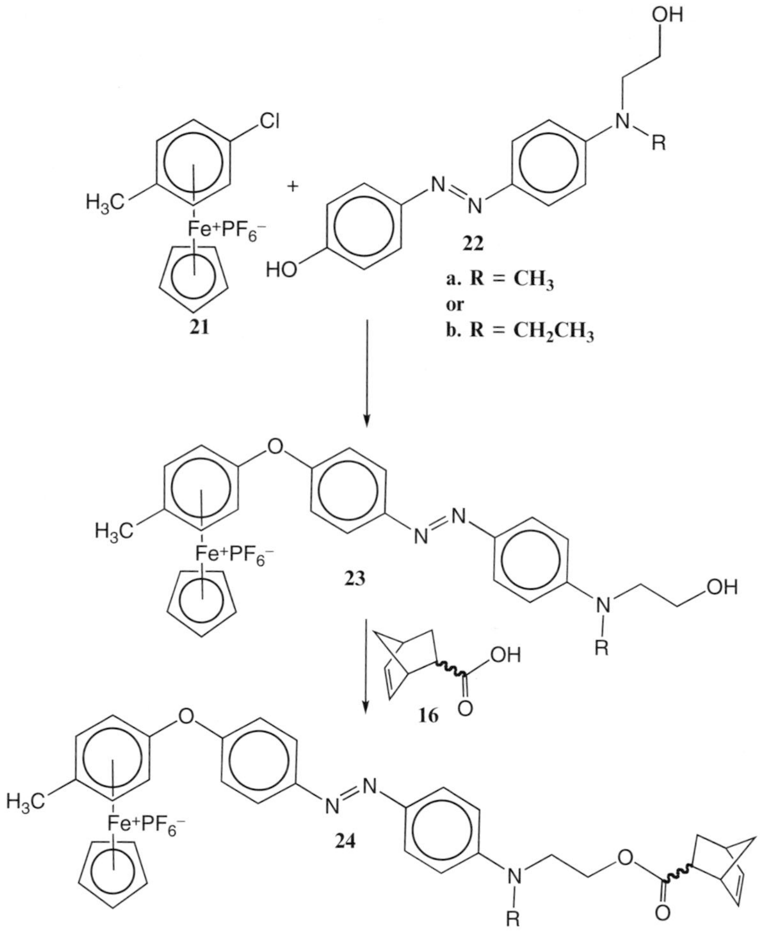

Scheme 7.7

polymer **33** in good yield (scheme 7.10). Polymer **33** displayed a λ_{max} at 511 nm which is much higher than those reported for aryl-azo dyes, the addition of HCl caused a bathochromic shift in the λ_{max} to 608 nm with a shoulder at 574 nm (Figure 7.4).

3 Synthesis of Organoiron Polymers with Azo Dyes in their Backbones

Polymers containing both netural and cationic iron moieties have been prepared from organoiron complexes containing ferrocene and azo dyes in the main chain with pendent cationic iron units. The monomers were prepared by first reacting bifunctional azo dyes **34** with the organoiron complex (**1**) (Scheme 7.11).

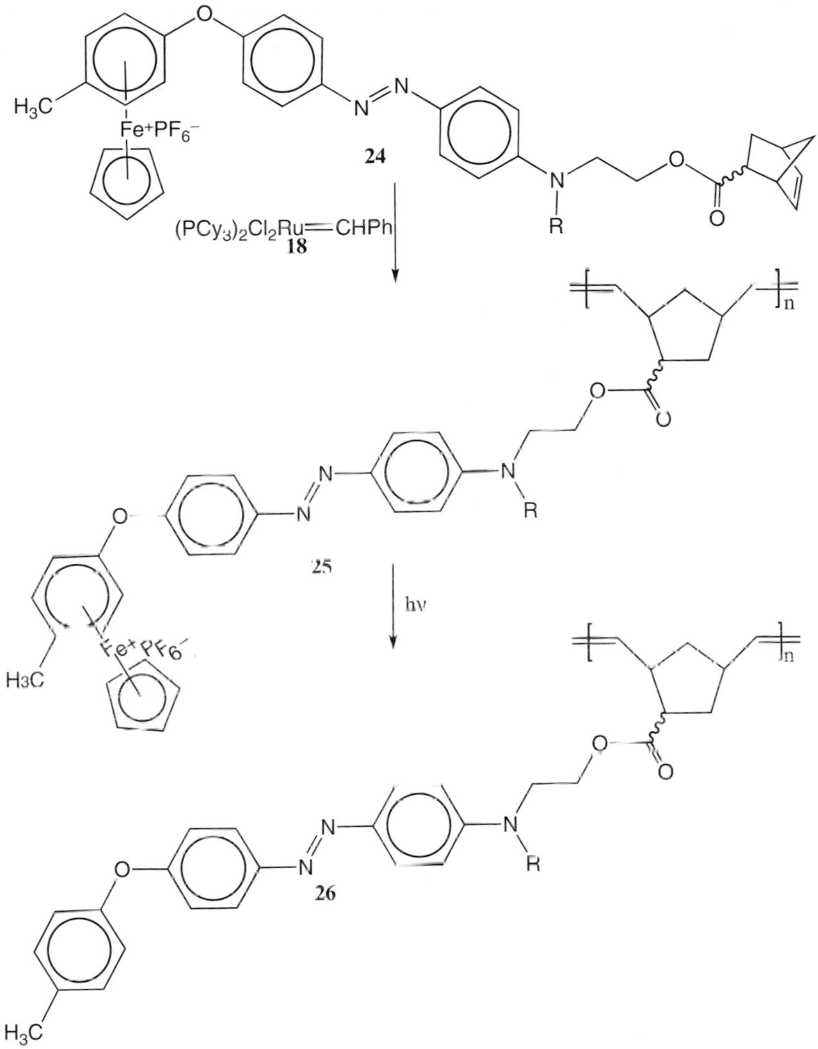

Scheme 7.8

The azo dye-functionalized organoiron complexes **35** were isolated in good yields and contained a hydroxyl group available for condenstation with 1,1′-ferrocene dicarbonyl chloride (**36**) (Scheme 7.12). The monomers **37** were isolated as vibrantly colored materials in high yields that showed good solubility in polar organic solvents.

Polymers **39** containing both neutral and cationic organoiron moieties as well as azo dyes in their backbone were isolated from the condensation polymerization of azo dye-containing organoiron monomers **37** with various *N*-, *S*-, and *O*-containing

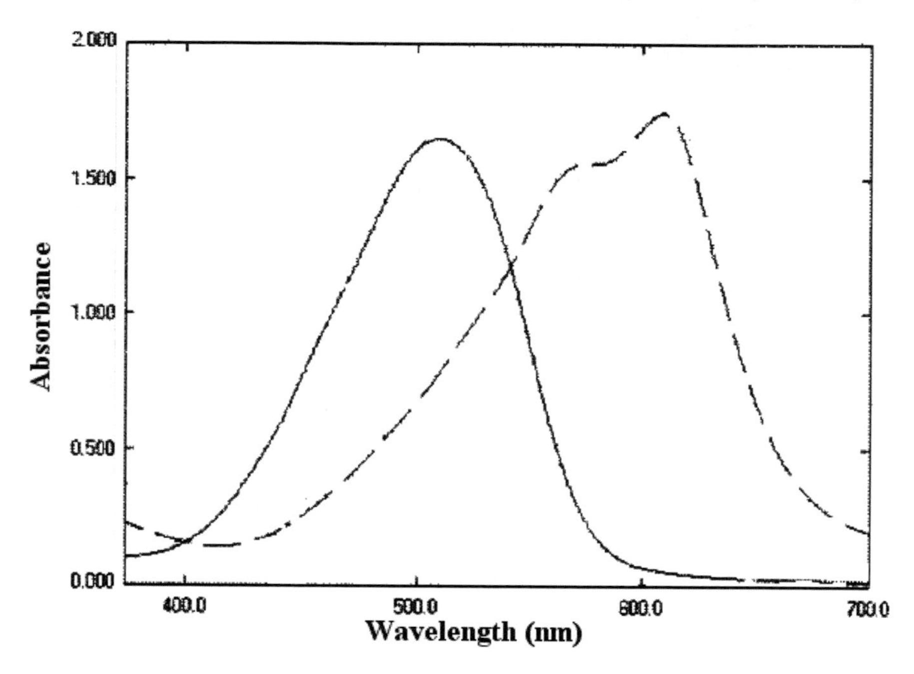

Figure 7.4 UV-vis spectrum of polymer **33**

dinucleophiles (Scheme 7.13). Polyferrocenes **40** containing azo dyes in the backbone were isolated by photolytic cleavage of the cationic cyclopentadienyliron moieties.

The M_w of polyferrocenes **40a–f** ranged from 8,000 to 21,000 with PDIs between 1.15 and 1.20 which corresponds to a M_w between 11,000 and 25,000 for polymers **39a–f**. Thermal gravimetric analysis indicated that polymers **39a–f** were thermally stable and showed two major weight-losses. The first weight loss occurred between 210 and 300 °C corresponding to the cleavage of the cationic organoiron moieties as well as, the degradation of the azo dye. The second weight loss between 400 and 500 °C was the result of decomposition of the polymer backbones. The glass transition temperatures of polymer **39a–f** were higher than those of their corresponding neutral polymers **40a–f**. The T_gs of polymers **39a–f** were between 126 and 164 °C whereas the T_gs for photolyzed polyferrocenes **40a–f** ranged between 85 and 92 °C. Electrochemical studies on polymers **39a–f** showed two distinct electrochemical processes. The reduction of the cationic organoiron centers occurred at –1.42 V and the neutral organoiron centres were oxidized at 0.89 V. UV-vis analysis showed that polyferrocenes **40a–f** displayed λ_{max} around 419 nm in a DMF solution, addition of hydrochloric acid caused a bathochromic shift to 530 nm, the spectrum of **40c** is given in figure 7.5 as an example.

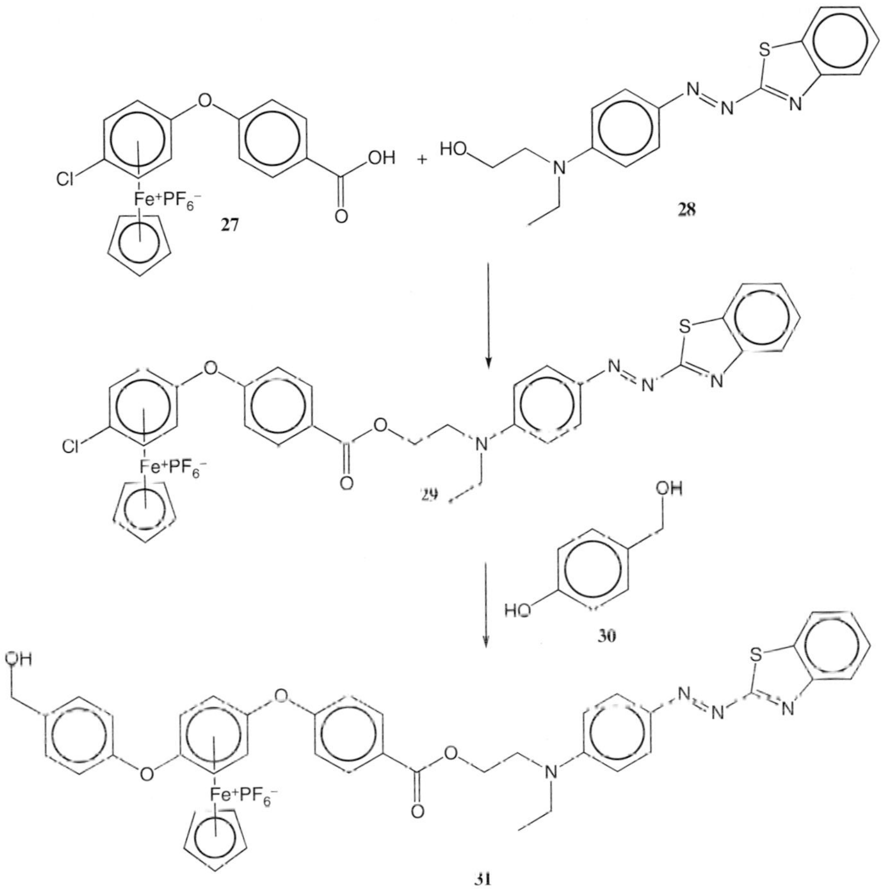

Scheme 7.9

4 Effects of Photoloysis on Organoiron Polymers Containing Azo Dyes

Azo dye-containing polymers must be able to resist decolouration by UV light to maintain their structural and mechanical integrity for any potential applications [24]. The affect of UV radiation on the photostability of polymer **7** was studied and compared to the photostability of the organic analogues **10**. Azo dyes are known to act as chelating ligands, forming complexes with various metals including iron. This metal-dye complex is formed by interaction of the metal *d*-orbitals, azo nitrogens, and if present a heteroatom in the ortho position. These complexes can form in a 1:1 Fe- and 1:2 Fe-azo dye complexes, which can facilitate azo dye degradation [25,26].

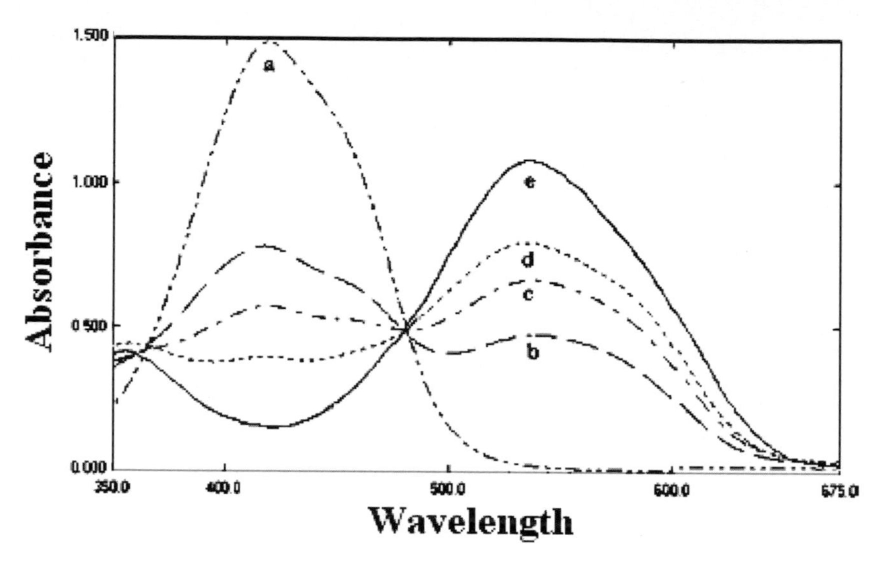

Figure 7.5 UV-vis spectra of polymer **41c** in: DMF (**a**), 1:9 DMF/HCl (20%) (**b**), 1:9 DMF/HCl (30%) (**c**), 1:9 DMF/HCl (40%) (**d**); 1:9 DMF/HCl (80%) (**e**)

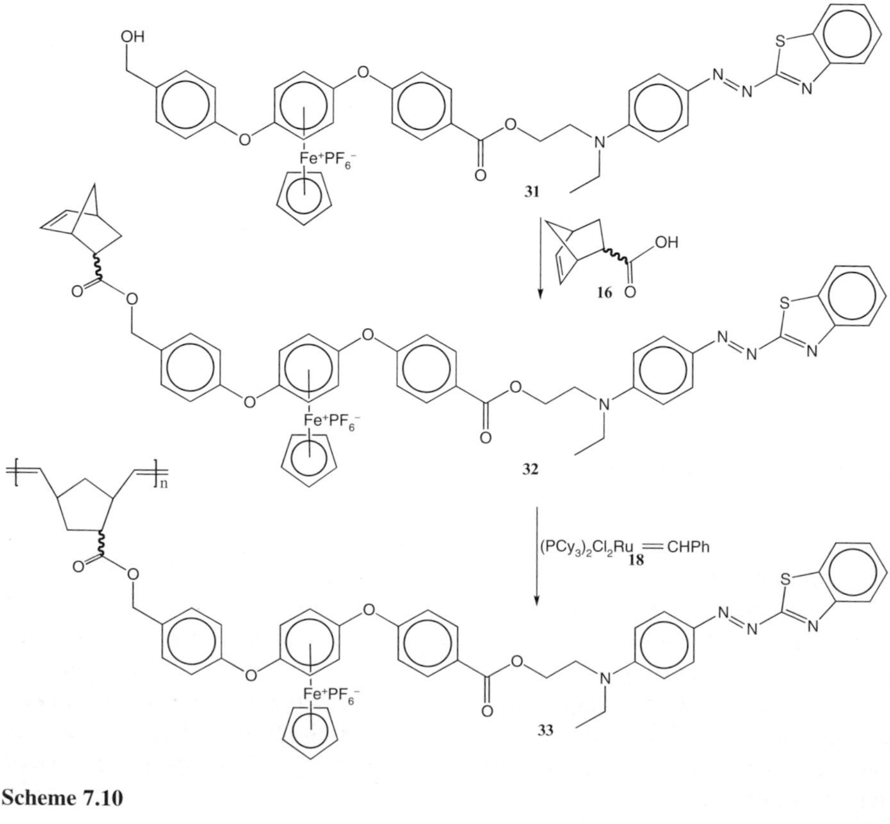

OH O O O N N=N S N $Fe^{+}PF_6^{-}$ **31**

16 OH O

O O O O O N N=N S N $Fe^{+}PF_6^{-}$ **32**

$(PCy_3)_2Cl_2Ru{=}CHPh$ **18**

n O O O O O N N=N S N $Fe^{+}PF_6^{-}$ **33**

Scheme 7.10

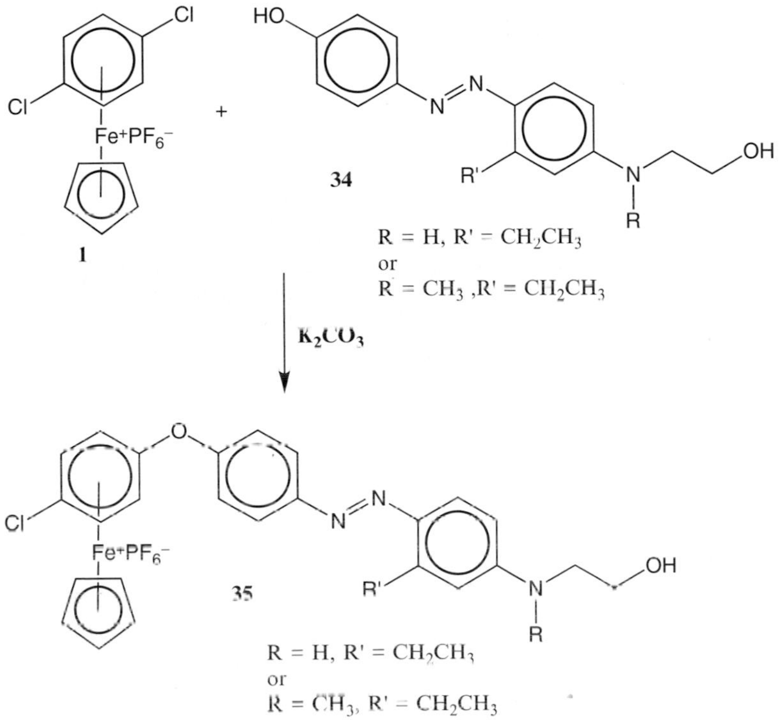

1 + **34**

R = H, R' = CH_2CH_3
or
R = CH_3, R' = CH_2CH_3

K_2CO_3

35

R = H, R' = CH_2CH_3
or
R = CH_3, R' = CH_2CH_3

Scheme 7.11

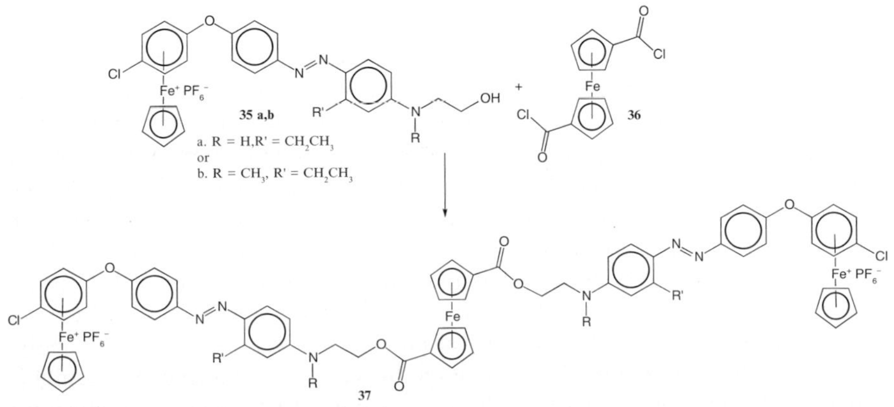

35 a,b + **36**

a. R = H, R' = CH_2CH_3
or
b. R = CH_3, R' = CH_2CH_3

37

Scheme 7.12

37

$H_2N-(CH_2)_x-NH_2$

38a-c
a:x = 3
b:x = 10
c:x = 12

XH—Spacer—HX

6a-c, 38a-c

39

hν

40

X	Spacer
S	$(CH_2)_8$
S	p-(C_6H_4)-S-p-(C_6H_4)
O	p-(C_6H_4)-$C(CH_3)_2$-p-(C6H$_4$)
N	$(CH_2)_3$
N	$(CH_2)_{10}$
N	$(CH_2)_{12}$

Scheme 7.13

The affect of the iron moiety on the photodegradation of the azo dye containing polymers **7** and **10** was studied. In this study [27] polymers **7** and **10** were dissolved in acetonitrile and exposed to UV radition and observed over time to determine the photo-discolorization. Figure 7.6 displays the discoloration of polymer **7**.

Polymer **7** showed a slow decrease in the absorbance between 400 to 550 nm, and complete discoloration after 2 hours at low concentrations. As shown in Table 7.1, the photo-discoloration follows pseudo first order kinetic rate constants for direct photolysis of the organometallic polymers at intial concentrations ranging from 0.01 to 0.2 mM. When the concentration was incresed 20-fold, a 16-fold decrease was observed in the rate constant under constant UV irradation. However, the darker the initial color of the original solution, the slower the discoloration occurred. The decrease in the discoloration rate of the darker solutions was attributed to increased light attenuation. As the solution becomes more instensely colored, the penetration and availability of photons decreases and causes the polymers to absorb less light [28]. Because of the increase in light attenuation at higher polymer concentrations the energy absorbed is only great enough to remove the iron moiety from the polymer, giving the organic analog with a negligible effect on

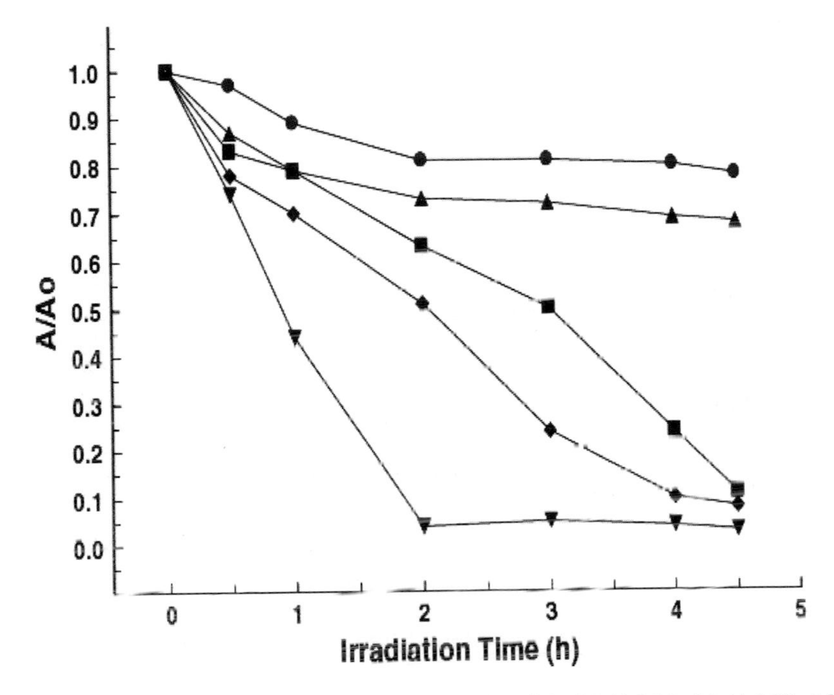

Figure 7.6 Direct photolysis in acetonitrile solution of: ( ) 0.2 mM, ( ) 0.15 mM, ( ) 0.05 mM, ( ) 0.03 mM, and ( ) 0.01 mM of metallated polymer

Table 7.1 Rate constants for the discoloration of metallated polymer solutions for direct photolysis

Concentration (mM)	K(h^{-1})
0.20	0.053
0.15	0.071
0.05	0.457
0.03	0.590
0.01	0.855

the color. The corelation between the discoloration rates and the initial polymer concentration is shown in Figure 7.7.

Figure 7.8 shows the comparison of direct photolysis of the various forms of the polymer in acetonitrile solutions over 9 hours of irradiation. The organometallic polymer shows only 30% discoloration during this time period and its organic analogue shows a very similar behavior. However, a polymer solution containing the organic analog with the decoordinated cyclopentadienyliron moiety still in solution displayed total discolouration in the 9-hour time frame. It is well documented that the photolytic cleavage of the cationic cyclopentadienyliron group in acetonitrile leads to the formation of $[(\eta\text{-}C_5H_5)Fe(CH_3CN)_3]^+$, which quickly decomposes into

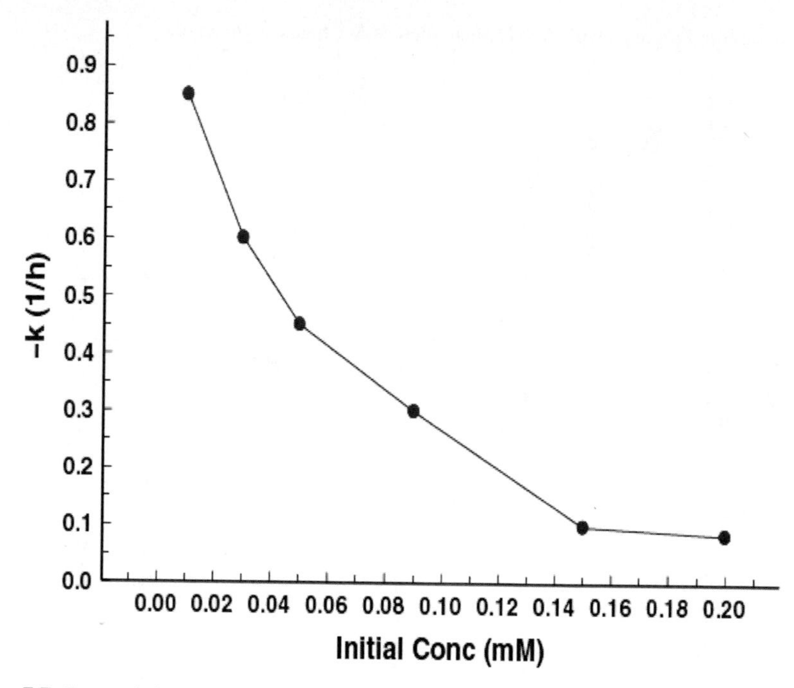

Figure 7.7 Rate relationship for direct photolysis of metallated polymer in acetonitrile

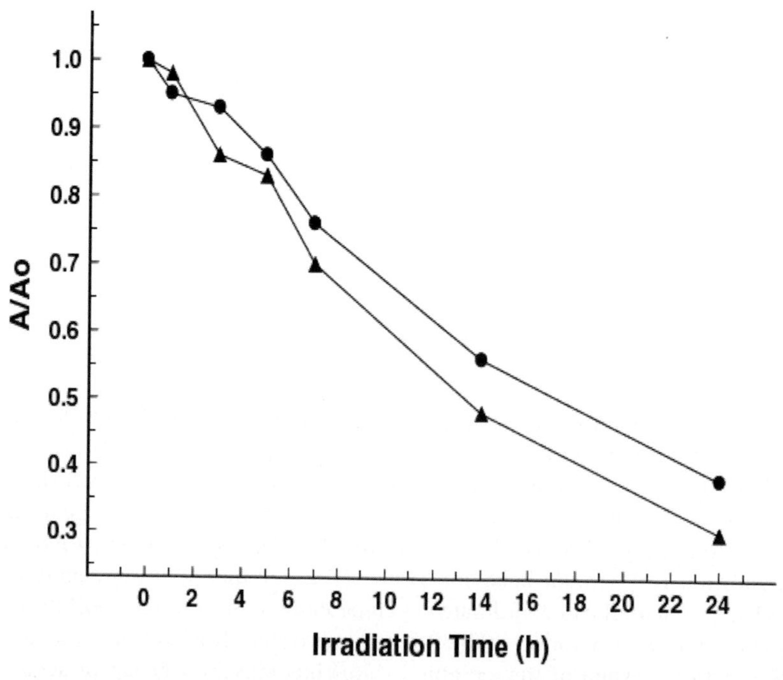

Figure 7.8 Photolysis of 0.2 mM demetallated polymer in acetonitrile with addition of: ( ) 0.1 mM Fe(III) and ( ) 0.1 mM Fe(II) perchlorate

ferrocene and iron (II) salts as the final demetallation products [28]. Therefore the rapid discoloration of the polymer solution containing the decoordiated metal is likely mediated by the presence of the ferrocene or the iron (II) salts. To verify which of the demetallation products resulted in the discolouration of the polymer solutions, ferrocene and Fe(II) salts were added to solutions of the organic analogues. The addition of ferrocene to the organic polymer solution showed no effect on the discoloration, indicating that the increased discolouration was due to the presence of the charged iron species.

Because Fe(II) salts are generated by the decoordination of the cyclopentadienyliron moiety, they are the most likely candidates responsible for the discoloration. However, because Fe(III) plays a role in the Photo-Fenton reaction, both Fe(II) and Fe(III) salts were added to the organic analogue polymer solutions. Solutions containing 0.1 mM iron (II) and (III) perchlorate gave similar rates in discoloration (Figure 7.8), indicating that both Fe(II) and Fe(III) take part in the discoloration.

Interestingly, at higher concentrations of Fe(III) perchlorate (≥1 mM), the rate of the discoloration was not affected. However the rates improved when low concentations of salts were used. Complete discoloration of the polymer solutions were seen after 3 hours with solutions containing 0.02 mM Fe(III). This behaviour suggests that a complex forms between the azo chormophore and the iron species. This complex is excited through UV radiation and causes an electron transfer from the azo group to the iron species [29,30]. However, as the iron concentration is increased, the complexes form agglomerates with arrangements that prevent the iron-azo dye interactions [29].

5 Summary

This chapter provides an overview of polymeric materials that contain azo dyes and cationic and/or neutral iron moieties. The azo groups are attached to the backbone or the side chains of the polymers. Changes in the azo dyes described in this article gave rise to materials that absorb in the 400 to 600 nm regin. Also, the protonation of the azo groups gave rise to a bathochromic shift of 50 to 120 nm. TGA analysis showed that these metal-containing macromelcules are thermally stable up to 200 °C. A number of organoiron polymers have been prepared that contained azo dyes either in the side chains or in the main chains. These polymers were found to be thermally stable and to possess unique UV-vis properties and electrochemical behavior.

References

1. Arimoto FS, Haven AC Jr. (1955) J Am Chem Soc. 77:6259.
2. Abd-El-Aziz AS, Manners I (eds). (2007) Frontiers in transition metal-containing polymers. John Wiley & Sons, Hoboken, New Jersey.
3. Manners I. (2004) Synthetic metal containing polymers. Wiley VCH, Weinheim, Germany.
4. Abd-El-Aziz AS, Carraher Jr CE, Pittman Jr CU. (2003) In: Sheats JE, Zeldin M (eds). Macromolecules containing metal and metal-like elements, Vol. 1. Wiley & Sons Inc, New Jersey.

5. Abd-El-Aziz AS. (2002) Macromol Rapid Commun. 23:995.
6. Archer RD. (2001) In: Inorganic and organometallic polymers. Wiley-VCH, New York.
7. Nguyen P, Gomez-Elipe P, Manners I. (1999) Chem Rev. 99:1515.
8. Abd-El-Aziz AS, Todd EK. (2003) Coord Chem Rev. 3:246.
9. Manners I. (2002) J Polym Sci Part A Polym Chem. 40:179.
10. Nguyen P, Gomez-Elipe P, Manners I. (1999) Chem Rev 99:1515.
11. Hudson RDA. (2001) J Organomet Chem. 47:637.
12. Natansohn A, Rochon P. (2002) Chem Rev. 102:4139.
13. Hasegawa M, Ikawa T, Tsuchimori M, Watanabe O. (2002) J Appl Polym Sci. 86:17.
14. Huang K, Qiu H, Wan M. (2002) Macromolecules. 35:8653.
15. Uznanski P, Pecherz J. (2002) J Appl Polym Sci. 86:1456.
16. Samyn C, Verbiest T, Persoons A. (2000) Macromol Rapid Commun. 21:1.
17. Iftime G, Labarthet FL, Natansohn A, Rochon P, Murti K. (2002) Chem Mater. 14:168.
18. Tawa K, Kamada K, Kiyohara K, et al. (2001) Macromolecules. 34:8005.
19. Liu X.-H, Bruce DW, Manners I. (1997) Chem Commun. 289.
20. Abd-El-Aziz AS, Okasha RM, Shipman PO, Afifi TH. (2004) Macromol Rapid Commun. 25:1497.
21. Abd-El-Aziz AS, Okasha RM, Afifi TH. (2004) J Inorg Organomet Polym. 14:269.
22. Abd-El-Aziz AS, Pereira NM, Boraie W, et al. (2006) Inorg Organomet Polym. 15:497.
23. Konstantinova T, Lazarova R, Bojinov V. (2003) Polym Degrad Stabil. 82:115.
24. Sokolowska-Gajda J. (1998) Dyes Pigments 36:149.
25. Rauf MA, Akhter Z, Kanwal S. (2004) Dyes Pigments 63:213.
26. Abd-El-Aziz AS, Elmayergi B, Asher B, Afifi TH, Friesen KJ. (2006) Inorganica Chimica Acta 359:3007.
27. da Silva CG, Faria JL. (2003) J Photochem Photobiol A. 155:133.
28. Gill TP, Mann KR. (1983) Inorg Chem. 22:1986.
29. Neamtu M, Yediler A, Siminiceanu I, Kettrup A. (2003) J Photochem Photobiol A. 161:87.
30. Park H, Choi W (2003) J Photochem. Photobiol A. 159:241.

Cisplatin Derivatives as Antiviral Agents

Michael R. Roner and Charles E. Carraher, Jr.

1 Introduction

The spotlight is again on the urgent need for the development of antivirals to treat both old and emerging virus outbreaks. The recent attempt to reinitiate the smallpox vaccination program to deal with biological terrorism has been halted, in part resulting from observation of adverse cardiac events following vaccination, and has resulted in the Centers for Disease Control recommending that people at risk for heart disease should be excluded from the vaccination program. Although the number and severity of adverse reactions associated with most if not all vaccines does not appear to be increasing, the percentage of the human population susceptible to these problems does appear to be increasing. Increased longevity and both infectious and environmental factors are producing a population increasingly harboring immunocompromised individuals for whom vaccinations are unworkable.

A.S. Abd-El-Aziz et al. (eds.), *Inorganic and Organometallic Macromolecules: Design and Applications.*

Diseases caused by old foes such as influenza always maintain the potential to initiate new pandemics of disease. Antivirals, unlike vaccines have the potential to act against new variants of old viruses and new viruses such as SARS [1,2], and the bird flu, and as interventions in biological terrorism. Additionally, a new climate appears to be emerging in which the acceptance of the risks inherent to vaccines is very low, the most recently casualty of this being the rotavirus vaccine [3].

Antivirals, however, all suffer from the problem of target specificity [4,5. Viruses, for the most part, utilize cellular machinery to replicate the viral genome and produce new virus particles. In an attempt to target viral replication, the cellular processes in uninfected cells are also undesirably affected. Polymeric drugs offer the opportunity to avoid some of these effects. This is reviewed in recent articles and briefly described following [6,7].

We have been surveying a number of metal-containing polymers as potential antiviral agents emphasizing both platinum and organotin-containing polymers [8–11]. Polymeric drugs offer many potential advantages over monomeric or small molecule drugs. These reasons have been described elsewhere and will be briefly described later in this chapter.

Researchers have suggested that at least some cancers have a viral relationship. Thus, we have begun testing polymers that show good anticancer activity against a variety of viruses.

Recently, we first tested a number of organotin products derived from known antibacterial drugs, namely ciprofloxacin, ampicillin (Scheme 8.1), and norfloxacin (Scheme 8.2). All of these products exhibited some antiviral activity against a number of viruses, that is reovirus ST3, vaccinia virus, herpes simplex virus (HSV-1), and varicella zoster virus (VZV) [8,9]. The organotin polymers from norfloxacin and ampicillin showed total inhibition to virus growth at concentrations of about 2 mg/ml, whereas norfloxacin and ampicillin, themselves, exhibited no viral inhibition. We also looked at organotin products derived from the known antiviral drug acyclovir [9]. Again, the organotin polymers showed

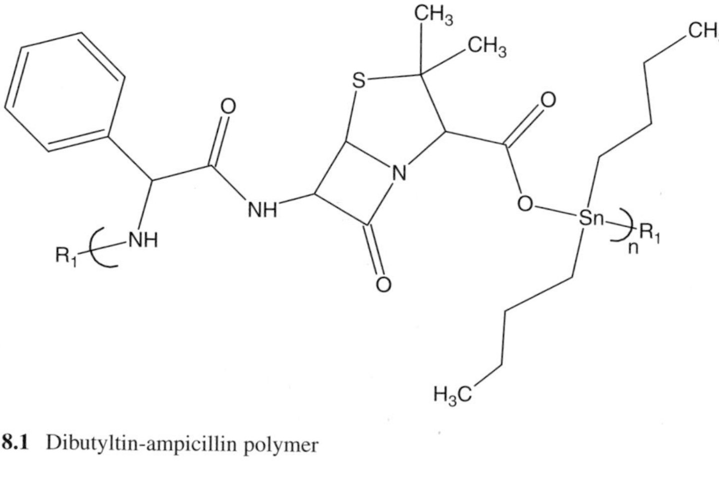

Scheme 8.1 Dibutyltin-ampicillin polymer

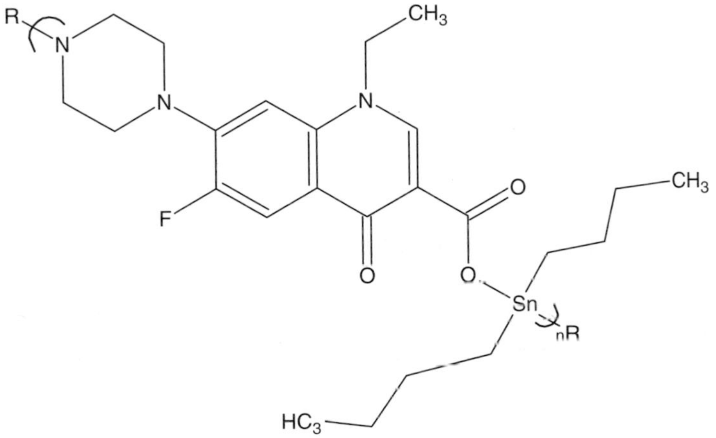

Scheme 8.2 Dibutyltin-norfloxacin polymer

inhibition of the viruses at lower concentration that found for acyclovir itself. Thus, the combination of known drugs with organotin moieties within polymers appear to be effective antiviral agents, more effective than either of the reactants themselves.

Our interest in these organotin-containing polymers is also derived from the fact that they are potent anticancer agents capable of inhibiting cancer cell growth at concentrations within the same range and less than that of cisplatin itself [12–19]. Further, these organotin polymers are much less toxic than cisplatin, the most widely used anticancer drug.

We have been investigating a number of polymeric derivatives of cisplatin as anticancer drugs for about 30 years [20]. Again, we have synthesized polymeric drugs that inhibit cancer cell growth within the same concentration range as cisplatin itself and these drugs, are again, much less toxic. We have also been investigating some of these polymeric cisplatin derivatives as antiviral agents. The use of polymeric derivatives of cisplatin as antiviral agents is the focus of this chapter.

2 Inhibition

The majority of known viruses are RNA viruses. It is not surprising that they also cause the majority of human diseases. The following are some of the more familiar diseases caused by RNA viruses:

- common cold
- poliomyelitis
- hepatitis
- encephalitis
- yellow fever

- rubella
- influenza
- measles
- mumps
- various hemorrhagic fevers

2.1 Features of an Ideal Antiviral Drug

Features of an ideal antiviral drug might include the following: effective inhibition of an essential viral process, mechanism to prevent drug-resistant viruses from developing, broad-spectrum activity against RNA and DNA viruses, and no negative effect on host cell processes.

2.2 Strategies for Antiviral Therapy

Most antivirals target one of five major viral processes: (1) attachment of the virus to the host cell, (2) penetration and/or uncoating of the virus to release the viral nucleic acid into the host cell, (3) replication of the viral genome, (4) viral gene expression to produce viral proteins, and (5) assembly and maturation of the virus structure and the release of the progeny virions with or without lysis of the host cell.

2.3 Attachment

Virus attachment can be inhibited by two broad methods which are described below.

1. Agents that mimic the viral attachment protein (VAP) can be introduced into the infected host. These VAPs will then bind to the cellular receptor and block the virus binding [21]. Anti-idiotypic antibodies can be produced that mimic the VAP. When these antibodies are introduced into the host they bind the cellular receptors that would normally be available to infectious virus. This effectively "blocks" the viral receptor on the cells and prevents the virus from attaching and infecting the cell [22]. Natural ligands of the viral receptor can be employed that bind the receptor and block virus use of this receptor. One example of this is vaccinia virus and the epidermal growth factor (EGF) receptor [23]. The fourth possibility is the use of synthetic ligands that resemble the receptor-binding domain of the VAP itself. These peptides will bind a cell receptor and render the receptor unable to bind the VAP, preventing virus infection [24,25].
2. Agents that mimic the viral receptor on the host cell and function by binding the VAP. Antibodies to the VAP, naturally produced as the humoral response to most

viruses, bind the VAP and prevent its interaction with host cell receptors [26,27]. Anti-idiotypic antibodies can be produced that mimic the cell receptor or extraneous receptors such as rsCD4 used by HIV [28]. When introduced into the host these antibodies serve as "binding targets" for the virus, but unlike the cells normally expressing these receptors, cannot be infected or support virus replication [28,29]. Additionally, synthetic receptor mimics can be produced to bind virus before it has an opportunity to bind cell receptors. One example of this approach is the use of sialic acid derivatives to bind influenza virus [30,31].

2.4 Penetration and Uncoating

These processes have proven difficult to examine at the molecular level for most viruses and therefore it has been difficult to specifically target these stages of the virus life cycle. Uncoating is for the most part mediated by cellular enzymes but like penetration, is often influenced by one or more virus proteins [32,33].

Pleconaril is a broad spectrum anti-picorna virus agent [34]. It is a small cyclic drug which binds to a canyon pore of the virus. In doing so it blocks the attachment and uncoating of the viral particle [35,36].

Amantadine (Scheme 8.3), and rimantadine (Scheme 8.4) are active against influenza A viruses. The action of these closely related agents is complex and incompletely understood, but they are believed to block cellular membrane ion channels [37,38]. Both drugs target the influenza A matrix protein (M2). Drug-treated cells are unable to lower the pH of the endosomal compartment (a function normally controlled by the M2 gene product), a process which is essential to induce conformational changes in the HA protein to permit membrane fusion.

2.5 Genome Replication

Many viruses have evolved their own specific enzymatic mechanisms to divert cellular energy to the replication of viral genomes. Very often there are sufficient differences between viral and cellular polymerases to provide a target for an antiviral agent without harming the unaffected agents. This strategy has yielded the majority of the antiviral drugs currently in use. Most of these drugs function as polymerase substrate, as nucleoside/nucleotide analogues. The toxicity of these drugs varies considerably from some which are well tolerated such as acyclovir [39] to others which are highly toxic such as IdU/TFT [40] and AZT [41,42]. There is a serious problem with the pharmacokinetics of these nucleoside analogs, as most typically have short serum half lives of 1 to 4 hours [43]. Nucleoside analogues are in fact pro-drugs, meaning that to be activated they need to be phosphorylated. Acyclovir [44] (Scheme 8.5), is phosphorylated by the herpes simplex virus (HSV) thymidine kinase 200 times more efficiently than by cellular enzymes. Gancyclovir [45] (Scheme 8.5) is 10 times more

3. Amantadine

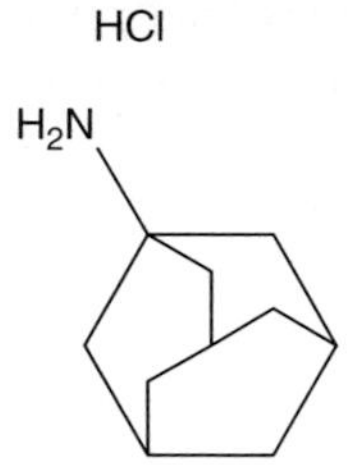

HCl

H_2N

Scheme 8.3 Amantadine

4 Rimantadine

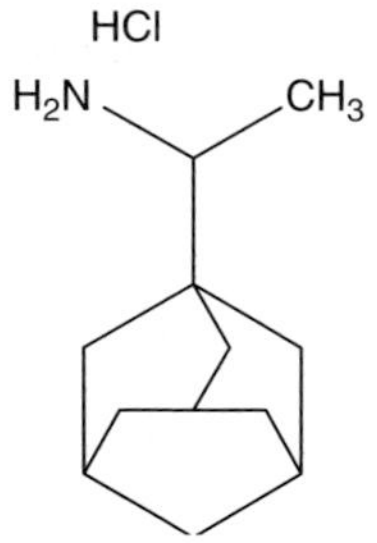

HCl

H_2N CH_3

Scheme 8.4 Rimantadine

5 Acyclovir

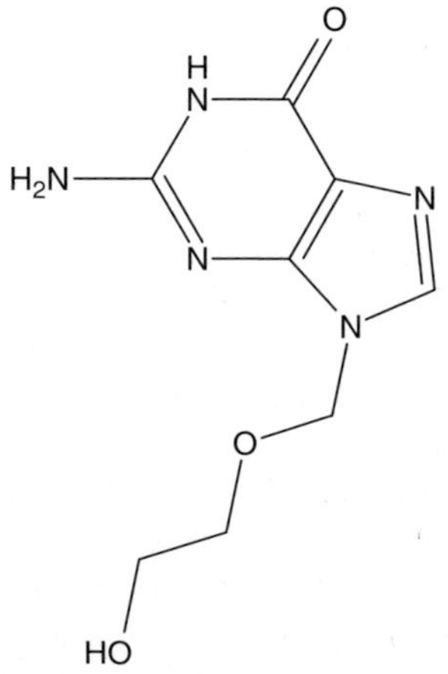

O H N N H_2N N N O HO

Scheme 8.5 Acyclovir

effective against cytomegalovirus (CMV) than acyclovir since it is specifically phosphorylated by a CMV-encoded kinase not present in HSV.

Other nucleoside analogs active against herpesviruses have been derived from Acyclovir and Gancyclovir and include **Schemes 8.6–8.14**. Additional nucleoside analogues with activity against HIV are shown in Schemes 8.15–8.17.

2.6 *Gene Expression*

Most viruses are heavily dependent upon the cellular machinery for transcription of viral genomes, mRNA splicing, translation, and protein transport. Unlike genome replication, uniquely viral proteins are not involved in these processes and to date none have been utilized as targets for antiviral therapy.

2.7 *Assembly, Maturation, and Release of Progeny Virus*

For the majority of viruses, assembly, maturation, and release of progeny virus processes are poorly understood. Two drugs with anti-influenza activity are available. These are Relenza [69,70] taken as an aerosol and Tamiflu [71,72] taken as a pill. Tamiflu is active against both influenza A and B strains. Both of these drugs function as neuraminidase inhibitors and prevent the release of budded viruses from the cell.

6 Gancyclovir

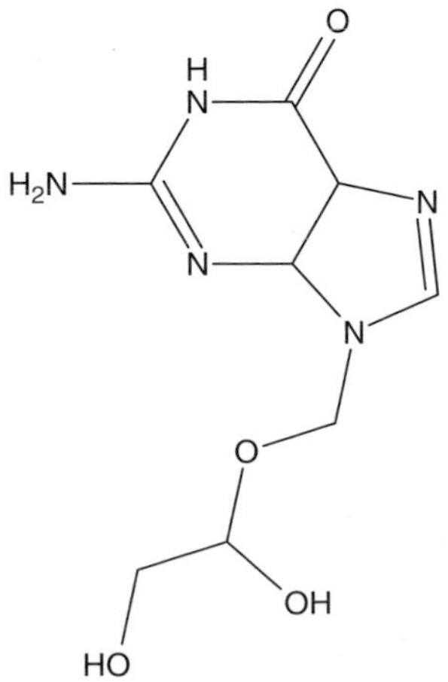

H_2N N H N N N O O OH HO

Scheme 8.6 Gancyclovir

7 Penciclovir[46-49]

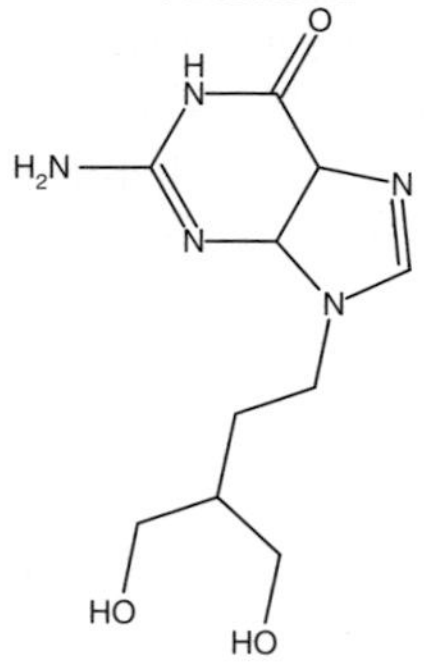

Scheme 8.7 Penciclovir [46–49]

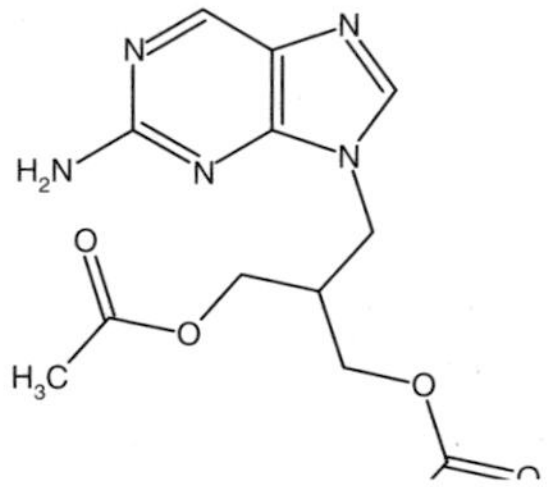

Scheme 8.8 Famciclovir [50,51]

9 BVDU[52-54]

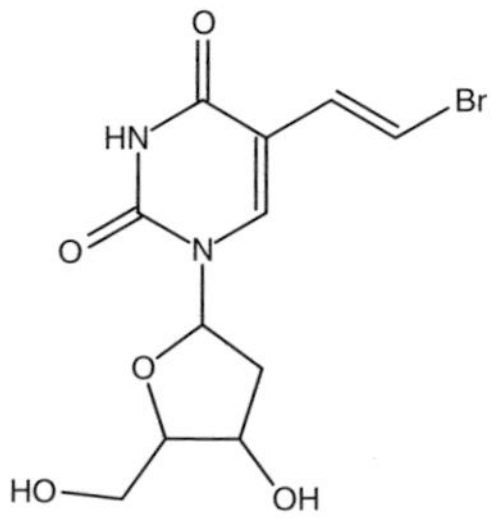

Scheme 8.9 BVDU [52–54]

2.8 Additional Antiviral Drugs

Foscarnet [73,74](Scheme 8.18) is a first line treatment for CMV retinitis and a treatment for CMV colitis if ganciclovir therapy is ineffective or not tolerated. Foscarnet crosses the blood–brain barrier and can treat susceptible infection in the

10 Brovavir[55]

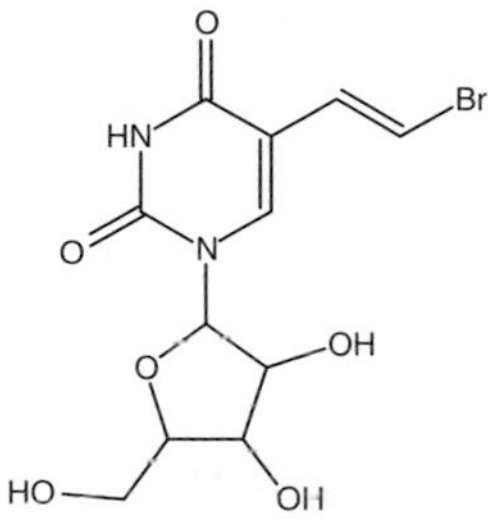

Scheme 8.10 Brovavir [55]

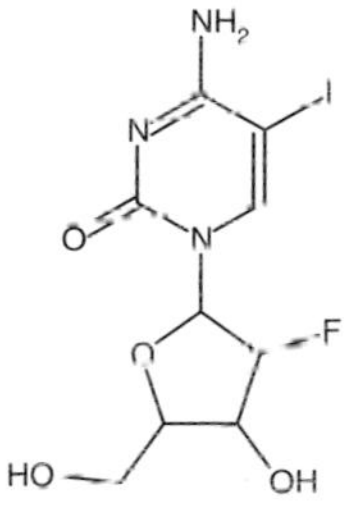

11.FIAC[56]

Scheme 8.11 FIAC [56]

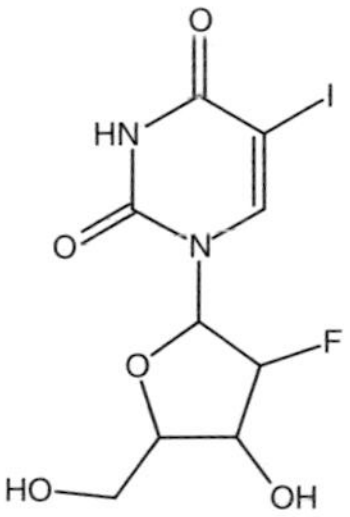

12 FIAU[57,58]

Scheme 8.12 FIAU [57,58]

brain. Strains of herpes that are resistant to treatment with acyclovir can be treated with foscarnet. Idoxuridine (IdU) [52] (Scheme 8.19) acts by irreversibly replacing thymidine in newly synthesized DNA and producing an abnormal and essentially nonfunctional DNA molecule. The drug acts on viral and host cell DNA and is highly toxic to host cells. Because of its high systemic toxicity, IdU has been

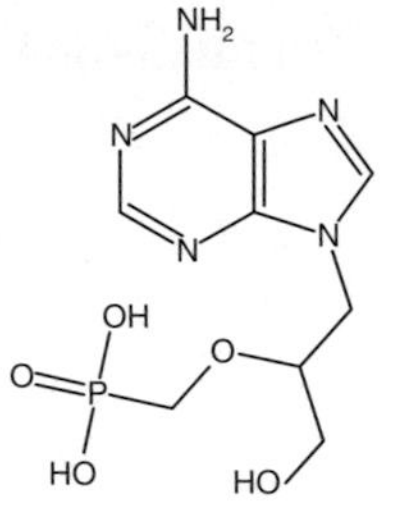

13 (S)-HPMPA [59]

Scheme 8.13 (*S*)-HPMPA [59]

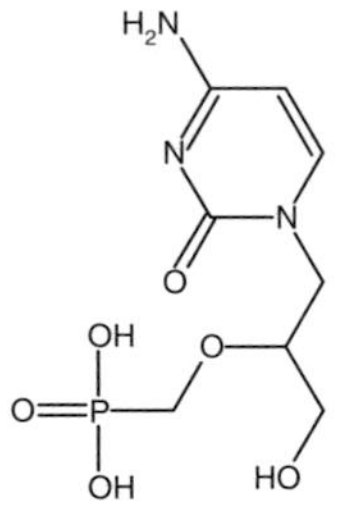

14 (S)-HPMPC[60-63]

Scheme 8.14 (*S*)-HPMPC [60–63]

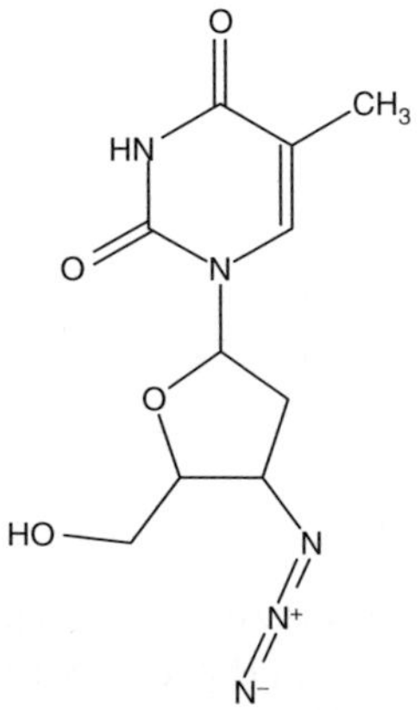

15 AZT[64,65]

Scheme 8.15 AZT [64,65]

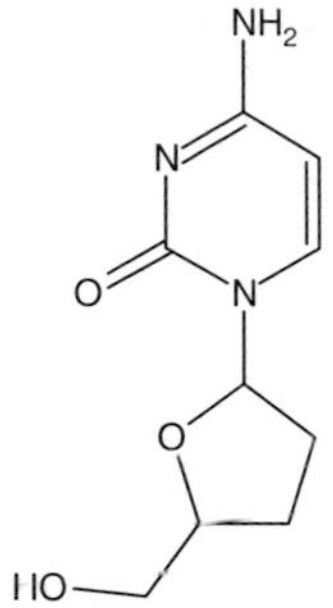

16 ddC[66,67]

Scheme 8.16 ddC [66,67]

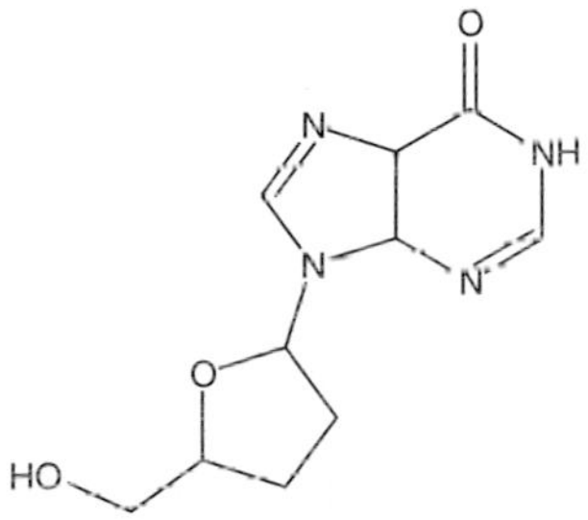

17 ddI[68]

Scheme 8.17 ddI [68]

18 Foscarnet

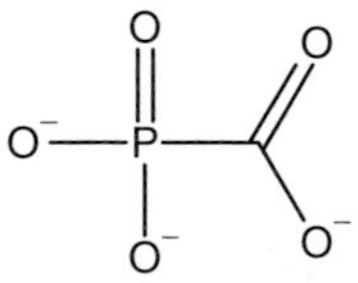

Scheme 8.18 Foscarnet

limited to topical therapy of herpes simplex keratoconjunctivitis. Ribavirin is a guanosine analog that inhibits the replication of many RNA and DNA viruses. Ribavirin (Scheme 8.20) is thought to inhibit messenger RNA formation. Ribavirin exhibits inhibitory activity in vitro against respiratory syncytial virus (RSV), influenza A and B, HSV-1, HSV-2, and many other viruses. Vidarabine (adenine arabinoside, ara-A) (Scheme 8.21) interferes with viral DNA synthesis and is effective in the treatment of HSV infections.

19 IdU

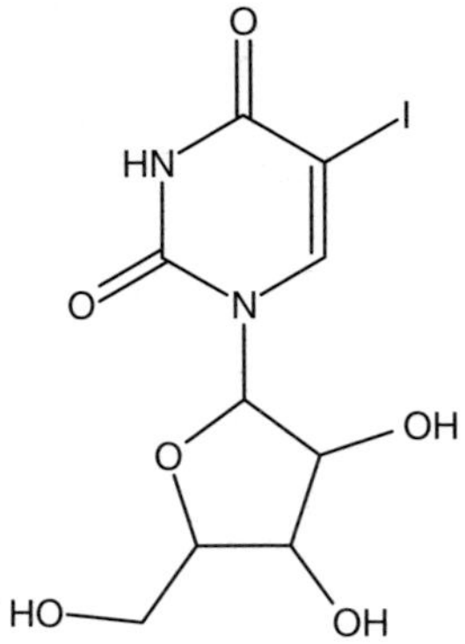

Scheme 8.19 IdU

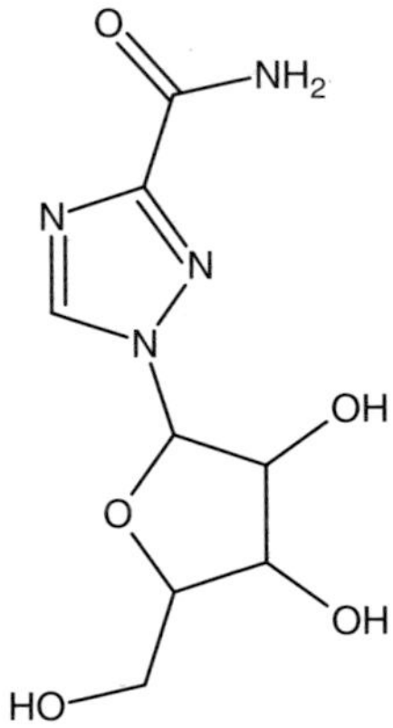

20 Ribavirin[75-78]

Scheme 8.20 Ribavirin [75–78]

3 Currently Approved Platinum-Containing Drugs

Although thousands of cisplatin analogues have been synthesized and screened only about 28 platinum compounds have entered clinical trials as anticancer agents [6,9]. Of these only four are currently approved. Those approved are cisplatin (Scheme 8.22), carboplatin (Scheme 8.23), oxaliplatin (Scheme 8.24), and nedaplatin (Scheme 8.25). Only the first two are commercially available for general use in the treatment of cancer.

Next to cisplatin (Scheme 8.22), carboplatin (Scheme 8.23) is the most widely used metal-containing anticancer drug. Although it is similar to cisplatin in its cell-killing ability, it shows moderate effectiveness with some malignancies that

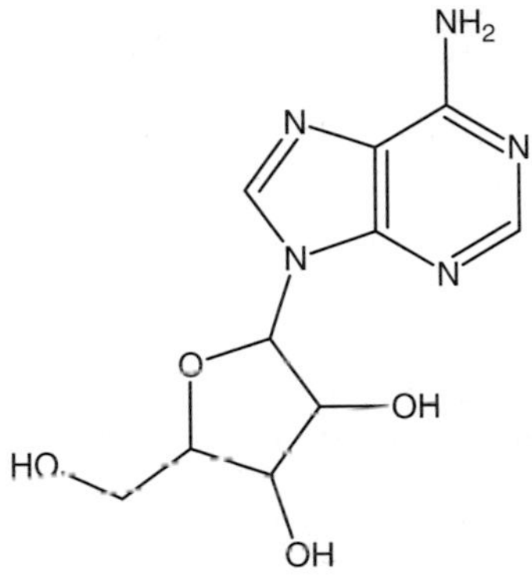

21 Vidarabine[79,80]

Scheme 8.21 Vidarabine [79,80]

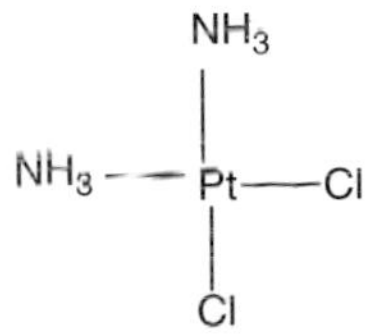

22 Cisplatin, Cis-DDP
(Cis-diaminedichloroplatinum(II)),

Scheme 8.22 Cisplatin, *Cis*-DDP (*Cis*-diaminedichloroplatinum(II))

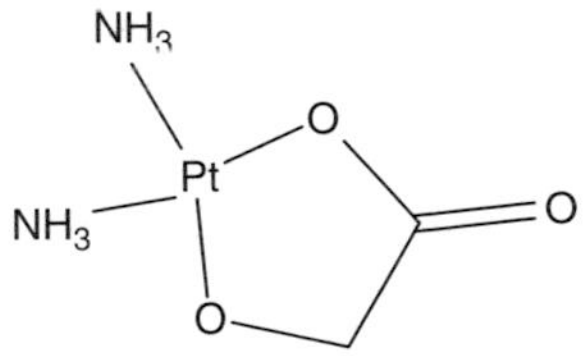

24 Nedaplatin
(Cis-diamineglycolatoplatinum (II))

Scheme 8.23 Carboplatin (Diamine cyclobutane dicarboxylate platinum(II)); *cis*-diamine-1,1-cy clobutanedicarboxylatoplatinum(II);*cis*-diamine-(cyclobutane-1,1-dicarboxylato)-platinum(II)

are less responsive to cisplatin such as non-small cell lung cancer. It also offers a different pharmacokinetic behavior. The presence of the bidentate carboxylate moiety gives decreased rates of reaction with the biological environment. Thus, it shows less nephrotoxicity and is preferred for patients suffering from kidney failure. It also shows a reduced rate of serum protein binding with only 10 to 20%

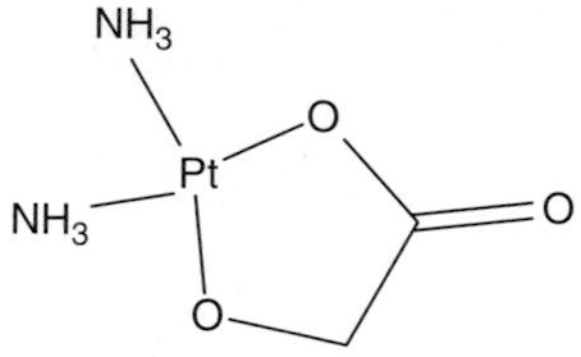

24 Nedaplatin
(Cis-diamineglycolatoplatinum (II))

Scheme 8.24 Nedaplatin (*cis*-diamineglycolatoplatinum (II))

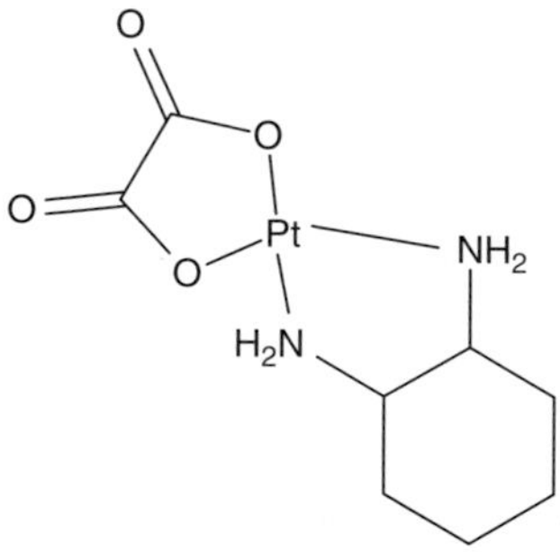

25 Oxaliplatin
(Oxalatoplatinum(II))

Scheme 8.25 Oxaliplatin (Oxalatoplatinum(II))

irreversibly bound to protein. This results in greater bioavailability and larger concentrations of the administered drug to about a five fold extent.

There are a number of good reviews covering carboplatin and related drugs [81–84].

4 Active Form of Cisplatin

While the active form of polymeric derivatives of cisplatin may be different, it may be informative to know how the active form of cisplatin itself as a beginning point in understanding how polymeric forms are active against both cancers and viruses.

In aqueous solution cisplatin is known to undergo spontaneous hydrolysis. The reaction produces species such as monoaquo platinum and diaqua platinum complexes, as shown in Eq. 8.1, arising from nucleophilic substitution in water.

```
H3N Cl              H3N Cl                 H3N   OH2+1
   \ /                 \ /                    \ /
   Pt  +  H2O  —>      Pt    +  H2O   —>       Pt          (8.1)
   / \                 / \                    / \
H3N Cl              H3N  OH2+1             H3N   OH2+1
```

Additionally, other aqueous species can exist including hydroxy complexes such as Schemes 26 and 27. The actual form of the hydroxyl species is pH dependent. At a pH of 7.4, 85% of the monohydrated complex will exist in the less reactive dihydroxy form. Lowering the pH to 6.0 results in the most common form (80%) being the monohydrate species. Thus, the number of possible aquated forms derived from cisplatin is high and dependent on pH, temperature, time, and the concentration of associated reactants such as the chloride ion and ammonia Figure 8.1 contains structures of some of these aquated forms of cisplatin including ones already noted.

The relatively high chloride concentration (about 100 mM) in blood minimizes hydrolysis and the formation of aquated species [9,13,85,86]. Once inside of the cell, where the chloride ion concentration is much lower (~4 mM), hydrolysis readily occurs giving a number of aquated species including the diaqua complex. At 37 °C the half-life for the completion of the formation of the diaqua complex is 1.7 hours with an activation energy of about 20 Kcal/mol (80 KJ/mol) [9,13,85,86].

Although the active form within the cell is believed to be the monohydrated structure (Scheme 8.28), the "preferred" extracellular species contains two *cis*-oriented leaving groups that are normally chloride ligands. As noted before,

```
H3N   OH2+1
   \ /
   Pt
   / \
H3N   OH
   26
```

Scheme 8.26

```
H3N   OH
   \ /
   Pt
   / \
H3N   OH
   27
```

Scheme 8.27

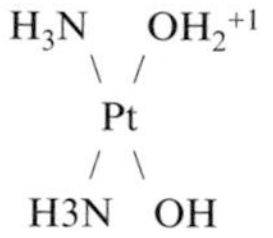

```
H3N   OH2+1          H3N   OH
   \ /                  \ /
   Pt                   Pt
   / \                  / \
H3N   OH             H3N   OH
  26                   27
```

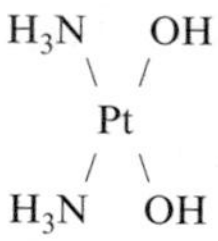

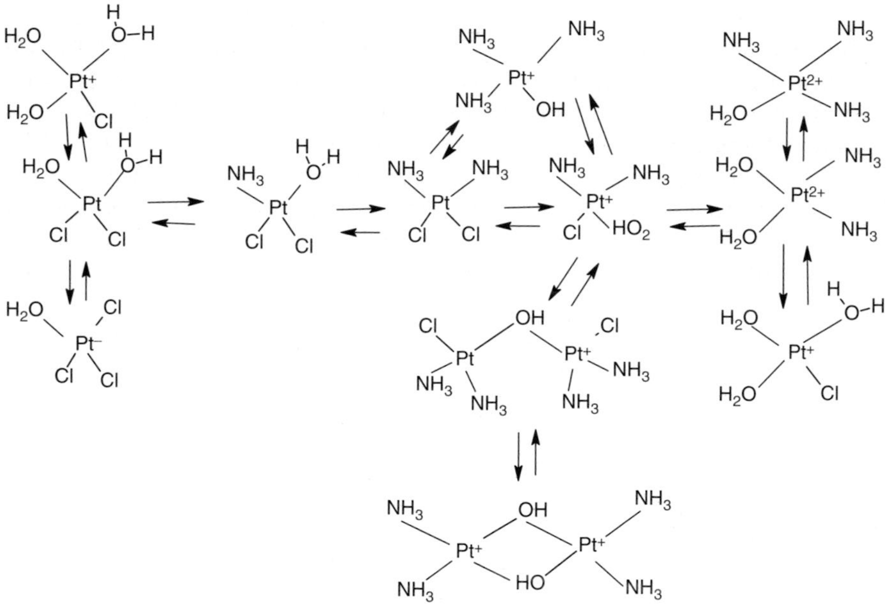

Figure 8.1 Selected aquated forms of cisplatin

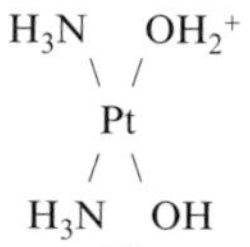

```
H3N   OH2+
   \ /
   Pt
   / \
H3N   OH
  28
```

Scheme 8.28

due to the high chloride ion concentration in blood these leaving groups will remain in position resulting in the molecule being electrically neutral until it enters the cell.

As noted above, *cis*-DDP enters cells by diffusion where it is converted to an active form. This results from the lower intracellular chloride concentration which promotes ligand exchange of chloride for water thus formation of the active aquated complex. Thus, the platinum-containing complex should be neutral to enter the cell and labile chloride groups need to be present to form the active species within the cell [87].

The antineoplastic activity of cis-DDP appears to be related to its interaction with DNA nucleotides as a monoaquo species [15]. The monohydrated complex reacts with the DNA nucleotides forming intra and interstrand crosslinks. Of the four nucleic acid bases, *cis*-DDP has been shown to preferentially associate with guanine. The most common are intrastrand crosslinks between adjacent guanines [6].

There are several possible crosslinks with DNA. One favored interstrand option occurs between the 6-NH groups of adenines on opposing strands in an A-T rich region [13,83–86,88–99]. This is because these groups are approximately 3.5 Å apart, close to the 3 Å distance between the *cis* leaving groups on the platinum. The second favored option is cross-linking occurring between the amino groups of guanine and cytosine in opposing strands. This is favored because the platinum is at right angles to the bases that in turn are coplaner with one another. This implies that the bases will have to either "bend down" or "turn edge" to achieve the necessary configuration to bind to the platinum complex. This binding pattern is believed to lead to perturbation of the secondary structure and minor disruption of the double helix. This is sufficient to cause inhibition of DNA replication and transcription with eventual cell death, yet too small to cause a response by damage recognition proteins and consequent excision of the affected segment and repair of the strand [13,83–86,88–103].

Although there are a variety of types of crosslinks that can be formed by cisplatin, the intrastrand crosslinks are the most common. Most of these crosslinks can be repaired but at least one type of interstrand crosslink may not induce response of cellular repair enzymes. Further, whereas intrastrand crosslinks are important in describing the activity of cisplatin, inducing apoptosis also appears to be a factor in the mechanism of cisplatin's anticancer activity [104,105].

Whereas *cis*-DDP is believed to act within the cell, some platinum-containing compounds appear to act on the cell membrane such as the so-called "platinum blues" [70,71]. Thus, the precise mode and site of activity may vary.

Despite the unquestionable success story of cisplatin, limitations remain including the powerful toxic side effects [103]. These toxic side effects include gastrointestinal problems such as acute nausea, vomiting and diarrhea; occasional liver dysfunction; myelosuppression involving anemia, leukopenia and thrombocytopenia; nephrotoxicity, and less frequently cases of immunosuppression, hypomagnesia, hypocalcemia, and cardiotoxicity [69]. The most serious side effect is damage to the kidney [13,106,107]. Much of the administered *cis*-DDP is filtered out of the body within a few hours, exposing the kidneys to bursts of high concentrations of platinum. The rapid rate at which the kidneys filter the platinum from the blood is believed to be responsible for the kidney problems. Another problem is the cumulative and irreversible hearing loss.

5 Structure-Activity Relationships

The topic of structure-activity relationships is complex and not fully defined for anticancer activity and is unknown for its ability to inhibit viruses. Most strategies employing polymeric derivatives of cisplatin are based in part on the assumption that the platinum-containing polymer will act similarly to cisplatin itself. However, it is likely that at least some of these polymeric drugs act in other ways to inhibit cancer growth. Although this complicates describing overall strategies, this may be an advantage because they would operate in another manner to complement cisplatin as a drug. Thus they may operate in several ways, allowing inhibition of cancer growth to occur via several mechanisms. Even so, it may be informative to briefly look at some structure-activity relationships established for cisplatin acting as an anticancer agent.

Although many promising products have been made, overarching structure-activity relationships are uncertain for a number of reasons, not the least of which is that different platinum-containing compounds may inhibit cancer growth in different ways. Thus, the following discussion should be considered as only one brief attempt at describing general structure-activity relationships. First, there should be two available anionic leaving groups such as chloride, bromide, or oxalate. Bidentate chelating groups such as dicarboxylate dianions, except for the especially labile malonato ligand, are often preferable to monodentate ligands because of their superior ability to remain intact in the bloodstream. Complexes with more labile groups such as the nitrate ion hydrolyze too rapidly for in vivo use, whereas ligands such as the cyanide ion bind the platinum too tightly impairing its activity [9,13,85,103]. Further, such complexes should have the *cis* geometry and be neutral with relatively inert amine or nitrogen donor groups. The neutrality of the molecule is believed to allow the platinum-containing drug to more easily traverse the cell membrane. The amines should be primary or secondary amines allowing for hydrogen-bonding to occur.

The relatively high Pt–N bond strength results in tight bonding of the amine, or related, ligands whereas the leaving groups such as chlorides and carboxylate anions are more weakly bonded and are readily replaced by other nucleophiles. A number of aqua complexes are formed with these aqua structures susceptible to replacement. When the pH is greater than 6.0, chloride displacement by the hydroxyl anion is favored leading to complexes containing the hydroxo group which is a relatively poor leaving group.

6 Arguments for Cisplatin-Derivative Drugs

Since the discovery by Rosenberg [88] that cisplatin is an effective anticancer drug, large synthesis and evaluation programs have aimed at the creation of cisplatin derivatives that show greater and more wide-spread activity against cancer but with lowered toxicity. More recent activity has focused on the construction of platinum-containing homing compounds that act specifically at the desired cancer site. For many decades now, oncologists worldwide have taken advantage of the cytotoxic action of a great variety of drug systems in the fight against cancer. However, despite undisputed successes in cancer chemotherapy, particularly in combination with surgery and other treatment modalities, numerous important pharmacological deficiencies of anticancer drugs have been well recognized in the medical fraternity. Most drugs lack cell specificity, failing to discriminate between normal and cancer cells. Thus, they tend to cause severe and dose-limiting systemic toxicity. As extraneous agents, they immediately expose themselves as targets for scavenger proteins or as substrates for glomerular filtration and first-pass liver metabolism. As a consequence, serum residence times are commonly short, predominant fractions of administered doses are prematurely excreted (and wasted in the process), and bioavailability (concentration in the target tissue) is generally low. Many drugs are polar, charged, or salt like. Therefore, they are poor substrates for membrane penetration, intracellular trafficking, and cell entry by the passive diffusion mechanism common to neutral and nonpolar compounds. In addition, drugs possessing poor solubility in aqueous media are sluggishly and incompletely dissipated in the central circulation and become easy targets for the reticuloendothelial system. Lastly, and most importantly, acquired drug resistance, which gradually builds up in the target cells after initially successful chemotherapy, is a relatively common phenomenon requiring premature treatment termination. The overall result of these shortcomings is a narrow therapeutic window and grossly limited overall chemotherapeutic effectiveness.

One approach to circumvent these deficiencies is to convert the active agent into some form of prodrug that will encounter minimal interference by scavenger mechanisms. This prodrug will be able to cross intercellular membranes and approach the target site, which in cancer chemotherapy means being able to breach the lysosomal compartment of the cancerous cell. One such group of prodrug involves the presence of the platinum-containing agent in polymers.

7 Arguments for Polymeric Drugs

Polymers can act as carriers or drugs or some combination of these two extremes. This topic is more fully discussed elsewhere. As carriers, polymers can be designed that contain many positive elements including so-called homing devices as well as the "bullet." The polymer can also be modified to achieve water solubility, desired control release kinetics, a balance of hydrophilic and hydrophobic character, and be nontoxic and nonimmunogenic. Its bulk helps to shield the drug temporarily from attack by serum proteins giving extended serum circulation half-lives. An often cited molecular weight range is from 25,000 to 80,000 as this will retard premature renal excretion while minimizing toxic effects as occasionally shown by high molecular weight polymers.

Using a pinocytotic cell entry mechanism, the carrier-attached drug, even if polar or charged, will be transported into the intracellular space, thereby overcoming possible influx inhibition, or efflux acceleration, mediated by certain well-defined drug resistance mechanisms. The process is expedited by the presence of potentially cationic moieties, such as *tert*-amino groups. Cationic sites in a polymer are known to facilitate pinocytosis while at the same time increasing affinity for the neoplastic cell, which in many cancers is negatively charged. Lastly, the enhanced permeation and retention (EPR) effect associated with macromolecules, in contrast to small compounds, provides preferred distribution of polymers to tumorous tissue. As a consequence, conjugate accumulation in the tumor is favored over that in healthy tissue. The overall result is reduced systemic toxicity and enhanced bioavailability. The drug-containing polymer can also act as a drug itself.

It is hoped that inclusion of the cisplatin-like moiety into a polymer will achieve the following:

1. It will limit movement of the biologically active drug. Because of their size, polymers are not as easily passed through membranes present in the body. Cisplatin itself is rapidly excreted from the body causing the kidney and other organs to be exposed to high concentrations of platinum. Polymers with chain lengths of about 100 units and greater typically are unable to move easily through biological membranes. Restricted movement may prevent a build-up in the kidneys and other organs thereby decreasing renal and other organ damage. Further, the platinum from polymers could be released slowly, reducing the exposure of organs to large concentrations of platinum-containing complexes.
2. It may enhance activity through an increased opportunity for multiple bonding interactions at a given site (e.g., chemical bonding, hydrogen bonding, hydrophobic interactions).
3. It should increase delivery of the bioactive moiety and decrease toxicity. In aqueous solutions, cisplatin hydrolyzes with a reaction half-life of nine hours at room temperature or 2.4 hours at 37 °C. Cisplatin hydrolyzes in the body forming a wide variety of platinum-containing agents, none of which is as active as cisplatin itself and most of which exhibit increased toxicity to the body. Formation of these hydrolysis products increases the amount of platinum complex that must be added to effect desired tumor reduction. Consequently, this increases the amount of platinum complexes that must be processed by the body. The polymeric structure should also shield the platinum moiety from unwanted hydrolysis increasing the concentration of platinum in the beneficial form that is retained in the body thus permitting lower effective doses of the drug to be used. The nature of the more hydrophobic polymer chain should also act to protect the platinum moiety from ready attack by water.
4. It should by-pass the cell's defense system. The cell's defensive response is armed as a result of invasion of other chemo drugs. Recent studies are indicating that introduction of chemo drugs into cells causes the buildup of "housekeeping" proteins that are rather general in their ability to select and remove foreign compounds present in the cell. This may be a principal reason why chemo treatments

lead to resistance to drugs, even to drugs that have not been previously used. It is possible that the polymeric nature of the platinum carriers will discourage the housekeeping proteins from removing them allowing the polymers to function as anticancer drugs under conditions where smaller platinum-containing drugs are not successful.

8 Polymer Synthesis

The polymers were synthesized by simply adding equal molar aqueous solutions of potassium tetrachloroplatinate II and the diamine-containing reactant with the resulting polymer captured as a precipitate. Polymer is formed in several minutes to several days depending on the reactivity of the Lewis base. The polymers can be made in gram to larger quantities as needed employing simple equipment and commercially available reactants. Thus, the polymers are ideally suited for commercialization.

9 Antiviral Activity

Some cancers are believed to have a viral relationship. As such it is informative to look at the viral response to some of the platinum polymers.

Tetramisole is an antihelmentic which acts on the cyclic nucleotide phosphodiesterases. It actually consists as a combination of optical isomers, the most active one being levamisole. Levamisole was the first synthetic chemical that exhibited immunomodulatory properties. It appears to restore normal macrophase and T-lymphocyte functions.

Cisplatin polymer analogs, made through reaction of tetrachloroplatinate with tetramisole, were tested for their ability to inhibit EMC-D viruses that are responsible for the onset of juvenile diabetes symptoms in ICR Swiss male mice [109]. Briefly, the mice were treated with 1, 5, and 10 mg/kg. Doses of 1 and 5 mg/kg decreased the severity and incidence of virus-induced diabetes in comparison to untreated mice. In another series of tests, doses of 1 and 10 mg/kg were administered 1 day prior to injection of the virus but here there was an increase in the severity and incidence of virus-induced diabetes. Other studies were undertaken showing that the polymer showed different activity profiles than the tetramisole (Scheme 8.29), itself.

Methotrexate is a folic acid antagonists that indirectly suppresses the synthesis of purine and is particularly effective in rapidly proliferating cell populations such as cancer. It depresses the primary and secondary antibody response, the homograft reaction, the graft-versus-host response, and the development of hypersensitivity. Methotrexate is used in the treatment of certain cancers. It is very toxic in long-term use, especially to the liver. The toxic effect of methotrexate is reversed by use of folinic acid. This "finding" led to the development of the "rescue technique" where the most rapidly dividing cells are killed and the other cells are left unchanged.

The polymers were synthesized by simply adding equal molar aqueous solutions of potassium tetrachloroplatinate II and the methotrexate with the resulting polymer (Scheme 8.30) captured as a precipitate.

In a study related to that carried out employing tetramisole, a methotrexate polymer (Scheme 8.30) was similarly tested. In the initial study female mice were treated. Generally, only male IRC Swiss mice develop diabetes-like symptoms. Female mice must first be treated with testosterone before they can develop diabetes-like symptoms. The mice were divided into 3 groups all of which received injections of the polymer (0.5 cc IP of polymer solution containing 6.4 mg/kg). Groups I and II were treated with testosterone 1 week later. On day 8, group 1 was again given a second 0.5 cc intraperitoneal (i.p.) inoculation of the polymer solution. On day 9, all groups received 1×10^4 pfu (plaque forming units) of EMC-D virus. On day 17, all mice were given a 1 hour glucose tolerance test. The glucose

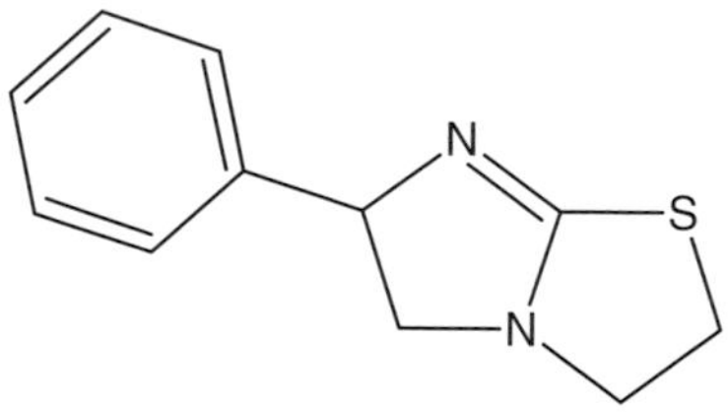

Tetramisole
29

Scheme 8.29 Tetramisole

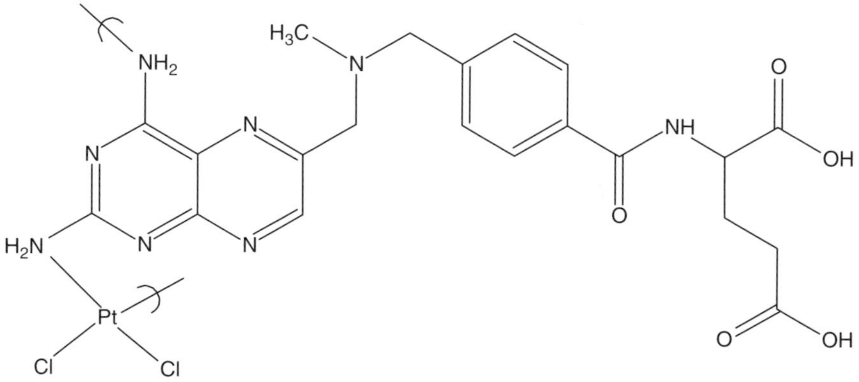

30 Methotrexate Polymer

Scheme 8.30 Methotrexate Polymer

level for groups I and III were similar and significantly below the level for the diabetic mice in group II. This is consistent with the polymer effectively blocking the diabetogenic effects of the virus. Further, other results from this study were consistent with this strain of female mice being susceptible to developing diabetes-like symptoms even without the testosterone treatment.

A related study was carried out except using male mice. Here, again, the polymer (Scheme 8.30) showed a greater positive effect on the control of diabetes than either of the reactants themselves. The glucose levels were near those of non-infected mice for the polymer-treated mice. Again, the incorporation of both the platinum and methotrexate into a polymer was the effective agent and not either of the drugs themselves. These two experiments are related to generation of a vaccine that can be employed to prevent onset of β cell damage by RNA viruses.

The third experiment focused on treatment subsequent to viral infection [108,109]. The polymer was 100% effective in viral control with delivery of the polymer (Scheme 8.30) 1 day after the mice were infected. In summary, the methotrexate polymer (Scheme 8.30) is an effective antiviral agent against at least the EMC RNA virus.

Recently, we looked at the ability of methotrexate, tetrachloroplatinate, a physical mixture of methotrexate and tetrachloroplatinate and a methotrexate-platinum polymers to inhibit various viruses. A related study employing tilorone and a tilorone derivative and cisplatin polymeric derivatives was carried out. The results will be briefly reported later. For these studies each cell line is especially chosen to be compatible to support growth for the particular virus. DSC-1 cells are African green monkey kidney epithelial cells, mouse L929 are fibroblast cells, vero cells are African green monkey kidney epithelial cells, and human 143 cells are fibroblast bone osteosarcoma cells.

The virus were chosen to represent a broad range of virus. The reovirus ST3 virus is a RNA virus that is currently being investigated because of its ability to inhibit certain cancer cells while leaving normal cells alone. Generally, drugs that are capable of inhibiting one RNA virus will be effective against other RNA viruses. The other viruses are all DNA virus and the activity against different DNA viruses must be studied individually. Vaccinia is responsible for small pox; herpes simplex is responsible for at least 45 million infections yearly in the United States, or 1 out of 5 adolescents and adults; and varicella zoster is responsible for chickenpox and shingles.

The cell lines chosen for viral replication studies are cancer cell lines so that a measure of the ability of the test agents to inhibit cell growth is obtained. Each cell line is especially chosen to be compatible to support growth for the particular virus. The cell lines are all transformed cell lines.

Table 8.1 contains the GI_{50} values for these compounds for the various cell lines in μg/ml. Viral replication studies are carried out at concentrations where cell death is less than 5%.

For comparison, the GI_{50} for cisplatin for L929 cells is 50 μg/ml. Thus, all of the tested compounds except for the tetrachloroplatinate exhibit GI_{50} values less than that of cisplatin, and with the exception to the tetrachloroplatinate exhibit similar abilities to inhibit cell growth.

Table 8.1 Toxicity of methotrexate-related compounds to cancer cell lines

	GI_{50} (microgram/mL)			
Compound	L929 Cells	143 Cells	Vero Cells	BS-C-1 Cells
K_2PtCl_4	375	275	225	225
Methotrexate	15	10	12	10
Polymer	10	10	10	12
Methotrexate Mix	12	8	10	10

Table 8.2 Inhibition concentrations (μg/ml) for the tested compounds

Tested Compound →	K_2PtCl_4	Methotrexate	Polymer mixture	
Reovirus ST3				
GI50	–	–	–	–
GI100	–	–	–	–
GI50*	–	–	–	–
GI100*	–	–	–	–
Vacinia WR				
GI50	–	–	–	–
GI100	–	–	8	–
GI50*	–	–	4	–
GI100*	–	–	–	–
HSV-1				
GI50	–	2	3	2
GI100	–	8	6	4
GI50*	–	2	3	2
GI100*	–	8	3	2
VZV				
GI50	–	3	4	4
GI100	–	8	4	6
GI50*	–	3	2	2
GI100*	–	8	2	3

*Based on the amount of methotrexate

Tables 8.1 and 8.2 give the results of preliminary studies involving the ability of the various polymeric drugs to inhibit the DNA and RNA associated viruses. The results are presented as means of four experiments using duplicate samples in each experiment.

Using a plaque reduction assay, the ability of each compound to prevent viral growth is summarized in Table 8.2.

The tetrachlorate is essentially inactive against the tested viruses. Methotrexate, the physical combination of tetrachloroplatinate, and the polymer all exhibit good activity against HSV-1 and VZV, both DNA viruses whose genome replication occurs in the nucleus with some activity against vaccinia virus—a DNA virus with cytoplasmic DNA replication—and no activity against reovirus—a dsRNA virus with cytoplasmic replication. The polymer and physical mixture show good activity at lower concentrations than the methotrexate itself. Table 8.3 also contains the GI_{50} and GI_{100} for the various compounds including columns based on the amount of methotrexate itself because much of the activity appears to come from the presence of the methotrexate and little if any from the tetrachloroplatinate. It is seen that the polymer generally exhibits the lowest concentrations for inhibition consistent with the greater ability of the polymer to inhibit the tested viruses. Polymer activity may result from controlled release of the methotrexate, through the polymer itself acting as a drug, or some combination of these two extremes.

In summary, methotrexate, the mixture, and the polymer are all active against HSV-1 and VZV viruses whereas the tetrachloroplatinate showed little or no activity against any of the tested viruses. The best inhibition, that is inhibition at the lowest concentration, was found for the polymer and the mixture consistent with there being some cooperative effect of the methotrexate and platinum moieties. It is possible that the mixture of tetrachloroplatinate and methotrexate may have formed polymer when mixed together in solution consistent with the similar findings found for the mixture and polymer.

A similar study was carried out employing tilorone (Scheme 8.31) and a tilorone derivative. It is known that a molecular complex of tilorone and RNA exhibit an antiviral effect similar to that of polynucleotide interferon (IFN) inducers such as poly(I)-poly(C), larifan, and ridostin [110]. It is possible that the current responses are related to this.

Tilorone, 2,7-bis[2-9diethylamino)ethoxy]-9H-fluorene-9-one, is the first recognized synthetic small molecule structure that is an orally active interferon inducer. Because of its potential importance a number of similar structures were synthesized. These derivatives are given various numbers that follow the name tilorone. Tilorone 11,567 is one of these derivatives.

Structures of the platinum polymers are presented in Schemes 8.32 and 8.33. Each is a *cis*-derivative obeying the *trans*-effect.

The toxicity of the tilorone polymers to the various cell lines is given in Table 8.3 in ng/ml. Both compounds show similar toxicities to the cancer cell lines tested but not to the Balb 3T3 cells which are partially transformed cells. Again, for comparison, cisplatin shows a GI_{50} of 50,000 ng/ml, over 250 times that of either platinum-tilorone polymer.

Table 8.4 gives the results of preliminary studies involving the ability of the Pt-tilorone polymeric drugs to inhibit DNA and RNA associated viruses.

Table 8.3 Toxicity of tillorone compounds to various cell lines

	GI_{50} (nanogram/mL)				
Compound	L929 Cells	143 Cells	Vero Cells	BS-C-1 Cells	Balb 3T3 Cells
Pt-Tilorone 11,567	225	200	175	200	10,000
Pt-Tillorone	180	125	175	200	500

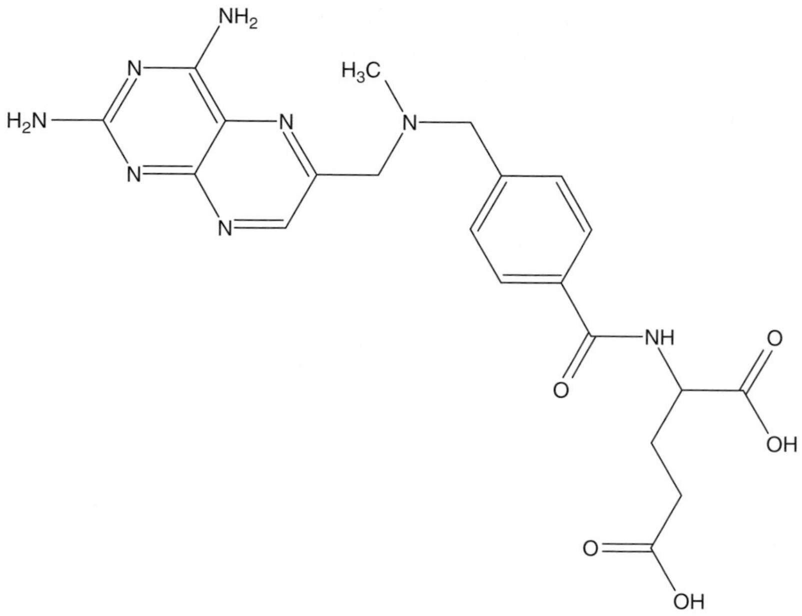

31 Tilorone

Scheme 8.31 Tilorone

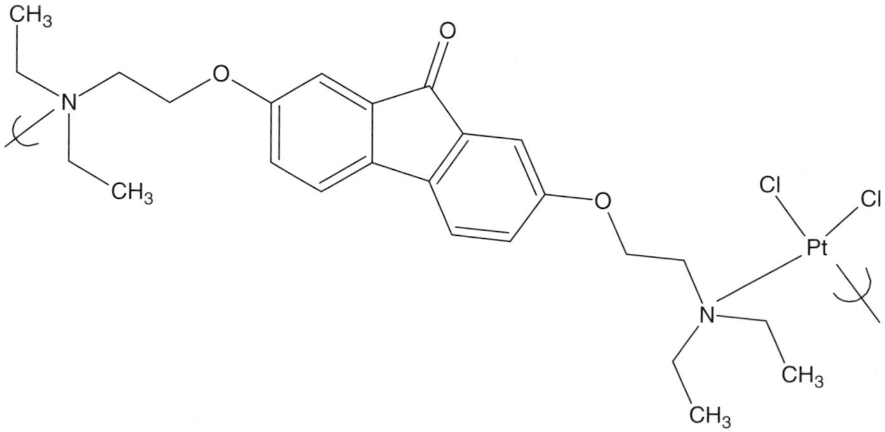

32 Product of potassium tetrachloroplatinate II and tilorone.

Scheme 8.32 Product of potassium tetrachloroplatinate II and tilorone

Both of the tillorone polymers exhibited good inhibition of all of the tested viruses at low concentrations. For comparison, the organotin polymers of norfloxacin and ampicillin showed good inhibition of these same viruses at a little higher concentrations,

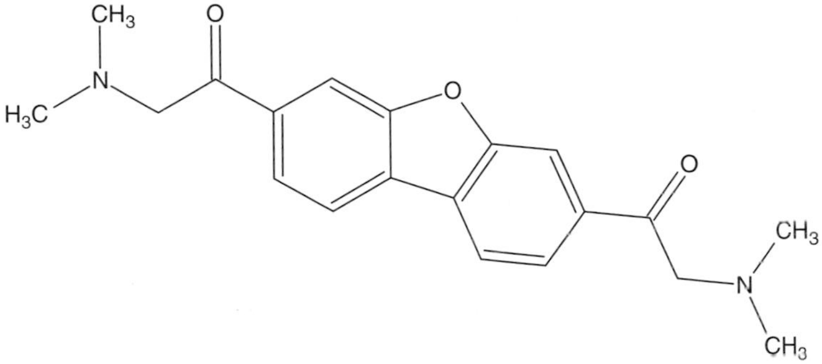

33 Product of potassium tetrachloroplatinate II and tilorone 11,567.

Scheme 8.33 Product of potassium tetrachloroplatinate II and tilorone 11,567

Table 8.4 Plaque-reduction assay results for the platinum tilorone polymers

	GI_{99+} Concentration (ng/ml)			
Compound	Reovirus ST3	Vaccinia WR	HSV-1	BS-C-1
Pt-Tilorone 11,567	200	150	150	150
Pt-Tilorone	125	100	125	100

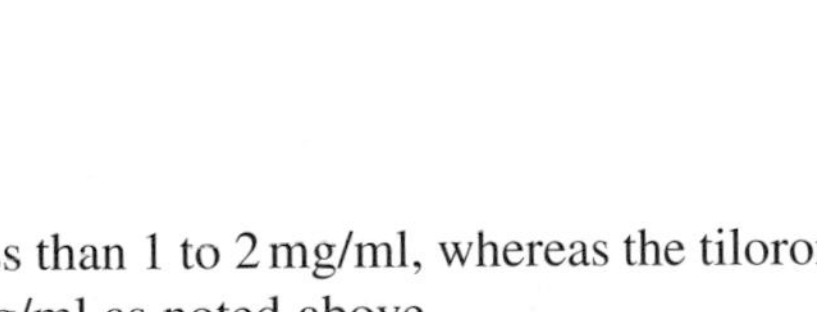

34

Scheme 8.34

within the range of less than 1 to 2 mg/ml, whereas the tilorone compounds showed good activity at 0.2 mg/ml as noted above.

In summary, the tilorone polymers inhibit both RNA and DNA viruses and deserve further consideration as antivirial agent in the war against viruses and possible bioterrorism involving viruses. Further, they show good inhibition of virus replication in both transformed cell lines but also in normal cell lines, a condition that better mimics antiviral therapy of humans.

A number of platinum polyamines were tested for antiviral activity in tumor cells [111]. For instance, the polymer from tetrachloroplatinate and 2,6-diamino-3-nitroso-pyridine (Scheme 8.34) which exhibited a cell differential ratio of 3.4, was tested at a concentration of 2.2 μg/ml on L929 cells infected with

Encephalomyocarditis (EMC) virus, strain MM. A virus reduction of about 25% was seen. This is considered to be a moderate antivirial response.

In general, agents capable of inhibiting one RNA virus will inhibit other RNA viruses but each DNA virus must be evaluated separately. The platinum polyamines were studied against RNA viruses. The behavior toward RNA viruses was varied with some showing little activity but the majority showing inhibition of viral replication at polymer concentrations below which tumoral inhibition is found (< 1 μg/ml).

The effect of platinum polyamines on the transformation of 3T3 cells by SV40 virus was also studied [111]. In summary, these polymers showed no effect on the transformation process.

10 Conclusions

Polymeric derivatives of cisplatin inhibit the replication of a number of important viruses. The results are consistent with this family of compounds showing promise as a new family of antiviral agents. Further, the results indicate that additional study of members of this family as antiviral agents is warranted.

References

1. Ksiazek TG, Erdman D, Goldsmith CS, et al. (2003) N Engl J Med. 348(20):1953.
2. Drosten C, Gunther S, Preiser W, et al. (2003) N Engl J Med. 348(20):1967.
3. Kapikian AZ. (2001) Novartis Found Symp. 238:153.
4. Piliero P. (2004) AIDS Read. 14(12):655.
5. Blower S, Wald A, Gershengorn H, Wang F, Corey L (2004) J Infect Dis. 190(9):1610.
6. Carraher C, Siegmann-Louda, D. (2004) Macromolecules containing metals and metalloids. Wiley, NY.
7. Birch K, Subr V. (2004) Adv Drug Delivery Rev. 56:1023.
8. Roner M, Carraher, C, Roehr, J, Bassett K, Siegmann-Louda D. (2004) Polym Mater Sci Eng. 91:744.
9. Roner M, Carraher C, Roehr J, Bassett K, Siegmann-Louda D, Zhao A. (2004) Polym Mater Sci Eng. 90:515.
10. Bleicher R, Carraher, C. (2002) Polym Mater Sci Eng. 86:289.
11. Carraher C, Bleicher, R. (2004) Macromolecules containing metal and metal-like elements, biomedical applications, Vol. 3. Wiley, Hoboken.
12. Siegmann-Louda D, Carraher, C, Pflueger F, Coleman J, Harless S, Luing H. (2000) Polym Mat Sci Eng. 82:83.
13. Siegmann-Louda D, Carraher, C, Ross, J, Li F, Mannke K, Harless, S. (1999) Polym Mat Sci Eng. 81:151.
14. Siegmann-Louda D, Carraher, C, Pfueger F, Nagy D. (2001) Polym Mat Sci Eng. 84:658.
15. Siegmann-Louda D, Carraher C, Chamely, D, Cardoso A, Snedden D. (2002) Polym Mat Sci Eng. 86:293.
16. Siegmann-Louda D, Carraher C, Graham M, Doucetter R, Lanz L. (2002) Polym Mat Sci Eng. 87:247.

17. Siegmann-Louda D, Carraher C, Snedden D., Komulainen A. (2004) Polym Mat Sci Eng. 90:512.
18. Doucette R, Siegmann-Louda D, Carraher C, Cardoso A. (2004) Polym Mat Sci Eng. 91:564.
19. Doucette R, Siegmann-Louda D, Carraher C. (2004) Polym Mat Sci Eng. 91:567 and 569.
20. Siegmann-Louda D, Carraher C. (2004) Macromolecules containing metal and metal-like elements, biomedical applications, Vol. 3. Wiley, Holboken.
21. Abraham G, Colonno RJ. (1988) J Virol. 62(7):2300. 22. Ludwig DS, Schoolnik GK. (1987) Med Hypotheses. 23(3):303.
23. Eppstein DA, Marsh YV, Schreiber AB, Newman SR, Todaro GJ, Nestor JJ Jr. (1985) Nature. 318(6047):663.
24. Ghosh JK, Shai Y. (1998) J Biol Chem. 273(13):7252.
25. Wild T, Buckland R. (1997) J Gen Virol. 78 (Pt 1):107.
26. Nybakken GE, Oliphant T, Johnson S, Burke S, Diamond MS, Fremont DH. (2005) Nature. 437(7059):764.
27. Galmiche MC, Goenaga J, Wittek R, Rindisbacher L. (1999) Virology. 254(1):71.
28. Chanh Tc, Dreesman GR, Kennedy RC. (1987) Proc Natl Acad Sci USA. 84(11):3891.
29. Dalgleish AG. (1991) Ann Ist Super Sanita. 27(1):27.
30. Suzuki T, Tsukimoto M, Kobayashi M, et al. (1994) J Gen Virol. 75 (Pt 7):1769.
31. Sauter NK, Bednarski MD, Wurzburg BA, et al. (1989) Biochemistry. 28(21):8388.
32. Chu JJ, Ng ML. (2004) J Virol. 78(19):10543.
33. Hong SS, Boulanger P (1995) EMBO J. 14(19):4714.
34. Florea NR, Maglio D, Nicolau DP. (2003) Pharmacotherapy. 23(3):339.
35. Reisdorph N, Thomas JJ, Katpally U, et al. (2003) Virology. 314(1):34.
36. Shia KS, Li WT, Chang CM, et al. (2002) J Med Chem. 45(8):1644.
37. Furuta Y, Takahashi K, Fukuda Y, et al. (2002) Antimicrob Agents Chemother. 46(4):977.
38. Donath E, Herrmann A, Coakley WT, Groth T, Egger M, Taeger M. (1987) Biochem Pharmacol. 36(4):481.
39. Laerum OD. (1985) Scand J Infect Dis Suppl. 47:40.
40. Verheyden JP. (1988) Rev Infect Dis. 10 Suppl 3:S477.
41. Lee H, Hanes J, Johnson KA. (2003) Biochemistry. 42(50):14711.
42. Zaretsky MD. (1995) Genetica. 95(1-3):91.
43. Brigden DWhiteman P. (1983) J Infect. 6(1 Suppl):3.
44. Elion GB. (1982) Am J Med. 73(1A):7.
45. Fletcher CV, Balfour HH Jr. (1989) Dicp. 23(1):5.
46. Fowles SE, Pierce DM, Prince WT, Staniforth D. (1992) Eur J Clin Pharmacol. 43(5):513.
47. Earnshaw DL, Bacon TH, Darlison SJ, Edmonds K, Perkins RM, Vere Hodge RA. (1992) Antimicrob Agents Chemother. 36(12):2747.
48. De La Fuente R, Awan AR, Field HJ. (1992) Antiviral Res. 18(1):77.
49. Vere Hodge RA, Sutton D, Boyd MR, Harnden MR, Jarvest RL. (1989) Antimicrob Agents Chemother. 33(10):1765.
50. Harrell AW, Wheeler SM, Pennick M, Clarke SE, Chenery RJ. (1993) Drug Metab Dispos. 21(1):18.
51. Ashton RJ, Abbott KH, Smith GM, Sutton D. (1994) J Antimicrob Chemother. 34(2):287.
52. De Clercq E, Descamps J, Ogata M, Shigeta S. (1982) Antimicrob Agents Chemother. 21(1):33.
53. De Clercq E. (1982) Antimicrob Agents Chemother. 21(4):661.
54. De Clercq E. (1982) Bull Soc Ophtalmol Fr. 82(6-7):913.
55. Kawai H, Yoshida I, Suzutani T. (1993) Microbiol Immunol. 37(11):877.
56. Lopez C, Watanabe KA, Fox Jj. (1980) Antimicrob Agents Chemother. 17(5):803.
57. Klein RJ, Friedman-Kien AE. (1984) J Invest Dermatol. 83(5):344.
58. Schat KA, Schinazi RF, Calnek BW. (1984) Antiviral Res. 4(5):259.
59. De Clercq E, Holy A, Rosenberg I, Sakuma T, Balzarini J, Maudgal PC. (1986) Nature. 323(6087):464.

60. Snoeck R, Lagneaux L, Delforge A, et al. (1990) Eur J Clin Microbiol Infect Dis. 9(8):615.
61. Kim Cu, Luh BY, Martin JC. (1990) J Med Chem. 33(6):1797.
62. Holy A, Votruba I, Merta A, et al. (1990) Antiviral Res. 13(6):295.
63. Li Sb, Yang Zh, Feng JS, Fong CK, Lucia HL, Hsiung GD. (1990) Antiviral Res. 13(5):237.
64. De Clercq E. (2002) Biochim Biophys Acta. 1587(2-3):258.
65. Nakashima H, Tochikura T, Kobayashi N, Matsuda A, Ueda T, Yamamoto N. (1987) Virology. 159(1):169.
66. Jeffries DJ. (1989) J Antimicrob Chemother. 23 Suppl A:29.
67. Sangkitporn S, Shide L, Klinbuayaem V, et al. (2005) Southeast Asian J Trop Med Public Health. 36(3):704.
68. Connolly KJ, Allan JD, Fitch H, et al. (1991) Am J Med. 91(5):471.
69. Schmidt RE. (2002) Med Microbiol Immunol (Berl). 191(3-4):175.
70. Colman PM. (2005) Expert Rev Anti Infect Ther. 3(2):191.
71. Montalto NJ. (2001) Am Fam Physician. 63(4):635.
72. Ward P, Small I, Smith J, Suter P, Dutkowski R. (2005) J Antimicrob Chemother. 55 Suppl 1:i5.
73. Oberg B. (1982) Pharmacol Ther. 19(3):387.
74. Rigsby RE, Rife CL, Fillgrove KL, Newcomer ME, Armstrong RN. (2004) Biochemistry. 43(43):13666.
75. Schoub BD, Prozesky OW. (1977) Antimicrob Agents Chemother. 12(4):543.
76. Migus DO, Dobos P. (1980) J Gen Virol. 47(1):47.
77. Yan Y, Svitkin Y, Lee JM, Bisaillon M, Pelletier J. (2005) RNA. 11(8):1238.
78. Parker WB. (2005) Virus Res. 107(2):165.
79. Sloan BJ, Kielty JK, Miller FS. (1977) Ann N Y Acad Sci. 284:60.
80. Miwa N, Kurosaki K, Yoshida Y, Kurokawa M, Saito S, Shiraki K. (2005) Antiviral Res. 65(1):49.
81. Dabrowiak JC, Bradner WT. (1987) Prog Med Chem. 24:129.
82. Kelland L. (1992) Crit Rev Oncol Hematol. 15:191.
83. Heim M. (1992) Metal complexes in cancer chemotherapy. VCH, NY.
84. Mckeage M, Kelland L. (1992) Molecular aspects of drug-DNA interactions. Macmillian, NY.
85. Neuse E. (1999) South African J Sci. 95:509.
86. Zwelling LA, Kohn KW. (1979) Cancer Treat Rep. 63(9-10):1439.
87. Zwelling LA, Kohn KW, Anderson T. (1978) Proc Am Assoc Res. 19:233.
88. Rosenberg B. (1973) Naturwissenschaften. 60(9):399.
89. Rosenberg B. (1979) Cancer Treat Rep. 63(9-10):1433.
90. Rosenberg B. (1971) Platinum Metals Rev. 15:42.
91. Roberts J, Pascoe J. (1972) Advances in antimicrobial and antineoplastic chemotherapy, Vol. 2. University Park Press, Baltimore, p 249.
92. Thomson A, Mansy S. (1972) Advances in antimicrobial and antineoplastic chemotherapy. University Park Press, Baltimore, p 199.
93. Drobnik J, Horacek P. (1973) Chem Biol Interact. 7(4):223.
94. Macquet JP, Theophanides T. (1976) Biochim Biophys Acta. 442(2):142.
95. Mansy S, Rosenberg B, Thomson AJ. (1973) J Am Chem Soc. 95(5):1633.
96. Goodgame D, Jeeves I, Phillips F. (1975) Biochim Biophys Acta. 378:153.
97. Dehand J, Jordanov J. (1976) J Chem Soc Chem Commun. 598.
98. Pegg AE. (1978) Nature. 274(5667):182.
99. Beck DJ, Fisch JE. (1980) Mutat Res. 77(1):45.
100. Cohen GL, Bauer WR, Barton JK, Lippard SJ. (1979) Science. 203(4384):1014.
101. Cohen GL, Ledner J, Baues W, Ushay H, Caravana C, Lippard SJ. (1980) J. Amer. Chem Soc. 102:2487.
102. Brouwer J, van de Putte P, Fichtinger-Schepman AM, Reedijk J. (1981) Proc Natl Acad Sci USA. 78(11):7010.
103. Lippard B. (1999) Coord Chem Revs. 182:263.
104. Sorenson CM, Barry MA, Eastman A. (1990) J Natl Cancer Inst. 82(9):749.

105. Chu G. (1994) J Biol Chem. 269(2):787.
106. Gottieb J, Drewinko B. (1975) Cancer Chemother Rep Part 1. 59:621.
107. Stadnicki S, Fleischman R, Schaeppi U, Merriman P. (1975) Cancer Chemother Rep Part 1. 59:467.
108. Carraher C, Lopez I, Giron D. (1985) Polym Mater Sci Eng. 53:644.
109. Carraher C, Lopez I, Giron D. (1987) Advances in Biomedical Polymers. Plenum, NY.
110. Roner M, Carraher C, Dhanji, S. (2005) Polymer Mater Sci Eng. 92:499.
111. Giron D, Espy M, Carraher C, Lopez I. (1985) Polymeric Materials in Medication. Plenum, NY.

Vanadocene-Containing Polymers

Theodore S. Sabir and Charles E. Carraher Jr.

1 Introduction

Essentially all vanadium-containing organometallics are derived from dicyclopentanylenevanadium (IV) dichloride, vanadocene dichloride. Vanadocene dichloride and its derivatives are members of a family of distorted sandwich compounds. Unlike ferrocene, vanadocene dichloride has a distorted tetrahedral geometry with the cyclopentadienes facing the vanadium atom. It is structurally related to a number of other metallocene dichlorides including those where the metal is Ti, Zr, Hf, Mo, and Nb. These compounds not only share related geometries, but also similar chemical reactions and biological activities.

The major interest in these metallocenes involves their use as catalysts in the synthesis of a number of stereoregular vinyl polymers such as polyethylene and polypropylene. They are also being investigated for their biological activities. Because one of the major driving forces for the synthesis of polymeric derivatives is their biological activity, we will briefly review some of this effort focusing on vanadocene dichloride, since that is the compound most often studied.

A.S. Abd-El-Aziz et al. (eds.), *Inorganic and Organometallic Macromolecules: Design and Applications.*

2 Anticancer Activity

The distorted tetrahedral metallocene dihalides form a class of hydrophobic organometallic anticancer agents that inhibit a wide range of cancer cell lines. They inhibit such cell lines as leukemias P388, and L1210, colon 38, Lewis lung carcinomas, B16 melanoma, and solid and fluid Ehrilich ascites tumors. Titanocene dichloride is active against xenografted tumors of rectum, stomach, colon, breast, neck, and stomach and is now undergoing phase II clinical trials [1–6].

Both titanocene and vanadocene dichloride inhibit DNA and RNA replication accumulating in nucleic acid rich portions of tumor cells [7,8]. Because of this it is believed that the formation of DNA-metallocene adducts in vivo is involved in their ability to inhibit cell growth. Further, it has been found that metallocene-protein interactions are also important and that metal-tranferrin compounds have been included in their mode of action [9].

Along with blood plasma proteins, proteins involved in cellular replication have also been implicated in the action of several metallocenes. Vanadocene dichloride inhibits both protein kinase C and bacterial topoisomerase activity [10].

Mokdsi and Harding recently studied the inhibition of human topoisomerase II employing a number of different metallocenes [11]. Their findings indicate that the mechanism of activity involves a complex pathway involving a different species that influences transport and delivery of the active compounds to the cancer cells. Subsequent interaction with nucleic acids and/or proteins is also involved. Whereas inhibition of toposiomerase II is a possible site of activity, other sites are not ruled out. Several recent reviews have been published in this area [12–14].

In contrast to cisplatin, vanadocene complexes did not increase the DEL recombination frequency in yeast nor encourage any of the DNA damage associated promoters in HepG2 cells [15]. Vanadocene compounds do promote the activation of the *c*-fos promoter without affecting the minimal promoter containing p53 response elementsor the GADD45 promoter. The results were consistent with the apopotic signal associated with the presence of vanadocene compounds not being triggered by primary DNA damage and not requiring p53 induction [15]. Thus, it appears to control cell growth differently than cisplatin and may complement its use.

It is important to note that many metallocene derivatives lack the biological activities cited before, so their presence does not automatically impart biological activity onto molecules containing this moiety.

3 Spermicidal Activity

Researchers at the Wayne Hughes Institute have pioneered work related to the use of metallocene complexes, focusing on vanadocene derivatives, as spermicidal drugs. Briefly, several vanadium salts were found to act as modulators of cellular redox potential and to exert pleiotropic effects in a number of biological systems by catalyzing the creation of active oxygen compounds [16,17]. Such reactive oxygen species affect sperm motility through peroxidation of membrane lipids and proteins [18,19]. Peroxidative damage to the sperm plasma membranes is also believed to be important in the onset of male infertility [20]. Sperm are susceptible to oxidative stress because of the high content of unsaturated fatty acids and the relative paucity of cytoplasmic enzymes to scavenge reactive oxygen compounds that initiate lipid peroxidation [21].

They have tested metallocenes containing Ti, Zr, Mo, and Hf and found that they did not have spermicidal activity [22]. In comparison, a wide range of vanadocene compounds exhibit good spermicidal activity. Some of these compounds are shown in Figure 9.1 in decreasing order of efficacy [22].

Vanadocene dichloride disrupts mitotic spindle formation in cancer cells [23]. Although the control cancer cells had organized mitotic spindles, organized as a bipolar microtubule array with the DNA organized on a metaphase plate, the vanadocene dichloride treated cancer cells had aberrant monopolar mitotic structures where the microtubules were found on only one side of the chromosomes with the chromosomes arranged in a circular pattern. This is evidence that the vanadocene dichloride can stop cell division though disrupting bipolar spindle formation causing an arrest in the G2/M phase of cell cycle reproduction [23].

Thus, results from studying vanadocenes as antiproliferative compounds have shown another potential avenue for their anticancer activity.

4 Model Compounds for Polymer Formation

Polymer formation is generally achieved through reaction of Lewis acid (the vanadocene dichloride) base reactions. The exception to this is reviewed first. Lee and Brubaker [23] described the attachment of various metallocenes, including vanadocene, to polystyrene-divinylbenzene beads for the purpose of creating stabilized catalysts. The reactions occurred by employing exchange reactions between Cp_4V, Cp_3V, or $(C_6H_5CH_2C_5H_4)Cp_2VCl$ and the polystyrene-divinylbenzene beads through exchange of the Cp groups resulting in the attachment of the various vanadocenes to the beads.

Tarasov and coworkers reported on the formation of 1:1 complexes from the reaction of vanadocene dichloride and tartaric acids.[25] The products probably include materials with structure (**1**) when in basic solution.

Chelated materials were formed from reaction of vanadocene dichloride with heterocyclic 1,2-dithiolates [26]. The product from 1,3-dithiolene-2-thione-4,5-ditholate is given in (**2**).

Cyclic phosphate compounds are formed when aqueous solutions of vanadocene dichloride and phosphates are brought together. The proposed structure is given as (**3**) [27].

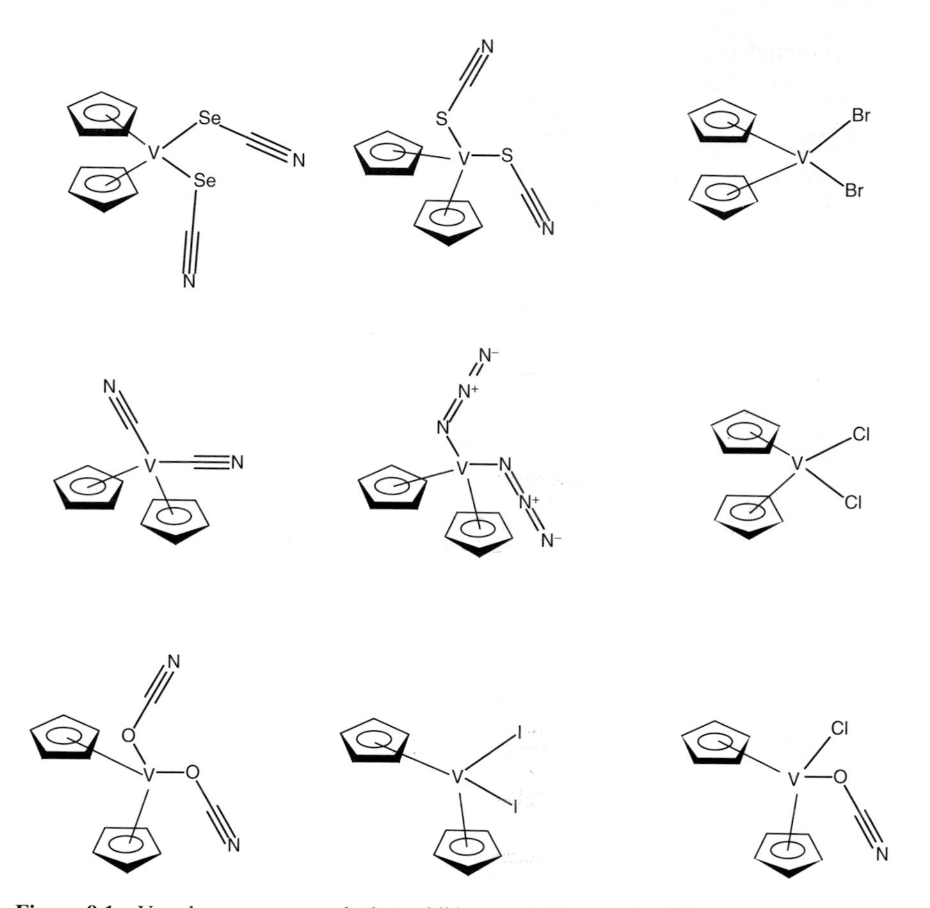

Figure 9.1 Vanadocene compounds that exhibit spermicidal activity [23]

(1)

This work is aimed at helping understand the ability of metallocene dichlorides to inhibit nucleic acid growth. These reactions are similar to ones that can be carried out to form polymers.

5 Dendrimers

Several examples of the formation of complexes from dendrites are present in the literature. The first report describes the formation of metallosilicates that are subsequently incorporated into a silica matrix using the sol-gel method. The metallosilates structure is given in (**4**). These dendrimeric silanes show good activity and selectivity towards the epoxidation of alkenes with *t*-butylhydroperoxide [28].

Malik, Duncan, Tomalia, and Esfand [29] reported the use of a variety of amine and acid-containing polymers that chelated platinum forming cisplatin derivatives. They also reported the synthesis of similar dendritic-antineoplastic drugs using vanadocene dichloride. The structure from the poly(propyleneimine) dendrimers contains units are shown in (5) and (6).

(**2**)

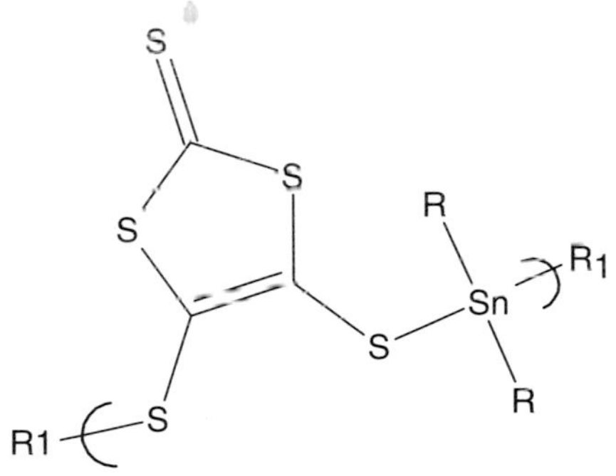

S S S S Sn R R R1 R1 S

(**3**)

O O P OH V O

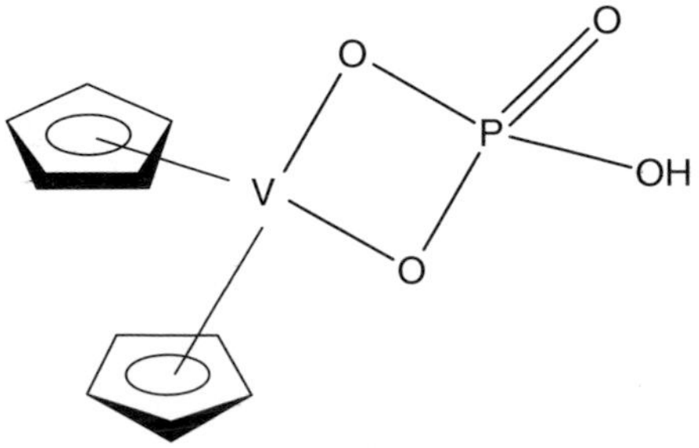

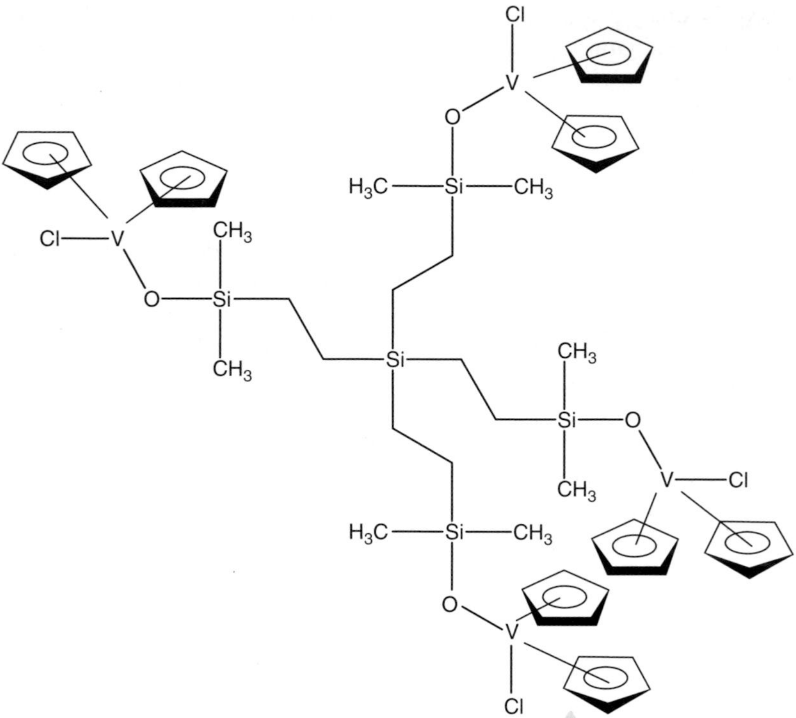

Cl
V
O
H_3C—Si—CH_3
Cl—V
CH_3
O—Si
CH_3
Si
CH_3
Si—O
CH_3
V—Cl
H_3C—Si—CH_3
O
V
Cl

(4)

(5)

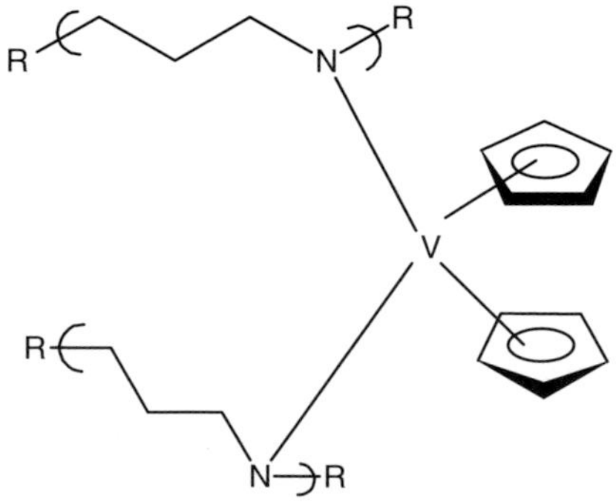

R
N
R
V
R
N—R

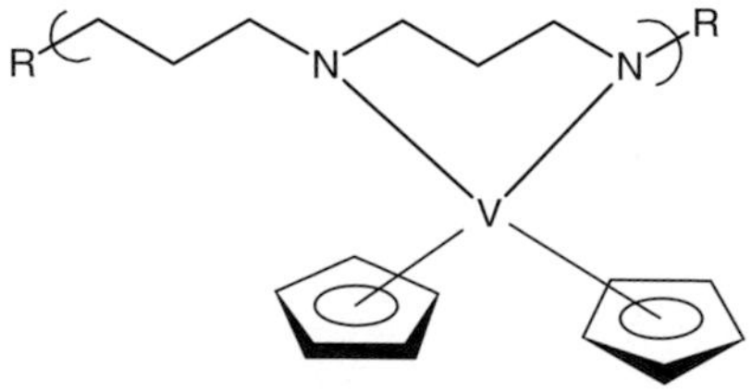

R
N
N
R
V

(6)

6 Linear Polymers

Recently we reported the synthesis of vanadocene polyethers [30] from the reaction of vanadocene with diols forming products with structure shown in (**7**).

We also synthesized the analogous polyesters (**8**) from reaction with salts of carboxylic acids [31].

7 Acyclovir Polymers

Acyclovir is widely used to inhibit several herpes viruses, particularly HSV-1 and HSV-2 [32–34]. It is also used to treat varicella-zoster virus (VZV), Epstein-Barr virus (EBV), and cytomegalovirus (CMV). The inhibitory activity of acyclovir is highly selective. Thus, acyclovir is a first line antivirial drug.

Recently, we reported the synthesis of organotin products, as shown in (**9**), containing acyclovir [35,36].

(**7**)

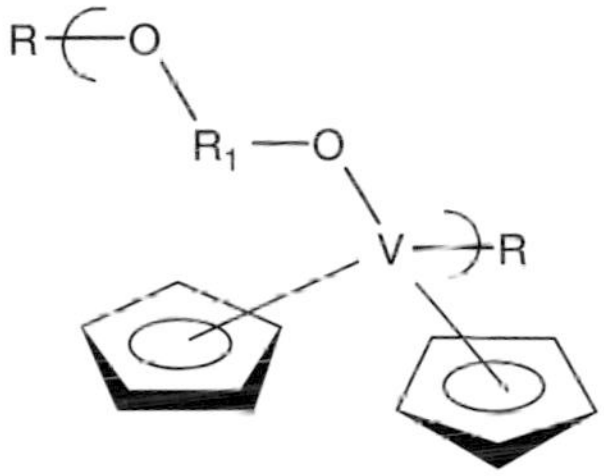

R O R1 O V R

(**8**)

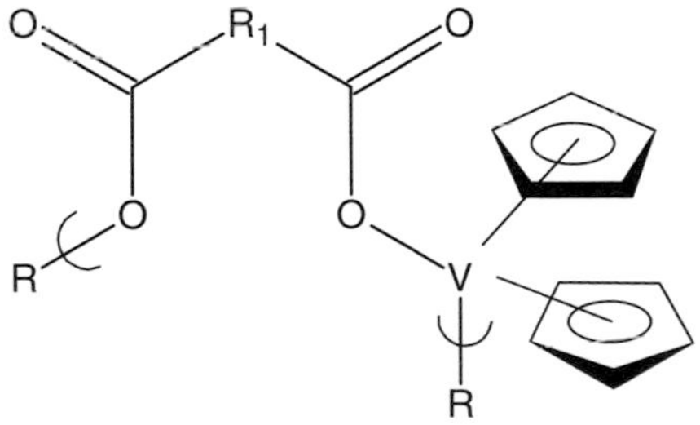

O R1 O O O R V R

HN O N N N O O Sn R R R1 n R1 NH

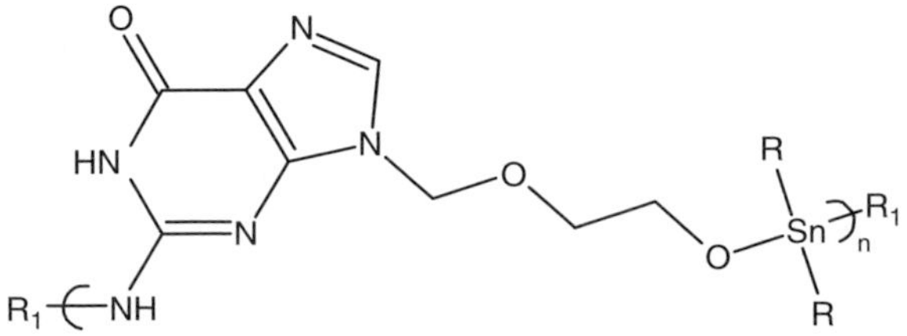

(**9**)

These materials are potent antivirial agents [37,38]. More recently we reported the synthesis of Group IVB metallocene products and the synthesis of Group IVB acyclovir polymers of the following structure from reactions with titanocene dichloride shown in (10) [39].

As part of our effort to create anticancer, antibacterial, and antiviral agents we recently synthesized the analogous product except employing vanadocene dichloride giving polymers [40] of the form shown in (11).

Our main reason for synthesizing the vanadocene polymers is for comparison with the organotin and Group IVB analogous polymers containing the acyclovir moiety with respect to biological activity.

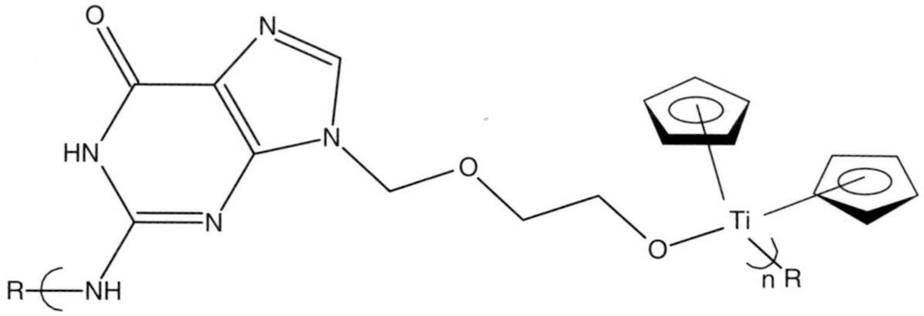

(10)

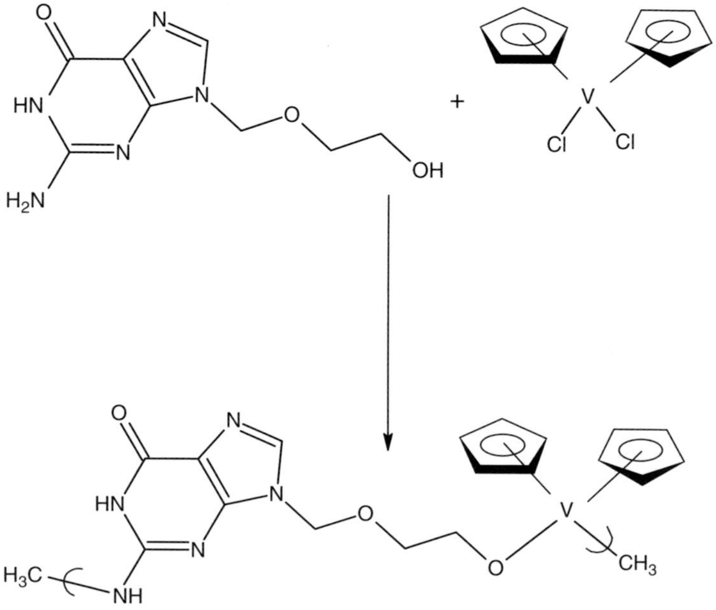

(11)

Polymers offer potential advantages compared to smaller, monomeric drugs. First, a polymer should be filtered out by the kidneys more slowly than small compounds, decreasing kidney damage and increasing the time the compound remains in the body to attack the cancer or other target organism. By selective targeting of the polymer it can be used to deliver local, high-dose chemotherapy simply through control of membrane penetration. Second, a polymer may be effective against tumors that have developed resistance to other chemotherapeutic agents because the polymer is not recognized by cellular resistance mechanisms. Third, the size and structure of the polymer provides more binding sites to cellular targets, increasing effectiveness. Fourth, a polymer can be designed to incorporate multiple anticancer agents within the same molecule that act against cancer cells by different mechanisms. Fifth, polymer structure can permit easy coupling to other molecules, such as those that specifically target cancer cells, allowing delivery of a polymeric drug to a particular site without impairing its effectiveness. Sixth, a polymer can be designed as either a large stable compound that enters the cell by pinocytosis and is active in a polymeric form, or as an unstable compound that slowly degrades into active monomeric units in a time-delivery fashion. Seventh, polymers accumulate in solid tumors more than in normal tissues because of the enhanced permeability and retention effect, resulting in high amounts of polymers in the interstitial space resulting from leaky vasculature and limited lymphatic drainage typical of tumors. Finally, polymers can be synthesized to fine-tune many structural characteristics, such as monomeric components, chain length, crosslinking, and polarity, to maximize anticancer activity and vary the spectrum of activity. Thus, polymers have great flexibility in their design and many potential benefits compared to monomeric drugs.

The reaction between acyclovir and vanadocene dichloride occurs almost instantaneously with a color change from green to a brown or brownish green. The product is formed in about 61% yield. It is a high polymer with a weight average molecular weight in HMPA of 2.1×10^6 (dn/dc = 0.3) corresponding to an average degree of polymerization of about 5,200. Chain length was followed as a function of time. Condensation polymers generally degrade though a random scission route [41,42]. The Schultz-Flory equation (Eq. 9.1) was developed to describe the change in chain length with time for such systems.

$$1/(DP)_t = 1/(DP)_0 + 0.5k[K]t \tag{9.1}$$

where DP_t is the weight average degree of polymerization at time t, $(DP)_0$ is the original weight average degree of polymerization, k is the rate of chain scission, and K is the molar concentration of the solvent which is generally taken to be a constant. Thus, a plot of $1/(DP)$ as a function of time should give a straight line. This is what is roughly found in Figure 9.2. Thus over approximately 7 weeks molecular weight decreases from 2.1×10^6 to 6.8×10^4.

Infrared spectroscopy was carried out on the products and reactants. Table 9.1 contains infrared bands and assignments for acyclovir and the vanadocene polymer. All locations are given in cm^{-1}.

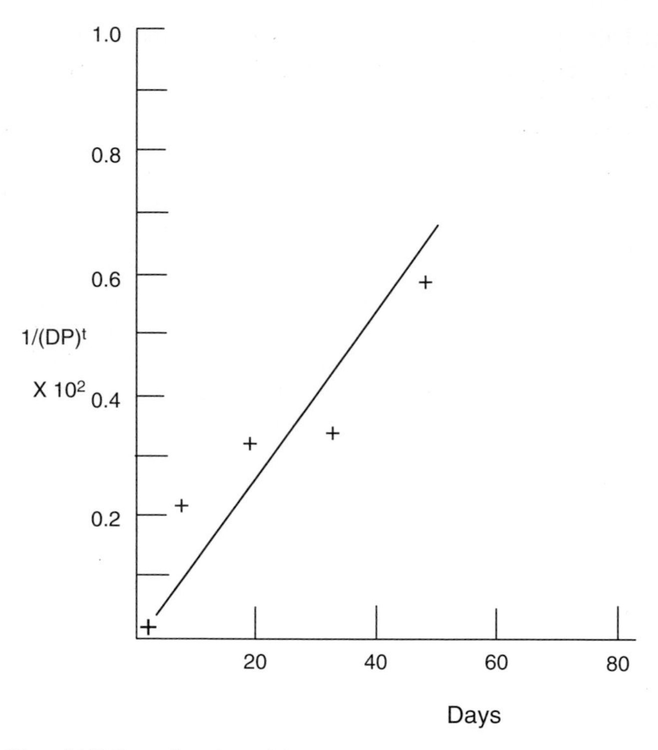

Figure 9.2 Plot of $1/DP$ as a function of time

Table 9.1 Infrared band locations and assignments for selected bands for the polymer derived from reaction of acyclovir and vadanocene polymer

Assign	Acyclovir	Cp_2VCl_2	Cp_2VCl_2 Polymer
CH St Arom.	3172	3095	3101
CH St Alip.	2930		2929
	2851; 2716		2851; 2,720
Cp CC st.		1445	1461
Cp CH ip Bending		1025	1016
Cp CH op Bending		850	850; 810
Purine Carbonyl	1716		1699
OH St.	3441; 3523		3450
NH St	3295		3300
COC Ether	1070		1067
NH	1632		1630
C–H; C–N Purine	1485		1480
C–N; N–C Purine	1183		1182

The metallocene monomers and polymers all show bands at about 1440 (CC stretch), 1015 (CH ip bending), and 820 (CH op bending) characteristic of various Cp bending and stretching modes. Consistent with other studies, these bands are less intense than most bands associated with most organic compounds. Even so, they are present in the products. The OH is absent or greatly diminished for the products consistent with the formation of the M–O and absence of the R–OH groupings.

The location of bands associated with the V–O was reported to be at about 950 for the symmetric stretch. The presence of a new band at about 979 was tentatively assigned to the V–O stretch. Acyclovir had a single band at 454 that was present in the product at 449. A new band about 419 was tentatively assigned as the asymmetric V–O stretching. The location of the V–N band was not know but believed to be in the general area of 400. The new band at 397 was tentatively assigned to the Zr–N stretch. The V–Cl stretching band occurs in the range of 385 to 300 [21] and is not identified in the present work.

8 F IIR Matrix-Assisted Laser Desorption/Ionization Time-of-Flight Mass Spectrometry (MALDI-TOF MS)

We have been developing procedures for the analysis of metal-containing polymers because of their general difficulty in achieving sufficient solubility to allow full advantage to be taken of analysis techniques that require the polymer to be decently soluble in appropriate solvents. One of these techniques is mass spectrometry. We recently investigated the use of matrix-assisted laser desorption/ionization time-of-flight mass spectrometry (MALDI MS) as a tool to better describe the composition of such metal-containing polymers. Because the focus is on the fragments rather than entire chains we have named this approach Fragmentation MALDI MS or simply F MALDI MS. Here we report some of the results of the acyclovir-vanadocene polymers. Table 9.2 contains the ion fragments created in the 100 to 1,000 amu range. A number of abbreviations are employed to describe ion fragment assignments. V represents Cp_2V, Ac represents acyclovir minus two protons; C represents a methylene moiety. The symbols "+" and "–" represent additions or deletions.

Table 9.2 Most abundant ion fragments generated in the 100 to 1,000 amu range for the product of vanadocene dichloride and acyclovir

m/e assignment	(Possible)	m/e assignment	(Possible)
152	Ac–OCCOC	226	Ac
248	Ac,Na	264	V,Na+OCCO
304	V+2OCCO	481	U+OCCOC
494	U,Na+OCCO	508	U,Na+OCCOC
522	UAc–OCC,OCCOC	537	U,Na–OCC,OCCOC

Although we have often somewhat arbitrarily divided the mass range for low (below 1,000 amu), medium (11,000–10,000 amu) and high (10,000–100,000 amu) range ion fragments, there is a difference that needs to be taken into account as the ion fragment mass increases. Briefly, the masses used to calculate proposed ion fragment structures must be more precise as the mass range increases.

To illustrate a strength of MALDI MS, Table 9.3 concentrates on only a small spectral range, from 3,680 to 3,910 amu, which corresponds to about 10 units. There is an orderly breakage at the expected sites. The only internal fragmentation was the loss of two cyclopentadienyl groups which gave rise to the ion fragments at 3,910, 3,888, and 3,753 amu. It was believed that most of the chain remains intact with little fragmentation except at the site of bond scission. Here, one end of the chain has a Cp_2V moiety that probably loses the Cp groups. Loss of Cp groups is common for EI-MS, which is less gentle than MALDI MS. Even so, the loss of two Cp groups is not unexpected and occurs only at the site of bond scission.

Only a few "inner-chain" breakages are believed to occur. Again, these scissions occur at terminal Cp_2V groups and involve the loss of Cp, whereas this is not prevalent, it does occur occasionally. Bond breakages are as expected and noted below by arrows in (**12**). It is interesting to note that there was not widespread breakage of the guanosine–NH bond, as there was evidence for the breakage of the Cp_2V–O and/or Cp_2V–NH linkage and breakage of the methylene-ether system.

Table 9.3 Ion fragments in the mass range of 3,680 to 3,910 daltons for the product of acyclovir and vanadocene dichloride

m/e	(Tentative) assignment	m/e	(Tentative) assignment
3910	10U-2Cp	3888	10U-2Cp,O,C
3860	9U+Ac	3849	9U+V,OC
3833	9U+V,O	3825	9U+9U+V
3800	9U+Ac–OCCOC	3786	9U+Ac–PN
3753	9U+V–Cp	3716	9U+OCCOC
3680	9U+OCC		

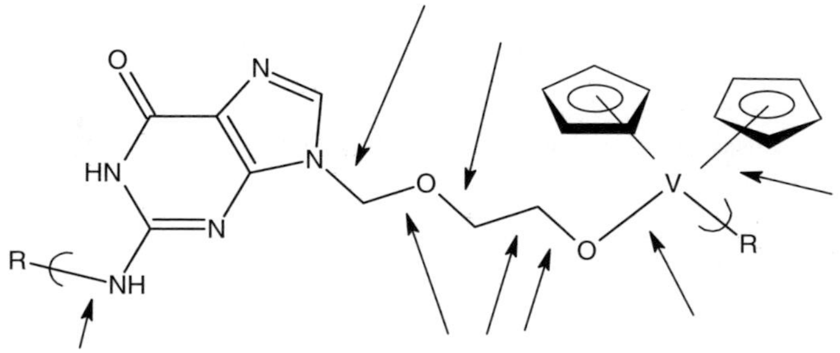

O HN N N N NH R O O V R

(**12**)

9 Fibers

In an unrelated study we found that the vanadocene polymers produce fibers when isolated. The fibers were mixed in with nonfiberous flat plate-like structures. Fibers have been found for a number of different metal-containing condensation polymers, normally those that have rigid backbones. The acyclovir backbone contains both the rigid guanosine portion and the more flexible ether-tail. Apparently, the presence of the guanosine ring system is sufficient to cause fiber formation. Some of the fibers were sufficiently small and therefore not readily visible with the naked eye but rather can be seens only with tenfold magnification. There appeared to be two types of fibers—one that was relatively short (with lengths of about 0.4 mm) and thicker fiber (diameter about 0.01 mm) with aspect ratios ranging from 40 to 100 (Figure 9.3) These fibers appeared to have many spurs coming off the main shoot and were very abundant.

The second fibers were much longer (4 mm) and somewhat thinner (0.01 mm), with aspect ratios ranging from 400 to 1,000 (Figure 9.4). They were more slick, more flexible, and clear. As noted before, this was not the only report of such anomalous fibers. Mechanical tests should be made of these fibers and their potential in composites determined.

Figure 9.3 Representative shorter fibers at ×60 magnification

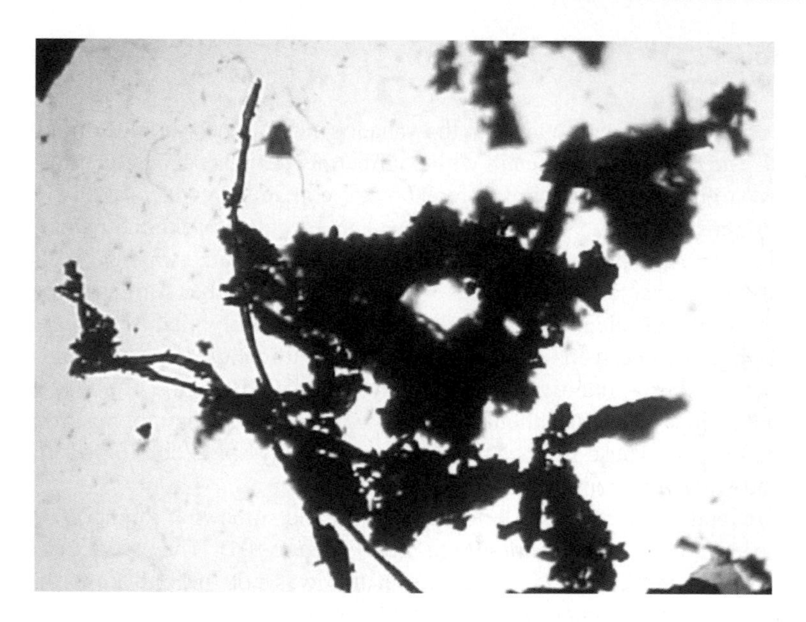

Figure 9.4 Fibers at ×10 magnification. Representative "longer" fibers

References

1. Kopf-Maier P, Kopf H. (1988) Struct Bonding. 70:103.
2. Kopf-Maier P. (1993) Metal complexes in cancer chemotherapy: antitumour bis(cyclopentadi enyl)metal complexes. VCH, Weinheim , Germany.
3. Kopf-Maier P, Kopf H. (1994) Metal compounds in cancer therapy: organometallic titanium, vanadium, niobium, molybdenum, and rhenium complexes: early transition metal antitumour drugs. Chapman and Hall, London.
4. Christodoulou C, Eliopoulos A, Young L, Hodgkins L, Ferry D, Kerr D. (1998) Br J Cancer. 77:2088.
5. Lummen G, Sperling H, Luboldt H, Otto T, Rubben H. (1998) Cancer Chemother Pharmacol. 42:415.
6. Krogeer N, Kleeberg U, Mross K, Sass G, Hossfeld D. (2000) Onkologie. 23:60.
7. Kopf-Maier P, Krahl D. (1981) Naturwissenschaften. 68:273.
8. Kopf-Maier P, Krahl D. (1983) Chem. Biol Interact. 44:317.
9. Sun H, Li H, Weir R, Sadler P. (1998) Angew Chem Ind Ed. 37:1577.
10. Kuo L, Li A,Marks T. (1996) Metal ions in biological systems: metallocene interactions with DNA and DNA-processing enzymes, Vol. 33. Marcel Dekker, New York.
11. Mokdsi G, Harding M. (2001) J Inorg Biochem. 83:205.
12. Harding M, Mokdsi G. (2000) Current Medicinal Chem. 7:1289.
13. Murthy M, Rao L, Kuo L, Toney J, Marks T. (1998) Inorg Chimica Acta. 152:117.
14. Djordjevic C. (1995) Metal Ions Bio Systems. 31:595.
15. Aubrecht J, Naria R, Ghosh P, Stanek J, Uckun F. (1999) Tox Applied Pharm. 154:228.
16. Sakurai H, Tamura H, Okatani K. (1995) Biochem Biophys Res Commun. 206:133.
17. Shi X, Wang P, Mao Y, Ahmed N, Dalal N. (1989) Ann Clin Lab Sci. 26:39.
18. Aitken R, Clarkson J, Fishel S. (1989) Biol Reprod 40:183.
19. Jones R, Mann T, Sherins R. (1979)Fertil Steril. 31:531.

20. Aitken J, Fisher H. (1994) Bioessays. 16:259.
21. Alvarez J, Touchstone J, Blaso L, Stoney B. (1987) J Androl 8:338.
22. D'Cruz O, Ghosh P, Uckun F. (1998) Biology Reproduction. 58:1515.
23. Navara C, Benyumov A, Vassilev A, Naria R, Ghosh P, Uckun F. (2001) Anti-Cancer Drugs. 12:369.
24. Lee J, Brubaker C. (1977) J Organometal Chem. 135:115.
25. Tarasov O, Glebov A, Lineva A, Latyaeva V, Sal'nikov YU. (1989) Metalloogranicheskaya Khimiya 2:627.
26. Zeltner S, Dietzsch W, Olk R, et al. (1994) Zeitschrift Anorgan. Allgemeine Chemie. 620:1768.
27. Vinklarek J, Honzicek J, Holubova J. (2004) Inorganic Chim Acta. 357:3765.
28. Juwiler D, Neumann R. (2001) Catalysis Lets. 72:241.
29. Malik N, Ducan R, Tomalia D, Esfand R. (2002) US Patent Appl. 20030064050.
30. Carraher C, Randolph E. (1997) Polym Mater Sci Eng. 77:497.
31. Carraher C, Randolph E. (1998) Polym Mater Sci Eng. 79:50.
32. Schaeffer HJ. (1976) US Patent No. 4199574.
33. Elion G, Furman P, Fyfe J, Miranda P, Beauchamp L, Schaeffer H. (1977) Proc Natl Acad Sci USA. 74;5716.
34. Physicians' Desk Reference, 55 Edition. (2001) Medical Economics, Thompson Heathcare, Montville, NJ.
35. Bleicher R, Carraher C. (2002) Polym Mater Sci Eng. 86:289.
36. Carraher C, Bleicher R. (2004) In; Macromolecules containing metal and metal-like elements, Vol. 3 Wiley, Hoboken, NJ.
37. Roner M, Carraher C, Zhao A, Roehr J, Bussett K, Siegmann-Louda D. (2004) Polym Mater Sci Eng., 90:515.
38. Roner M, Carraher C, Zhao A, Roehr J, Bassett K, Siegmann-Louda D. (2003) Polym Mater Sci Eng. 89:525.
39. Sabir T, Carraher C. (2005) Polym Mater Sci Eng. 92:485.
40. Sabir T, Carraher C. (2005) Polym Mater Sci Eng. 93:396.
41. Elias H. (1984) Macromolecules, Vol. 2. Plenum, New York, p 834.
42. Carraher C. (2003) Polymer Chemistry, 6th ed. Marcel, Dekker, New York.

Hafnium-Containing Nanocomposites

A.D. Pomogailo, A.S. Rozenberg, G.I. Dzhardimalieva, A.M. Bochkin, S.I. Pomogailo, and N.D. Golubeva

1 Introduction

In recent years, a great deal of attention has been paid to various approaches to produce metal–polymer nanocomposites containing ultrafine particles of metals or their compounds, in polymer matrices through polymer synthesis [1]. The development of effective processes for the fabrication of high-temperature composites based on refractory compounds, such as hafnium oxides and carbides, is of considerable interest for the preparation of thermally stable, high-strength coatings. For example, the melting point of hafnium carbide is as high as 3928 ± 20 °C [2] and is only exceeded by that of tantalum carbide (3,983 °C).

Hafnium compounds are widely used in the fabrication of composites capable of withstanding high temperatures and aggressive attack and also in basic organic

A.S. Abd-El-Aziz et al. (eds.), *Inorganic and Organometallic Macromolecules: Design and Applications.*

synthesis, in particular as Ziegler–Natta catalysts. Hafnium carbide composites are perspective candidates as oxidation-resistant high-temperature materials which can, for example, survive surface temperatures of 4,800 °F for more than 10 minutes [3]. It is known that some missile components are exposed to temperatures exceeding 5,000 °F and for their protection high-temperature materials are needed.

The most simple methods for preparation of hafnium carbide and its composites are the follows. The powder of HfO_2 was thermally treated with Mg in molar ratio 5:4 under a CH_4 flow ranging from 800 to 950 °C [4]. The effective high temperature coating for carbon fiber reinforced carbon and carbon fibre reinforced silicon carbide was prepared with use of HfC [5]. For this purpose hafnium carbide layers were obtained in a thermally simulated chemical vapor deposition (CVD) reactor on nonporous substrates by reaction of hafnium tetrachloride, methane and addition of hydrogen (Eq. 10.1):

$$HfCl_4 + CH_4 + H_2 \rightarrow HfC + HfCl_{4\text{-}X} + HCl \quad (10.1)$$

Thermodynamic modeling of the reaction shows the possibility of depositing a nearly carbon-free hafnium carbide layer. The carbon-to-hafnium ratio of the deposited layer closely agreed with the thermodynamic predictions. The same method is described in reference [6]. Such an approach allowed the production of hafnium carbide as a wear resistant coating for hard metal cutting tools as well as a barrier for oxygen in multilayer coatings to protect carbon fiber reinforced carbon composite materials at high temperatures [7]. The CVD process was carried out 850 °C, at temperature above 1,050 °C HfC-whiskers are formed [8]. The thermally induced CVD technique was used to produce multilayer coatings containing HfC for carbon materials protection [9]. It is important that such coatings show high adhesion properties. Problems concerning the formation of coatings composed of the carbides of metals of the IV–V groups on fiber under conditions of chemical transport through the gas phase are considered in reference [10].

For the CVD processes organometallic hafnium compounds are often used. Thus, thin films of hafnium carbide were obtained in a planar reactor from bis(cyclopentadienyl)dimethylhafnium, $(\eta^5\text{-}C_5H_5)Hf(CH_3)_2$ as precursor [11]. The carbon content ranged from 11 to 40 weight % and increased with the deposition rate. The film hardness varied between 1,300 and 2,000 HK.

There are currently three basic approaches to the preparation of metal-containing polymers [12] as shown in Structure 10.1 (I) reactions of metal compounds with functionalized linear polymers in which the main polymer chain remains unchanged (so-called polymer-analogous conversions); (II) polycondensation of appropriate precursors, a process in which a metal ion is incorporated into the main chain, and its removal leads to destruction of the polymer; (III) polymerization and copolymerization of metal-containing monomers (recently devised approach):

Transition-metal-containing polymers prepared by process I (Structure 10.1, I) are, as a rule, characterized by a low content of combined metal and are used primarily in ion-exchange extraction, preconcentration, and separation of metals by selective extraction [13–15].

Unfortunately, little or no data have been reported on the preparation of Hf-containing polymers. Most effort has been concentrated on polymers containing Ti and, to a lesser extent, Zr, electronic analogs of hafnium [12]. Thus, even the recent reviews [16,17], devoted to macromolecules contained the triads of these metals have almost no information about hafnium-containing polymers. Also, no data exist on the feasibility of the synthesis and polymerization of Hf-containing monomers—compounds containing unsaturated ligands capable of polymerization. The same refers to the pyrolysis of Hf-containing polymers.

This chapter details the synthesis of appropriate Hf-containing precursors, their polymerization, the composition and structure of the resulting Hf-containing polymers, approaches to controlled thermolysis of the polymers, and the composition of the Hf-containing thermolysis products. Hf-containing polymers were prepared by the three processes I–III.

2 Results and Discussion

2.1 *Preparation of Hf-Containing Polymers Via Polymer-Analogous Conversions (Process I)*

By reacting methanolic solutions of PAA and $HfOCl_2$, up to 34 wt% Hf can be introduced into the polymer. The IR spectra of such macrocomplexes show a strong, narrow absorption band at 1,720 cm^{-1}, which points to a significant concentration of

$$\sim CH_2-\underset{\displaystyle L}{\underset{|}{CH}}-CH_2-\underset{\displaystyle L}{\underset{|}{CH}}-CH_2-\underset{\displaystyle L}{\underset{|}{CH}}\sim \; + MX_n \longrightarrow \sim CH_2-\underset{\displaystyle L}{\underset{|}{CH}}-CH_2-\underset{\underset{\displaystyle MX_n}{\downarrow}}{\underset{\displaystyle L}{\underset{|}{CH}}}-CH_2-\underset{\displaystyle L}{\underset{|}{CH}}\sim, \qquad (I)$$

$$HL\text{–}R\text{–}LH + MX_n \xrightarrow[-2HX]{} \sim L\text{–}R\text{–}L\text{–}MX_{n-2}\sim, \qquad (II)$$

$$\underset{\underset{\displaystyle MX_{n-m}}{|}}{\underset{\displaystyle Z}{\underset{|}{CH_2=CH}}} \xrightarrow{\text{Initiation}} \underset{\underset{\displaystyle MX_{n-m}}{|}}{\underset{\displaystyle Z}{\underset{|}{\sim CH_2-CH\sim}}}, \qquad (III)$$

where *L*, *X*, and *Z* are functional groups, and *R* is a bridge group.

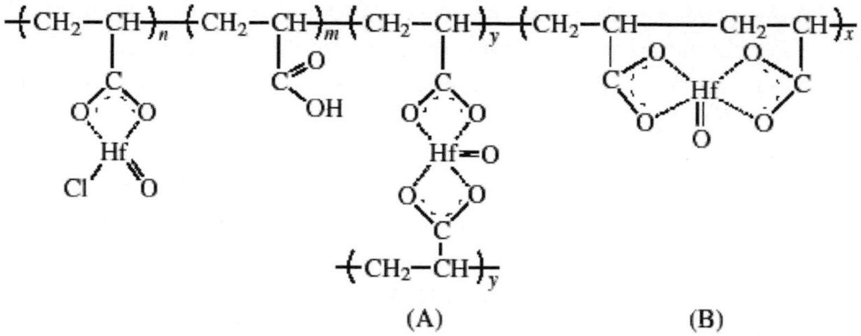

(A) (B)

unreacted carboxyl groups. At the same time, the presence of totally symmetric (1,328 and 1,457 cm^{-1}) and asymmetric (1,751 cm^{-1}) stretching modes of COO^- confirm that the hafnium ions are in bidentate and bridge coordination with carboxylate ions. Moreover, the presence of residual chlorine (up to 2.6%) in the synthesized macrocomplex indicates that the PAA–Hf(IV) (polymer **1**) chain contains monosubstituted hafnium oxychloride units. The general structural scheme of the oxypolyacrylate complex is represented in Structure 10.2.

In addition to interchain coordination (Structure 10.2, A), intramolecular cyclization (Structure 10.2, B) cannot be ruled out. Moreover, the formation of crosslinked structures may also be contributed by oxobridges with the participation of hafnyl groups, as will be shown below.

2.2 Preparation of Hf-Containing Polymers through Polycondensation (Process II)

To prepare polymers containing metals (e.g., Sn, Pb, Ni, Pd, Pt, or Ti) in the main chain, use is commonly made of the linear condensation of metal dihalides and bifunctional Lewis bases [16], such as diols, diamines, dicarboxylic acids, hydrazines, oximes, and dithiols. The preparation of such metal-containing polymers can be represented in Scheme 10.1.

Although this scheme includes the titanium triad, no case of the preparation of Hf-containing polymers has come to our notice.

Consider the most characteristic condensation reactions for the preparation of Hf-containing polymers.

2.2.1 Diol-Based Hf-Bontaining Polymers

Polymers containing Hf atoms in the main chain were prepared by the reactions in Scheme 10.2.

Reaction 1 in Scheme 10.2 was used to prepare polymers based on 1,4-butanediol ($m = 4$) and 1,9-nonanediol ($m = 9$) (Table 10.1, polymers **2** and **3**, respectively).

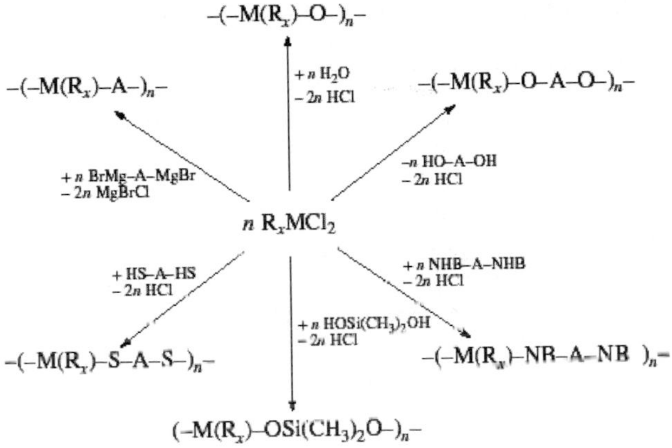

$$-(-M(R_x)-O-)_n-$$

$$-(-M(R_x)-A-)_n- \quad -(-M(R_x)-O-A-O-)_n-$$

$+ n$ H_2O, $- 2n$ HCl; $+ n$ BrMg–A–MgBr, $- 2n$ MgBrCl; $-n$ HO–A–OH, $- 2n$ HCl

$$n\ R_xMCl_2$$

$+$ HS–A–HS, $- 2n$ HCl; $+ n$ NHB–A–NHB, $- 2n$ HCl; $+ n$ $HOSi(CH_3)_2OH$, $- 2n$ HCl

$$-(-M(R_x)-S-A-S-)_n- \quad -(-M(R_x)-NB-A-NB\ )_n-$$

$$(-M(R_x)-OSi(CH_3)_2O-)_n-$$

M = Si, Ge, Sn, Pb
As, Sb
Ni, Pd, Pt
Ti, Zr, Hf
R = H, alkyl, cyclopentadiene, phosphine ($x = 1, 2$)
B = H, alkyl, aryl
A – alkylene, arylene, or $-M(R_x)$

Scheme 10.1

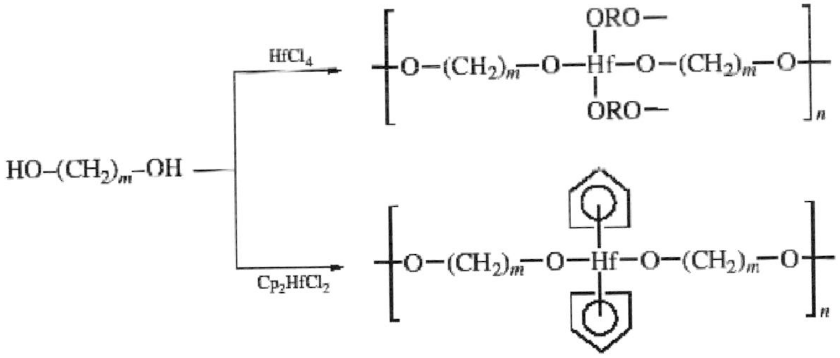

$$HO-(CH_2)_m-OH \xrightarrow{HfCl_4} \left[-O-(CH_2)_m-O-\underset{ORO-}{\overset{ORO-}{Hf}}-O-(CH_2)_m-O-\right]_n \quad (1)$$

$$HO-(CH_2)_m-OH \xrightarrow{Cp_2HfCl_2} \left[-O-(CH_2)_m-O-\underset{Cp}{\overset{Cp}{Hf}}-O-(CH_2)_m-O-\right]_n \quad (2)$$

Scheme 10.2

Table 10.1 Elemental analysis data for Hf-containing polymers (process I)

	Weight percent (assay/nominal)		
Polymer	Hf	C*	H
2	42.5/43.03	24.89/26.51	5.19/5.31
3	43.4/43.34	18.06/26.30	4.52/5.84
4	45.41/43.0	24.82/34.7	5.41/4.59
5	44.94/43.20	22.54/42.23	5.31/4.61
6	42.42/38.23	31.07/47.71	5.35/6.92

* The low carbon content may be the result of the formation of hafnium carbide in the course of analyses. The carbon content was higher if analyses were carried out in the presence of PbO.

Table 10.2 Molecular weight of Hf-containing polymers

Polymer	M_n	M_w	γ	t_g, Y°C
2	4,100	5,600	1.37	90
3	150,000	314,000	2.09	90
4	33,000	46,000	1.4	–
5	42,000	57,000	1.36	–
6	41,000	68,000	1.66	–
9	132,000	224,000	1.70	56

Note: M_n and M_w are the number-average and weight-average molecular weights, respectively; $\gamma = M_w / M_n$; t_g is the glass transition temperature.

Reaction 2 in Scheme 10.2 was used to prepare polymers based on 1,4-butanediol ($m = 4$, polymer **4**), 1,6-hexanediol ($m = 6$, polymer **5**), and 1,9-nonanediol ($m = 9$, polymer **6**). The Hf content of these polymers is 42–45 wt %. Their molecular weight was determined by analyzing the shape of the thermomechanical curve (Table 10.2).

The polymers have the form of white, fine powders (specific surface of up to 120 m^2/g) which dissolve very little in organic solvents and water.

The IR spectra of the polymers show absorptions characteristic of diol residues and free hydroxyls: 3,369, 2,929, 2,854, and 1,626 cm^{-1} (Figure 10.1).

The formation of three-dimensional networks of the form shown in Structure 10.4 is hindered sterically and is accompanied by side reactions, e.g., diol cyclization, as in reactions of 1,4-butanediol with organoaluminum compounds [18], or dehydrocyclization, as in reactions of 1,4-butanediol and other diols with Cp_xMCl_{4-x} [19]. Such processes prevent conversion from reaching 100%, as also evidenced by the presence of residual chlorine in the polymers.

Characteristically, Hf(IV) compounds contain oxo-bridges [20] of the form shown in Structure 10.5.

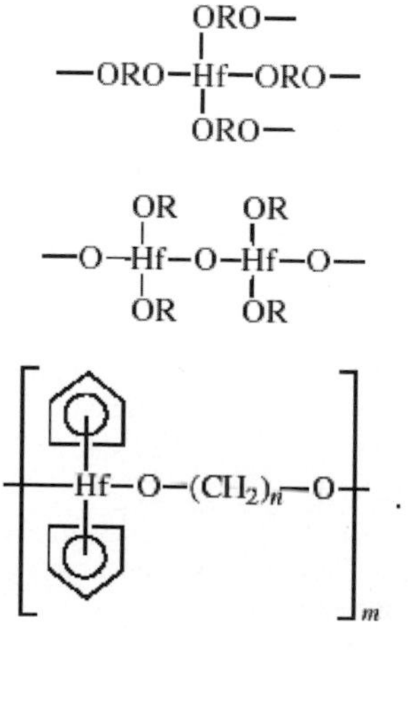

ORO—
—ORO—Hf—ORO—
ORO—

OR OR
—O—Hf—O—Hf—O—
OR OR

$$\left[-\mathrm{Hf(Cp)_2}-\mathrm{O}-(\mathrm{CH_2})_n-\mathrm{O}- \right]_m .$$

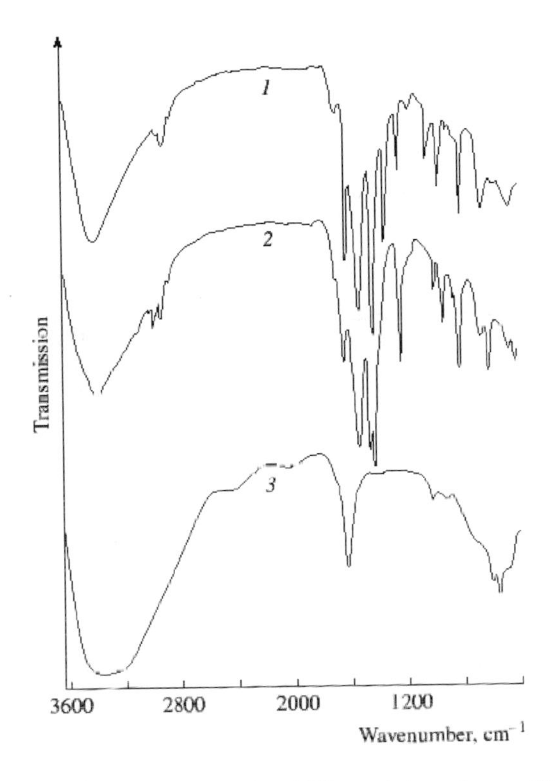

Figure 10.1 IR spectra of (*1*) 1,9-nonanediol and Hf-containing polymers (2) **2** and (*3*) **3**

Chemical analysis data indicate that, on the average, 3.0–3.8 Cl atoms per Hf atom are replaced in reactions of diols with $HfCl_4$, and 1.5–1.7 Cl atoms, in reactions with Cp_2HfCl_2. Given this, we took into account the possible formation of oxobridges in calculating the compositions of the condensation-type Hf-containing polymers from chemical analysis data (Table 10.1).

The basic approach to the preparation of Hf(IV)-containing organic polymers is to react readily available cyclopentadienyl derivatives of hafnium (most frequently, hafnocene dichloride) with the ligands used in reactions (1) and (2) (Scheme 10.2). This approach was used to prepare polymers of the type shown in Structure 10.6. The hafnium content of these polymers is in the range of 44 to 46 wt%, and their molecular weight is between 46,000 and 68,000. Their IR spectra show absorption bands arising from the Cp ring and diol fragments (Figure 10.2).

2.2.2 *n*-Diethynylbenzene-Based Hf-Containing Metal- Organic Polymers

Compounds containing M–C (M = transition metal) bonds are known to be unstable and to decompose at low temperatures. Polymerization of such monomers is

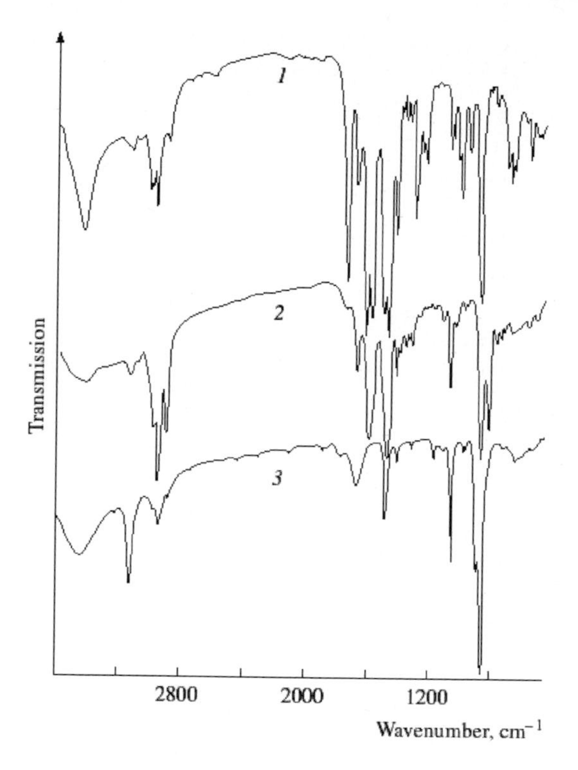

Figure 10.2 IR spectra of the condensation-type Hf-containing polymers (*1*) **4**, (*2*) **5**, and (*3*) **6**

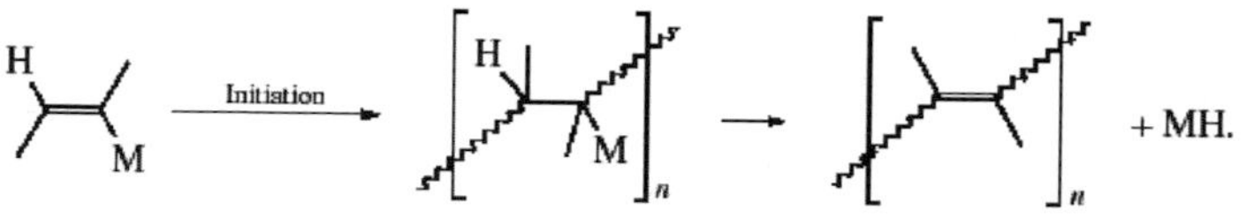

H, M — Initiation → [H, M]$_n$ → [ ]$_n$ + MH. (3)

Scheme 10.3

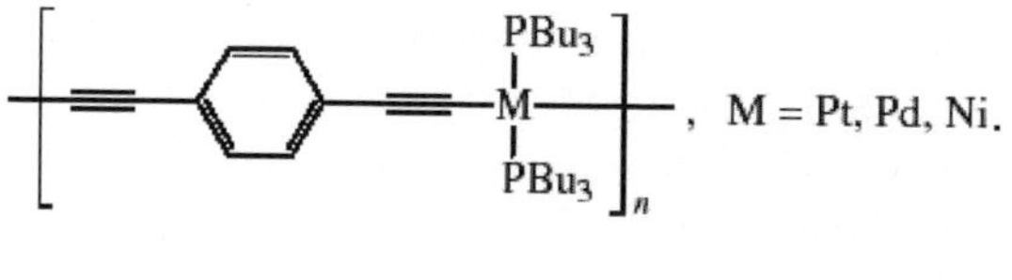

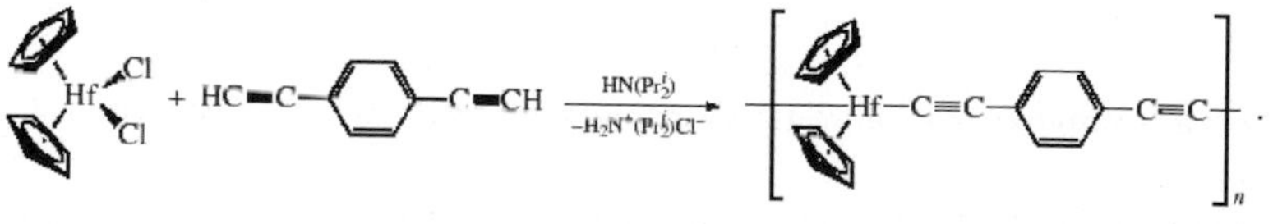

[–C≡C–C$_6$H$_4$–C≡C–M(PBu_3)$_2$–]$_n$, M = Pt, Pd, Ni.

Cp_2HfCl_2 + HC≡C–C$_6$H$_4$–C≡CH $\xrightarrow[-H_2N^+(Pr^i_2)Cl^-]{HN(Pr^i_2)}$ [Cp_2Hf–C≡C–C$_6$H$_4$–C≡C–]$_n$. (4)

Scheme 10.4

$$Cp_2HfCl_2 + LiC{\equiv}C{-}C_6H_4{-}C{\equiv}CLi \longrightarrow [{-}Cp_2Hf{-}C{\equiv}C{-}C_6H_4{-}C{\equiv}C{-}]_n \quad (5)$$

Scheme 10.5

$$HC{\equiv}C{-}C_6H_4{-}C{\equiv}CH \xrightarrow[NH_3]{LiNH_2} LiC{\equiv}C{-}C_6H_4{-}C{\equiv}CLi. \quad (6)$$

Scheme 10.6

accompanied by metal elimination (in hydride form) and the formation of metal-free polymers (Scheme 10.3):

One way of preventing the undesirable β hydrid transfer is by eliminating the hydrogen atom bonded to the β-carbon atom of the metalorganic compound. To this end, the hydrogen can be replaced by CH_3 or another group. Even more effective is the use of aceylenic metalorganic compounds. Such monomers were synthesized for Ni, Pd, and Pt (with stabilizing ligands) (Scheme 10.4) [21–23]:

It seems likely that there are no fundamental limitations on the preparation of Hf-containing polymers using this strategy. Among possible stabilizing ligands are cyclopentadienyl rings. Indeed, using *n*-diethynylbenzene as an acetylenic monomer containing active hydrogen for the formation of Hf–C–C bonds, one can achieve polymer chain growth through condensation. This acetylenic precursor is often used in the preparation of metal-containing polymers; its synthesis was described in detail by Hay [24].

Polycondensation was carried out according to Scheme 10.5 (polymer **7**).

Synthesis was run under an inert atmosphere (Ar or N2) in a mixed solvent (DMSO + diisopropylamine, $(i\text{-}Pr)_2NH$) at 20 °C by adding equivalent increments of Cp_2HfCl_2 or *n*-diethynylbenzene. The molecular weight of the polymers thus synthesized was ~10,000 and depended on the reaction conditions. The yield and molecular weight of the polymer can be raised using the dilithium salt of *n*-diethynylbenzene as a precursor in the polycondensation process shown in Scheme 10.6 (polymer **7a**).

Only general information regarding the synthesis of this dilithium salt is available in the literature [25].

We devised a special process for quantitative synthesis of the dilithium salt from BuLi and diethynylbenzene in toluene at –40 °C. The synthesized Hf-containing polymers are light yellow to yellow-brown in appearance, depending on the process conditions, and do not dissolve in toluene. The low-molecular fraction is THF-soluble (M_n =12970, M_w = 37530, polydispersity index γ = 2.89) (Table 10.3).

The molecular weight of the THF-insoluble fraction is 100,000 to 200,000. The IR spectra of these polymers show the C≡C stretching band (2110 cm^{-1}) and the bands arising from C=C stretches in the benzene (1,600 cm^{-1}) and Cp (855, 1,024, and 1,440 cm^{-1}) rings. According to scanning electron microscopy results, the polymers have a characteristic filamentary texture; the fibers are several tens of microns in length and about 1 μm in diameter (Figure 10.3).

Table 10.3 Characteristics of diethynylbenzene-based Hf-containing polymers

Polymer	wt % Hf	ν C=C (Ar), cm^{-1}	ν C≡C, cm^{-1}	ν C=C (Cp), cm^{-1}	π CH (Cp), cm^{-1}	M_n	M_w
7	36	1,600	2,110	1,024; 1.440	855	10,000	
7a	43	1,600	2,110	1.024; 1,440	855	12,970	37,530

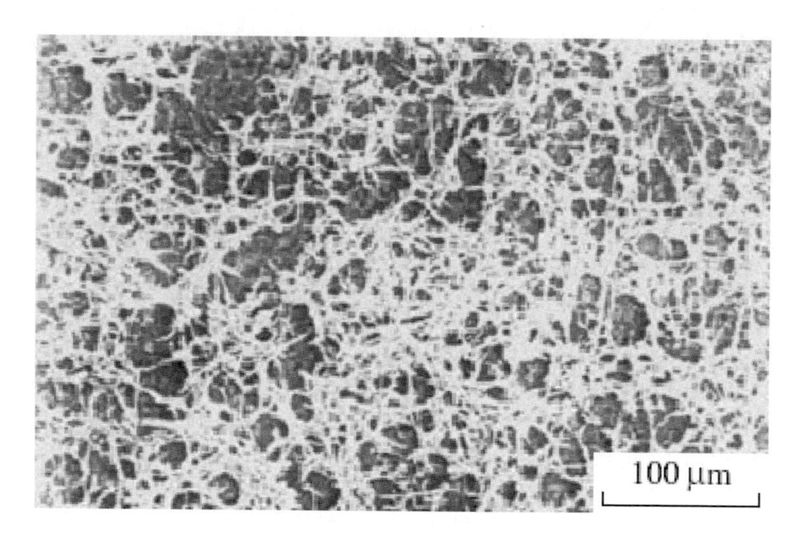

Figure 10.3 Scanning electron micrograph illustrating the typical particle shape of polymer **6**

2.3 Preparation of Hf-Containing Polymers Through Polymerization of Metal-Containing Monomers (Process III)

Reacting $HfOCl_2$ wlith the sodium or potassium salt of acrylic or methacrylic acid, we obtained for the first time Hf(IV)-containing monomers. Hafnium oxoacrylates and methacrylates were syn-thesized according to the Scheme 10.8 (polymer **7**).

$$HfOCl_2 + 2CH_2{=}C(R)\text{-}COONa \longrightarrow HfO(CH_2{=}C(R)\text{-}COO)_2 + 2NaCl \quad (7)$$

Scheme 10.7

$$HfOCl_3 + 2CH_2{=}C(R)\text{-}COOH \longrightarrow HfO(CH_2{=}C(R)\text{-}COO)_2 + CO_2{+}H_2O \quad (8)$$
$R{=}H_2\ CH_3$

Scheme 10.8

$HfOCl_2$ reacts with fumaric acid to form the coordination polymer **8.**

$$HfOCl_2 + HOOC\text{-}CH{=}CH\text{-}COOH \rightarrow \text{-(-Hf(O)-OCOCH=CH-COO-)-}_n + 2H\mu l \quad (9)$$

Scheme 10.9

Table 10.4 Elemental analysis data for Hf-containing polymers (process III)

	Weight percent (assay/nominal)		
Polymer	Hf	C*	H
8	50.45/51.80	14.82/13.94	2.01/1.76
9	53.6/50.40	14.86/20.28	2.44/2.25
10	47.0/46.86	22.40/25.00	3.21/3.15
11	41.01/39.81	29.04/42.47	4.11/3.57
12	36.62/37.48	40.74/44.99	4.91/4.20

* The low carbon content may be the result of the formation of hafnium carbide in the course of analyses. The carbon content was higher if analyses were carried out in the presence of PbO.

Table 10.5 Characteristic frequencies (cm^{-1}) in the IR spectra of Hf-containing polymers

Polymer	$\nu C{=}C$	ν_s^{COO}	ν_{as}^{COO}	νCH (Cp)	δCH (Cp) I	δCH (Cp) II	νHfO	νOH (H_2O)
9	1,641	1,449, 1,374	1,544				667, 470	3,429
10	1,645	1,461; 1374	1,548				473, 621	3,402
11	1,638	1,446, 1,374	1,549	3,104	1,018	814	676, 452	
12								

Note: I, in-plane modes; II, out of-plane modes; ν, stretches; δ, bends; as, asymmetric modes; s, symmetric modes; Cp = cyclopentadienyl group.

The elemental analysis data for $HfO(CH_2{=}CHCOO)_2\ H_2O$ (Scheme 10.9, **9**) and $HfO(CH_2{=}C(CH_3)COO)_2 \cdot H_2O$ (polymer **10**) are listed in Table 10.4.

The IR spectra of these polymers show absorption bands due to carboxylate groups (ν_s^{COO} = 1,373, 1,449, 1,461 cm^{-1} ; ν_{as}^{COO} = 1,544, 1,548 cm^{-1}) (Table 10.5; Figure 10.4), which indicates that the complexes have bidentate and bridge structures.

The strong absorptions in the range 800 to 1,000 cm^{-1} are attributable to the $\delta_{OH}(H_2O)$, $\pi(-CH{=}CH_2)$, and ν_{cc}(=CH-C)modes [26]. At the same time, the broad bands between 700 and 900 cm^{-1}, characteristic of …HfOHfO … chains, are missing. The absorption bands at 473, 621 (polymer **10**), 470, and 667 cm^{-1} (polymer **9**) are the result of Hf–O stretching vibrations. Zirconium and hafnium oxyacetates, formates, and propionates [27] and zirconium methacrylate [28] are known to exist as dimers and tetramers. Carboxylates of cyclopentadienyl derivatives of hafnium, $Cp_2Hf(CH_2{=}CHCOO)_2$ (Scheme 10.10, **11**) and $Cp_2Hf(CH_2{=}C(CH_3)COO)_2$ (Scheme 10.10, **12**), were obtained by reacting Cp_2HfCl_2 with the potassium salt of acrylic or methacrylic acid. The composition and structure of these compounds were determined by elemental analysis and IR spectroscopy. Their IR spectra show bands arising from the carboxylate group (ν_s^{coo} = 1,373, 1,446, 1,458 cm^{-1}; ν_s^{coo} = 1,538, 1,549, 1,567 cm^{-1}) and also (νMO (432, 452, 492, 640, 676 cm^{-1}) and (νCC (1,636, 1,638 cm^{-1}) (Table 10.5, Figure 10.5). In addition, the spectra of these compounds show strong absorptions at 3,004, 3,005 (CH stretches), 1,421, 1,446 (CC stretches), 1,018, 1,020 (in-plane CH bends), 814, and 816 cm^{-1} (out-of-plane CH bends), which result from vibrations of the Cp ring [29]. Most of the

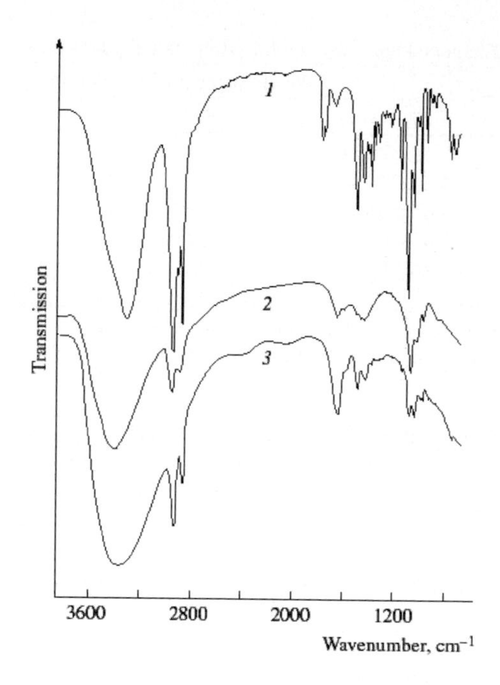

Figure 10.4 IR spectra of the Hf-containing polymers (*1*) **9** and (*2*) **10** and (*3*) HfOCl2 • *n*H2O

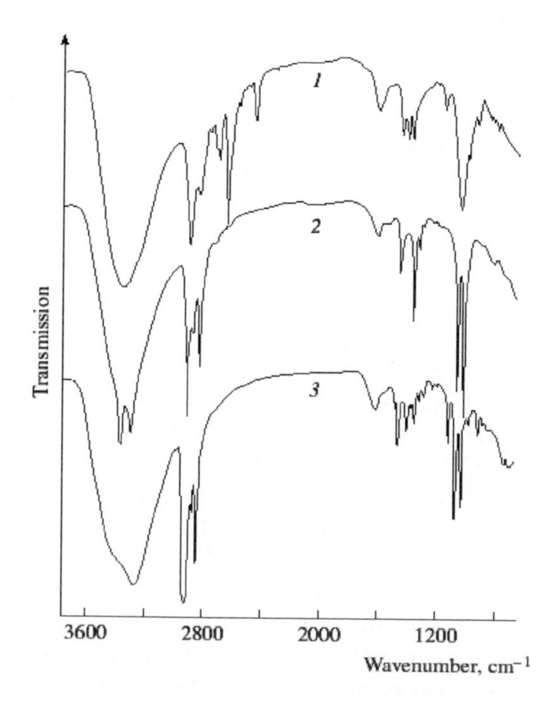

Figure 10.5 IR spectra of the Hf-containing polymers (*1*) **12** and (*2*) **11** and (*3*) Cp2HfCl2

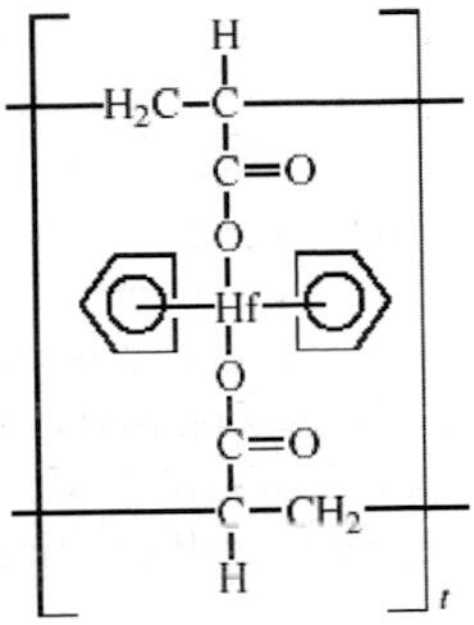

Polymer **11** 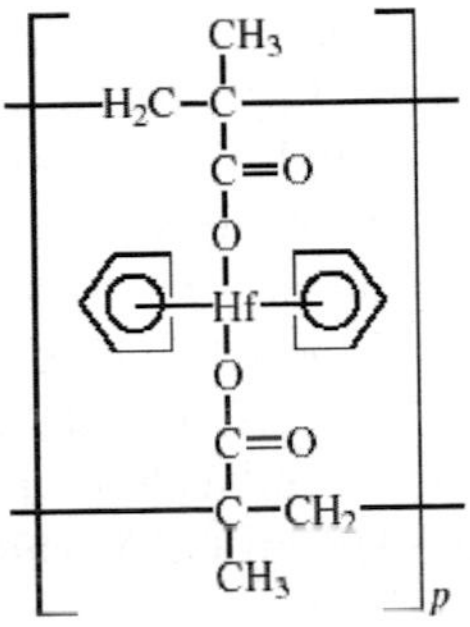 Polymer **12**

Scheme 10.10

synthesized monomers are insoluble in organic solvents. Their thermal polymerization yields high-molecular polymers (based on $HfO(OCOCH{=}CH_2)_2 \cdot H_2O$ (**9**), $HfO(OCOC(CH_3){=}CH_2)_2\ H_2O$ (**10**), $Cp_2Hf(OCOCH{-}CH_2)_2$ (**11**), and $Cp_2Hf(OCOC(CH_3){=}CH_2)_2$ (Scheme 10.10, **12**)(Table 10.2, polymer **9**). These polymers range in Hf content from 37 to 51 wt%, just as the corresponding monomers. IR spectroscopy results confirm that the cyclopentadiene structure is retained during polymerization:

3 Thermodynamic Analysis of the Hf–C–H–O System

Hafnium oxide and carbide are known to be among the highest melting point materials: the melting point of hafnium carbide (fcc Hf sublattice) is 3,928 ± 20°C.The only hafnium compounds with carbon are monocarbides, $HfC_y\square_{1-y}$ ($\square$ is a carbon vacancy), with the basic B (NaCl) structure. The monocarbides have broad homogeneity ranges. The carbon-rich phase boundary of the cubic monocarbide coincides with the stoichiometric composition $HfC_{1.0}$. In order to reinforce carbon foams for high temperature applications, the possibility of covering such a substrate by a few micrometers thick deposit of hafnium carbide was investigated [30]. The conditions of HfC deposit are studied by means of thermodynamical calculations which allows for selection of the appropriate gaseous product, experimental parameters and associated various range. Evaporation of cubic non-stoichiometric HfC_y ($0.6 < y \leq 1.00$) and its enthalpies of both atomization ($\Delta H°_{s298}$ = (672–998) ± 65 kJ/mol) and formation, depending on composition were studied [31]. The phase diagram was compiled taking into account the regulating character of these compounds [32].

A key issue in studies of thermolysis kinetics in hitherto unexplored systems, including Hf-containing polymers, is the optimization of the experimental procedure. Preliminary thermodynamic analysis of the Hf–C–H–O system offers the possibility of establishing the temperature stability limits of HfC and HfO_2. Earlier, thermodynamic studies of the Hf–C–H–O system were carried out in relation to hafnium acetylacetonate, which is used to prepare HfC [33]. The most convenient approach to producing hafnium-containing nanocomposites with the aim of optimizing the formation of refractory hafnium oxide or carbide nanoparticles is controlled thermolysis of polymers **1–12**. Note that polymer-mediated synthesis is an effective means of producing nanoparticles without aggregation during thermolysis, in contrast with other processes [34]. Thermodynamic analysis results for the Hf–C–H–O system lead us to the following conclusions: In the absence of hydrogen (system Hf–C–O), the solid phases existing in the range of 0 to 3,000 °C at an Hf:C:O atomic ratio of 1:3:2 are HfC, HfO_2, and C, and the gas phase consists of CO. Below 1,700 °C, the solid phases present are Hfo2 and C; above 1,700 °C, the phase composition is HfC(s) + CO, which corresponds to the overall reaction HfO_2 (*s*)+3C(*s*)=HfC(*s*) + 2CO.

Increasing the O and C contents (e.g., Hf:C:O = 1:5:4) causes no changes in the content of the solid phase or in the temperature of the HfO_2 (*s*) →HfC(s) conversion (~1,700–1,750 °C), but the composition of the gas phase changes to CO + CO_2.

The addition of hydrogen (bonded) (e.g., Hf:C:O:H = 1:3:2:4) (Figure 10.6) which corresponds, in particular, to the initial composition of the system HfO_2 + 3C +$2H_2$, causes little or no reduction of HfO_2 and produces insignificant changes in the content of the solid phase and the temperature of the HfO_2 (s) → HfC(s) conversion. At the same time, below 800 °C hydrogen reacts with C(s) to form a number of volatile hydrocarbons, primarily CH_4. In the temperature range of 800 to 1,700 °C, the major phases in the system are HfO_2 (s), C(s), CO, and H_2. Above 1,700 °C, the phases present are HfC(s), CO, and H_2. Thus, at thermodynamic equilibrium, the HfO_2 (s)→HfC(s) conversion in the $HfC_xO_yH_z$ polymer system occurs between 1,700 and 1,750 °C. One should keep in mind that, under real thermolysis conditions, the initial stages of the HfO_2 (s)→HfC(s) conversion involve highly reactive nanoparticles with an increased heat content, and the reaction system is far from equilibrium. At Hf:C:O:H = 1:3:2:4, increasing the heat content of HfC(s) by just 40 kJ/mol (Figure 10.7) reduces the temperature of the HfO_2 (s)→HfC(s) conversion to between 1,450 and 1,500 °C [35] or probably to an even lower level.

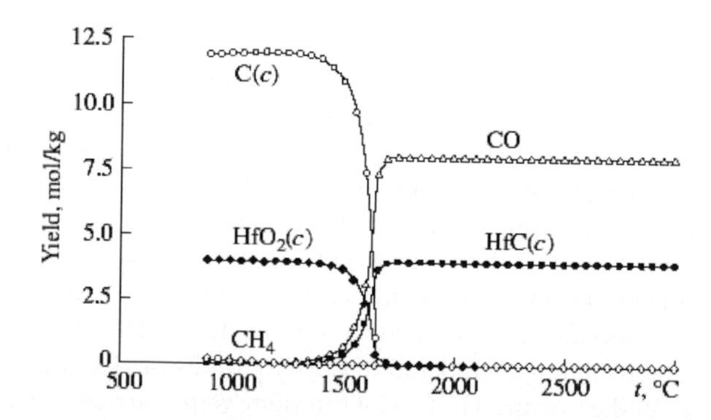

Figure 10.6 Phase composition of the Hf:C:O:H = 1:3:2:4 system at thermodynamic equilibrium (*p* = 0.1 Mpa, *c* denotes condensed phases)

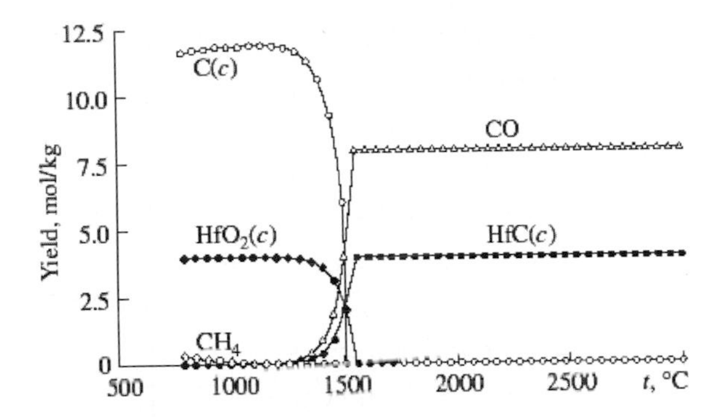

Figure 10.7 Phase composition of the Hf:C:O:H = 1:3:2:4 system at thermodynamic equilibrium (p = 0.1 Mpa, c denotes condensed phases)

3.1 Thermal Transformations of Hf-Containing Polymers

The solid-state thermolysis under consideration can be initiated by reactions taking place around structural inhomogeneities (e.g., point defects, dislocations, cracks, vacancies, and others) and propagating to structurally homogeneous regions. From this viewpoint, solid-state polymerization initiates thermal decomposition of the metal-containing polymer, which is a product common to these coupled processes. The free energy liberated during thermolysis exceeds that absorbed by the forming metal-containing polymer. Moreover, the concept of coupled processes is an important step in describing self-organization as being due to feedback processes. Polymer-mediated synthesis plays a key role in the formation of nanoparticles: impregnation processes lead to severe aggregation of hafnium particles during thermolysis.

The thermolysis of organohafnium compounds has been studied very little. According to studies of $Cp_2Hf(BH_4)$ thermolysis in the range of 300 to 600 °C at 20 Pa by Hanko et al. [36], the product of thermolysis at 600 °C is HfC. Thermolysis of mixed-ligand bis(cyclopentadienyl) complexes Cp_2HfR_2 (R = Cl, $N(C_2H_5)_2$, C_4H_9) leads to Cp elimination and the formation of HfC film at temperatures as low as 250 to 350 °C [37]. In light of this, it was expected that the thermal transformations of the Hf-containing polymers and monomers would be accompanied by the formation of refractory hafnium compounds, in particular HfC. This led us to investigate the thermolysis of the synthesized Hf-containing compounds (Scheme 10.11).

The H:O atomic ratio in these compounds decreases in the order **6** (14.0) > **5** (11.0) > **3** (6) > **12** (5.0) > **11** (4.0) = **2** (4.0) > **9** (1.2) > **8** (0.4). At 370 °C the gas release from these compounds reaches completion in 60 to 300 minutes, with $\beta_{\Sigma}^{370°C} \times 10^3$ = 2.62 (**2**), 0.905 (3), 1.48 (5), 2.09 (6), 1.74 (9), 2.42 (8), and 3.37 mol/g (**11**) (here, $\beta = \frac{P_{\Sigma}^{T_r} V}{m_0 R T_r}$, where $P_{\Sigma}^{T_r}$ is the total pressure of the released gas at 20 °C V is the

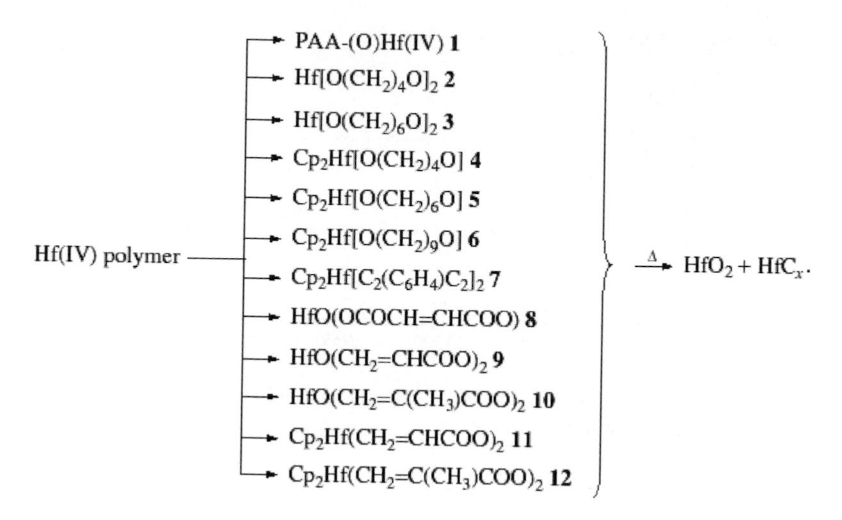

volume of the reaction vessel, m_o is the initial mass of the sample, R is the gas constant, $\alpha^{\Sigma} = \beta_{\Sigma}\mu_i$ is the total number of moles of the gaseous reaction products per mole of the sample, and μ_i is the molecular mass of substance, i. The weight loss Δm under these conditions is 7.3 (**2**), 31.8 (**3**), 30.6 (**5**), 35.7 (**6**), 14.1 (**9**), 26.1 (**8**), and 29.0% (**11**). At 600 °C gas evolution becomes insignificant as $\beta_{\Sigma}^{600} \times 10^3$ reaches 2.62 (**2**), 6.95 (**3**), 4.7 (**5**), 7.4 (**6**), 12.3 (**9**), 12.55 (**8**), or 9.52 mol/g (**11**). The associated weight loss Δm is 28.4 (**2**), 37.9 (**3**), 40.05 (**5**), 42.25 (**6**), 28.4 (**9**), 33.7 (**8**), and 42.3% (**11**). Thermolysis at 1,000 °C for 150 minutes results in $\beta_{\Sigma}^{1,000°C} \times 10^3 = 19.7$ (**2**), 24.65 (**3**), 11.85 (**5**), 36.95 (**6**), 16.65 (**9**), 8.0 (**8**), and 34.1 mol/g (**11**), and Δm = 30.3 (**3**), 40.15 (**4**), 38.9 (**6**), 31.2 (**9**), 36.5 (**8**), and 43.3% (**11**).

The thermolysis kinetics of **1–12** have the following features in common: In all of the samples, the gas evolution rate decreases steadily with time, the considerable gas evolution was observed already on the initial stage of the heating of sample. Figure 10.8 illustrates typical gas evolution kinetics during thermolysis.

At a given conversion time, increasing the thermolysis temperature increases the total gas volume evolved and, accordingly, the weight loss. The slope of the gas release and weight loss curves is a nonlinear function of time, which indicates that the contribution of the pyrolysis of the volatile reaction products increases with temperature (Table 10.6).

It should be mentioned that such kinetic behavior of gas evolution (Figure 10.8) is typical for multistage successive reactions which are the first order ones with decreasing rate constants for the each subsequent stage [38]. At constant temperature the dependence of conversion on time $\eta^{\Sigma}(t) = \alpha^{\Sigma}(t)/ \alpha^{\Sigma}$(&infinity;) in coordinates $[-\ln(1-\eta^{\Sigma}), t]$ is linear plots following one by one. Indeed, if there are a few successive reactions accompanied by gas evolution on each stage (Eq. 10.2).

Table 10.6 Thermolysis of Hf-containing polymers.

Polymer	[H]:[O]	$T_{exp.}$, °C	τ_{term},min	α_f^{Σ}	$\frac{\Delta m}{\pi o}$	$\left(\frac{\Delta m}{\pi o}\right)_{HfC_2}$	$\left(\frac{\Delta m}{\pi o}\right)_{HfC}$
2	4.0	370	270	1.03	147		
		600	180	6.50	154	160.3	182.2
		1000	150	7.74	–		
3	9.0	370	200	0.37	130.5		
		600	120	2.76	155.5	199.7	219.6
		1000	130	10.10	125.0		
5	11.0	370	135	0.59	121.5		
		600	150	1.89	161.5	163.8	185.8
		1000	150	4.63	158.5		
6	14.0	370	122	0.97	166.8		
		600	150	3.18	197.4	256.4	276.5
		1000	160	17.2	181.8		
9	1.2	370	380	0.61	47.7		
		600	80	4.03	95.5	125.9	146.0
		1000	150	5.91	–		
8	0.4	370	160	0.67	80.7		
		600	85	3.54	104.0	98.0	118.05
		1000	95	2.52	112.0		
11	4.0	370	290	1.52	131.0		
		600	150	4.35	194.0	240.0	260.5
		1000	150	5.76	195.5		
12	5.0	370	370	1.78	159.0		
		600	145	5.17	263.9	268.0	288.0
		1000	200	6.88	260.5		

$$A_{1(s)} \xrightarrow{k_1} A_{2(s)} + m_1C_{1(g)}$$

$$A_{2(s)} \xrightarrow{k_2} A_{3(s)} + m_2C_{2(g)}$$

$$A_{n-1(s)} \xrightarrow{k_{n-1}} A_{n(s)} + m_nC_{n-1(g)}, \tag{10.2}$$

on the assumption of the rate constants $k_1 >> k_2 >>..>> k_{n-1}$ and $\eta^{\Sigma}(\infty) = 1$ one may assume from Eq. 10.3,that:

$$\eta^{\Sigma}(t) \approx \Sigma_{i=1}^{n-1} N_i[1-\exp(-k_i t)] \tag{10.3}$$

where $N_i = \alpha_i^{\Sigma}(\infty)/\Sigma_{i=1}^{n-1}\alpha_i^{\Sigma}(\infty)$, $\Sigma_{i=1}^{n-1}\alpha_i^{\Sigma}(\infty) = \alpha_i^{\Sigma}(\infty)$, and $\Sigma_{i=1}^{n-1}N_i = \eta^{\Sigma}(\infty)=1$ (hear $\alpha_i^{\Sigma}(\infty)$ is the maximum molar gas evolution in the i-reaction).

If for the successive stages i = 1, 2....n−1 both $t >> 1/k_j$ for all i = 1, 2…j and $t \approx 1/k_{j+1}$ for all i = j+1, j+2… n−1, taking into account $\Sigma_{i=1}^{n-1}N_i=1$, then

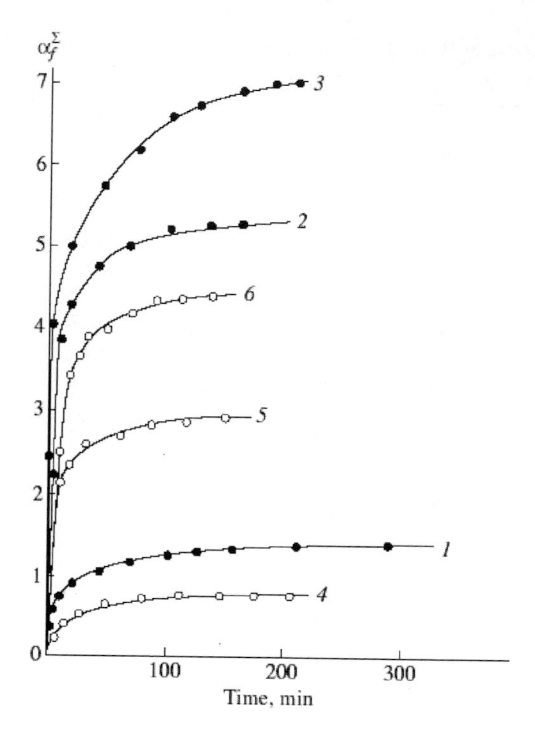

Figure 10.8 Gas evolution kinetics during thermolysis of (*1–3*) **12** and (*4–6*) **2** at (*1*, *4*) 370, (*2*, *5*) 600, and (*3*, *6*) 1,000°C

$$\eta^{\Sigma}(t)\approx\Sigma_{i=1}{}^{j}N_i+\Sigma_{j+1}{}^{n-1}N_{j+1}[1-\exp(-k_{j+1}t)]=\Sigma_{i=1}{}^{j}N_i+\Sigma_{j+1}{}^{n-1}N_{j+1}-\Sigma_{j+1}{}^{n-1}N_{j+1}\exp(-k_{j+1}t)=\Sigma_{i=1}{}^{j}N_i-\Sigma_{j+1}{}^{n-1}N_{j+1}\exp(-k_{j+1}t)=1-\Sigma_{j+1}{}^{n-1}N_{j+1}\exp(-k_{j+1}t). \quad (10.4)$$

Then,

$$1-\eta^{\Sigma}(t)\approx\Sigma_{j+1}{}^{n-1}N_{j+1}\exp(-k_{j+1}t) \quad (10.5)$$

In the case when $i = j + 1, j + 2 \ldots n-1$, the values

$$1/k_{j+1} << 1/k_{j+2}, \ldots 1/k_{n-1}, \text{ then } \Sigma_{j+2}{}^{n-1}N_{j+2}[1-\exp(-k_{j+2}t)]0, \text{ and}$$
$$1-\eta^{\Sigma}(t)\approx N_{j+1}\exp(-k_{j+1}t) \quad (10.6)$$

Thus, for two-stage reaction

$$\eta^{\Sigma}(t)\approx N_1[1-\exp(-k_1t)]+N_2[1-\exp(-k_2t)]. \quad (10.7)$$

If during heating $t \approx 1/k_2 >> 1/k_1$, then

$$\eta^{\Sigma}(t)\approx N_2[1-\exp(-k_2t)]=1-N_2\exp(-k_2t) \quad (10.8)$$

and $-n(1 - \eta^{\Sigma}) = \ln N_2 + k_2t = \theta_1 + k_2t$, that is observed experimentally (Figure 10.9).

At thermolysis of the sample 3 at 600 °C in semilogarithmic coordinates:

$$[F(\eta) = -\ln(1 - \eta^{\Sigma}), t],$$

$$F(\eta) = 0{,}07 + 5{,}36 \cdot 10^{-3}t \text{ for the first stage,}$$
$$F(\eta) = 0{,}18 + 1 \cdot 10^{-3}t \text{ for the second stage;}$$

and at 1,000 °C, respectively

$$F(\eta) = 0.28 + 3.13 \cdot 10^{-3}t \text{ for the first stage,}$$
$$F(\eta) = 0.62 + 2.02 \cdot 10^{-2}t \text{ for the second stage.}$$

The total gas volume evolved at 600 and 1,000 °C is lower for the polycondensation polymers (**2–6**) compared with the carboxylate polymers (**8**, **9**, **11**, **12**) (Figure 10.10). The net weight loss is, as a rule, lower than would be expected for decomposition of the polymers to HfO_2 or HfC (%): 40.6 (46.2) for **2**, 48.7 (53.6) for **3**, 50.4 (55.1) for 5, 54.8 (59.2) for **6**, 37.4 (43.4) for **9**, 31.8 (38.3) for **8**, and 53.3 (57.7) for **11**. Only in the case of **8** (hafnyl fumarate) is the weight loss at 600 and 1,000 °C close to that for decomposition to HfC. According to mass spectrometric analysis of the gas phase, the major gaseous product is CO_2 during thermolysis of the Hf-containing carboxylate polymers (**8**, **9**, **11**) at 370 °C and cyclopentadiene vapor in the case of **11**. Increasing the thermolysis temperature to between 600 and 1,000 °C leads to a considerable release of H_2, a thermolysis product of C–H–O compounds. As shown

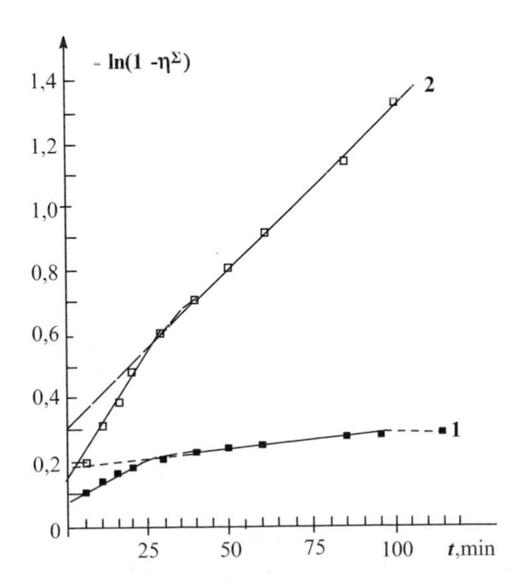

Figure 10.9 The kinetics of gas evolution at thermolysis of **3** in semilogarithmic coordinates at: *1* – 600, *2* – 1,000 °C (for $-0.5 \ln(1 - \eta^{\Sigma})$)

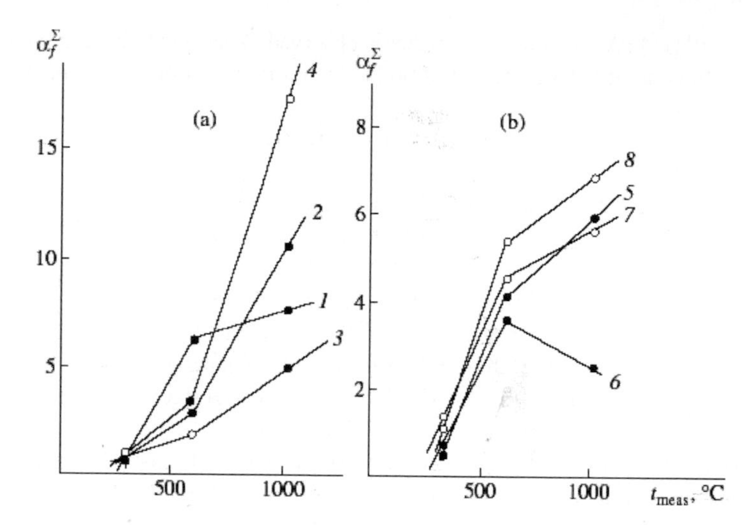

Figure 10.10 Yield of thermolysis products as a function of *t*meas for diol-based Hf compounds (**a**) and carboxylate derivatives (**b**): *(1)* **2**, *(2)* **3**, *(3)* **5**, *(4)* **6**, *(5)* **9**, *(6)* **8**, *(7)* **11**, *(8)* **12**

by mass spectrometric analysis, the volatile thermolysis products of the Hf–diol polymers include a wide variety of diols and their oxygen-free fragments, even at 370 °C. With increasing temperature, the mass spectrum becomes richer in the pyrolysis products of these compounds.

3.2 Composition of Solid Pyrolysis Products of Hf-Containing Polymers

Optical examination shows that the as-prepared polymers have the form of fine powders consisting of colorless, glassy particles transparent in transmitted light. Comparison with specific surface data indicates that the particles have rather high porosity. Optical examination of 370 °C thermolysis products indicates that all of the powders are close in morphology and consist of irregularly shaped particles containing opaque inclusions. At 600 and 1,000 °C we obtained opaque, coke-like substances. The solid thermolysis products were also characterized by X-ray diffraction (Table 10.7). All of the 300 °C thermolysis products, except sample **8**, were amorphous. Sample **8** consisted of an amorphous phase and crystalline HfO_2. The solid products of the 600 °C thermolysis of **9**, **2–4**, and **6** consisted of well-crystallized *m*-HfO_2 (monoclinic) and trace amounts of HfC with different degrees of crystallinity (sample **9** after thermolysis consisted of a mixture of *m*-HfO_2 and *t*-HfO_2). The product of the 600 °C thermolysis of **11** was amorphous. In the case of **8** (hafnyl fumarate), the solid phase consisted of HfC and trace levels of HfO_2. In the thermolysis product of **7**, the dominant phase was cubic hafnium carbide,

Table 10.7 Effect of thermolysis temperature on the phase composition of solid thermolysis products

Polymer	Phase composition		
	370 °C	600 °C	1,000 °C
2	Amorphous	*m*-HfO_2 + HfC*	*m*-HfO_2 + HfC*
3	Amorphous	*m*-HfO_2 + *t*-HfO_2	*m*-HfO_2 + *t*-HfO_2*+ HfC*
5	Amorphous	*m*-HfO_2 + t-HfO_2*+ HfC*	*m*-HfO_2 + HfC*
6	Amorphous	*m*-HfO_2 + t-HfO_2* + HfC*	*m*-HfO_2 + *t*-HfO_2* + HfC*
9	Amorphous	*m*-HfO_2 + t-HfO_2 + HfC*	*m*-HfO_2 + *t*-HfO_2 + HfC*
11	Amorphous	Amorphous	*m*-HfO_2 + *t*-HfO_2 + HfC*
12	Amorphous	Amorphous	*m*-HfO_2 + *t*-HfO_2 + HfC*
8	Amorphous + t-HfO_2 + HfC*	Amorphous + HfC + *t*-HfO_2 *	*m*-HfO_2 + HfC*

Note: *m* and *t* denote monoclinic and tetragonal phases, respectively.

* Heavily disordered phase.

c-HfC. The *d*-spacings of the strongest reflections from *c*-HfC and the relative intensities for stoichiometric *c*-HfC (in parentheses) are 2.68 (100), 2.32 (90), 1.64 (70), 1.40 (80), 1.34 (30), 1.065 (50), 1.038 (50), 0.95 (40), and 0.89 Å (50).

Typical X-ray diffraction patterns of the thermolysis products are displayed in Fig. 10.11. Note that the diffraction patterns show a number of relatively strong diffraction peaks which cannot be attributed to *m*-HfO_2 or *o*-HfO_2 (orthorhombic). One possible reason is the formation of ordered HfC_y phases. It should also be taken into account that the pyrolysis of organic ligands may lead to the formation of carbon (graphitization). This seems to be evidenced by the diffraction peaks close in position to those from hexagonal carbon (*h*-C):3.36 (100), 2.03 (50), 1.67 (80), 1.16 (50), 0.99 (40), and 0.83 Å (40). The products of thermolysis at 1000 °C consist mainly of m Hf_{O2} and trace levels of HfC, just as at 600 °C. Sample **5** consists of well-crystallized *m*-HfO_2 and imperfect HfC. For **8**, the percentage of HfC decreases with increasing temperature, and the sample consists of a mixture of m HfO2 and trace levels of HfC.

In the case of **11**, the product of 1,000 °C thermolysis consists of HfC and *m*-HfO_2. The observed significant broadening of diffraction peaks from the thermolysis products suggests that thermal transformations of the Hf-containing precursors lead to the formation of fine HfO2 and HfC particles stabilized by the polymer matrix. According to preliminary electron-microscopic examination, the particle size of the forming nanocomposites ranges from 10 to 50 nm, and the particles have a core–shell structure: the core consists of hafnium carbide or oxide and is surrounded by a pyrolyzed polymer shell (Figure 10.12).

Polymers containing Hf(IV) in the main chain or side chains can be prepared by any of the known processes for synthesis of metal-containing polymers. An attractive approach is the polycondensation of dicyclo-pentadienyl derivatives of hafnium(IV) with diols or oxygen-free diacetylenic ligands. An alternative approach is polymerization of the Hf(IV) monomers with (meth)acrylate and fumarate groups, synthesized for the first time in this study. We investigated the thermolysis of the synthesized

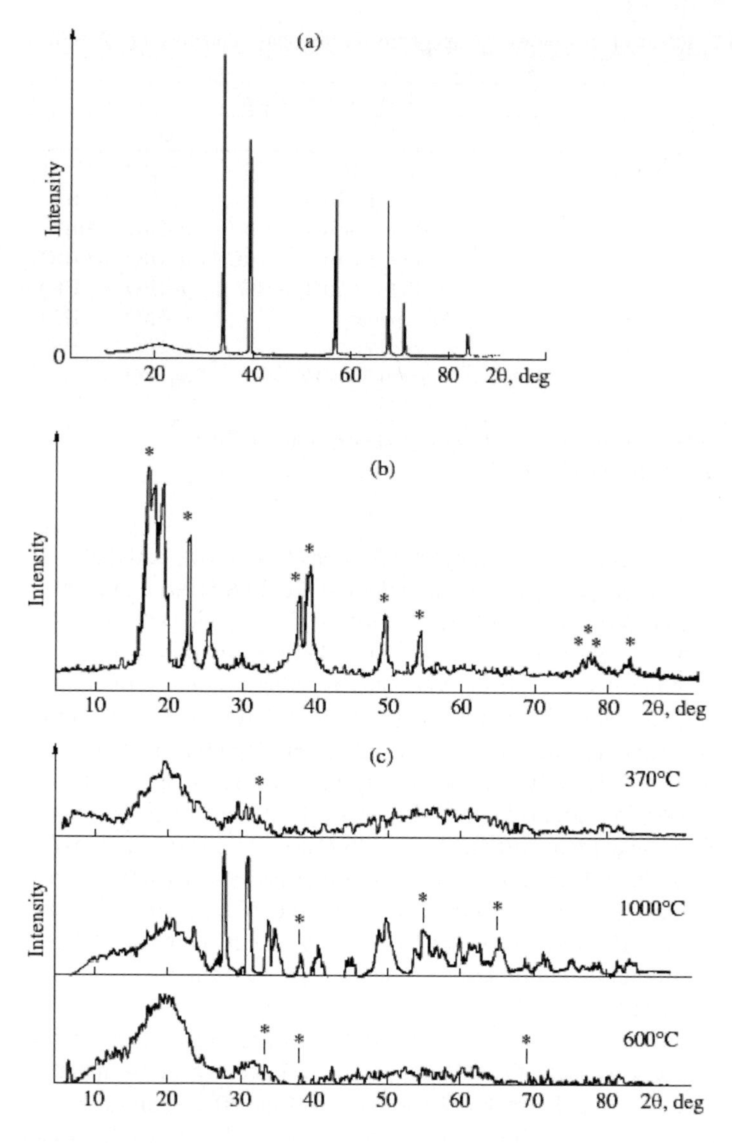

Figure 10.11 X-ray diffraction patterns of HfC (**a**) and thermolysis products of Hf-containing polymers based on **7** and **8** (**b**,**c**), respectively

organohafnium polymers in a self-generated atmosphere between 300 and 1,000 °C and determined the compositions of the gas and solid phases that were formed in the static nonisothermal system. This analysis, in conjunction with X-ray diffraction results, indicates that pyrolysis leads to the formation of metal–polymer nanocomposites consisting of crystalline HfO_2 and HfC stabilized by the polymer

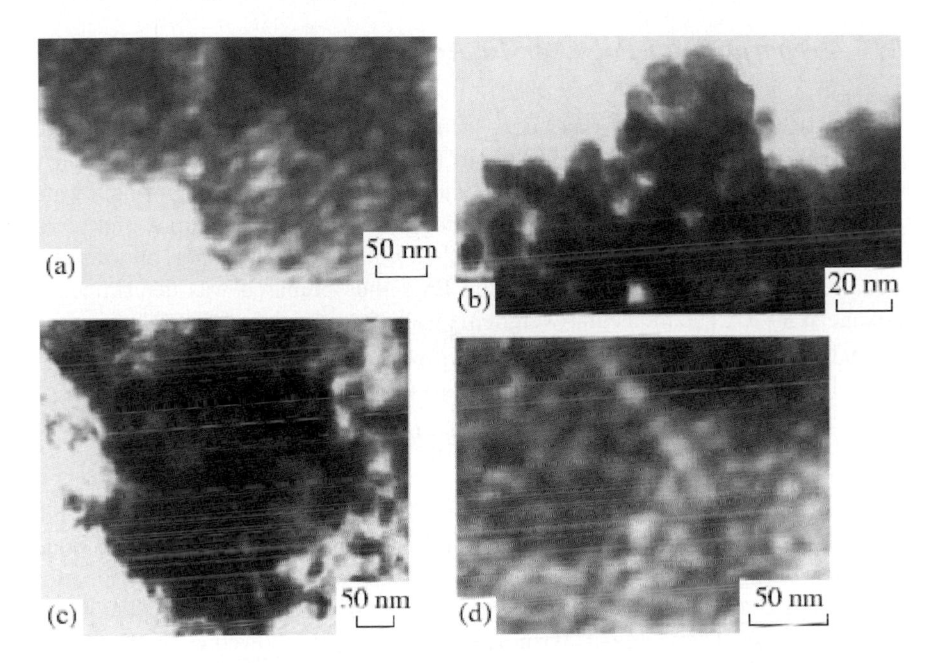

Figure 10.12 Electron micrographs of thermolysis products of polymers **11** (**a**,**b**), **1** (**c**), and **4** (**d**); thermolysis temperature of 370 (**a**), 1,000 (**b**), and 1,350 °C (**c**,**d**)

matrix. The hafnium carbide content depends on the ligand shell of Hf(IV) and thermolysis conditions. Kinetically, the polymer-mediated synthesis of nanoparticles is a two-step process involving polymerization transformations of the metal-containing precursor, which initiate subsequent transformations of the polymer formed. In this process, the free energy liberated during thermolysis exceeds that absorbed by the forming metal-containing polymer [39]. In other words, the free energy liberated through chemical reactions was consumed by other concurrent processes—an approach to describing self-organization (i.e., feedback processes) [40]. Solid-state reactions, including thermolysis, can be initiated by transformations taking place around structural inhomogeneities (point defects and dislocations) and propagating to structurally homogeneous regions. We hope that further research will make it possible to optimize both the synthesis and thermolysis of Hf-con-taining polymers for the preparation of HfC-based nanocomposites.

4 Experimental

We used standard procedures to synthesize $(\eta^5\text{-}C_5H_5)_2HfCl_2$ (Cp_2HfCl_2, where Cp = cyclopentadienyl), $HfOCO_3$, diethynylbenzene, its dilithium derivative, and LiBu and to purify solvents and inert gases.

4.1 *Preparation of Poly(Acrylic Acid) (PAA)*

$K_2S_2O_8$ (0.36 g, 1.33 mmol) was added dropwise, while stirring to a solution of acrylic acid (6.0 g, 0.08 mol) in H_2O (22 ml) at 80 °C over a period of 1 hour. The resulting viscous solution was diluted with a mixture of acetone (140 ml) and water (40 ml), filtered, and reprecipitated by adding a 4:1 mixture of acetone and petroleum ether. The polymer was filtered, washed with petroleum ether, and dried in vacuum. The intrinsic viscosity η of the PAA thus prepared was determined in dioxane at 30 °C. The molecular weight M_η was determined from the relation $\eta = K M^\alpha$ ($K = 8.5 \cdot 10^{-4}$, $\alpha = 0.5$), $M_\eta = 330000$.

4.2 *Preparation of Hf(IV) Polyacrylate*

A solution of $HfOCl_2$ (1.33 g, 0.005 mol) in methanol was added dropwise, while stirring to a solution of PAA (1.0 g, 0.01 mol) in methanol. After precipitation of the resultant macromolecular complex, the reaction was heated at 50 °C for 90 minutes. The desired product was isolated by centrifugation, washed with methanol, and dried in vacuum. The yield was 0.6 g. Elemental analysis showed C 26.59, H 3.92, Cl 2.62, and Hf 34.4 wt %.

4.3 *Polycondensation of Hf-Containing Precursors*

4.3.1 Polycondensation Through Diethynylbenzene in a Mixture of DMSO and Diisopropylamine, (i-Pr)2NH

Freshly distilled diethynylbenzene (0.77 g) in DMSO (6 ml) and Cp_2HfCl_2 (1.054 g) in DMSO (8 ml) was added to a three-neck flask fitted with a magnetic stirrer and inert-gas inlet. Next, $(i\text{-Pr})_2NH$ (0.58 g) and Cu_2Cl_2 (0.02 g) were added dropwise, while heating, and the mixture was stirred for 3 hours at room temperature. In the second step, an excess of Cp_2HfCl_2 (1.52 g) in DMSO was added, and then $(i\text{-Pr})_2NH$ (2 ml) was added dropwise at room temperature over a period of 3 hours. In the third step, an excess of diethynylbenzene (0.52 g in 4 ml of DMSO) was added. In subsequent steps, we added Cp_2HfCl_2 (1.39 g), (*i*-Pr)2 NH (2 ml), and Cu_2Cl_2 (0.016 g) in DMSO (4 ml) over a period of 3 hours. The polymer was filtered off, dried to remove DMSO, and extracted with toluene between 50 and 60 °C. Distilling off the toluene, we obtained 3.29 g of the polymer. Its molecular weight was 56100.

4.3.2 Polycondensation Through the Dilithium Derivative of Diethynylbenzene

BuLi (3 ml, 5.3 mmol) at 45°C was added to diethynylbenzene (0.36 g) in tetrahydrofuran (THF). After stirring, Cp_2HfCl_2 (1.99 g) was added to the suspension of the dilithium derivative. The mixture was slowly heated to 50 °C, the reaction medium turned yellow. After LiCl removal, the polymer was yellow-brown. The number-average molecular weight of the THF-soluble fraction was $M_n = 12,970$, and its weight-average molecular weight was $M_w = 37,530$. The molecular weight of the THF-insoluble fraction (20–40%) ranged from 100,000 to 200,000.

4.3.3 Polycondensation of Hf-Containing Precursors With Diols

The best results were achieved when a mixture of ether and toluene was used as the solvent, the reaction temperature was 70 °C, and the resulting HCl was removed from the reaction zone in order to shift the chemical equilibrium toward the formation of the desired product. Attempts to remove HCl by passing Ar had limited success. The most effective approach was to combine HCl with triethylamine, $(C_2H_5)_3$ N, to give a quaternary ammonium salt. At a reaction time of 1 h, the reaction yield was between 56 and 66%. This process, however, requires subsequent removal of the ammonium salt from the polymer (for example, through extraction with chloroform in a Soxhlet apparatus).

4.4 Preparation of Hf-Containing Monomers

4.4.1 Hafnium(IV) Oxocarboxylates

These compounds were prepared by reacting $HfOCl_2$ with the sodium or potassium salt of acrylic or methacrylic acid. The reaction mixture in methanol was stirred between 40 and 50 °C for 5 hours. The resulting precipitate was filtered off, washed, and dried in vacuum.

4.4.2 Carboxylates of Cyclopentadienyl Derivatives of Hafnium(IV)

These compounds were prepared by reacting Cp_2HfCl_2 with the potassium salt of acrylic or methacrylic acid. An excess of the salt was added to a Cp_2HfCl_2 solution in benzene, and the reaction was stirred at 50 °C. The unreacted potassium methacrylate and KCl were separated by filtration. Removing benzene in vacuum, we obtained powder products.

4.5 Thermal Transformations of Hf-Containing Precursors

Thermolysis was performed isothermally, at 370, 600, and 1,000 °C and occasionally in the range of 1,350 to 1,400 °C. The tube containing the sample was placed in an isothermally heated zone, whose volume was approximately 5% of the total volume of the reactor.

The process was run in a self-generated atmosphere, under dynamic vacuum, or in argon (p = 72–74 kPa). The kinetics of gas evolution were followed using a membrane manometer. The morphology of the starting solid materials and reaction products was examined by optical microscopy in transmitted—in particular, with crossed polarizers,—and reflected light, using the thermal stage of an MBI-6 microscope and an MBI-15 microscope. IR absorption spectra of both precursors and reaction products, gaseous and condensed, were measured in the range 400 to 4,000 cm^{-1} (Specord 75 IR spectrophotometer). Gaseous reaction products were also analyzed by mass spectrometry using an MS-3702 quadrupole mass spectrometer (75 eV). X-ray diffraction studies of powder samples were performed on a DRON UM-2 diffractometer with Cu $K\alpha$- radiation. In electron-microscopic examination, we used JEOL JSM-840 and JEM-100CX instruments.

4.6 *Thermodynamic Data*

The equilibrium composition of the Hf–C–O–H system for Hf-containing polymers was evaluated by thermodynamic analysis at a pressure of 0.1 MPa, temperatures from 0 to 3,000 °C and a varied initial composition of the reaction system, using the Terra program (Bauman State Technical University).

Acknowledgments This work was supported by the Russian Foundation for Basic Research grant nos. (07-03-00113 and 07-03-91582-ACP-a), the Programs of Chemistry Sciences and Materials Sciences Departments of RAS nos 3 and 9 and the Lawrence Livermore National Laboratory (project B 324 143).

References

1. Pomogailo AD, Rozenberg AS, Uflyand IE. (2000) *Nanochastitsy metallov v polimerakh*: metal nanoparticles in polymers). Khimiya, Moscow.
2. Toth LE. (1971) *Transition metal carbides and nitrides*. Academic, New York.
3. Newman RW. (1993) Jons Hopkins Apl Technical Digest. 14:24.
4. Kobayashi H, Shimosaka K, Kamiyama Y, Mitamura TJ. (1993) Cheram Soc 101:342.
5. Ache HF, Goschnick J, Sommer M, Emig G, Schoch G, Wormer O. (1994) Thin Solid Film. 241:356.
6. Emig G, Schoch G,Wormer OJ. (1993) Physique IV. 3:535.
7. Wunder VK, Popovska N, Emig GJ. (1999) Physique IV. 9:509.
8. Motojima S, Kawashima Y. (1996) J Mater Sci. 31:3697.
9. Wunder V, Popovska N, Wegner A, Emig G, Arnold W. (1998) Surface & Coating Technol. 101:329.
10. Baklanova NI, Kulyukin VN, Korchagin MA, Lyakchov NZ. (1998) J Mater Synth Proc. 6:15.
11. Spatenka P, Suhr H, Erker G, Rump M. (1995) Appl Phys A. 60(3):285.

12. Wöhrle D, Pomogailo AD. (2003) *Metal complexes and metals in macromolecules: synthesis, structures, and properties*. Wiley, Weinheim.
13. Hubicki Z. (1998) *Ion Exch*. 6(1):188.
14. Vibhute CP, Khopkar SM. (1987) *Anal Chim Acta*. 193:387.
15. Hubicki Z. (1987(*Przem Chem*. 66(6):290.
16. Wohrle D. (1992). In: Kricheldorf HR (ed) *Handbook of polymer synthesis, part* B. Marcel Dekker, New York, p. 1133.
17. Carraher CE Jr. (2006) Transition metal-containing polymers. In: Abd-El-Aziz AS, Carraher CE Jr., Pittman CU, Jr, Zeldin M (eds) Macromolecules containing metal and metal-like elements. (eds.). Wiley, Hoboken, NJ, p. 111.
18. Ziemkowska W, Pasynkiewicz S. (1996) *J Organomet Chem*. 508:243.
19. Wali A, Ganeshpure PA, Pillai SM, Satish S. (1994) *Eng Chem Res*. 33(2):444.
20. Kharitonov YA, Zaitsev LM. (1968)*Zh Neorg Khim*. 13(3):902.
21. Takahashi S, Morimoto H, Murata E, *et al*. (1982) *J Polym Sci. Polym Chem*. 20(2):565.
22. Takahashi S, Murata E, Kataoka S, *et al*. (1980) *J Polym Sci Part A: Polym Chem*. 18(2):661.
23. Sonogashira K, Ohga K, Takahashi S, Hagihara N. (1980) *J Organomet Chem*. 188(2):237.
24. Hay AS, (1960) *J Org Chem*. 25(4):637.
25. Talalaeva TV, Kocheshkov K. (1971) *Metody elementoorganicheskoi khimii. Litii, natrii, kalii, rubidii, tsezii* (Techniques of Organoelement Chemistry: Lithium, Sodium, Potassium, Rubidium, Cesium). Nauka, Moscow.
26. Nakamoto K. (1986) *Infrared and raman spectra of inorganic and coordination compounds* Wiley, New York.
27. Prozorovskaya ZN, Petrov KI, Komissarova LN. (1968) *Zh Neorg Khim*. 13(4):965.
28. Besecke S, Schoder G, Ganzler W. FRG Patent 3137840, 1981.
29. Wailes PC, Coutts RSP, Weigold DH. (1974) *Organo-metallic chemistry of titanium, zirconium, and hafnium*. Academic, New York.
30. Sourdiaucourt P, Derre A, Delhaes P, David P. (1999) J Phys IV. 9:Pr8-373.
31. Gusev AI, Zyryanova AN. (1997) Dokl Akad Nauk. 354: 493.
32. Gusev AI. (1992) Dokl Akad Nauk 322:918.
33. Timofeev AN, Filatov IY, Sevast'yanov VG, Marushkin KI. (1994) *Vysokochist Veshchestva*. 5:45.
34. Pomogailo AD, Rozenberg AS, Dzhardimalieva GI. (2005) In: Nicolais L, Carotenuto, GH (eds) Controlled pyrolysis of metal-containing precursors as a way for synthesis of metallopolymer nanocomposites, *metal–polymer nanocomposites*. Wiley, New York, p. 75.
35. Toth LE. (1971) *Transition metal carbides and nitrides*. Academic Press, New York.
36. Hanko K, Vass G, Szepes L. (1995) *J Organomet Chem*. 492:235.
37. Grafov AV, Mazurenko EA, Mel'nik OV, Kofman,VY. (1993) *Ukr Khim Zh*. 59(12):1235.
38. Rozenberg AS, Nechiporenko GN. (1999) Chem Phys Report. 18(5):905.
39. Emanuel' NM, Knorre AG. (1962) *Kurs khimicheskoi kinetiki* (course in chemical kinetics). Vysshaya Shkola, Moscow. 1962.
40. Tret'yakov YD. (2003) *Usp Khim*. 72(8):731.

Nanoscale Dendrimer-Platinum Conjugates as Multivalent Antitumor Drugs

Bob A. Howell, Daming Fan, and Leela Rakesh

1 Introduction

1.1 Development of Conventional Polymers as Drug Carriers for Platinum Anticancer Drugs

The observation that platinum compounds inhibit bacterial cell division brought about a rather massive examination of the biological activity of these compounds [1]. Platinum(II) compounds which contain 2 inert *cis* ligands and 2 labile ligands (chloride displays a near optimum hydrolysis rate under physiological conditions; a half-life of about 1 hour at 37 °C) display remarkable antitumor activity. Cisplatin [*cis*-dichlorodiammineplatinum(II)], the first of these to be synthesized and commercialized, is a broad-spectrum cancer drug effective against a wide range of tumors [2]. Cisplatin is currently the most widely used anticancer drug. It is often used in combination with organic antitumor compounds or with carboplatin [1,1-cyclobutanedicarboxyalato(diammine)platinum(II)] the second platinum anticancer drug to gain widespread commercial use. Carboplatin displays a somewhat different toxicity profile than does Cisplatin, making it an attractive compliment to Cisplatin [3–6].

A.S. Abd-El-Aziz et al. (eds.), *Inorganic and Organometallic Macromolecules: Design and Applications.*

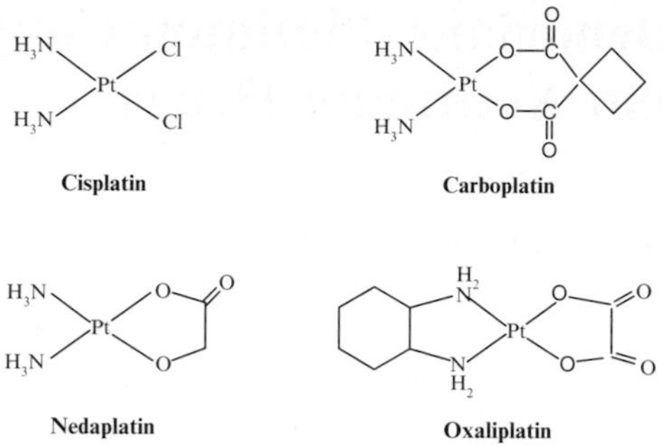

Cisplatin

Carboplatin

Nedaplatin

Oxaliplatin

Figure 11.1 Structures of cisplatin, carboplatin, nedaplatin, and oxaliplatin

The potential of these drugs has been limited because of the severe side effects, which accompany their administration. Among the most debilitating side effects induced by organoplatinum drugs are severe kidney damage [7] and extreme nausea (as a class the platinum compounds are among the most effective nausea producing agents known. to the point that some patients refuse to complete the treatment regimen [8]). In an attempt to identify active but less toxic drugs, literally hundreds of platinum compounds in which the structure of the amine ligand has been varied have been synthesized and evaluated for antitumor activity. For the most part, this has been a fruitless undertaking. Whereas some ligands impart better solubility, activity, or toxicity than similar properties associated with compounds derived from simple ammonia ligands, no compounds with clearly superior performance have been found. Of the hundreds of compounds synthesized, fewer than 30 have entered clinical trials as antitumor agents [9,10]. Of these, two in addition to those noted above, Oxaliplatin and Nedaplatin (Figure 11.1), have received approval for use in the treatment of cancer.

The generation of Oxaliplatin makes use of *trans*-1,2-diaminocyclohexane as the amine ligand. This ligand has proven to be an excellent inert component of organoplatinum antitumor agents. In an alternative approach we have, for some time, been using water soluble polymers as platforms on which a platinum drug or prodrug might be supported and from which it might be slowly released into the extracellular fluid [11–19]. This approach has several major potential advantages over the traditional forced hydration therapy currently practiced. Firstly, the solubility of the drug formulation may be dramatically enhanced such that the volume of the fluid required to introduce a satisfactory dose of drug is strongly diminished [Cisplatin has a solubility of about 10 mg/l in aqueous saline]. More importantly, if the release rate is optimal, the drug is released into the blood stream at a level that is beneath the toxicity threshold such that side effects may be mitigated [20]. Early attempts involved the formation of noncovalent complexes with poly(*N*-vinylpyrollidone) [11–14]. A strong positive with the respect to the use of this polymer is its long history in biological applications and its approval for use in food and drug applications. More recently platform polymers have been poly(acrylamide)s in which the amine portion of the amide is derived from a 1,2-oxazine [21,22]. These polymers are versatile materials and may be readily modified in a number of ways for covalent attachment of organoplatinum species [23]. The advent of dendritic polymers, particularly the poly(amidoamine) (PAMAM) dendrimers, has provided a water-soluble, nontoxic base for the preparation of multivalent organoplatinum drugs [24–28].

1.2 The Development of Dendrimer Chemistry

Since Herman Staudinger proposed the macromolecular hypothesis in 1926 [29], the 20th century has witnessed significant development of macromolecular chemistry. Three major macromolecular architectures have evolved since then, namely linear (class I), crosslinked (class II), and branched types (class III), as shown in Figure 11.2. These three classes of traditional synthetic polymers are produced by largely statistical polymerization processes.

Dendritic polymers have recently been recognized as the fourth major class of macromolecular architecture consisting of four subclasses, namely: (1) random hyperbranched (IV(a)), (2) dendrigrafts (IV(b)), (3) dendrons (IV(c)), and (4) dendrimers (IV(d)), (Figure 11.3). Among the four subclasses of dendritic polymers, dendrons and dendrimers are the most intensely studied subset and have a high degree of structural definition. The term *dendrimer* was coined by Tomalia from the Greek term dendro for tree-like. Figure 3a shows that dendritic polymers

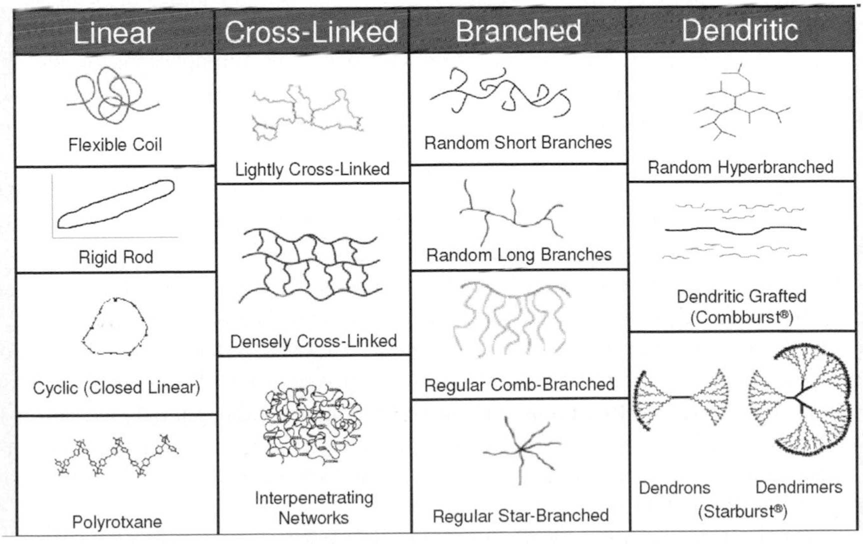

Figure 11.2 Four major classes of macromolecular architecture

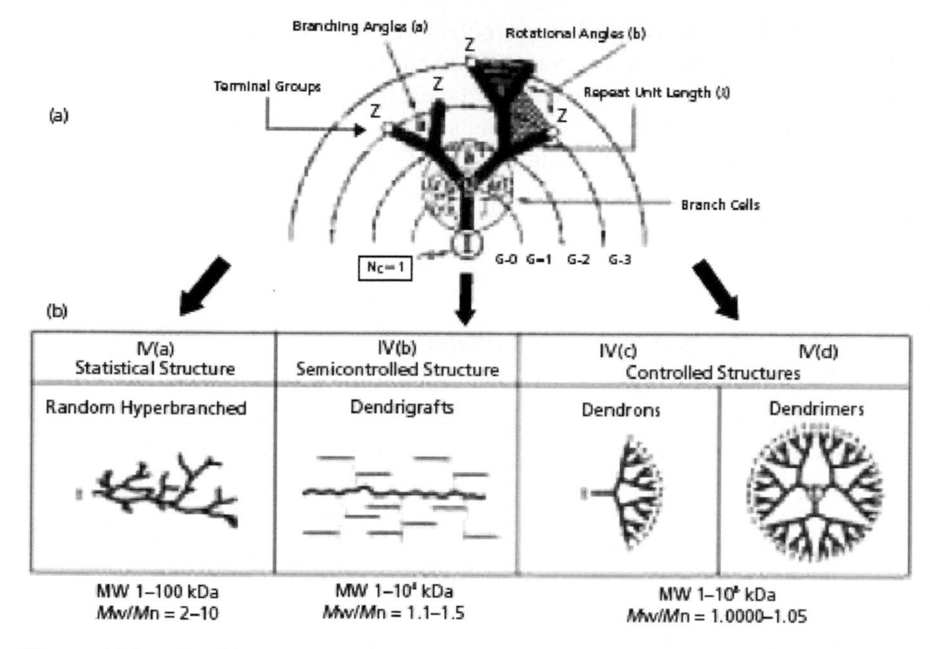

Figure 11.3 a Dendritic polymers as open, covalent assemblies of branch cells. **b** Dendritic polymers: subclasses of the 4th major new class of macromolecular architecture. (Courtesy of reference [27]).

are generally open, covalent assemblies of branch cells. The level of structure control, such as symmetry and polydispersity, of respective subclasses are dependent on the propagation methodology as well as the branch-cell (BC) construction parameters. The BC parameters are determined by the compositions of the BC monomers and the nature of excluded volume defined by the BC, i.e. the length of the arms, the symmetry, rigidity or flexibility, and the branching and rotation angles within each of the BC domains.

The structure of dendrimer depends on the core multiplicity (N_c), the branch-cell multiplicity (N_b), and the critical molecular design parameters (CMDPs) [30]. A typical branching structure is given below [31].

For simulation purposes the PAMAM dendrimers can be constructed by fixing the branch-cell lengths (*l*) and the branching angles (a) and assigning random angles for the rotational angles (b). The initial stage of a PAMAM dendrimer, an initiator core with three polymeric strands, is known as generation 0. When each of these strands branches out to two more strands, the dendrimer is designated generation 1. Subdividing each polymeric strand of the dendrimer produces higher generations. The two-dimensional projections of these dendrimers help to illustrate the geometric progression (Figure 11.4).

The polymeric strands expand symmetrically until there is no room for additional terminal groups (z). As seen in Figure 11.5, it is necessary to control the

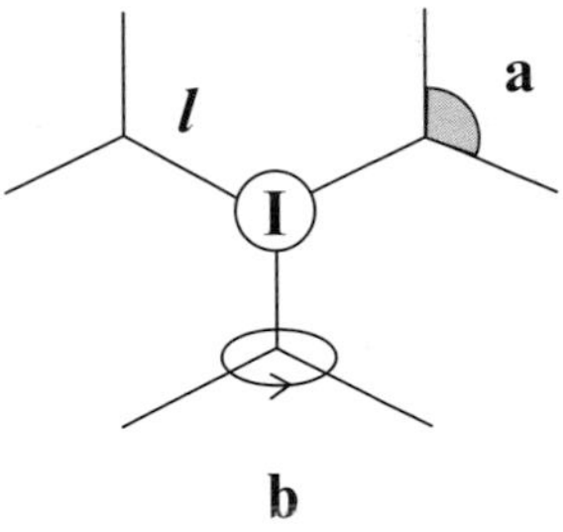

Figure 11.4 Control parameters: I, initiator core; l, branch-cell length; a, branching angles; and b, rotational angles (Courtesy of reference [27]).

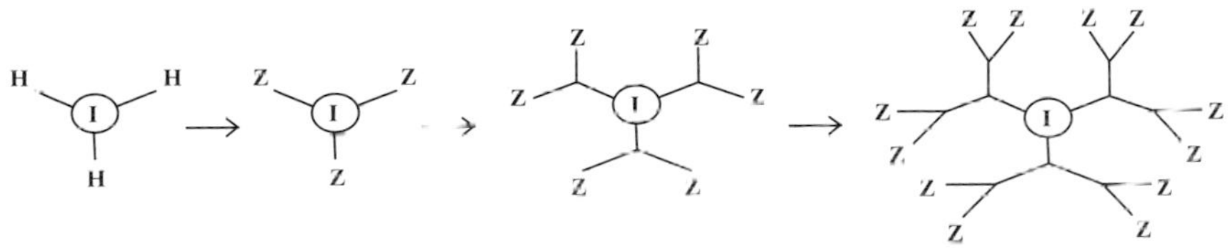

Figure 11.5 Two-dimensional projections of Starburst dendrimers with initiator core multiplicity ($N_c = 3$) and branch-cell multiplicity ($N_b = 2$).

branch-cell length (l) and the branching angles (a) in order to create the two-dimensional projections of the dendrimers. The tree-like structures are called Bethe Lattices or Cayley Trees. The branching process and the means by which polymeric strands occupy space are reflected by mathematical concepts. Cantor used such concepts to create two-dimensional graphics. Cantor Dust was developed by removing the middle third of a line segment and repeating this process on the two remaining segments and so forth (Figure 11.6). The pattern becomes more congested in higher generations. The same observation applies to the two-dimensional projections of PAMAM dendrimers in Figure 11.6.

The initiator-core multiplicity (N_c) and the branch-cell multiplicity (N_b) directly affect the number of terminal groups, the number of repeating units (branch cells), and the molar mass of the dendrimers. The percolation threshold for a Cayley tree having z branches is $p_c = 1/(Z - 1)$ [32].

Dendrimers have attracted considerable attention in the polymer field over the past two decades as they have been recognized as the most important macromolecules possessing tunable internal packing density, void volumes, solvent-dependent size, branching dimensions, and surface functionalities. Since the first report of a dendrimer-like molecule in 1978 [33], significant progress has been made in the dendrimer chemistry. A large number of dendrimer compositions (families) and dendrimer surface modifications have been reported. A plethora of applications related to controlled release of pharmaceuticals have been reported. Currently, there are two widely studied dendrimer families, namely the Tomalia-type polyamidoamine (PAMAM) dendrimers and the Fréchet-type polyether dendrimers. PAMAM dendrimers are the first complete dendrimer family to have been synthesized,

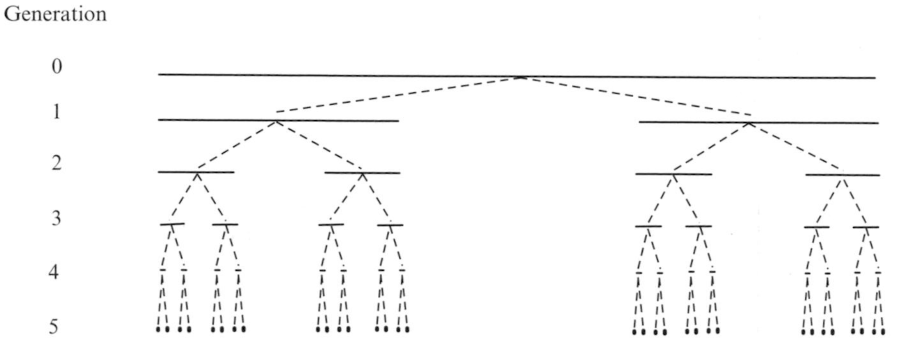

Figure 11.6 Cantor Dust

characterized, and commercialized and are the best-understood series at the present time [26]. Based on these extensive studies, dendrimers have been recognized as a unique new class of nanostructures, which are being studied and developed for many industrial and medical applications such as: (i) size or mass standards (ii) medical diagnostics (iii) controlled delivery of pharmaceuticals or other guest molecules (iv) new materials.

The merger of dendrimer chemistry and coordination chemistry lead to the creation of the metallodendritic regime. Incorporation of metal ions into the framework of dendrimers was initiated in the early 1990s, either through the use of metal branching centers or by internal metal complexation or encapsulation at specific binding site(s) [34–37]. Recently, several groups have focused on the incorporation or complexation of (transition) metal components onto the dendrimer branches or periphery and currently a broad spectrum of metallodendritic species are known. A detailed review of metallodendrimers, in which the metals serves as branching centers, building block connecters, terminal groups, and structural auxiliaries, has been presented [38]. Metallodendrimers represent novel supermolecular architectures possessing new physical, optical, electrochemical, photochemical, biological, and catalytic properties. In general, these metallodendrimers reflect advantages of both homogeneous and heterogeneous catalysts because the metal sites in these well-defined spherical polymeric assemblies are easily accessible to substrate molecules and reagents, and therefore exhibit characteristics usually encountered in homogeneous catalysis such as fast kinetics, specificity, and solubility. Several reports of specific metallodendrimers and applications have been appeared [39–62].

1.3 Dendrimer Synthesis: Divergent and Convergent Methods

Dendrimers are highly branched macromolecules with precisely controlled size, shape and end-group functionality. In contrast to conventional synthetic polymers, dendrimers are unique core-shell structures consisting of three basic architectural

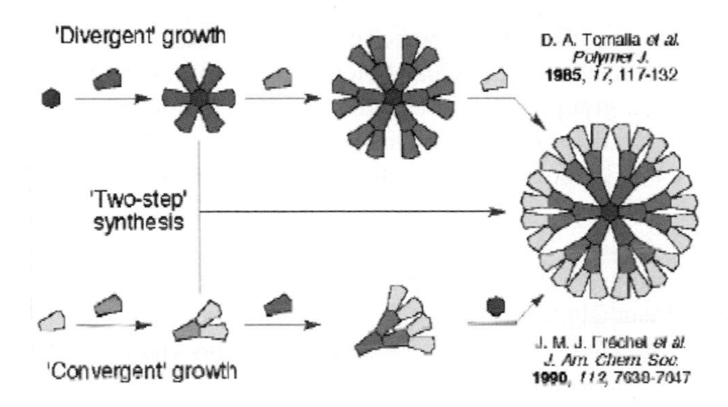

Scheme 11.1 Dendrimer synthesis methods

components: a core, an interior of shells (generations) having repeating branch-cell units, and terminal functional groups (the outer shell or periphery).

The synthetic methodologies used for the construction of dendrimers generally involve divergent or convergent hierarchical assembly strategies, which require relevant construction components with multiple functionalities and need the repetitious alternation of a growth reaction and an activation reaction, as shown in Scheme 11.1. Recently, a new "double-stage convergent" approach has been reported, in which the combination of divergent and convergent growth allows the synthesis of the fourth generation acetonide-protected monodendron with a reduced number of purification steps [63]. Within each of these major methods, there may be variable selections for branch-cell construction and dendron construction. For details of the chemistry employed in the synthesis of dendrimers, several comprehensive reviews may be consulted [64–66].

The divergent method can be envisioned by considering the sequential addition of repeating units (building blocks, monomers, or dendrons) to a starting core site, thus forming the layers or generations of repeating units within the superstructure. This protocol is exemplified by two commercial products, PAMAM dendrimers [27,28,67] and poly(propylene imine) (PPI) dendrimers [68,69]. During the 1980s, virtually all dendritic polymers were produced by construction from the root of the molecular tree. In the case of PAMAM dendrimer synthesis, the divergent approach involves in situ branch-cell construction in stepwise, iterative stages around a desired core to produce mathematically defined core-shell structures. Normally, ethylenediamine [core multiplicity (N_c) = 4], ammonia (N_c = 3), or cystamine (N_c = 4) may be used as cores and allowed to undergo repeating, two-step reaction sequences. These sequences involve: (1) an exhaustive alkylation of primary amines (Michael addition) with methyl acrylate, and (2) amidation of amplified ester groups with a large excess of ethylenediamine to produce primary amine terminal groups. The alkylation step produces ester terminated subshells that are referred as "half-generations" and are designated as (Gn.5). The second step involves amidation of the ester terminated (Gn.5) intermediates with substantial excesses of ethylenediamine to generate amine terminated full generations, referred as (Gn). The iteration of alkylation-amidation reaction produces amplifications of branch cells (generations) as well as surface functional groups. The mathematical calculation of molecular weight (MW), branch cells (BC), and terminal groups (Z) has been demonstrated for the example of PAMAM (cystamine core) dendrimers, as shown in Figure 11.2. It is apparent that both the core multiplicity (N_c) and branch-cell multiplicity (N_b) determine the precise number of terminal groups (Z) and mass amplification as a function of generation (G). It is generally noted that the molecular weights approximately double as one proceeds from one generation to the next. The surface groups (Z) and branch cells (BC) amplify mathematically according to a power function, thus producing discrete, monodispersed structures with precise molecular weights and a nanoscale diameter enhancement as described in Figure 11.7.

The convergent method essentially constructs dendritic architectures starting from the periphery inward to a reactive focal point at the root (Scheme 11.1). This leads to the formation of a single reactive dendron. In other words, unlike the divergent approach used for PAMAM, the growth involves a limited number of reactive sites. The dendrimer assembly is obtained by the connection of several dendrons through the single functional focal point. Notable features of the convergent method include simplicity and great precision, ease of purification, and functional versatility enabling the preparation of dendrons with differentiated functionalities located respectively at the focal point and at the chain ends of dendrons [70,71]. Two well-known dendrimer families are prepared by the convergent method: the polyether dendrimers derived from the 3,5-dihydroxybenzyl alcohol moiety [70], and the aliphatic polyester dendrimers derived from the 2,2-bis-hydroxymethylpropionic acid repeat unit [72].

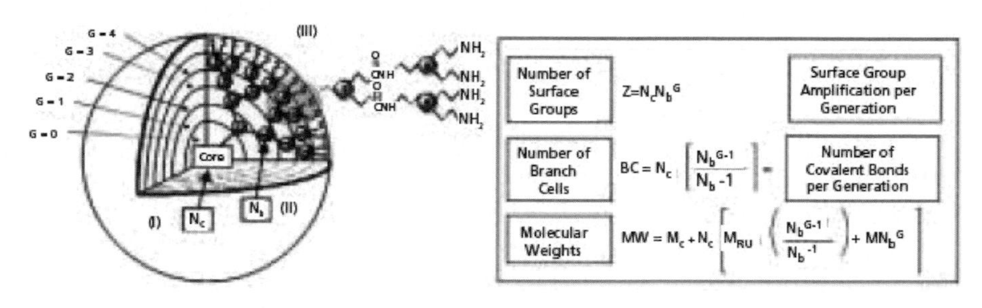

Gen	No. of NH_2 Surface Groups	Molecular Formula	MW	Hydrodynamic Diameter (nm)
0	4	$C_{24}H_{52}N_{10}O_4S_2$	609	1.5
1	8	$C_{64}H_{132}N_{26}O_{12}S_2$	1,522	2.2
2	16	$C_{144}H_{292}N_{58}O_{28}S_2$	3,348	2.9
3	32	$C_{304}H_{612}N_{122}O_{60}S_2$	7,001	3.6
4	64	$C_{624}H_{1252}N_{250}O_{124}S_2$	14,307	4.5
5	128	$C_{1264}H_{2532}N_{506}O_{252}S_2$	28,918	5.4
6	256	$C_{2544}H_{5092}N_{1018}O_{508}S_2$	58,140	6.7
7	512	$C_{5104}H_{10212}N_{2042}O_{1020}S_2$	116,585	8.1

Z = monomer-shell-saturation level, N_c = core (cystamine) multiplicity, N_b = branch cell (BC) multiplicity, G = generation.

Figure 11.7 Mathematical expressions for calculating the theoretical number of surface groups (Z), branch cells (BC), and molecular weights (MW) for [cystamine Core]-PAMAM dendrimers as a function of generation. Approximate hydrodynamic diameters (Gen = 0–7) based on gel electrophoretic comparison with the corresponding [Ethylenediamine Core]-PAMAM dendrimers. (Courtesy of reference [27])

1.4 Properties of Dendrimers

One of the important properties of dendrimers compared with those of traditional polymers is monodispersity, which has been verified by mass spectroscopy, size-exclusion chromatography, gel electrophoresis, and electron microscopy (TEM) [73]. Generally, convergent methods produce the most nearly monodisperse dendrimers as a result of purification at each step of dendron synthesis. Mass spectrometry studies have shown that PAMAM dendrimers produced by divergent methods are remarkably monodisperse and have masses consistent with predicted values for earlier generations (G0-5) [30,74–75]. Because of their molecular architecture, dendrimers show some significantly improved physical and chemical properties when compared to traditional linear polymers. In solution, linear chains exist as flexible coils; in contrast, dendrimers form a tightly packed ball. This has a great impact on their rheological properties. Dendrimer solutions have significantly lower viscosity than those of linear polymers [43]. When the molecular mass of dendrimers increases, their intrinsic viscosity goes through a maximum at the fourth generation and then begins to decline [76]. Such behavior is unlike that of linear polymers. For classical polymers the intrinsic viscosity increases continuously with molecular mass. The presence of many chain-ends is responsible for high solubility and miscibility and for high reactivity [43]. The solubility of dendrimers is strongly influenced by the nature of surface groups. Dendrimers terminated with hydrophilic groups are soluble in polar solvents, while dendrimers having hydrophobic end groups are soluble in nonpolar solvents.

The unique branched architecture and the multifunctional end groups make dendrimers important as unimolecular nanocontainer and scaffolding molecules. These properties are determined by the size, shape, and multiplicity of the construction components that are used for the core, interior, and surface of dendrimers. As discussed above, the unique core-shell structure of the dendrimer is manifest as a specific function of each architectural component. The core is thought of as the molecular information center from which size, shape, directionality, and multiplicity are expressed via the covalent bonding with outer shells. The branch-cell moieties within the dendrimer define the type and volume of interior void space that may be enclosed by terminal groups as the dendrimer grows. The interior composition and volume of solvent-filled void space determines the extent and nature of the guest-host (endo-receptor) properties that are possible within a particular dendrimer family and generation. Meanwhile, the terminal groups may serve as template polymerization regions for dendrimer growth and may also function as passive or reactive gates controlling entry or departure of guest molecules from the dendrimer interior. These three architectural components essentially determine the physical and chemical properties, as well as the overall size, shape, and flexibility of a dendrimer. There have been many efforts to describe theoretically the maximum core-shell filling and to predict the architecture to be generated in megamer synthesis. The most successful is reflected in the Mansfield-Tomalia-Rakesh equation (Figure 11.8) [27,77]. Many of the interesting properties of dendrimers—from design and synthesis to applications—have been reviewed [3–41,54,55,58,62,78–80].

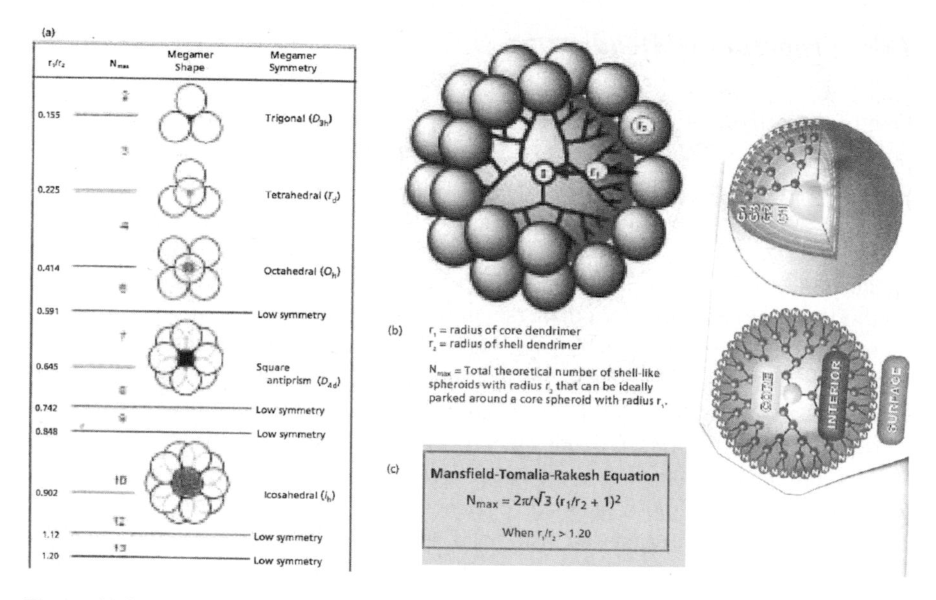

Figure 11.8 (a) Symmetry properties of a core-shell structure where $r_2 / r_1 < 1.20$. (b) Sterically induced stoichometry based on the respective radii of core and shell dendrimers (c) Mansfield-Tomalia-Rakesh equation for calculating the maximum shell filling when $r_2/r_1 > 1.20$

The unique properties of dendrimers have lead to many promising applications in the field of host-guest chemistry and pharmaceuticals, such as their use as molecular carriers, enzyme mimics [81], and drug-delivery vehicles [82–85]. The concept of dendrimer as drug delivery agent is principally based on the dualistic role of dendrimers as either endo-receptor (nano-container) or exo-receptor (nano-scaffolding). The interior void space within dendrimer was first noted in 1989 [86] and subsequently has been demonstrated by NMR studies and computer-assisted simulations [87], molecular inclusion work [34], and extensive photochemical probe experiments [88]. These studies showed that symmetrically branched dendrimers can be viewed as unimolecular micelles (nanoscale containers) and can manifest behavior reminiscent of either regular [89] or inverse micelles [90] depending on the nature of the surface groups and interiors. This "unimolecular encapsulation" behavior has been demonstrated via the concept of a "dendritic box," in which the surface of a generation-5 poly(propylene imine) (PPI) dendrimer was modified by conjugation with 1-phenylalanine or other amino acids [91]. Such modifications induced dendrimer encapsulation properties by the formation of dense, hydrogen-bonded surface shells with solid-state character. Small guest molecules could be entrapped in such dendrimer interiors and were unable to escape even after extensive dialysis [92]. A simple and general synthesis concept for the selective shell functionalization of dendritic polymers (e.g., PG and PEI) to generate molecular nanocarriers for the encapsulation and transport of polar guest molecules has been presented [93]. Dendritic polymers with a pH-responsive shell (PG-acetal/ketal and PEI-imine bonds) have been prepared which can selectively release the encapsulated guest molecules in a physiologically relevant pH range. The maximum number of captured guest molecules was determined by the shape and size of the guest molecules as well as by the amount, shape, and size of the available internal dendrimer cavities. In addition, hydrophobic and hydrophilic properties, as well as other supramolecular (i.e., noncovalent binding, acid-base reaction) properties, of these spatial binding-sites are expected to strongly influence the host-guest relationship.

2 PAMAM Dendrimer-Platinum Conjugates and Applications

2.1 PAMAM Dendrimer-Encapsulated Platinum Nanoparticles

PAMAM dendrimers have drawn considerable interest in recent years because of their well defined structures and number of potential applications. In particular, PAMAM dendrimer interior tertiary and secondary nitrogen atoms and surface amine groups provide versatile binding sites for metal ions through coordination chemistry or ligand exchange reactions. Higher-generation PAMAM dendrimers have spherical core-shell structures with closely packed surface groups surrounding interior voids. These attributes make PAMAM dendrimers useful as a template and stabilizer for synthesizing metal nanoparticles with controlled size, shape, and composition. Recently, there have been a number of reports which have described the synthesis, characterization, and applications of dendrimer-encapsulated nanoparticles (DENs) [94–111]. DENs are synthesized by a template approach in which metal ions are incorporated into the interior of dendrimers and then chemically reduced to yield nearly size-monodispersed particles having dimensions of less than 3 nm, as shown in Scheme 11.2. The dendrimer component hereby serves not only as a template for preparing metal nanoparticles but also as a platform to stabilize the nanoparticles. This makes it possible to tune solubility and provides a means for immobilization of the nanoparticle. DENs have a number of potential applications, but the current focus is on catalysis, such as homogeneous catalytic reactions in various solvents with the advantage of recycle for the catalytic DENs.

In the instance of dendrimer-encapsulated platinum nanoparticles, a general template-based approach for preparing monodisperse platinum nanoparticles suitable for use as electrocatalysts for oxygen reduction [99] and as homogeneous hydrogenation catalysis has been demonstrated [107]. Dendrimer-encapsulated platinum nanoparticles are prepared via a two-step process (*see* Scheme 11.2). First, platinum(II) ions are loaded into a PAMAM (G4) dendrimer containing hydroxyl surface groups. The platinum ions coordinate in fixed stoichiometries with interior functional groups. Second, the Pt^{2+}/dendrimer composites are reduced with borohydride to yield encapsulated platinum nanoparticles. Such encapsulated nanoparticles are soluble in water and are stable, either dry or solvated, for at least several months as a result of the porous stabilizer function of the dendrimer. High-resolution transmission

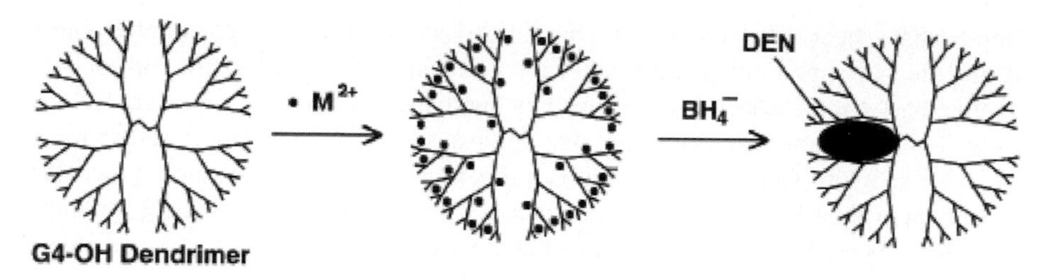

Scheme 11.2 Dendrimer-encapsulated nanoparticle synthesis (Reprinted with permission from reference [112]

electron microscopy (HRTEM) images show that the nanoparticles are quite monodisperse and that their shape is roughly spherical with average diameters of 1.4 ± 0.2 nm (G4-OH(Pt_{40})), 1.6 ± 0.2 nm (G4-OH(Pt_{60})), respectively. The size of the platinum naonparticles is generally dependent on the type and generation of the dendrimer used as the template, as well as on the number of metal ions preloaded into the dendrimer interior prior to reduction. The nanoscale, monodisperse, and stable encapsulated platinum nanoparticles may act as electrocatalysts for oxygen reduction. The encapsulated nanoparticles could be attached to a gold electrode surface [99] or coupled to glass carbon electrodes *via* an electrooxidation reaction [111]. The results conclusively demonstrate that the surface of the dendrimer-encapsulated platinum nanoparticles is accessible to reactants in solution and can exchange electrons with the underlying electrode surface. Both platinum and palladium DENs display high catalytic activity for the hydrogenation of alkenes in aqueous solution. These DENs are quite stable before, during, and after hydrogenation reactions [107]. Importantly, the hydrogenation rate can be controlled though use of dendrimers of different generations by virtue of selective nanoscopic filtration by higher generation dendrimers. Platinum and bimetallic platinum-gold nanoparticles prepared by using (G4–OH) PAMAM dendrimers have been adsorbed onto a high surface area silica support and thermally activated to remove the dendrimers to generate highly active nanoscale catalysts [113,114]. Utilizing dendrimer-encapsulated nanoparticles as catalyst precursors offers the opportunity to exert a degree of control over the metal size and composition, while varying the carrier or substrate. These platinum and bimetallic platinum-gold nanoparticles are active for both toluene hydrogenation and carbon monoxide oxidation.

Despite the considerable work on dendrimer-encapsulated nanoparticles, the mechanisms by which PAMAM dendrimers function as the template to form DENs and the feasibility for controlling the shape and size of nanoparticles using dendrimer templates are not fully understood. The platinum(II) ion uptake by PAMAM (G2 and G4) dendrimers containing terminal hydroxyl groups was studied using ^{1}H, ^{13}C, and ^{195}Pt NMR and AFM measurements [115]. For the PAMAM(G2) dendrimer, it was found that half of the platinum(II) precipitated after 2 days. The remaining platinum(II) species were coordinated to 1, 2, or 3 nitrogen ligands in the open structure of the PAMAM(G2) dendrimer. In the case of the G4-OH dendrimer, 80% of platinum(II) was taken up deep within the dendrimer in a very slow process but without precipitation. Reduction of the G4-OH-Pt(II) complexes with borohydride generated platinum nanoparticles 80% of which were less than 1.1 nm in height; small enough to be within individual dendrimers. Thus, the dendrimer not only acts as a template for the formation of small nanoparticles within it but also more generally acts to prevent precipitation of platinum nanoparticles that are larger than 1.1 nm. In a more recent study employing AFM measurements, the size of dendrimer-stabilized platinum nanoparticles deposited on mica surfaces provide insights into the mechanism of PAMAM-mediated templating of metal nanoparticles in solution [116]. If the complexation of a platinum precursor with PAMAM (G4-OH) is prematurely terminated, AFM images and feature height distributions show evidence of arrested precipitation of platinum colloids. In contrast, sufficient Pt-PAMAM complexation time leads to AFM images and height distributions that have relatively narrow, normal distributions with mean values that increase with the nominal Pt:PAMAM ratio. The surface density of the features in the AFM images suggests that these platinum nanoparticles reside on the mica surface as two-dimensional surface aggregates. These observations are consistent with an intradendrimer templating mechanism proposed for platinum and other metal nanoparticles [117,118].

2.2 *PAMAM Dendrimer as an Exo-Receptor for Platinum Drugs*

2.2.1 Antitumor Activity of PAMAM(G3.5)-Cisplatin

Organoplatinum antitumor drugs, particularly Cisplatin, are widely used for the treatment of human testicular, ovarian, bladder, head and neck carcinomas. However, the severe side effects accompanying administration of these drugs as well as the development of Cisplatin resistant diseases have spurred continued activity in this area. Both new compounds and new delivery vehicles continue to be formulated and evaluated [20]. New platinum compounds containing *trans*-1,2-diamminocyclohexane as an inert ligand such as Oxaliplatin offer promise for the treatment of Cisplatin resistant tumors such as colorectal tumors [119]. There has been a continuing interest in the development of hydrophilic, biocompatible polymeric materials that could serve as drug carriers to achieve site-specific and time-controlled delivery of therapeutics both to alleviate undesired side effects and enhance the efficacy of treatment [120,121]. For drug carriers, high molecular weight materials (>20,000 Da) provide advantages over low molar mass compounds (<1,000 Da) including enhanced targeting of tumor tissue and efficiency of treatment [122–124]. The use polymeric prodrugs takes advantage of the endocytotic cell uptake mechanism and also may overcome problems of water solubility and lower the toxicity of the drug until it is delivered to the target tissue. As noted above, the advent of dendrimers, which are highly branched macromolecules with precisely controlled shape, nanoscale size and end-group functionality, has provided an excellent opportunity to design novel drug carriers.

The first preclinical study of a PAMAM dendrimer-platinate conjugate for the delivery of antitumor agents was reported in 1999 [125]. A generation 3.5 PAMAM dendrimer containing carboxylate surface groups was treated with Cisplatin to generate a dendrimer-platinate (20–25 wt% Pt, about 22 Cisplatin units) which was highly water soluble and had a much higher platinum loading than observed for *N*-(2-hydroxypropyl)methacrylamide (HPMA) copolymer platinates (3–8 wt% platinum) [126]. Size exclusion chromatography (GPC) and particle size photon correlation spectroscopy revealed that the PAMAM(G3.5)-cisplatin conjugate consisted of a number of species including those arising from monodentate and bidentate to carboxylate groups as well as cross-linked dendrimer via platinum bridges, which caused an increase in particle size from 3 to 4 nm in the parent dendrimer to between 30 and 40 diameter of the dendrimer-Cisplatin adduct. Thus, a variety of platinum species, including some physically entrapped Cisplatin, were present in the polymer-drug conjugate. In vivo the dendrimer-Cisplatin and Cisplatin administered intraperitoneally (i.p.) were equally active against L1210, and at high dose dendrimer-cisplatin displayed activity against B16F10 whereas Cisplatin did not. Additionally, when administered intravenously (i.v.) to treat a palpable squamous cell (s.c.) B16F10 melanoma, the PAMAM(G3.5)-Cisplatin adduct displayed antitumor activity whereas Cisplatin was inactive. Measurement of platinum levels in blood and tissues after i.v. injection of Cisplatin (1 mg/kg) or dendrimer-Cisplatin (15 mg/kg)—the maximum tolerated dose (MTD) of these compounds—showed selective accumulation of the dendrimer-Cisplatin in solid tumor tissue by the enhanced permeability and retention (EPR) effect. The improved activity in the s.c. solid tumor model versus the i.p. ascites is indicative of the importance of the EPR effect in tumor targeting. This PAMAM(G3.5)-Cisplatin conjugate was also less toxic (3- to 15-fold) than Cisplatin and thus has potential for further investigation as a novel antitumor approach. However, the release of active platinum species was not detected by atomic absorbance spectrometry (less than 1% of the total platinum released) in pH 7.4 and pH 5.5 buffer solutions over a period of 72 hours. It should also be noted that the exact proportion of the dendrimer-Cisplatin made available as the active diaquo species is not yet known and indeed the time course of platinum liberation is yet to be determined.

2.2.2 Synthesis and Characterization of PAMAM(G4.5)-[(DACH)Pt]

Although the activity of the above polymer-platinum conjugate is impressive, the ill-defined nature of the multiple species present as well as the variable release rates for these entities makes this less than an ideal formulation for the treatment of disease. More recently, a generation 4.5 PAMAM dendrimer nanoconjugate containing (1,2-diaminocyclohexane)platinum(II) [(DACH)Pt] moieties covalently bound to surface carboxylate groups has been prepared [127]. For the preparation of a useful drug formulation, the PAMAM(G4.5) dendrimer has several positive features. Generally, high generation symmetrical dendrimers (G≥4) have globular structure with peripheral densely-packed terminal groups [127]. In the case of

PAMAM (G4.5), 128 carboxylate groups are closely packed on the dendrimer surface. This strongly facilitates the interaction of carboxylate groups with platinum species to construct a dendrimer-based multivalent platinum conjugate. In addition, the carboxylate functionality should serve as labile ligands for platinum moieties such that release of active platinum species [(DACH)Pt] should occur at a sustained rate over a period of time. Biological studies using PAMAM dendrimers have also demonstrated that high generation PAMAM dendrimers are nonimmunogenic and display low mammalian toxicity, while anionic PAMAM dendrimers (surface groups with carboxylate or hydroxylic functionalities) are non-toxic in vitro. [128,129] These intrinsic properties of PAMAM (G4.5) dendrimer make it an interesting multivalent macromolecule that can serve as an exo-receptor for the construction of novel dendrimer-based platinum anticancer agents. The *trans*-1, 2-diaminocyclohexane platinum(II) moiety was selected as the active species for the dendrimer-platinum nanoconjugate because 1,2-diaminocyclohexane is known to serve as a superior inert ligand for the preparation of platinum antitumor compounds and to contribute to enhanced antitumor activity. Furthermore, the relative bulky size of [(DACH)Pt$(OH_2)_2$]$(NO_3)_2$ and the hydrophobic nature of DACH inhibit the guest drug from penetrating the sterically-crowded surface to access the interior of the dendrimer.

The synthesis of the PAMAM(G4.5)-[(DACH)Pt] nanoconjugate and a diagram of a structural model are presented in Scheme 11.3. [(DACH)PtCl_2] was prepared from tetrachloroplatinate as previously described [16]. This, in turn, was treated with aqueous silver nitrate to generate the corresponding diaquo species. Treatment of this intermediate with a pH 5.0 aqueous solution of PAMAM(G4.5) dendrimer produced the dendrimer-based platinum conjugate with carboxylate groups as the labile ligands at the surface of the dendrimer. The PAMAM(G4.5)-[(DACH)Pt] conjugate was purified by dialysis against deionized water (3500 Da cut-off). The water solubility of the resulting PAMAM(G4.5)-[(DACH)Pt] is extremely good. The resultant sample was checked for purity using thin layer chromatography and dried by lyophilization.

This nanoscale multivalent PAMAM(G4.5)-[(DACH)Pt] conjugate was fully characterized using a variety of spectroscopic, chromatographic and thermal methods. The complexation of surface carboxylate groups by platinum is apparent from the downfield chemical shift of the carboxylate group in the ^{13}C NMR spectrum of the PAMAM(G4.5)-[(DACH)Pt], as shown in Figure 11.9. Generally, the chemical shift of surface carboxylate of a PAMAM half-generation is smaller than that of interior carbonyl groups [64]. Here the strong peak at δ 175.2 ppm in Figure 11.9a corresponds to the 128 surface carboxylates of PAMAM (G4.5). Upon formation of the PAMAM(G4.5)-[(DACH)Pt] conjugate, this absorption is shifted downfield to 177.7 ppm (Figure 9b), reflecting coordination of the surface carboxylate groups with platinum.

The ^{1}H NMR spectrum of the conjugate unambiguously shows two characteristic regions representing the ethylene groups (3.60–2.40 ppm) of the PAMAM dendrimer and the cyclohexyl portion of the inert ligand (1.80–1.20 ppm). It is also important to note that the integration of the signal in the 2 regions indicates that there are about 40 [(DACH)Pt] units coordinated to each dendrimer. That this is a maximum possible loading of [(DACH)Pt] units per dendrimer is apparent from

$$[(DACH)PtCl_2] \xrightarrow[H_2O]{AgNO_3} [(DACH)Pt(OH_2)_2](NO_3)_2$$

$$[(DACH)Pt(OH_2)_2](NO_3)_2 + PAMAM(G4.5) \longrightarrow PAMAM(G4.5)[(DACH)Pt]$$

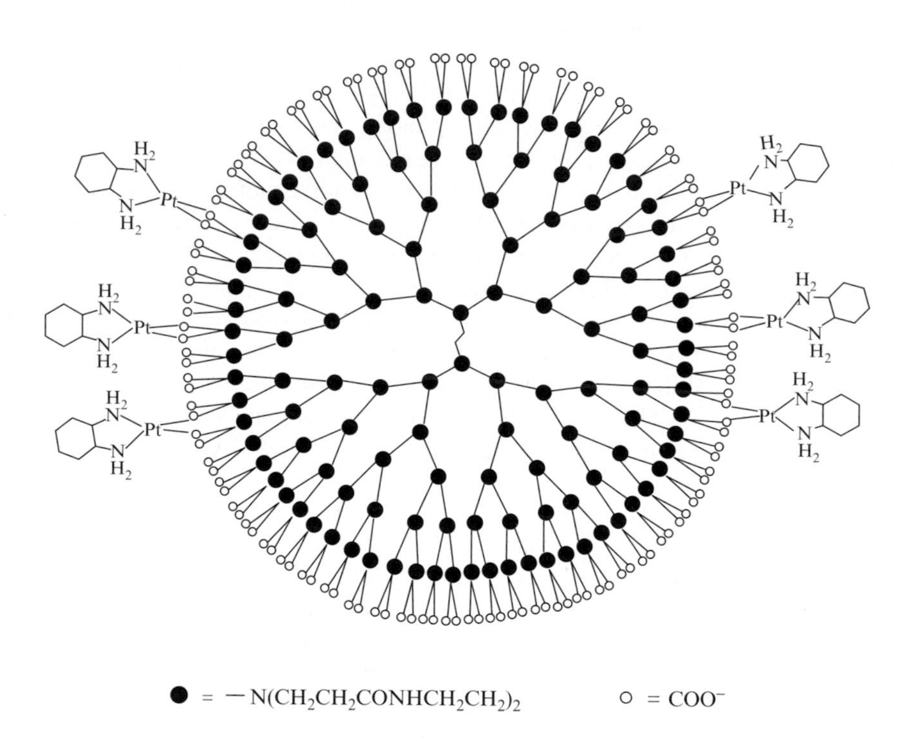

Scheme 11.3 Synthesis of PAMAM(G4.5)-[(DACH)Pt] nanoconjugate and a diagram of its structure (other [(DACH)Pt] units omitted for clarity)

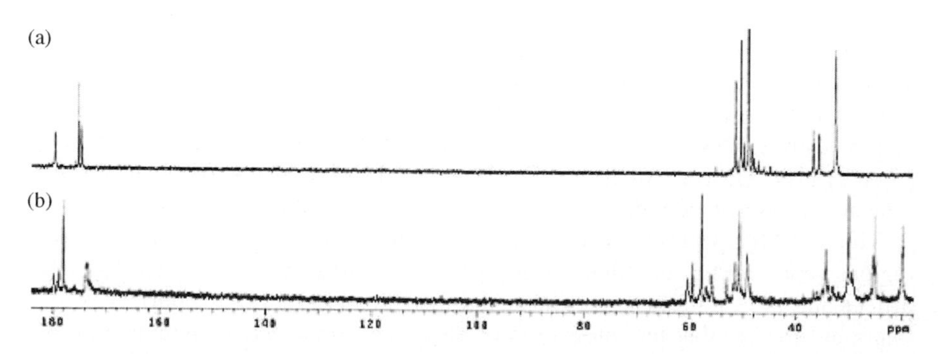

Figure 11.9 ^{13}C NMR spectra of (a) PAMAM(G4.5) and (b) PAMAM(G4.5)-[(DACH)Pt] conjugate in D_2O

Table 11.1 MALDI-TOF mass values for six preparations of PAMAM(G4.5)-[(DACH)Pt]

Batch	Molar Ratio of G4.5:[(DACH)Pt]	Observed Mass Value	No of [(DACH)Pt]*
1	1:48	29271	33
2	1:64	30706	38
3	1:64	31075	40
4	1:72	31538	41
5	1:128	32318	44
6	1:192	30935	39
	Pure PAMAM (G4.5)	20665	–

* No. of [(DACH)Pt] is calculated on the basis of mass value of pure PAMAM(G4.5)

a series of experiments in which the ratio of [(DACH)Pt$(H_2O)_2$]$(NO_3)_2$ to dendrimer was increased well beyond the theoretical maximum of 64 [(DACH)Pt] units. For example, even when the ratio of [(DACH)Pt$(H_2O)_2$]$(NO_3)_2$ to dendrimer was 192 to 1 (well above that required to saturate the carboxylate surface), the dendrimer-Pt conjugate generated contained 39 (DACH)Pt units. The matrix assisted laser desoprtion/ionization time-of-flight mass spectroscopy (MALDI-TOF MS) data in Table 11.1 support this observation and indicate an average of 40 [(DACH)Pt] moieties per dendrimer are bound to the PAMAM(G4.5) surface. This is in close agreement with the results from ^{1}H NMR spectra and clearly supports the observation that the maximum loading is approximately 40 [(DACH)Pt] moieties per dendrimer molecule. The loading capacity of 40 [(DACH)Pt] units onto the nanocarrier PAMAM(G4.5) is consistent with a recent report of a similar limitation for the interaction of PAMAM (G4, amine surface) with an organic ligand [131]. Thermogravimetric (TGA) analyses of the PAMAM(G4.5)-[(DACH)Pt] conjugate was carried out to further ascertain this value. The conjugate begins to decompose at 173 °C and a stable residue of platinum oxide is obtained at 910 °C, as shown in Figure 11.10. The mass of the residual oxide corresponds to 24.1% of the initial sample mass. This is in excellent agreement with that expected (24.7%) for a dendrimer-Pt conjugate containing 40 [(DACH)Pt] moieties. In poly(acrylamide) gel electrophoresis (PAGE), the migration difference between PAMAM(G4.5) and its [(DACH)Pt] conjugate is clearly a result of loading of [(DACH)Pt] moieties onto dendrimer, leading to the increment of the mass value of the conjugate. Remarkably, the drug loading capacity about 40 [(DACH)Pt] units on PAMAM(G4.5) dendrimer is significantly higher than that observed for the PPI "dendritic-box" (4 large or 8–10 small guest molecules)[91] and in traditional *N*-(2-hydroxypropyl)methacrylamide (HPMA) copolymer (3–8 wt% platinum) [126].

Molecular dynamic simulation is a powerful tool for the theoretical evaluation of the energetic and structural properties of this new class marcomolecules—dendrimers. There have been numerous molecular dynamics studies for dendrimers which are in good correlation with experimental results. Molecular dynamic simulation has been carried out to investigate the various binding sites for interac-

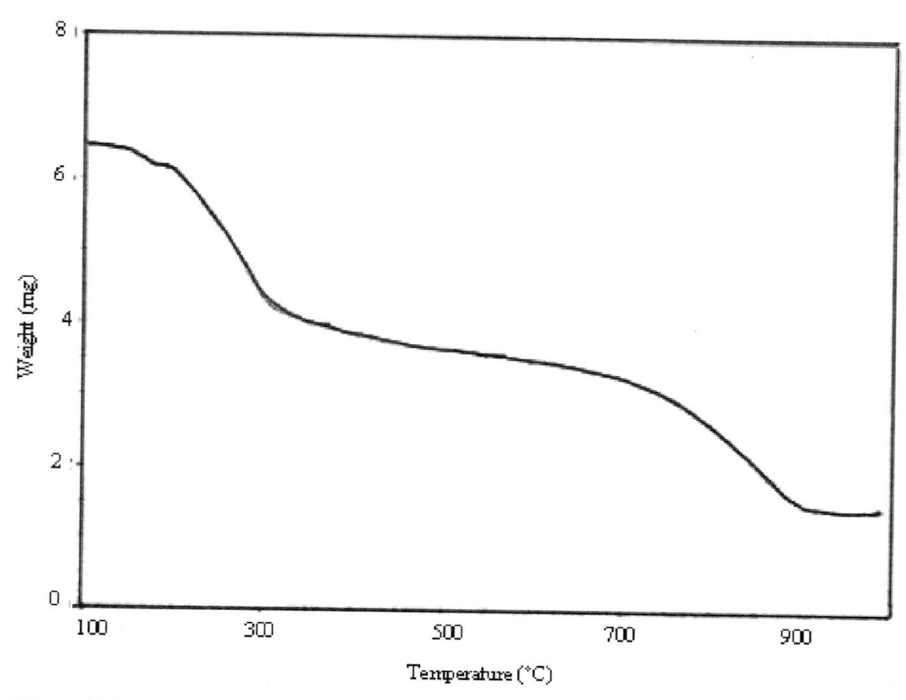

Figure 11.10 Thermogravimetric analysis for PAMAM(G4.5)-[(DACH)Pt]

tion of surface carboxylate groups of PAMAM (G4.5) dendrimers with [(DACH)Pt] species. To some extent, available experimental results have been used to form a base for the computational study. For example, preferred binding sites, in some cases, could be inferred from the results of NMR studies of the binding of platinum moieties to similar polymers, DNA in particular. The available NMR data are not always as useful as might be expected. However, this has proven to be a useful starting point for molecular simulation. Computational simulation has been used to probe the attachment of active organoplatinum fragments to dendrimers, which would serve as a nanocarrier from which the active species might be released over an extended period of time.

The molecular dynamic simulations for the PAMAM(G4.5)-[(DACH)Pt] conjugate were performed using the Forcite module of Accelyrs with universal forcefield [Accelrys Software Inc., San Diego] for 500 psec with 1 femto-second time intervals at 298 K in vacuum after global minimization of 5000 iterations. The minimum energy was then computed for each addition of [(DACH)Pt] to 2 carboxylates at the PAMAM(G4.5) surface. Various methods have been employed to authenticate the total number of [(DACH)Pt] units which can be attached to the surface carboxylate groups of the dendrimer. The simulation for PAMAM(G4.5)-[(DACH)Pt] shows that the minimum energy for the conjugate is reached when 44 [(DACH)Pt] units are attached to PAMAM(G4.5) dendrimer surface (Figure 11.11). This simulation

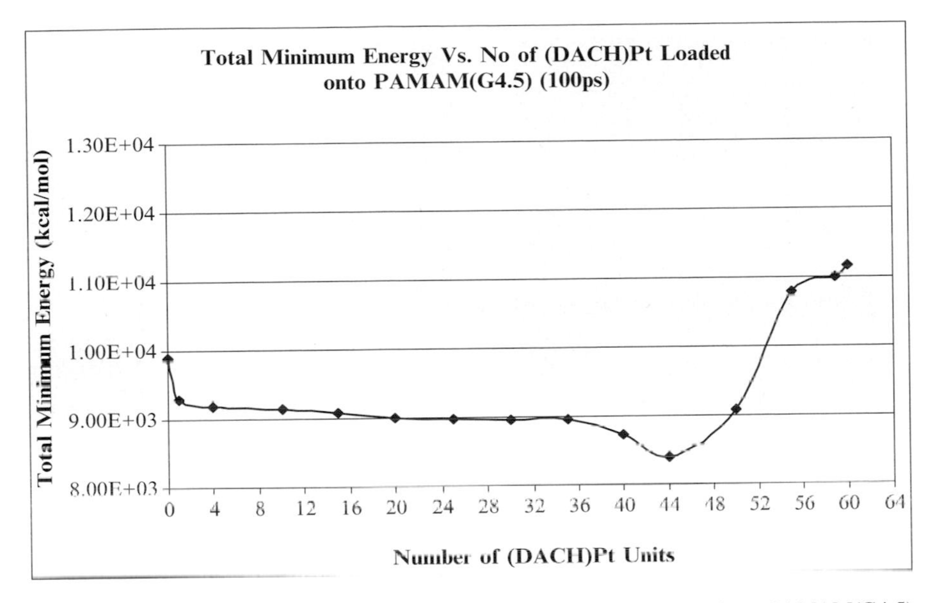

Figure 11.11 Total minimum energy versus number of (DACH)Pt units added onto PAMAM(G4.5) (using 500 ps simulation)

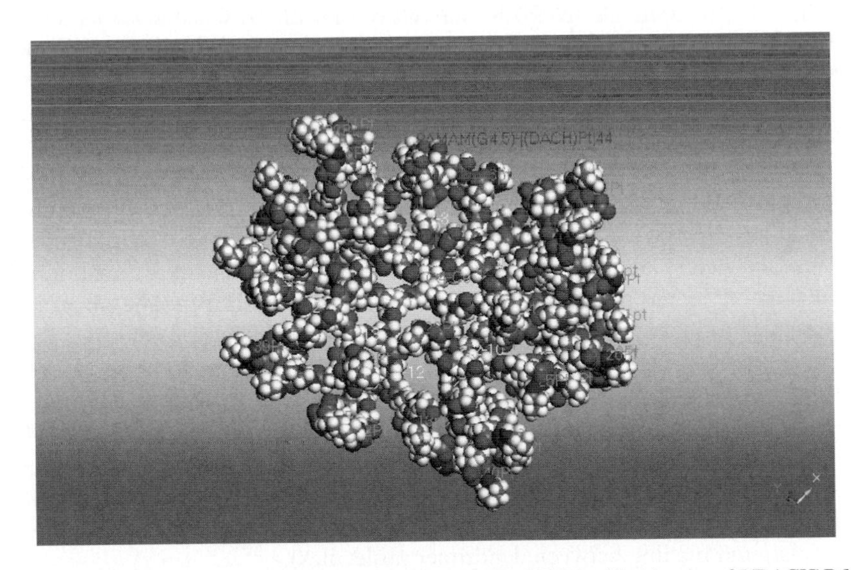

Figure 11.12 Molecular model of PAMAM(G4.5)-[(DACH)Pt] with 44 units of [(DACH)Pt]

result is in good agreement with the experimental data which demonstrate that about 40 [(DACH)Pt] units are attached to the surface carboxylate groups of the PAMAM(G4.5) dendrimer. A molecular model of the PAMAM(G4.5)-[(DACH)Pt] conjugate is presented in Figure 11.12.

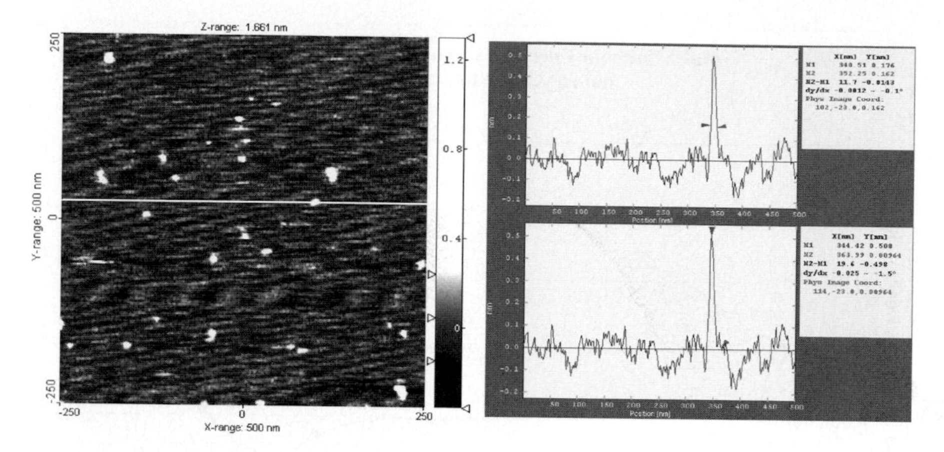

Figure 11.13 AFM images of PAMAM(G4.5)-[(DACH)Pt] nanoconjugate *(left)* and profile data for a dendrimer-Pt molecule *(right)*

Dendrimers are viewed as a unique new class of synthetic nanodevices and PAMAM dendrimers have been generally used as nanoscale building blocks to construct more complex dendritic topologies [132]. The dimensions of this PAMAM(G4.5)-[(DACH)Pt] conjugate have been determined using MAC mode atomic force microscopy (AFM). The AFM images of the conjugate on a freshly cleaved mica surface clearly show both individual single molecules and a few aggregates, as shown in Figure 11.13. Detailed cross-sectional measurements on 150 isolated features give an average height of 0.40 (± 0.16) nm and an average diameter of 7.83 (± 1.62) nm. The right images of Figure 11.13 exhibit a cross-sectional profile for this particular dendrimer-Pt molecule with a diameter of 11.7 nm and height of 0.49 nm. The AFM images document the formation of a novel nanoconjugate of the PAMAM(G4.5) and [(DACH)Pt]. The size of the Pt-conjugate is in the nanoscale range as has been observed for other dendrimer species [73]. The particles depicted in the AFM image appear to be substantially uniform in size and globular in shape except for a few aggregates present. The aggregates represent dimmers and trimmers present in the PAMAM(G4.5) dendrimer as received from a commercial source (Dendritic Nanotechnologies, Inc.). This was confirmed by MALDI-TOF MS. There was no change in the spectrum on conversion of the dendrimer to the conjugate (i.e., there would appear to be no platinum binding between dendrimer molecules).

2.2.3 In Vitro Release of [(DACH)Pt]

The release profile for the active component [(DACH)Pt] was investigated in pH 7.4 phosphate buffer and pH 5.0 phosphate-citrate saline solutions at 37 °C by

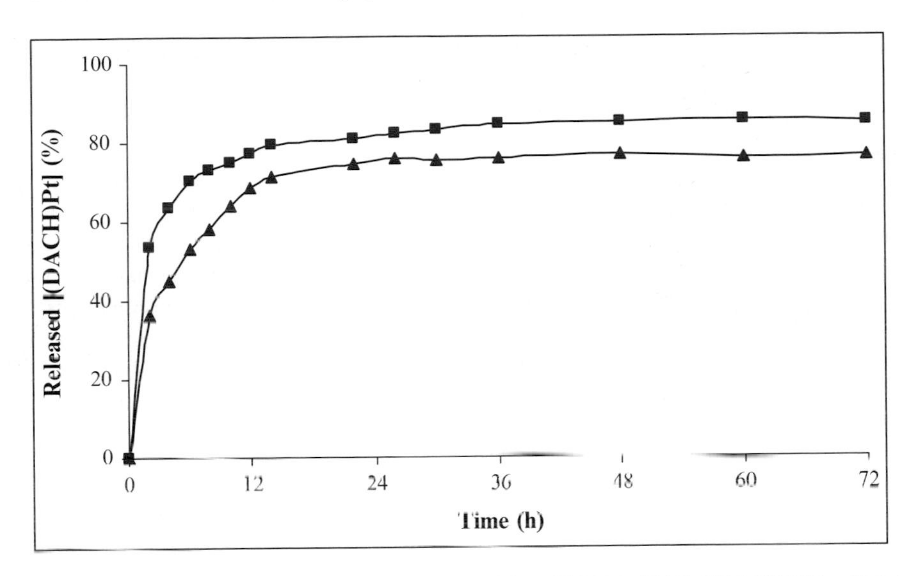

Figure 11.14 In Vitro release profile of [(DACH)Pt] from PAMAM(G4.5)-[(DACH)Pt] in pH 7.4 ( ) and pH 5.0 ( ) buffer saline

measuring the UV-vis absorbance of the dialysis saline at 292 nm, which is the characteristic UV-vis absorbance peak of the [(DACH)Pt] species. The preliminary in vitro release tests showed that most of the active species [(DACH)Pt] is smoothly released from the dendrimer nanocarrier over a period of 24 hours. The profile for the release of [(DACH)Pt] from the substrate PAMAM(G4.5) is shown in Figure 11.14. Overall, the release rate is very good with 76% in pH 7.4 buffer and 85% of the available platinum species in pH 5.0 buffer released in 24 hours, respectively. This behavior suggests that sustainable release should occur inside endosomes at lower pH. The mode of release is probably similar to that observed for other classic platinum drugs in which the labile ligands are carboxylate groups. Generally, under physiological conditions, hydrolysis occurs in a stepwise fashion to form first the monohydrated and subsequently the diaquo platinum species. A study of hydrolysis of Oxaliplatin showed that the ring-opening step has a half-life 16 min and the loss of the oxalate ligand occurs with a half-life of 92 min at 37 °C [133]. In this case, the [(DACH)Pt] units are released from PAMAM(G4.5)-[(DACH)Pt] nanoconjugate with a half-life of 105 min in pH 5.0 and 310 min in pH 7.4 saline. Therefore, the loss of labile ligand is considerably slower for the PAMAM(G4.5)-[(DACH)Pt] conjugate than that of the similar process for Oxaliplatin and suggests a sustainable release of active drug. These observations suggest that the use of PAMAM(G4.5) nanocarrier for [(DACH)Pt] may be used to generate a drug formulation with water solubility, dosage limitations, and response characteristics superior to those of classical platinum drugs.

Clearly, dendritic polymers may be utilized as nanocarriers for the improved delivery of antitumor agents. They offer several advantages over conventional polymers.

Most notably, the nanoscale size, uniform shape, and high surface functionality of these polymers offer the potential for the generation of "multivalent" drugs which permit the administration high dosage at low volume and which display enhanced delivery of active agent to the tumor site. In this case, a PAMAM(G4.5) dendrimer has been utilized as a nanocarrier for the generation of a dendrimer-platinum conjugate containing [(DACH)Pt] as the active agent. The dendrimer-platinum conjugate is well-defined with approximately forty [(DACH)Pt] units bound through surface carboxylate groups. This is distinctly unlike an earlier dendrimer-platinum formulation generated from the interaction of a PAMAM(G3.5) dendrimer with cisplatin in which some platinum species are bound at surface groups, some are bound at interior tertiary nitrogen atoms, and some are physically trapped as unchanged cisplatin. There are probably at least two reasons for this difference. The first is the more open structure of the smaller generation 3.5 dendrimer. In addition, the large difference in reactivity toward nucleophilic ligands of Cisplatin and diaquo(1,2-diaminocyclohexane)platinum(II) probably limits the effectiveness of the interaction of Cisplatin with surface groups. The loading of active platinum species is much larger than that achieved for earlier PAMAM(G3.5) dendrimer conjugates and considerably higher than that normally observed for linear polymer platinum conjugates. At the same time, the water solubility and release characteristics are superior to those observed for most polymeric platinum drugs. This conjugate displays sustained release of active platinum species over a period of 24 hours under physiological conditions. The PAMAM(G4.5)-[(DACH)Pt] represents a considerable improvement over previous dendrimer-platinum formulations and offers significant opportunity for clinical use.

3 Perspectives on Dendrimer Platinum Conjugates

The availability of nontoxic, water-soluble dendritic polymers containing high surface functionality offer great potential for the development of novel, highly effective organoplatinum antitumor formulations of toxicity much lower than that characteristic of simple platinum drugs. For the development of well behaved dendrimer-platinum conjugates, the dendrimer should be large enough so that surface crowding prevents entry of reactive species into the interior of the molecule and the platinum reagent should be sufficiently reactive so as to be readily bound by ligands at the surface of the dendrimer. These conditions seem to be well met by a generation 4.5 PAMAM dendrimer (128 surface carboxylate groups) and diaquo(1,2-diaminocyclohexane)platinum(II). Presumably, because of the steric crowding, the theoretical maximum number of platinum moieties (64) cannot be attached to the dendrimer surface. In the case of (1,2-diaminocyclohexane)platinum(II) approximately 40 units may be placed at the surface of the dendrimer. This dendrimer-platinum conjugate displays sustained release of active platinum species over 24 hours under physiological conditions. These results offer considerable optimism for the use of dendrimers in the development of more effective but less toxic organoplatinum antitumor agents.

Acknowledgement Financial support for this work by the Army Research Laboratory (ARL DAAL-01-1996-02-044) is gratefully acknowledged.

References

1. Rosenberg B, Van Camp L, Krigas T. (1968) Nature. 205: 698.
2. (a) Rosenberg B. (1985) Cancer. 55: 2303. (b) Rosenberg B. (1980) In: Nucleic acid-metal ion interactions, Spiro TG (ed),. John Wiley and Sons, Inc, New York, p 1–29.
3. Dabrowiak J, Bradner W. (1987) Progr Med Chem. 24: 129.
4. Kelland L. (1992) Crit Rev Oncol Hematol. 15. 191.
5. Heim M. (1993) Metal Complexes in Cancer Chemotherapy, B. Keppler, Ed., VCH, NY.
6. McKeage M, Kelland L. (1992) Molecular Aspects of Drug-DNA Interactions, S. Neidle, M. Waring, Eds., Macmillian, NY.
7. Jones TW, Chopra S, Kaufman JS, Flamenbaum W, Trump BF. (1985) Lab Invest. 52: 363.
8. (a) Rosenberg B. (1971) Inorganic and Nutritional Aspects of Cancer, 91; 129. (b) Aggrarwal SK, Menon GK. (1981) J Clin Hematol Oncol. 11: 73.
9. Neuse E. (1988) Polym Adv Technol. 9: 786.
10. Lebwohl D, Canetta R. (1998) Europ J Cancer. 34: 1522.
11. Howell BA, Walles EW. (1983) Macromolecular complexes of amidocarbonylic water-soluble polymers and square planar platinous and equivalent organometalics. US Patent No. 4405757.
12. Howell BA, Walles EW. (1985) Macromolecular complexes of amidocarbonylic water-soluble polymers and square planar platinous and equivalent organometalics. US Patent No. 4551502.
13. Howell BA, Walles EW, Rahidianfar R, Glass JR, Hunchison BJ, Johnson DA. (1988) In: Platinum and other metal coordination compounds in cancer chemthorapy, Nicholi M (ed). Martinus Nijhoff Publishing, Boston, MA, p 670.
14. Howell BA, Walles EW, Rashidianfar R. (1988) Makromol Chem Macromal Symp. 19: 329.
15. Howell BA, Walles EW. (1986) Polym Prepr. 27(1): 460.
16. Howell BA, Richards RM. (1996) Polym Mater Sci Eng. 74: 274.
17. Howell BA, Sastry BBS. (1996) Polym Mater Sci Eng. 74: 276.
18. Dyksterhouse RM, Howell BA, Squattrito PJ. (2000) Acta Cryst. C56: 64.
19. Saltmarch JA, Howell BA, Squattrito PJ. (2000) Acta Cryst. C56: e335.
20. Siegmann DW, Carraher CE, Jr. (2004) In: Marcomolecules containing metal and metal-like elements, Vol. 3, Biomedical applications, Abd-El-Aziz AA, Carraher CE, Jr, Pittman CU, Jr, Sheets JE, Zeldin M (eds). Wiley-Interscience, John Wiley and Sons, Inc., New Yorkp. 119.
21. Howell BA, Han G, Pakyz DJ. (1996) Polym Prepr. 37(1): 601.
22. Howell BA, Southwell JE. (1997) Polym Prepr. 38(2): 355.
23. Southwell JE. (1998) Water-Soluble Polymers Containing Coordinating Pendant Groups Derived from 2-Acryloyltetrahydro-1,2-oxazine", M.S. Thesis, Central Michigan University, Mt. Pleasant, MI.
24. Tomalia DA, Dewald JR, Hall MJ, Martin SJ, Smith PB. (1984) Preprints of the 1st SPSJ International Polymer Conference, Soc. of Polym. Sci., Japan, Kyoto, p 65.
25. Tomalia DA, Baker H, Dewald JR, et al. (1985) Polym J Tokyo. 17: 117.
26. Tomalia DA, Fréchet JMJ (eds.). (2001) Dendrimers and other dendritic polymers. John Wiley and Sons, New York, NY, p xxvii.
27. Tomalia DA. (2004) Aldrichimica Acta. 37(2): 39.
28. Tomalia DA. (2005) Materials Today. 8: 34.
29. Staudinger H.(1970) From organic chemistry to macromolecules. Wiley-Interscience: New York, 1970.
30. Tomalia DA. (1993) Aldrichimica Acta. 26: 91.
31. Kosugi J. (1994) MS Thesis. Central Michigan University.
32. Tomalia DA, Hedstrand DM, Wilson LR. (1990) Encyclopedia of polymer science and engineeringWiley, New York, p 46.

33. Buhleier E, Wehner W, Vögtle F. (1978) Synthesis 15: 155.
34. Newkome GR, Moorefield CN, Baker GR, Johnson AL, Behera RK. (1991) Angew Chem Int Ed Engl. 30: 1176.
35. Denti G, Serroni S, Campagna S, Ricevuto V, Balzani V. (1991) Inorg Chim Acta. 182: 127.
36. Campagna S, Denti G, Serroni S, Ciano M, Juris A, Balzani V. (1992) Inorg Chem. 31: 2982.
37. Denti G, Campagna S, Serroni S, Ciano M, Balzani V. (1992) J Am Chem Soc. 114: 2944.
38. Newkome GR, He E, Moorefield CN. (1999) Chem Rev. 99: 1689.
39. Archut A, Fischer M, Issberner J, Vögtle F. (1997) Dendrimere: hochverzweigte moleküe mit neuen eigenschaften. GIT Labor-Fachz., 41: 198.
40. Ardoin, N Astruc D. (1995) Bull Soc Chim Fr. 132: 875.
41. Bosman AW, Janssen HM, Meijer EW. (1999) Chem Rev. 99: 1665.
42. Chow HF, Mong TKK, Nongrum MF, Wan CW. (1998) Tetrahedron. 54: 8543.
43. Fréchet JMJ. (1994) Science 263: 1710.
44. Gorman C. (1998) Adv Mater. (Weinheim Fed Repub Ger). 10: 295.
45. Issberner J, Moors R, Vögtle F. (1994) Angew Chem Int Ed Engl. 33: 2413.
46. Majoral JP, Caminade AM. (1998) Topics in current Cchemistry. Springer-Verlag, Berlin, p 79.
47. Newkome GR, (ed). (1994) Advances in dendritic macromolecules, Vol. 1, JAI, Greenwich, CT, p1.
48. G. R. Newkome, Ed. (1995) Advances in dendritic macromolecules, Vol. 2. JAI, Greenwich, CT, p 1.
49. Newkome GR, (ed). (1996) Advances in dendritic macromolecules, Vol. 3. JAI, Greenwich, CT, p 1.
50. Newkome GR, (ed). (1999) Advances in dendritic macromolecules, Vol. 4. JAI, Greenwich, CT, p 1.
51. Newkome GR, Moorefield CN. (1996) Comprehensive supramolecular vhemistry, Pergamon, New York, p 777.
52. Schlüter AD. (1998) In: Topics in current chemistry. Vögtle F, (ed). Springer-Verlag, Berlin, p 165.
53. Seebach D, Rheiner PB, Greiveldinger G, Butz T, Sellner H. (1998) In: Topics in current chemistry, Vögtle F, (ed). Springer-Verlag, Berlin, p 125.
54. Smith DK, Diederich F. (1998) Chem Eur J. 4: 1353.
55. Tomalia DA, Naylor AM, Goddard WA III. (1990) Angew Chem Int Ed Engl. 29: 138.
56. Vögtle F, Plevoets M, Nachtsheim G, Wörsdörfer U. (1998) J Prakt Chem. 340: 112.
57. Venturi M, Serroni S, Juris A, Campagna S, Balzani V. (1998) In: Topics in current chemistry, Vögtle F, (ed). Springer-Verlag, Berlin, p 193.
58. Zeng F, Zimmerman SC. (1997) Chem Rev. 97: 1681.
59. Narayanan VV, Newkome GR. (1998) Topics in current chemistry, Vögtle F, (ed). Springer-Verlag, Berlin/Heidelberg, p 19.
60. Trollsås M, Hedrick JL. (1998) Macromolecules. 31: 4390.
61. Boulas PL, Gómez-Kaifer M, Echegoyen L. (1998) Angew Chem Int Ed Engl. 37: 216.
62. Matthews OA, Shipway AN, Stoddart JF. (1998) Prog Polym Sci. 23: 1.
63. Ihre H, Hult A, Fréchet JMJ, Gitsov I. (1998) Macromolecules. 31: 4061.
64. Esfand R, Tomalia DA. (2001) In: Dendrimers and other dendritic polymers, Fréchet JMJ, Tomalia DA, (eds). John Wiley and Sons, New York, p 587.
65. Fréchet JMJ, Ihre H, Davey M. (2001) Dendrimers and other dendritic polymers, Fréchet JMJ, Tomalia DA, (eds). John Wiley and Sons, New York, p 569.
66. Van Genderen MHP, Mak MHAP, De Brabander-van den Berg EMM, Meijer EW. (2001) In: Dendrimers and other dendritic polymers, Fréchct JMJ, Tomalia DA, (eds). John Wiley and Sons, New York, p 605.
67. Tomalia DA, Baker H, Dewald JR, et al. (1985) Polym J Tokyo. 17: 117.
68. Newkome GR, Yao Z, Baker GR, Gupta VK. (1985) J Org Chem. 50: 2003.
69. De Brabander-van den Berg EMM, Meijer EW. (1993) Angew Chem. 105: 1370.
70. (a) Hawker CJ, Fréchet JMJ. (1990), J Chem Soc Chem Commun. 1010-. (b) Hawker CJ, Fréchet JMJ. (1990) J Am Chem Soc. 112: 7638.
71. Miller TM, Neenan,TX. (1990) Chem Mater. 2: 346.
72. Ihre H, Hult H, Soderlind E. (1996) J Am Chem Soc. 118: 6388.

73. Jackson CL, Chanzy HD, Booy FP, et al. (1998) Macromolecules. 31: 6259.
74. Tomalia DA. (1994) Adv Mater. 6: 529.
75. Peterson J, Allikmaa V, Subbi J, Pehk T, Lopp M. (2003) Eur Polym J. 39: 33.
76. Mourey TH, Turner SR, Rubenstein M, Fréchet JMJ, Hawker CJ, Wooley KL. (1992) Macromolecules. 25: 2401.
77. Masfiled ML, Rakesh L, Tomalia LDA. (1996) J Chem Phys. 105: 3245.
78. Tomalia DA, Durst HD. (1993) Top Curr Chem. 165: 193.
79. Fréchet JMJ, Hawker CJ, Gitsov I, Leon JW. (1996) J Macromol Sci Pure Appl Chem A. 33: 1399.
80. Kim Y, Zimmerman SC. (1998) Curr Opin Chem Biol. 2: 733.
81. Brady PA, Levy EG. (1995) Chem Ind (London), 18: 21.
82. Langer R. (1995) Chem Eng Sci. 50: 4109.
83. Duncan R. (1999) Abstr Pap Am Chem Soc. 217: 141.
84. Duncan R, Kopecek J. (1984) Adv Polym Sci. 57: 51.
85. Peppas NA, Nagai T, Miyajima M. (1994) Pharm Tech Jpn. 10: 611.
86. Tomalia DA. (1989) In: Frontiers of macromolecular science: proceedings from the 32nd IUPAC . Saegusa T, et al. (eds). Blackwell, Boston, MA, p 207.
87. Naylor AM, Goddard WA, Kiefer GE,. Tomalia DA. (1989) J Am Chem Soc. 111: 2339.
88. Moreno-Bondi MC, Orellana G, Turro NJ, Tomlalia DA. (1990) Macromolecules. 23: 910.
89. Watkins DM, Sayed-Sweet Y, Klimash JW, Turro NJ, Tomalia DA. (1997) Langmuir. 12: 3136.
90. Sayed-Sweet Y, Hedstrand DM, Spinder R, Tomalia DA. (1997) J Mater Chem. 7: 1199.
91. Jansen JFGA, Brabander-van den Berg EMM, Meijer EW. (1994) Science. 265: 1226.
92. Jansen JFGA, Meijer EW, Brabander-van den Berg EMM, (1995) J Am Chem Soc. 117: 4417.
93. Krämer M, Stumbé J-F, Türk H, et al. (2002) Angew Chem Int Ed. 41: 4252.
94. Crooks RM, Zhao M, Sun L, Chechik V, Yeung LK, (2001) Acc Chem Res. 34: 181.
95. Esumi K. (2003) Top Curr Chem. 227: 31
96. Balogh L, Tomalia DA. (1998) J Am Chem Soc. 120: 7355.
97. Zhao M, Sun S, Crooks RM. (1998) J Am Chem Soc. 120: 4877.
98. Zhao M, Crooks RM. (1999) Chem Mater. 11: 3379.
99. Zhao M, Crooks RM. (1999) Adv Mater. 11: 217.
100. Sooklal K, Hanus LH, Ploehn HJ, Murphy CJ. (1998) Adv Mater. 10: 1083.
101. Huang J, Sooklal K, Murphy CJ, Ploehn HJ. (1999) Chem Mater. 11: 3595.
102. Sooklal K, Huang J, Murphy CJ, Hanus L, Ploehn HJ. (1999) Mater Res Soc Symp Proc. 576: 439.
103. Lakowicz JR, Grycynski I, Grycynski Z, Murphy CJ. (1999) J Phys Chem B. 103: 7613.
104. Hanus LH, Sooklal K, Murphy CJ, Ploehn HJ. (2000) Langmuir 16: 2621.
105. Lemon BI, Crooks RM. (2000) J Am Chem Soc. 122: 12886.
106. Esumi K, Suzuki A, Yamahira A, Torigoe K. (2000) Langmuir. 16: 2604.
107. Zhao M, Crooks RM. (1999) Angew Chem Int Ed Engl. 38: 364.
108. Yeung LK, Crooks RM. (2001) Nano Lett. 1: 14.
109. Niu Y, Yeung LK, Crooks RM. (2001) J Am Chem Soc. 123: 6840.
110. Scott RWJ, Wilson OM, Crooks RM. (2005) J Phys Chem B. 109: 692.
111. He H, Crooks RM, (2005) J Am Chem Soc. 127: 4930.
112. American Chemical Society. (2003) Chem. Mater. 15:3873.
113. Lang H, May RA, Iversen BL, Chandler BD. (2003) J Am Chem Soc. 125: 14832.
114. Lang H, Maldonado S, Stevenson KJ, Chandler BD. (2004) J Am Chem Soc. 126: 12949.
115. Pellechia PJ, Gao J, Gu Y, Ploehn HJ, Murphy CJ. (2004) Inorg Chem. 43: 1421.
116. Gu Y, Xie H, Gao J, Liu D, Williams CT, Murphy CJ, Ploehn HJ. (2005) Langmuir 21: 3122.
117. Crooks RM, Buford IL, Sun L, Leung LK, Zhao M. (2001) Top Curr Chem. 212: 81.
118. Niu Y, Crooks RM, (2003) Chim. 6: 1049.
119. (a) Connors TA, Jones M, Ross WC, Braddock PD, Khokhar AR, Lobe ML. (1972) Chem Biol Interact. 5 415. (b) Burchenal JH, Kalaher K, O'Toole T, Chisholm J. (1972) Cancer Res. 5: 415. (c) Tashiro T, Kawada Y, Sakurai Y, (1989) Biomed Pharmacother. 43: 251. (d) E. Raymond, S. Faivre, J.M. Woynarowski, (1998) Semin Oncol. 25: 4. (e) Weiss RB,

Christian MC. (1993) Review Drugs. 46: 360. (f) Raymond, E Faivre S, Chaney S, Woynarowski J, Cvitkovic C. (2002) Mol. Cancer Ther. 1: 227.
120. Langer RL. (1998) Nature. 392: 5.
121. (a) Allen TA, Cullis PC. (2004) Science 303: 1818. (b) Langer RL. (1992) Science. 293: (2001) 58.
122. Duncan R. (1992) Cancer Res. 46: 175.
123. Maeda H, Seymour LW, Miyamoto Y. (1992) Bioconjugate Chem. 3: 351.
124. Seymour LW, Miyamoto Y, Maeda H, Brereton M, Strohalm J, Ulbrich K, Duncan R. (1995) Eur J Cancer 31A: 766.
125. Malik N, Evagorou EG, Duncan R. (1999) Anti-Cancer Drugs 10: 767.
126. Gianasi E, Wasil M, Evagorou EG, Keddle A, Wilson G, Duncan R. (1999) Eur J Cancer. 35: 994.
127. Fan, D Howell BA, Rakesh L. (2005) Polym Mater Sci Eng. 93: 946.
128. Caminati G, Turro NJ, Tomalia DA. (1990) J Am Chem Soc. 112: 8515.
129. Roberts JC, Bhalga MK, Zera RT. (1996) J Biomed Mater Res. 30: 53.
130. (a) Malik N, Wiwattanapatapee R, Klopsch R, et al. (2000) J Cont Rel. 65: 133. (b) Wiwattanapatapee R, Carreno-Gomez G, Malik N, Duncan R. (2000) Pharm Res. 17: 991.
131. Deng S, Locklin J, Patton D, Baba A, Advincula RC. (2005) J Am Chem Soc. 127 1744.
132. (a) Tomalia DA, Mardel K, Henderson SA, Holan G, Esfand R. (2003) In: Handbook of nanoscience, engineering and technology, Goddard WA III, Brenner DW, Lyshevski SE, Iafrate GJ, (eds). CRC Press, Boca, Raton, FL, p 2021.. (b) Uppuluri S, Swanson DR, Piehler LT, Li J, Hagnauer GL, Tomalia DA. (2000) Adv Mater. 12: 796.
133. Jerremalm E, Videhult P, Alvelius G, et al. (2002) J Pharm Sci. 91: 2116.

Mössbauer Spectroscopy and Organotin Polymers

Anna Zhao, Charles E. Carraher Jr., Tiziana Fiore, Claudia Pellerito, Michelangelo Scopelliti, and Lorenzo Pellerito

1 Introduction

Mössbauer spectroscopy allows the structural analysis of certain metal atoms situated in complex structures. Briefly, Mössbauer spectroscopy is a resonant absorption spectroscopy that is observed best in isotopes having long lived, low-lying excited nuclear energy states. The largest recoil-free resonant cross-section is found for ^{57}Fe. Currently, Mössbauer spectroscopy is being used on Mars to identify iron compounds that are present in the Martian landscape. There are over 20,000 entries in SciFinder for Mössbauer spectroscopy, of which the two largest entries are for iron and tin-containing compounds. There are only about 100 entries for organotin compounds, with only a handful related to organotin-containing polymers. Mössbauer spectroscopy is an extremely powerful structural characterization tool that has been greatly overlooked because each Mössbauer spectrometer must be dedicated to a single element and measurements generally take hours to days to complete.

This chapter deals with the study of a particular potentially important set of organotin polymers that illustrates the strength of this spectral technique.

A.S. Abd-El-Aziz et al. (eds.), *Inorganic and Organometallic Macromolecules: Design and Applications.*

2 Basics

2.1 Mössbauer Spectroscopy

Mössbauer spectroscopy is a powerful technique that may give information on electronic distribution on about 44 different nuclei as a consequence of their structural environment [1–9]. The effects of interaction between the nuclear magnetic moment, an external magnetic field, electric charges and moments of the absorbing and surrounding atoms are known as hyperfine interactions [10].

The three main hyperfine interactions are:

1. Electric monopole interaction, detectable as a line shift (isomer shift δ, mm s^{-1});
2. Electric quadrupole interaction, detectable as a line splitting (nuclear quadrupole splitting Δ, mm s^{-1}), and;
3. Magnetic dipole interaction, detectable as a line splitting (nuclear Zeeman effect).

This third interaction, which arises from the interaction between the nuclear magnetic moment and local magnetic field created by electronic spins, or external applied magnetic field, completely removes the degeneracy of nuclear levels. For the sake of simplicity, and because it does not apply to ordinary organotin compounds, it will not be considered in this brief summary of the Mössbauer effect.

The first two interactions allow the extraction of two important Mössbauer parameters, isomer shift, δ, and nuclear quadrupole splitting, Δ, by considering the energy of the electrostatic interaction of the nucleus (*W*) (supposed to have a nonspherical symmetry), and the electric field, having a symmetry lower than cubic, created by the charge distribution around the Mössbauer nucleus.

$$W = \int \rho(x_1, x_2, x_3) \cdot U(x_1, x_2, x_3) dV \qquad (12.1)$$

where $\rho(x_1,x_2,x_3)$ is the nuclear charge density at coordinates x_1,x_2,x_3, and $U(x_1,x_2,x_3)$ is the electric potential exerted on the same point, by the charges distributed around the Mössbauer nucleus. Equation 12.1 is integrated over the whole nuclear volume (Figure 12.1).

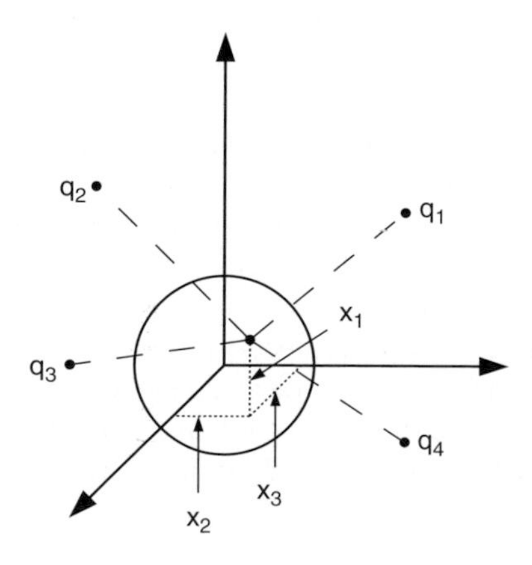

Figure 12.1 Interaction of the nucleus, supposed to have a nonspherical symmetry, and the electric field, having a symmetry lower than the cubic

The γ rays responsible of Mössbauer effect originate from a transition between excited and ground nuclear states in the emitter:

$$E_s = (E_e + W_e^s) - (E_g + W_g^s) = (E_e - E_g) + (W_e^s - W_g^s) = E_0 + W_T \qquad (12.2)$$

where E_0 is the nuclear transition energy in absence of interaction with surrounding electric fields, while W_T is the difference between W_e^s and W_g^s, the interaction energies for the source nucleus in the excited and in the ground state respectively, which may be calculated according to equation 12.1. After a Maclaurin series expansion and some simplifications, it is possible to get

$$W_I = \frac{1}{2}\sum_{\alpha=1}^{3} U_{\alpha\alpha} \int \rho(x_1, x_2, x_3)\left(x_\alpha^2 - \frac{r^2}{3}\right) dV + \frac{2}{3}\pi e|\Psi(0)|^2 \int \rho(x_1, x_2, x_3) r^2 dV \qquad (12.3)$$

where $U_{\alpha\alpha}$ is the appropriate component of the electric field gradient along an x_α axis, calculated at $x_\alpha = 0$; $|\psi(0)|^2$ is the probability to find electrons around the Mössbauer nucleus in the (0,0,0) point; $r^2 = x_1^2 + x_2^2 + x_3^2$.

The first term of Eq. 12.3 represents the interaction of the nuclear quadrupole moment with a non-cubic external electric field in the (0,0,0) coordinates and determines the nuclear quadrupole splitting, the consequence of the degeneracy removal of the excited state. The second term describes the interaction of the nuclear charge with the electronic cloud around the nucleus, and determines the shift of the nuclear level energies.

2.1.1 Isomer Shift

If we consider a γ emitting source as a spherical nucleus in a electric field with cubic symmetry, the first term of Eq. 12.3 is equal to 0, and

$$W_T = \frac{2}{3}\pi e|\Psi(0)|^2 \int \rho(x_1.x_2.x_3) r^2 \,\mathrm{dv} \qquad (12.4)$$

and the only effect shown will be the shift of both the excited and ground energy levels. The energy of the emitted γ ray will be:

$$E_s = E_0 + \left(W_e^s - W_g^s\right) = E_0 + \frac{2}{3}\pi Z e^2 |\Psi(0)|_s^2 \left(\langle r_e^2\rangle - \langle r_g^2\rangle\right) \qquad (12.5)$$

where

$$\langle r_e^2\rangle = \frac{\int \rho(x_1, x_2, x_3) \cdot r_e^2 dV}{\int \rho(x_1, x_2, x_3) dV}$$

and

$$\langle r_g^2\rangle = \frac{\int \rho(x_1, x_2, x_3) \cdot r_g^2 dV}{\int \rho(x_1, x_2, x_3) dV}$$

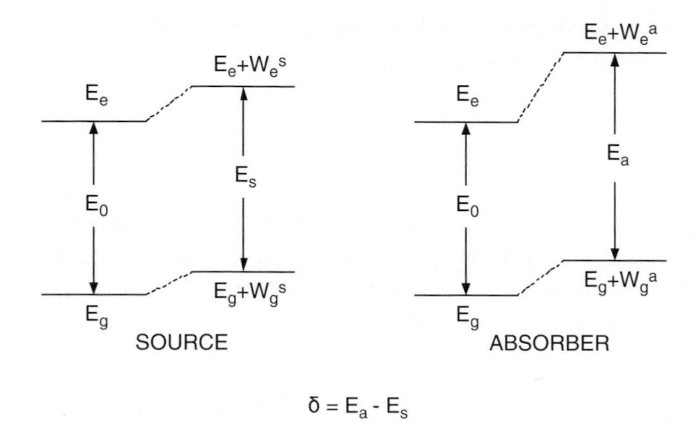

Figure 12.2 Shifts of the energy levels of the source and absorber nuclei

For an absorbing spherical nucleus, in the same electric field with cubic symmetry, it is possible to write an analogous relation for the energy of the absorbed γ rays:

$$E_a = E_0 + \left(W_e^a - W_g^a\right) = E_0 + \frac{2}{3}\pi Ze^2 \left|\Psi(0)\right|_a^2 \left(\left\langle r_e^2\right\rangle - \left\langle r_g^2\right\rangle\right) \tag{12.6}$$

so that if $|\Psi(0)|_s^2$ is different from $|\Psi(0)|_a^2$ and $<r_e^2>$ from $<r_g^2>$, then E_s will be different from E_a (Figure 12.2).
From Eq. 12.6 we have

$$\delta = \left(E_a - E_s\right) = \frac{2}{3}\pi Ze^2 \left(\left|\Psi(0)\right|_a^2 - \left|\Psi(0)\right|_s^2\right)\left(\left\langle r_e^2\right\rangle - \left\langle r_g^2\right\rangle\right) \tag{12.7}$$

which, owing to the fact that $<r_e^2> = 3/5\ R_e^2$ and $<r_g^2>= 3/5\ R_g^2$, leads to

$$\delta = \frac{2}{5}\pi Ze^2 (\left|\Psi(0)\right|_a^2 - \left|\Psi(0)\right|_s^2)(R_e^2 - R_g^2) \tag{12.8}$$

Replacing $(R_e^2 - R_g^2)$ with $2R^2(\Delta R/R)$, it is finally possible to describe the isomer shift as

$$\delta = \left(E_{\gamma a} - E_{\gamma s}\right) = \frac{4}{5}\pi Ze^2 R^2 \frac{\Delta R}{R}\left(\left|\Psi(0)\right|_a^2 - \left|\Psi(0)\right|_s^2\right) \tag{12.9}$$

The isomer shift, δ (mm s^{-1}) depends from an atomic factor, (ΔR/R), and on a nuclear factor $[(|\psi(0)|_a^2 - |\psi(0)|_s^2)]$ [2,3,6,10,11].

(ΔR/R) is positive for ^{119}Sn and, as a consequence, δ reflects the increase of the electronic density in the 5*s* orbital. The increase of the electronic density in the 5*p* orbitals produces a decrease in *d* orbitals following the deshielding action of the *p* electrons. Such an effect is less important of that of *s* electrons, so that covalent tin(IV) ($5s5p^3$) containing compounds will have a more positive

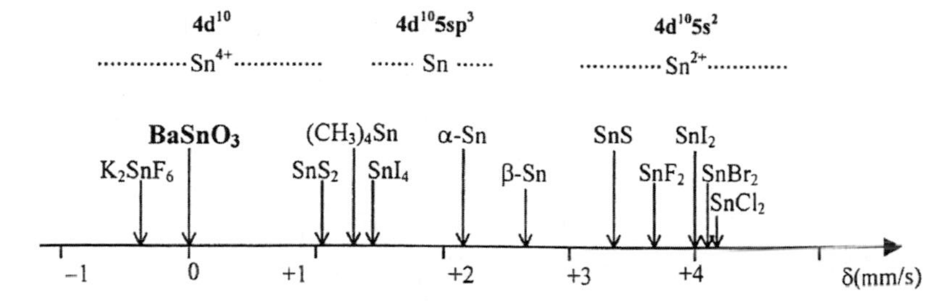

Figure 12.3 Scale of ^{119}Sn isomer shifts of some selected compounds, at liquid nitrogen temperature, with respect to $BaSnO_3$

Table 12.1 Limiting δ values, mm s^{-1}, for organotin(IV) derivatives

Absorber[a]	δ^b range,mms^{-1}	References
RSn^{IV}	0.52 – 1.80	12, 13
R_2Sn^{IV}	0.75 – 1.81	14, 15
R_3Sn^{IV}	0.86 – 1.72	16, 17
R_4Sn^{IV}	0.82 –1.40	18, 19

[a] R = organic radical

[b] Isomer shift with respect to R.T. $CaSnO_3$, $BaSnO_3$, SnO_2, referred to solid state absorbers at liquid nitrogen temperature, unless otherwise stated in the cited Refs. Low and high δ values in each class are reported, inherent to literature data 1986–2005.

δ than those containing "ionic" $Sn^{4+}(4d^{10})$. Increase of the tin-ligand bonds polarity will be followed by decrease towards to the limit value of a perfect Sn^{4+} ion (Figure 12.3). Characteristic δ, mm s^{-1}, values for organotin(IV) derivatives are reported in Table 12.1.

2.1.2 Nuclear Quadrupole Splitting

The interaction between a non-spherical nucleus and the electric field generated by the ligands contains a term that describes the interaction between the electric field and the nuclear quadrupole moment:

$$W_T = \frac{1}{2}\sum_{\alpha=1}^{3} U_{\alpha\alpha}\int p(x_1,x_2,x_3)\left(x_\alpha^2 - \frac{r^2}{3}\right)dV \qquad (12.10)$$

Nuclei having I > 1/2 possess a quadrupole moment which interacts with the electric field gradient (efg) generated by the surrounding charges. Such an interaction provokes the splitting of the nuclear levels due to the partial removal of the degeneracy of the $(2I + 1)^{th}$ nuclear level. The splitting (of the nuclear levels) arises because a nucleus not possessing spherical symmetry may assume different orientations respect to the electric field, each corresponding to a different energy content. The efg can be defined as:

$$grad.\vec{E} = \begin{bmatrix} \frac{\partial^2 V}{\partial x^2} & \frac{\partial^2 V}{\partial x \partial y} & \frac{\partial^2 V}{\partial x \partial z} \\ \frac{\partial^2 V}{\partial y \partial x} & \frac{\partial^2 V}{\partial y^2} & \frac{\partial^2 V}{\partial y \partial z} \\ \frac{\partial^2 V}{\partial z \partial x} & \frac{\partial^2 V}{\partial z \partial y} & \frac{\partial^2 V}{\partial z^2} \end{bmatrix} \tag{12.11}$$

Diagonalizing and choosing the corresponding axis system, the efg is completely defined as

$$grad.\vec{E} = \left(\frac{\partial^2 V}{\partial x^2}, \frac{\partial^2 V}{\partial y^2}, \frac{\partial^2 V}{\partial z^2} \right) = \left(V_{XX}, V_{YY}, V_{ZZ} \right) \tag{12.12}$$

The three components V_{XX}, V_{YY}, V_{ZZ} are not independent, since they must obey the Laplace relation:

$$V_{XX} + V_{YY} + V_{ZZ} = 0 \tag{12.13}$$

Moreover, by convention, $|V_{ZZ}| \geq |V_{YY}| \geq |V_{XX}|$.

Thus, only two components are needed, in particular V_{ZZ}, (often indicated as **eq**), and the asymmetry parameter η, defined as:

$$\eta = \left| \frac{V_{XX} - V_{YY}}{V_{ZZ}} \right| \tag{12.14}$$

For ^{119}Sn, I is equal to 1/2 for the ground level, which consequently is not split, whereas I = 3/2 for the excited level, so that it will be split into two levels having $m = \pm 3/2$ and $\pm 1/2$ (Figure 12.4).

The energy difference, Δ, between the two permitted nuclear transitions is (Table 12.2):

$$\Delta = \frac{1}{2} \mathrm{eQV_{zz}} (1 + \frac{\eta^2}{3})^{1/2} = \frac{1}{2} e^2 \mathrm{qQ} (1 + \frac{\eta^2}{3})^{1/2} \tag{12.15}$$

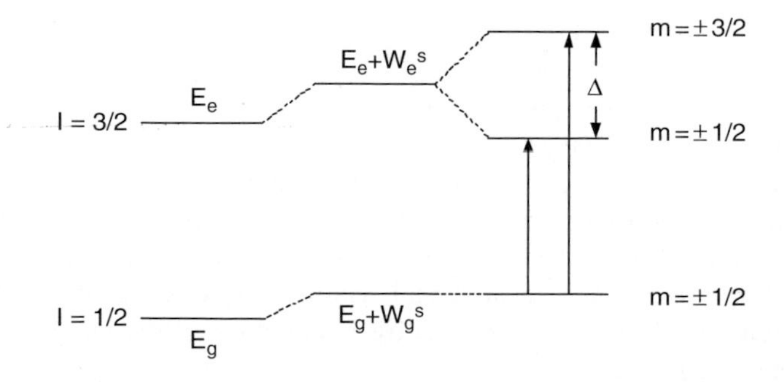

Figure 12.4 Permitted nuclear transitions for tin nuclei

Table 12.2 Limiting Δ values (mm s^{-1}) for several tetra-, penta-, hexa- and hepta-coordinated organotin(IV) derivatives

Structure	Absorber[a]	Δ^b range,mm s^{-1}	References
tet	$R_nSn(IV)$ (n=1–3)	1.19–3.21	[22,23]
tbp	RSn(IV)	2.02–2.45	[24]
	cis-$R_2Sn(IV)$	2.21–3.88	[25,26]
	trans-$R_2Sn(IV)$	2.15–4.16	[27,28]
	eq-$R_3Sn(IV)$	2.47–4.11	[29,30]
	fac-$R_3Sn(IV)$	1.82–2.55	[16,31]
	mer-$R_3Sn(IV)$	2.99–4.08	[32]
oct	RSn(IV)	1.48–2.08	[33,34]
	cis-$R_2Sn(IV)$	1.76–2.58	[35,36]
	trans-$R_2Sn(IV)$	2.31–5.70	[12,37]
pbp	RSn(IV)	1.92–2.23	[24]
	$R_2Sn(IV)$	3.12–4.66	[38,39]

[a] R = organic radicals

[b] Nuclear quadrupole splittings, referred to solid state absorbers at liquid N_2 temperature, unless otherwise stated in the cited Refs. Low and high Δ values are generally reported for each class, when available, inherent to literature data 1986–2005

oct, octahedral; tbp, trigonal bipyramidal; tet, tetrahedral; eq, equatorial; fac, facial; mer, meridional; pbp, pentagonal bipyramidal

where e is the electron charge; Q the nuclear quadrupole moment; V_{ZZ}, the efg component along the Z axis; and η the asymmetry parameter [2,3,6,20,21].

2.1.3 Molecular Dynamics of Mössbauer Nuclei

Mössbauer spectroscopy is a powerful technique that may give information on the occurrence of tin and organotin polymeric structures. This is made possible through the investigation of molecular dynamics of ^{119}Sn nuclei embedded in molecular fragments and in particular through the well established correlation between the recoil free fraction of γ rays (Debye-Waller-Mössbauer or Lamb-Mössbauer factor) and the mean square displacement $<x^2>$ of the nuclei (or of the masses bound to the Mössbauer nuclei). This correlation appears to be a linear dependence of the logarithm of the resonant peaks area, $\ln A_{tot}$ (proportional to Lamb Mössbauer factor, f_a) and the absolute temperature, T.

A number of researchers have reported [2,3,5,6,40–44] results on polymeric organotin derivatives based on reticular dynamics investigated through the Mössbauer effect and the related theory of lattice vibrations according to the Debye model.

However, it must be pointed out that the dynamics of tin nuclei embedded into polymeric materials, as investigated by variable temperature ^{119}Sn Mössbauer spectroscopy, does not give information on the polymericity of the material because it depends only upon the bonding situation at tin atoms. In fact, monomeric and polymeric di-*p*-tolyl-di-*p*-styryltin(IV) and tetra-*p*-styryltin(IV), as well as polymeric *p*-styryl-SnR_3 (R = Me, Ph) and 1,4-divinylbenzene-$SnPh_2$, all show $\ln A_{tot}/T$ slopes corresponding to those for monomeric Me_4Sn and Ph_4Sn [11,21].

Reticular dynamics of [119]Sn atoms have been cataloged on the basis of fingerprint criteria [5,6].

Using the Herber notation, for T > $\theta_D/2$, (where θ_D is Debye temperature for the sample), f_a can be expressed as:

$$f_a = \exp(-3 E_\gamma{}^2 T / Mc^2 K\theta_D^2) \tag{12.16}$$

where E_γ is the Mössbauer transition energy; M is the effective vibrating mass (EVM); c, is the velocity of light; K, is the Boltzmann constant; and θ_D, is the Debye temperature.

Taking into account, for a thin absorber, the relationship between the total area, A_{tot}, and absolute recoil free fraction, f_a, Eq. 12.17 is obtained as:

$$\frac{d\ln A_{tot}}{dT} = \frac{d\ln f_a}{dT} = \frac{-3E_\gamma^2}{Mc^2 K\theta_D^2} \tag{12.17}$$

thus the relative and absolute factors, f_a^{rel}, $f_a(T)$ may be obtained.

Once f_a values have been calculated, the mean square displacements of [119]Sn nuclei, $<x^2>(T)$, could be extracted from the equation:

$$f_a = \exp(-k^2 < x^2 >) \tag{12.18}$$

where k is the wave vector of γ rays.

The Debye temperatures, θ_D, are calculated from Eq. 12.17, assumed to be temperature independent; as a consequence, the Debye cut-off frequencies, ν_D, can be obtained from:

$$\nu_D = \frac{K\theta_D}{h} \tag{12.19}$$

where h and K are, respectively, Planck and Boltzmann constants.

Absolute f_a values, f_a^{abs}, may be calculated according to the following procedure. The Lamb-Mössbauer factor of the source, f_s, may be calculated from the area of Mössbauer spectra, A, at liquid nitrogen temperature, of 0.1 mm thin β-tin, whose effective thickness, t_i, is

$$t_i = (\pi/2)\Gamma_{nat} N\sigma_0 f_a \tag{12.20}$$

in which Γ_{nat} is the [119]Sn natural line width; N, the number of absorbing resonant atoms for cm^2; and σ_0, the [119]Sn resonant cross section [45,46], whereas the f_a values are those reported in the literature. Once t_i has been calculated, together with the corresponding tabulated saturation functions $L(t_i)$ [46], it is possible to get f_s:

$$f_s = \frac{A}{(\pi/2)\cdot\Gamma_{nat}\cdot L(t_i)} \tag{12.21}$$

The absolute recoilless fraction, f_a^{abs}, may be calculated with a reverse procedure.

$L(t_i)$ values for the tin(IV) and organotin(IV) derivatives are calculated from the areas A_1 and A_2 of the two absorption peaks according to:

$$L(t_i) = \frac{A}{(\pi/2)\cdot\Gamma_{nat}\cdot f_s} \tag{12.22}$$

Once t_1 and t_2 are interpolated from $L(t_1)$ and $L(t_2)$ values, f_a^{abs} may be calculated according to Eq. 12.23:

$$f_a^{abs} = \frac{t_1 + t_2}{(\pi/2)\Gamma_{nat}\sigma_0 N} \tag{12.23}$$

$<x^2>(T)$ functions, θ_D and ν_D parameters may be subsequently calculated from $f_a^{abs}(T)$ by applying Eqs. 12.16, 12.18, and 12.19. Fingerprint criteria have been applied based on the temperature dependence of $<x^2>$ [47], of the magnitude of "parameter of intermolecular interaction" $M\theta_D^2$ (M = molecular mass) [48]; and of the slope of functions ln $A(T)$ [and, consequently, of ln$f_a^{rel,abs}(T)$].

Solid state organotin(IV) polymers are characterized with low $<x^2>$ increments with increasing T, high θ_D (and $M\theta_D^2$) values (from 58.6 to 132.2 K) [48], and low slopes $d(\ln f_a^{rel,abs})/dT$, whereas monomers and monodimensional polymers are distinguished from bi- and tri-dimensional polymers by the borderline $<x^2> \approx 0.8\cdot10^{-2}$ (77.3 K); $2.0\cdot10^{-2}$ (200 K); $2.8\cdot10^{-2}$ (280 K), Å^2 and θ_D ranging from 21.6 to 68.5 K [48].

Ample collections of slopes $d(\ln A)/dT$ for tin(IV) and organotin(IV) derivatives have been reported [5,6,24,49–51]. In particular slopes $d(\ln A)/dT$ ranged from $-1.73\cdot10^{-2}$ to $-2.87\cdot10^{-2}$ and from $-2.11\cdot10^{-2}$ to $-2.73\cdot10^{-2}$, respectively, for diorganotin(IV) and triorganotin monomeric derivatives; from $-0.49\cdot10^{-2}$ to $-1.35\cdot10^{-2}$ and from $-0.68\cdot10^{-2}$ to $-1.40\cdot10^{-2}$, respectively, for diorganotin(IV) and triorganotin(IV) polymeric derivatives (Figure 12.5a,b).

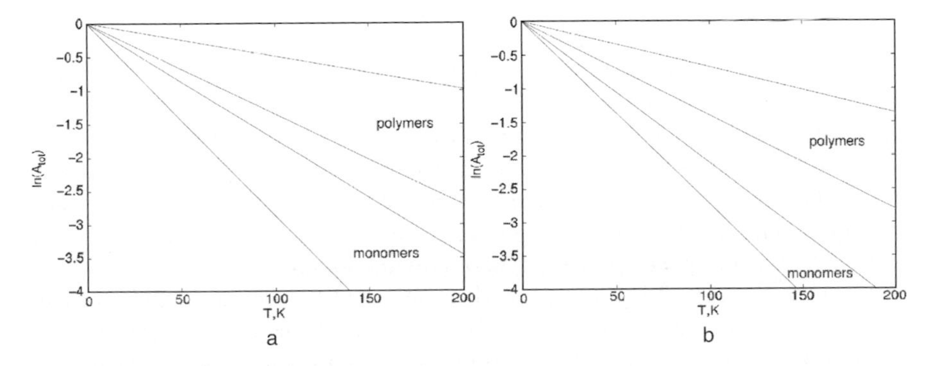

Figure 12.5 Estimates of the extent of intermolecular interaction in organotin(IV) derivatives, from ^{119}Sn molecular dynamics, determined through the slopes of ($d\ln A_{Tot}/dT$) for monomeric and polymeric diorganotin(IV) (**a**) and triorganotin(IV) (**b**) derivatives. (A_{Tot} = total area under the resonant peaks at T, K). Slope zones and limiting values include data extracted from references [5,6,24,50–55]

3 Research with Organotin Compounds

A number of groups are employing Mössbauer spectroscopy as a structural analysis tool. This includes groups lead by Lorenzo Pellerito. As noted before, little of this work is associated with polymers. Work related to polymers will be briefly reviewed in this chapter. Before we move to polymers, the work of Pellerito and co-workers will be discussed because of the structural similarities of the compounds studied by then to the ciprofloxacin products described in this chapter.

The use of Mössbauer spectroscopy for assisting in structure determination of polymers has been practiced since the 1960s with an early review dated in 1971 [56].

Mössbauer spectroscopy is typically coupled with infrared (IR), nuclear magnetic resonance (NMR), and/or X-ray evidence, the latter for only monomeric model compounds. Even so, we will concentrate on the Mössbauer spectroscopy results.

3.1 Noncarbonyl-Containing Polymers

As with many lead-containing organometallic compounds, some organotin products have a different structure in the solid in comparison with their structure in solution. Many of these are viewed as being polymeric in the solid state but monomeric in solution. These determinations are generally made based on the bond lengths that connect the various units. If the bond lengths connecting the various units are within those found for typical covalent bonds, then the material is considered to be polymeric. This tendency for some organotin compounds to be polymeric in the solid state but monomeric when dissolved is true for both carboxylate and noncarboxylate-containing compounds. This section focuses on noncarbonyl-containing polymers. Apperley and co-workers [57] report the following solid-state structures (**1**) and (**2**) for products containing a number of metal atoms including cobalt and gold. In the solid state the bond lengths are sufficient so that the products are considered polymeric.

deMello and co-workers formed products from the reaction of organotin compounds as well as tin(IV) tetrachloride from rutin (quercetin-3-rutinoside) (**3**) [58]. The products probably consist of a variety of structures but they reported by the authors as being six-coordinated (octahedral) in the solid and solution about the tin. Presumably the diorganotin atoms are complexed with four oxygen atoms for each tin atom. While referred to as polymeric, the molecular weights are only on the order of $7 \cdot 10^3$ corresponding to a degree of polymerization of about 8.

Recently, Deacon, Price, and coworkers reported the synthesis and study of polymers containing triphenyl and tributyltin as side chains and of polystannanes produced using the Wurtz polymerization [59]. The polystannanes were characterized employing Mössbauer spectroscopy. The polystannane shown below exhibits some liquid-crystal behavior. The products have weight average molecular weights of $2.5 \cdot 10^5$. The coordination number for tin is 4 with the tin present as a distorted tetrahedral geometry (**4**).

(1)

R
R_1 — N$^+$ — Sn — R
N$^+$
Co
R_1

(2)

R_1
N$^+$
R — Sn — N$^+$ — Au$^-$
R
R1

Mössbauer spectroscopy was carried out on polymers of the following form (**5**). As expected, the organotin is present in a distorted tetrahedral arrangement [60].

Another solid state example is the formation of organotin complexes with tetraethyl ethylene- and propylene-diphosphates as shown below (**6**). The complexes were described as being polymer chains with bridging bidentate ligands and an octahedral tin environment. Again, these complexes should dissociate in solution [61].

3.2 Carbonyl-Containing Polymers

Much of the effort has focused on the use of Mössbauer spectroscopy in determining the geometry of carbonyl-containing structures with respect to the arrangement about the organotin moiety.

Molloy, Blunden, and Hill studied 24 triphenyltin derivatives of substituted benzoic acids. The products were either tetrahedral monomers or trans chain polymers [62].

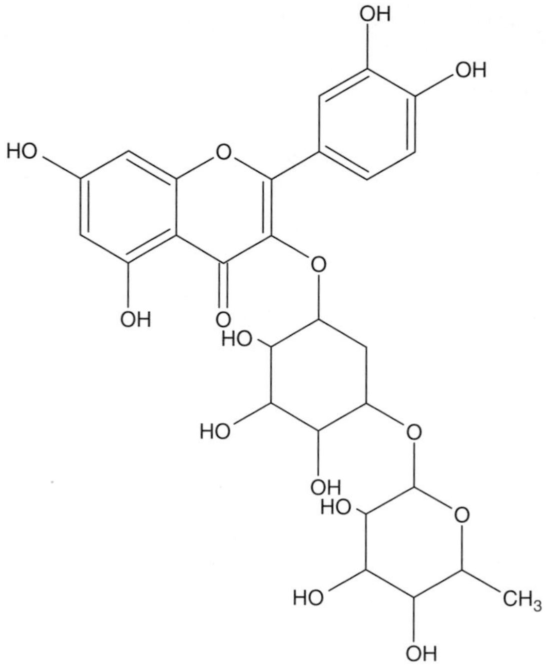

3

(3)

(4)

A prior study by Ford and Sams found similar results except employing triphenyltin haloacetates. In the solid state, many of the products are pentacoordinate polymers with O–C–O bridging between the tin atoms. Suggested repeat units are given below (**7**) where R = phenyl [63].

(5)

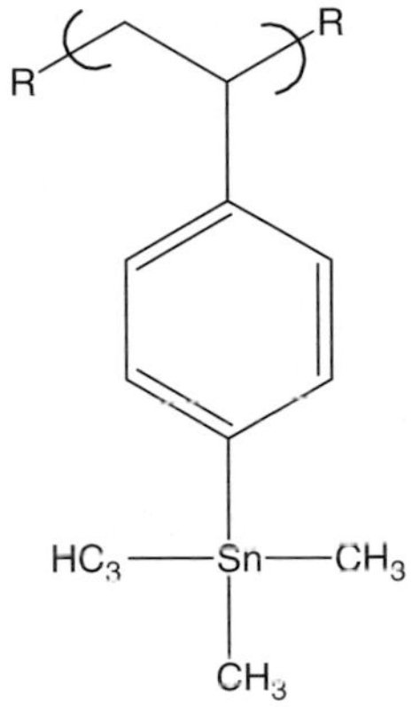

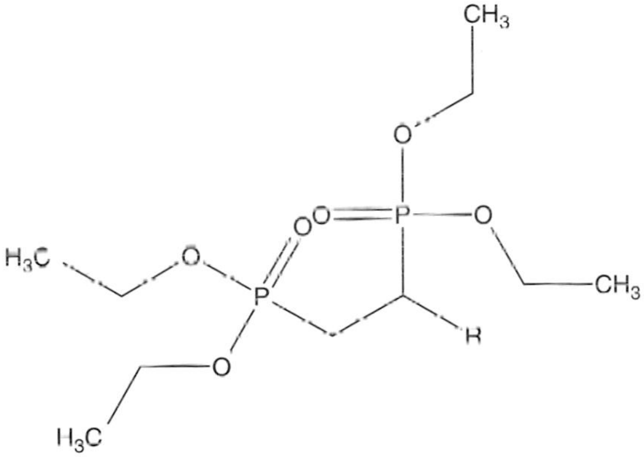

(6)

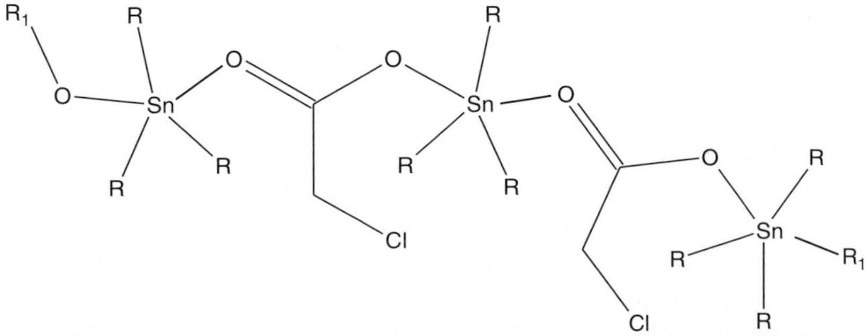

(7)

Organotin mercaptocarboxylates were formed and studied employing Mössbauer spectroscopy. Some of the compounds were linear polymers involving chelation by both oxygen atoms of the carboxyl group [64].

Another solid-state study involves the formation of organotin esters of 3-ureidopropionic acid with triphenyltin. The compound takes on a trigonal bipyramidal geometry with the axial positions occupied by the ester oxygen and the ureido-oxygen of an adjacent molecule with the polymer units held together by intermolecular hydrogen bonding between the NH group of the ureido fragment and the oxygen of the carboxylate carbonyl [65].

The description of diorganotin complexes with penicillamine enantiomers and 3-thiopropanoic acid in the solid state was undertaken. The 3-thiopropanoic acid complexes derivatives contain both trigonal bipyramidal and tetrahedral structures about the tin. These structures are maintained initially after solution. Again, cyclic formation with chelation occurring with the carbonyl oxygen of the carboxylic group occurs [66].

Organotin esters of 3-(2-furanyl)-2-propenoic acid were synthesized and studied employing Mössbauer spectroscopy. In the solid state, most of the products were polymeric with bridging carboxylate groups generally forming trigonal bipyramidal structures. In solution, the organotin complexes became monomeric with the organotin having a tetrahedral geometry (**8**) [67].

Recently, a number of compounds have been synthesized and analyzed employing Mössbauer spectroscopy by the group of Nath and Eng. Many of these compounds are monomeric but some are polymeric. Here we will look at one of these studies. Products were formed from reaction with ascorbates and various triorgano- and diorganotin halides. Below is the suggested structure about the dibutyltin-ascorbic acid product emphasizing the octahedral nature about the tin atom (**9**) [68–71].

For comparison, the product from trialkyltin halides take on a trigonal bipyramidal structure as pictured below (**10**) again emphasizing the geometry about the tin atom forming linear polymers in the solid state.

In summary, carbonyl-containing compounds often complex with the organotin moiety forming dative coordinate bonds. Carboxyl-containing compounds typically chelate the organotin moiety employing both oxygen atoms forming both trigonal bipyramid and octahedral structures. This chelation may have the two oxygen atoms bonded to the same tin atom or bonded to two different tin atoms.

4 Experimental

The ^{119}Sn Mössbauer spectra were measured at liquid nitrogen temperature with a multichannel analyzer [TAKES Mod. 269, Ponteranica, Bergamo (Italy)] and the following Wissenschaftliche Elektronik system [MWE, München (Germany)]: an MR250 driving unit, an FG2 digital function generator and an MA250 velocity transducer, moved at linear velocity, constant acceleration, in a triangular waveform. The

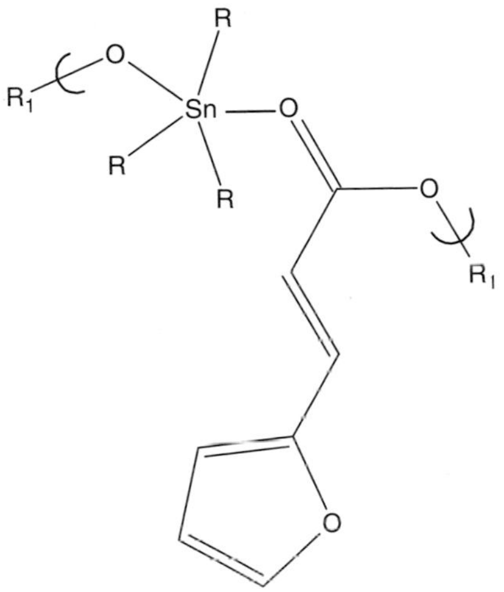

(8)

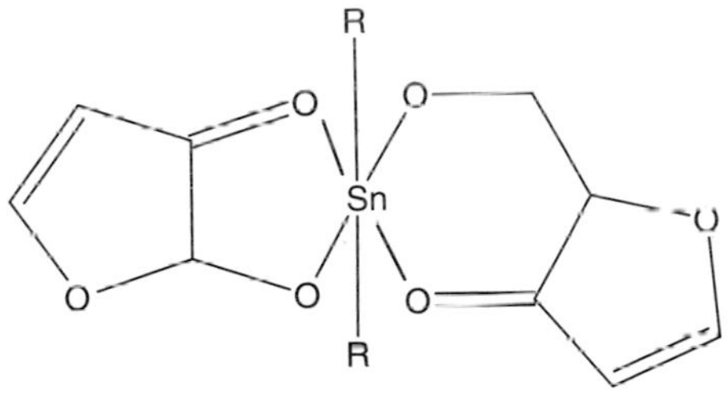

(9)

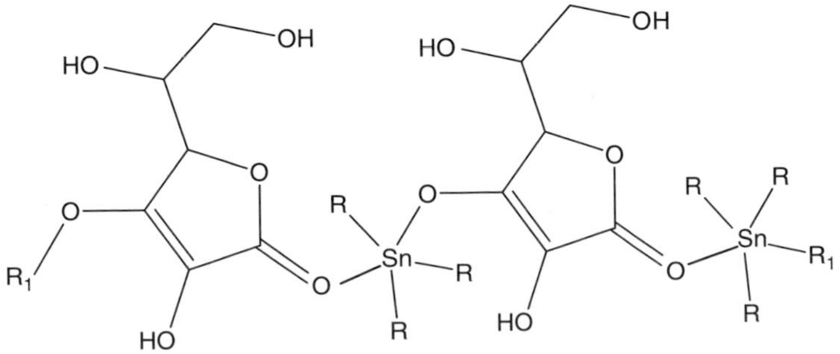

10

(10)

organotin(IV) samples were maintained at liquid nitrogen temperature in a model NDR-1258-MD Cryo liquid nitrogen cryostat (Cryo Industries of America, Inc; Atkinson, NH) with a Cryo sample holder. The 77.3 ± 0.1 K temperature was controlled with a model ITC 502 temperature controller of the Oxford Instruments (Oxford, England). The multichannel calibration was performed with an enriched iron foil (^{57}Fe = 95.2%, thickness 0.06 mm) (Dupont, MA), at room temperature, by using a ^{57}Co-Pd source (10 mCi, Ritverc GmbH) (St. Petersburg, Russia), while the zero point of the Doppler velocity scale was determined, at room temperature, through absorption spectra of natural $CaSnO_3$ (^{119}Sn = 0.5 mg/cm^2) and a $Ca^{119}SnO_3$ source (10 mCi, Ritverc GmbH) (St. Petersburg, Russia). The obtained $5 \cdot 10^5$ count spectra were refined, to obtain the isomer shift, δ (mm s^{-1}), and the nuclear quadrupole splitting, $|\Delta_{exp}|$ (mm s^{-1}).

5 Structural Possibilities

The present study focuses on the characterization of organotin polymers formed from the reaction of organotin(IV) with antibiotics forming polymers with the following general repeat units (Figures 12.6–12.11).

$$-O-Sn-O- \qquad -N-Sn-N- \qquad -N-Sn-O-$$

Whereas for R_2Sn(IV) ciprofloxacinate (Figure 12.11) this is the general repeat unit, there are several structural variations. These variations involve the precise structure about the organotin moiety. The organotin moiety can be connected through two oxygen atoms, two nitrogen atoms, or one nitrogen atom and one oxygen atom. The structures with two nitrogen and two oxygen atoms are referred to as the symmetric structures and the structure containing one oxygen atom and one nitrogen atom connected to the organotin is referred to as the asymmetric structure. These possibilities are shown below.

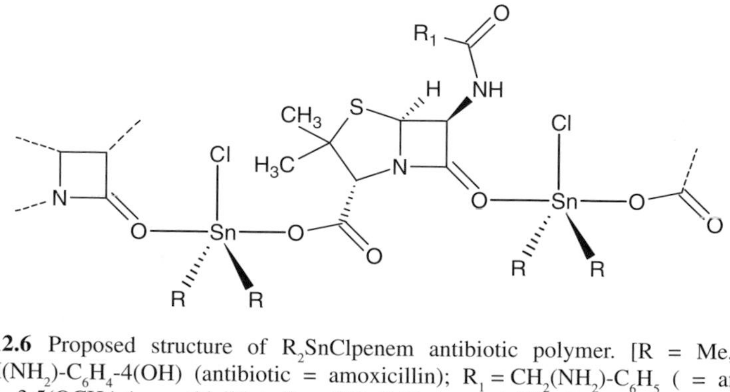

Figure 12.6 Proposed structure of R_2SnClpenem antibiotic polymer. [R = Me, Bu, Ph; R_1 = CH(NH_2)-C_6H_4-4(OH) (antibiotic = amoxicillin); R_1 = CH_2(NH_2)-C_6H_5 (= ampicillin); R_1 = C_6H_3–3,5(OCH_3) (= methicillin); R_1 = CH_2-C_6H_5 (= penicillin G)]

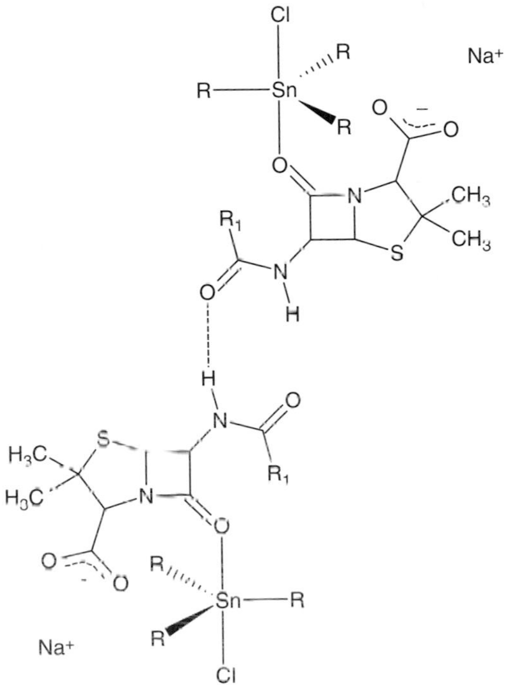

Figure 12.7 Proposed structure of R_3SnClpenem antibioticNa polymer. [R = Me, Bu, Ph; R_1 = $CH(NH_2)$-C_6H_4-4(OH) (antibiotic = amoxicillin); R_1 =$CH_2(NH_2)$-C_6H_5 (= ampicillin); R_1 = C_6H_3-3,5(OCH_3) (= methicillin); R_1 = CH_2-C_6H_5 (= penicillin G)]

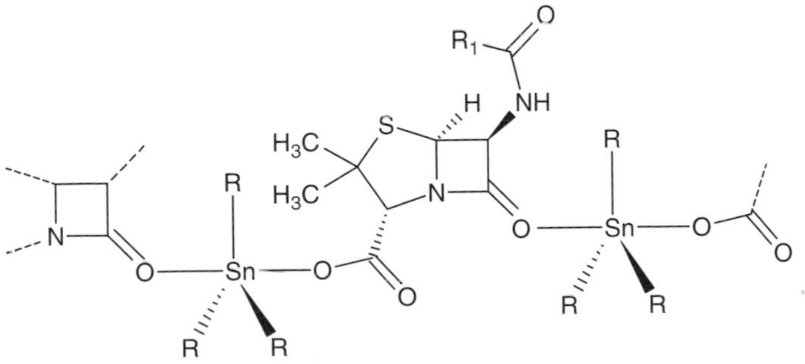

Figure 12.8 Proposed structure of R_3Snpenem antibiotic polymer. [R = Me, Bu, Ph; R_1 = $CH(NH_2)$-C_6H_4-4(OH) (antibiotic = amoxicillin); R_1 = $CH_2(NH_2)$-C_6H_5 (= ampicillin)]

The carbonyl group can be attached by what is referred to as bridging and nonbridging. The bridging structure forms a distorted octahedral arrangement about the organotin moiety whereas the nonbridging structure forms a distorted tetrahedral structure. Infrared and Mössbauer spectroscopy can be used to assign these structures. Furthermore, the bridging structures can be either cis or trans. Mössbauer spectroscopy is capable of making this assignment.

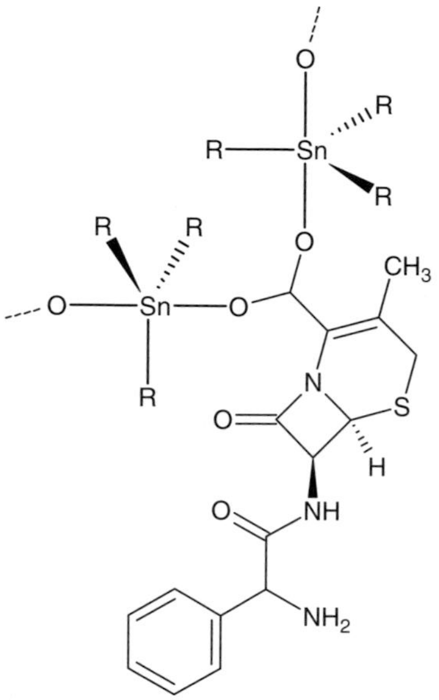

Figure 12.9 Proposed structure of R_3Sncephem antibiotic polymer. [R = Me, Bu; cephemic antibiotic = cephalexin]

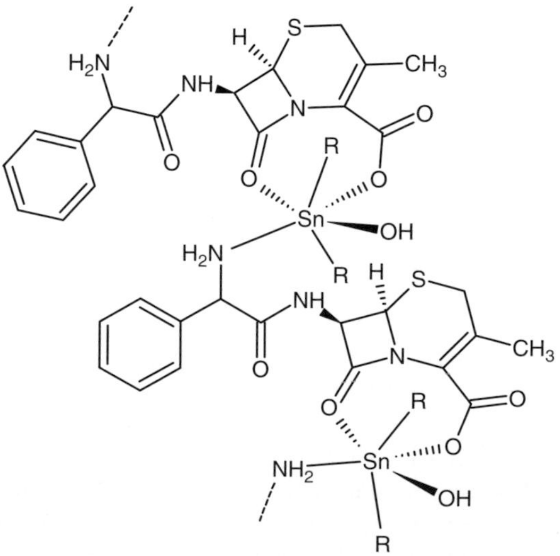

Figure 12.10 Proposed structure of R_3Sncephem antibiotic polymer. [R = Me, Bu; cephem antibiotic = cephalexin]

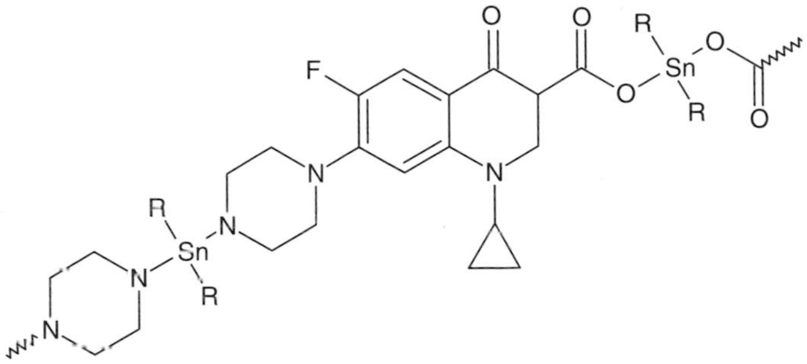

Figure 12.11 Proposed structure of R_2Snciprofloxacinate polymer

6 Mössbauer Data and Analysis

The experimental Mössbauer parameters, isomer shift, δ, and nuclear quadrupole splitting, $|\Delta_{exp}|$, mm s^{-1} of organotin antibiotic derivatives, Tables 12.3–12.5, are characteristic of organotin(IV) derivatives [2,5,72]. The isomer shift, δ, increased within the organotin(IV) antibiotics series from diphenyl to dibutyltin(IV) compounds, and from triphenyl to tributyltin(IV) for the triorganotin(IV) derivatives [2,5,72–74]. The small differences could result from small differences in the covalency of the tin-carbon bonds in the complexes.

The $|\Delta_{exp}|$ data for diorganotin(IV)chloro and triorganotin(IV)chloro antibiotic derivatives have been then rationalized according to the point charge model formalism [72–74] applied to the proposed trigonal bipyramidal structures of Figures 12.6–12.9 and to the skew trapezoidal of Figure 12.10.

The partial quadrupole splittings, p.q.s., mm s^{-1}, [$\{Alk\}^{tbe} = -1.13$; $\{Ph\}^{tbe} = -0.98$; $\{Alk\}^{tba} = -0.95$; $\{Ph\}^{tba} = -0.89$; $\{COO\}^{tba}_{unid} = -0.10$; $\{Cl\}^{tbe} = 0.20$; $\{Cl\}^{tba} = 0.00$; $\{C=O\}^{tba}_{lactamic} = \{C=O\}^{tba}_{DMA} = 0.10$; $\{C=O\}^{tbe}_{lactamic} = \{C=O\}^{tbe}_{DMA} = 0.40$; DMA = dimethylamide], which have been used in the application of the p.c.f., were literature or calculated values [72–75].

The resulting Δ_{calcd} values differ from the experimental data (Tables 12.3 and 12.4) less than ± 0.4 mm s^{-1}, the maximum difference allowed between experimental and calculated Δ for accepting the proposed geometry [74].

Furthermore, from the experimental nuclear quadrupole splittings of the previously mentioned complexes, $|\Delta_{exp}|$, C-Sn-C angles have been calculated as $(180 - 2\theta)$, where θ has been extracted by solving the equation $\Delta = 4\{R\} \times [1 - 3\cos^2\theta\sin^2\theta]^{1/2}$, by applying the Sham and Bancroft model [76] and reported in Tables 12.3 and 12.4.

The C–Sn–C angles, as calculated, are in good agreement with those expected for *cis*-R_2 trigonal bipyramidal structures around the tin(IV) atom.

The analysis of the experimental spectra (+ +) of the Et_2Sn(IV)ciprofloxacinate gives two doublets, the first attributable to N-Sn-N environment (——), the second to OCO–Sn–OCO (- -) (Figure 12.12).

Table 12.3 Experimental Mössbauer parameters, isomer shift, δ, mm s^{-1}, nuclear quadrupole splittings $|\Delta_{exp}|$, mm s^{-1}, measured at liquid N_2 temperature, calculated nuclear quadrupole splittings, Δ_{calcd}, and C–Sn–C angles for $R_2Sn(IV)$ Cl antibiotic and $R_3Sn(IV)$ C1 antibiotic derivatives [antibiotic = amoxicillin, ampicillin, methicillin and penicillin] [78–80]

Compound[a]	δ	$\|\Delta_{exp}\|$	Δ_{calcd}	Figure	C–Sn–C angles
$Me_2SnClamox{\cdot}2H_2O$	1.21	2.99	3.17	12.6	120
$Bu_2SnClamox{\cdot}2H_2O$	1.27	2.86	3.17	12.6	117
$Ph_2SnClamox{\cdot}2H_2O$	0.98	2.69	2.78	12.6	123
$Me_2SnClamp{\cdot}H_2O$	1.18	3.12	3.17	12.6	123
$Bu_2SnClamp{\cdot}H_2O$	1.35	2.98	3.17	12.6	120
$Ph_2SnClamp{\cdot}H_2O$	1.07	2.64	2.78	12.6	121
$Me_2SnClmet{\cdot}H_2O$	1.27	3.20	3.17	12.6	125
$Bu_2SnClmet{\cdot}H_2O$	1.37	3.19	3.17	12.6	125
$Ph_2SnClmet{\cdot}H_2O$	1.19	2.73	2.78	12.6	124
$Me_2SnClpenG$	1.26	3.29	3.17	12.6	128
$Bu_2SnClpenG$	1.30	3.22	3.17	12.6	126
$Ph_2SnClpenG$	1.21	2.79	2.71	12.6	126
$Me_3SnClamoxNa{\cdot}2H_2O$	1.26	3.24	−3.77	12.7	n.c.
$Bu_3SnClamoxNa{\cdot}2H_2O$	1.40	3.42	−3.77	12.7	n.c.
$Ph_3SnClamoxNa{\cdot}2H_2O$	1.16	2.87	−3.26	12.7	n.c.
$Me_3SnClampNa{\cdot}H_2O$	1.31	3.52	−3.77	12.7	n.c.
$Bu_3SnClampNa{\cdot}H_2O$	1.43	3.33	−3.77	12.7	n.c.
$Ph_3SnClampNa{\cdot}H_2O$	1.17	2.73	−3.26	12.7	n.c.
$Me_3SnClmetNa{\cdot}H_2O$	1.32	3.53	−3.77	12.7	n.c.
$Bu_3SnClmetNa{\cdot}H_2O$	1.44	3.53	−3.77	12.7	n.c.
$Ph_3SnClmetNa{\cdot}H_2O$	1.28	2.82	−3.26	12.7	n.c.
$Me_3SnClpenGNa$	1.36	3.48	−3.77	12.7	n.c.
$Bu_3SnClpenGNa$	1.44	3.30	−3.77	12.7	n.c.
$Ph_3SnClpenGNa$	1.30	2.93	−3.26	12.7	n.c.

[a] Sample thickness ranged between 0.50–0.60 mg ^{119}Sn cm^{-2}; Isomer shift, δ, ± 0.03, mm s^{-1} with respect to $BaSnO_3$ values; Nuclear quadrupole splittings, $|\Delta_{exp}|$ ± 0.02, mm s^{-1}.

amox, amoxicillinate; amp, ampicillinate; met, methicillinate; penG, penicillinate G; R, Me, Bu, Ph.

n.c., not calculated

The structures of six-coordinate R_2SnCh_2 (Ch = chelating ligand) species were considered to be simple octahedral. Most early studies interpreted results on the basis of simple *trans* or *cis* structures [83–89]. It is true that some structures are trans in solution [90,91] and the solid state [93], and others are certainly *cis* [93,94]. In 1977 Kepert reported that many octahedral organometallic complexes, including several tin complexes, are of neither regular *trans* (**11**) nor regular *cis* (**12**) geometry, but that an intermediate geometry, skew or trapezoidal bipyramidal, is more stable (skew structures have C–Sn–C angles of 135–155 °C) [95].

As a rule, R_2SnCh_2 complexes prefer a trans arrangement when the ligand bite (distance between the two coordinating atoms) is large and tend to be *cis* when

Table 12.4 Experimental Mössbauer parameters, isomer shift, δ, mm s^{-1}, nuclear quadrupole splittings $|\Delta_{exp}|$, mms^{-1}, measured at liquid N_2 temperature, calculated nuclear quadrupole splittings, Δ_{calcd}, and C-Sn-C angles for and $R_3Sn(IV)$antibiotic derivatives (antibiotic = amoxicillin, ampicillin and cephalexin) [78,79,81,82]

Compound[a]	δ	$\lvert\Delta_{exp}\rvert$	Δ_{calcd}	Figure	C–Sn–C angles
$Me_3Snamox{\cdot}H_2O$	1.34	3.44	−3.51	12.8	n.c.
$Bu_3Snamox{\cdot}H_2O$	1.44	3.32	−3.51	12.8	n.c
$Ph_3Snamox{\cdot}H_2O$	1.27	2.70	−3.06	12.8	n.c
$Me_3Snamp{\cdot}H_2O$	1.33	3.42	−3.51	12.8	n.c.
$Bu_3Snamp{\cdot}H_2O$	1.43	3.34	−3.51	12.8	n.c.
$Ph_3Snamp{\cdot}H_2O$	1.24	2.78	−3.06	12.8	n.c.
$Me_3Snceph{\cdot}H_2O$	1.33	3.46	−3.69	12.9	n.c.
$Bu_3Snceph{\cdot}H_2O$	1.42	3.22	−3.69	12.9	n.c.
$Me_2SnOHceph{\cdot}H_2O$	1.21	3.47	n.c.	12.10	141
$Bu_2SnOHceph{\cdot}H_2O$	1.36	3.76	n.c.	12.10	152

[a] Sample thickness ranged between 0.50 and 0.60 mg ^{119}Sn cm^{-2}; Isomer shift, δ, ± 0.03, mm s^{-1} with respect to $BaSnO_3$ values; Nuclear quadrupole splittings, $|\Delta_{exp}|$ ± 0.02, mm s^{-1}; amox, amoxicillinate; amp, ampicillinate; ceph, cephalexinate; R, Me, Bu, Ph; n.c. = not calculated

Table 12.5 Experimental Mössbauer parameters for diorganotin(IV)ciprofloxacinate[a,b]

Compound	δ_1	$\lvert\Delta_{1exp}\rvert$	Δ_{tet}	δ_2	$\lvert\Delta_{2exp}\rvert$
$Et_2Sncipro_2$	0.97	1.96	−1.82	1.08	2.29
$Bu_2Sncipro_2$	0.98	2.03	−1.82	1.20	2.66
$Ph_2Sncipro_2$	0.82	1.61	−1.61	0.92	2.04

[a] cipro = ciprofloxacinate; sample thickness ranged between 0.50 and 0.60 mg ^{119}Sn cm^{-2}; isomer shift, δ ± 0.03, mm s^{-1}, with respect to $BaSnO_3$; nuclear quadrupole splittings, $|\Delta_{exp}|$ ± 0.02, mm s^{-1}

[b] The partial quadrupole splittings, mm s^{-1}, used in the calculations are: {Alk} = −1.37; {Ph} = −1.26; {N} = −0.564.

the bite is small [86]. For example, acetylacetonate (acac)-type ligands form a six-membered chelate-metal ring, and trans configurations are expected. The calculated and experimental C–Sn–C angles range from 175 to 178° for the benzoylacetonates and dibenzoylmethanates trans structures [96–98]. Both picolinates and tropolonates (trop) have smaller bites than the acac family ligands and the structural assignment was described as skew or *cis*-skew configurations for $R_2Sn(trop)_2$ (119–143° for C–Sn–C angle) [99]. In the quinolinolate group of complexes, the oxinates, with less steric crowding about the central atom, have structures that are nearly *cis* (109–120°) [87].

The analysis of the Mössbauer spectra of the three diorganotin(IV) ciprofloxacinate complexes allowed the calculation of the Mössbauer parameters, isomer shifts, δ, and quadrupole splittings, $|\Delta_{exp}|$, reported in Table 12.5. Each complex showed the occurrence of a characteristic two doublets spectrum. This indicates that two different tin(IV) environments exist in the organotin polymers. The values of Mössbauer

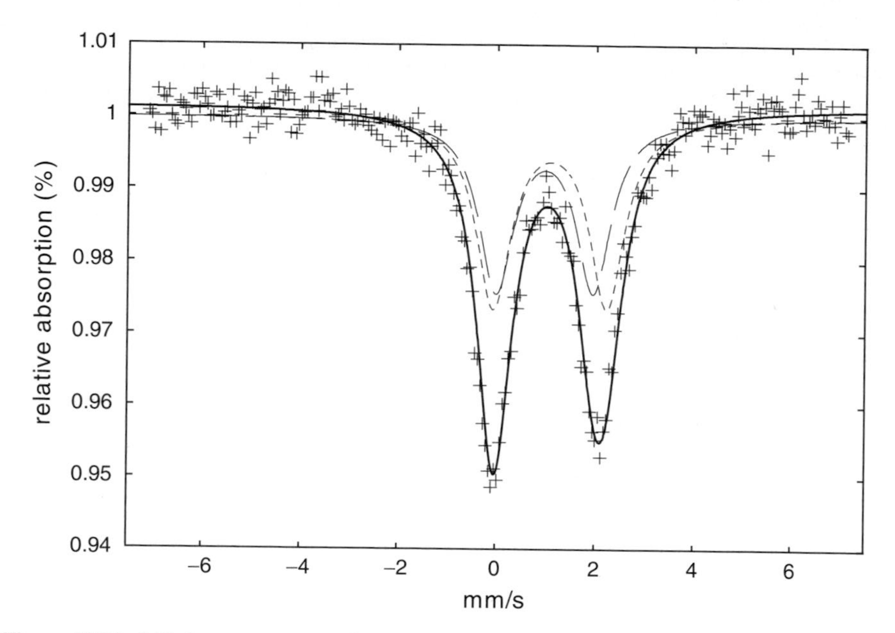

Figure 12.12 Mössbauer spectrum of the product of diethyltin dichloride and ciprofloxacin

11

(11)

12 **(12)**

parameters are in the range found for other organotin (IV) derivatives. As mentioned before, because of the electro-withdrawing property, the isomer shift of the diphenyltin derivative, δ (0.82, 0.92), is lower than that of the dialkyltin(IV) compounds (0.97,1.08; 0.98,1.20) [72–74,100]. As far as the experimental $|\Delta_{1exp}|$ values are

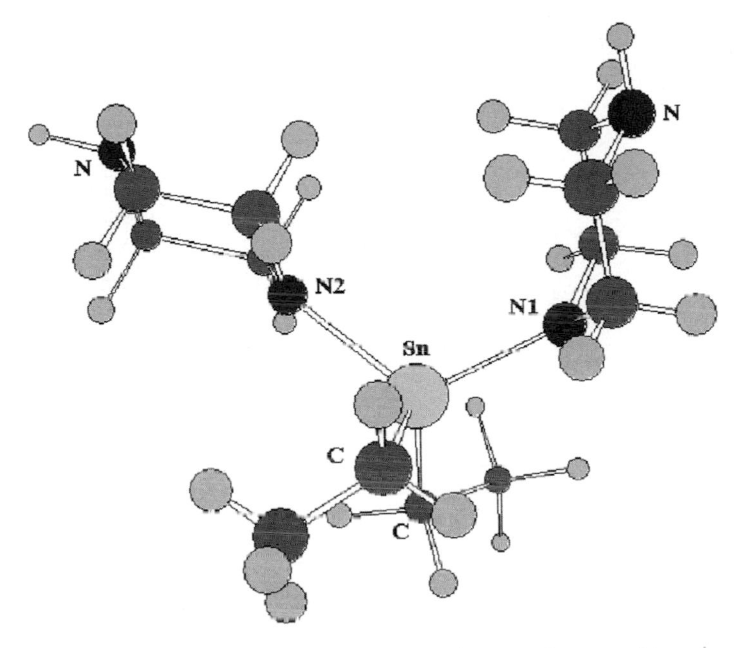

Figure 12.13 Tetrahedral configuration of tin connected to two nitrogens of two ciprofloxacin-derived moieties

concerned, they ranged from 1.61 for $Ph_2Sn(IV)$ciprofloxacinate to 2.03 mm s^{-1} for $Bu_2Sn(IV)$ciprofloxacinate, whereas $|\Delta_{2exp}|$ ranged from 2.04 to 2.66 mm s^{-1} for these organotinciprofloxacinates (Table 12.5).

The $|\Delta_{1exp}|$ values are consistent with a R_2SnN_2 tetrahedral configuration (Figure 12.13). The $|\Delta_{2exp}|$ values are consistent with a tetrahedral environment around the tin (IV) atoms, presumably distorted towards a skew trapezoidal trans-R_2SnO_4 configuration, with C–Sn–C angles <<180°.

6.1 Molecular Dynamics

Molecular dynamics of $R_2SnClamox{\cdot}2H_2O$, $R_3SnClamoxNa{\cdot}2H_2O$ have been investigated by variable temperature ^{119}Sn Mössbauer spectroscopy in an appropriate temperature range (Table 12.6) [101]. The lines representing the $\ln A_{tot}(T)$ functions for $R_2SnClamox{\cdot}2H_2O$ and $R_3SnClamoxNa{\cdot}2H_2O$ derivatives are definitely inside the polymeric zones, (Figure 12.14a, *lines A–C*; Figure 12.14b, *lines D–F*).
Once the absolute recoil free-fraction functions, f_a^{abs} (T), have been obtained for the previously mentioned complexes, sets of mean square displacements, $<x^2>(T)$, of the Debye temperatures, θ_D, and of the frequency cut-off, ν_D were calculated using Eqs. 12.6, 12.18, and 12.19 of Basics and the data reported in Table 12.6, whereas the functions $\ln f_a^{abs,\ rel}(T)$ and $<x^2>(T)$ are shown in (Figure 12.15). The slopes

Table 12.6 ^{119}Sn parameters and molecular dynamics of $R_2Sn(IV)Clamox{\cdot}2H_2O$ and $R_3Sn(IV)ClamoxNa{\cdot}2H_2O$ derivatives

Compounds[a]	Sample Thickness	Δ^{b}_{av} (mm s^{-1})	Δ^{c}_{av} (mm s^{-1})	Γ^{d}_{av} (mm s^{-1})	No. of data points	Temperature range (K)	Corresponding area range[e]	$d\ln A/dT$ $10^2 \cdot K^{-1}$)[f]	$\theta_D{}^{g}$(K)	ν^{h}(cm^{-1})
A. Me_2Sn Clamox·$2H_2O$	0.480	1.21	2.98	0.94	10	77.3–135	0.085–0.043	−1.29 (0.99)	53.1 (53.7 ± 0.2)	36.9 (37.3 ± 0.2)
B. Bu_2SnClamox·$2H_2O$	0.319	1.26	2.88	1.00	11	77.3–150	0.066–0.028	−1.12 (0.99)	53.2 (53.3 ± 0.2)	37.0 (37.0 ± 0.2)
C. Ph_2SnClamox·$2H_2O$	0.623	0.96	2.68	1.04	9	77.3–150	0.128–0.067	−0.829 (0.99)	54.3 (54.9 ± 0.3)	37.4 (38.1 ± 0.2)
D. Me_3SnClamox Na·$2H_2O$	0.315	1.18	3.12	1.05	13	77.3–170	0.060–0.018	−1.38 (0.99)	41.1 (41.9 ± 0.2)	28.6 (29.1 ± 0.1)
E. Bu_3SnClamox Na·$2H_2O$	0.451	1.36	3.16	1.01	11	77.3 – 160	0.076–0.029	−1.34 (0.99)	39.9 (39.4 ± 0.1)	27.7 (27.3 ± 0.06)
F. Ph_3SnClamox Na·$2H_2O$	0.309	1.15	2.66	1.00	11	82.4 – 140	0.069–0.032	−1.18 (.99)	41.7 (41.8 ± 0.1)	29.0 (29.1 ± 0.1)

[a] amox, Amoxicillin, 6-[D(–)-*β*-amino-*p*-hydroxyphenylacetamido]penicillinate

[b] Isomer shift, with respect to R.T. $CaSnO_3$ averaged over the investigated temperature range

[c] Nuclear quadrupole splittings averaged over the investigated temperature range

[d] Averaged values over the investigated temperature range of the full widths at half height of the resonant peaks, respectively at lower and larger velocity on respect to the spectrum centroid

[e] The total area under the resonant peaks of the Lorentzians have been calculated according to the equation Area = $(\pi/2)\varepsilon\Gamma$, [ε = percentage of the resonant effect]

[f] Slopes of the normalized total areas under the resonant peaks as functions of the temperature; in parentheses are the correlation coefficients of the straight lines lnA versus T

[g] Debye temperature obtained from $d(\ln A)/dT$ and from f_a^{abs}, in parentheses, together with standard errors. Effective vibrating masses have been assumed equal to the molecular weights for all the compounds

[h] Cut-off frequency calculated from $d(\ln A)/dT$ and from f_a^{abs}, in parentheses together with standard errors

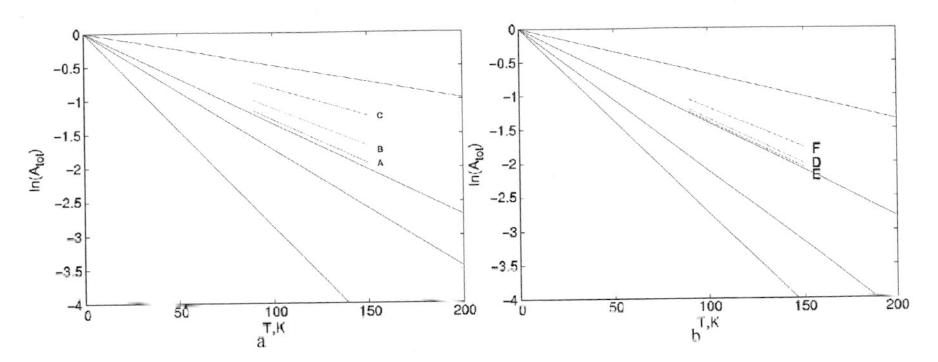

Figure 12.14 Estimates of the extent of intermolecular interaction in organotin(IV)amoxicillin derivatives, from ^{119}Sn molecular dynamics, determined through the slopes of ($d\ln A_{tot}/dT$) (A_{tot}, total area under the resonant peaks at T, K)(**a**) A = $Me_2SnClamox{\cdot}2H_2O$; B = $Bu_2SnClamox{\cdot}2H_2O$; C = $Ph_2SnClamox{\cdot}2H_2O$. (**b**) D = $Me_3SnClamoxNa{\cdot}2H_2O$; E = $Bu_3SnClamoxNa{\cdot}2H_2O$; F = $Ph_3SnClamoxNa{\cdot}2H_2O$

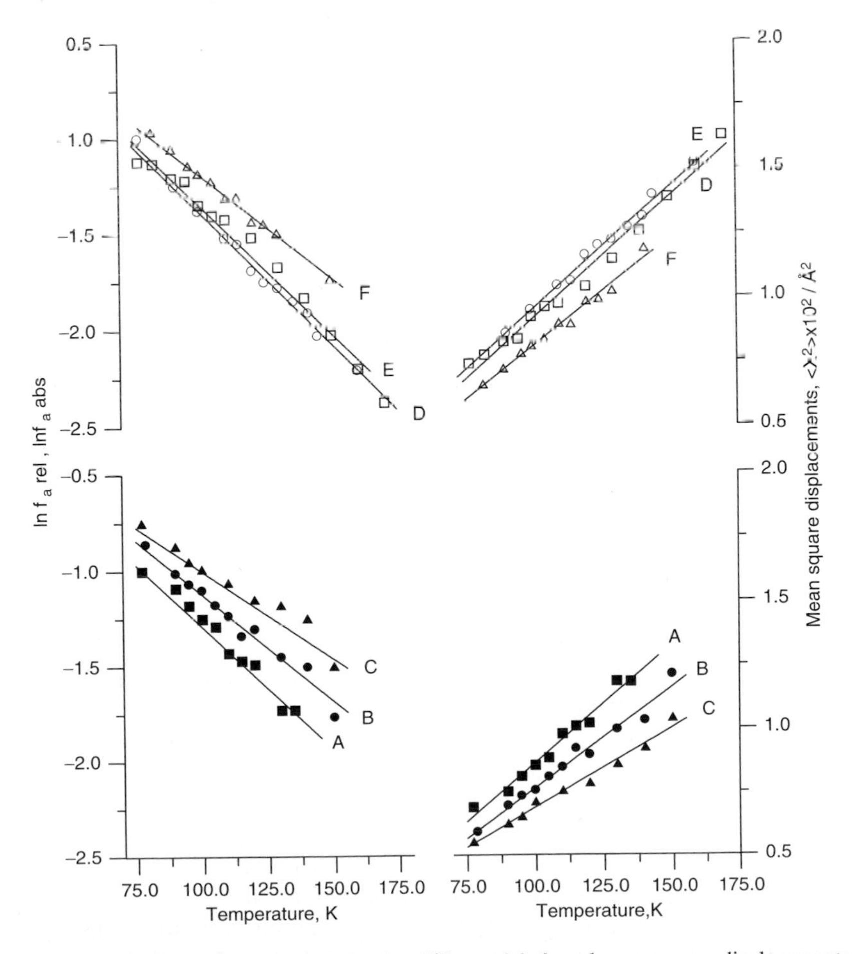

Figure 12.15 Recoil free fraction of absorber ^{119}Sn nuclei, f_a, and mean square displacements of ^{119}Sn, $<x^2>$, as functions of T for $R_2SnClamox{\cdot}2H_2O$ *(A–C)*, $R_3SnClamoxNa{\cdot}2H_2O$ *(D–F)*. Lines are $f_a^{rel}(T)$ and $<x^2>(T)$ functions obtained from slopes $d\ln A_{tot}/dT$, whereas data points are f_a^{abs} and related $<x^2>$ values

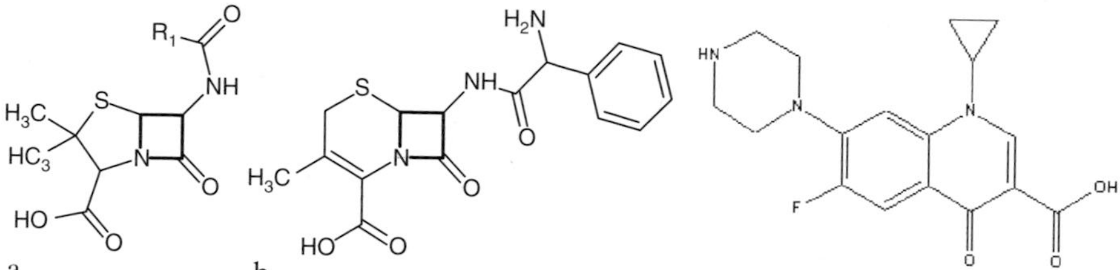

Figure 12.16 (**a**) General formula for penemic antibiotics: R_1 = CH(NH_2)-C_6H_4-4(OH) (antibiotic = amoxicillin); CH_2(NH_2)-C_6H_5 (= ampicillin); C_6H_3-3,5(OCH_3) (= methicillin); CH_2-C_6H_5 (= penicillin G)]; (b) Formula of cephalexin (cephemic antibiotic) and (c) of ciprofloxacin

$d\ln f_a^{a\,bs,\,rel}/dT$ are almost coincident with those obtained from ln $A_{tot}(T)$. The lines of $<x^2>$ as function of the temperature, both for R_2SnClamox·$2H_2O$ and R_3SnClamoxNa·$2H_2O$, lay in the typical zone for tin containing multi-dimensional polymers.

7 Infrared Data

A well characterized series of polymeric complexes, in solid state, is represented by organotin(IV) complexes with β-lactamic antibiotics, both penemic and cephemic.

The difference between the two classes of antibiotics is in the condensed (non-lactamic) ring, pentaatomic for the first class, hexaatomic for the second. This difference constraints different geometries, and thus different way of generating polymers. Polymeric behavior is strongly dependent on reagent used in complex synthesis.

In complexes with antibiotics [amoxicillin, ampicillin, methicillin, penicillinG, cephalexin and ciprofloxacin (Figure 12.16, *a–c*)], polymeric compounds may be originated by direct reaction between dialkyltin(IV) chlorides and antibiotic sodium salt, or by direct reaction between trialkyltin(IV) hydroxides and free antibiotic acid. Other combinations (diorganotin oxide/free acid ligand; triorganotin(IV) chloride/antibiotic salt) produce monomeric species.

The coordinating mode of the penemic antibiotic ligands towards the tin(IV) atom, in the isolated diorgano and triorganotin(IV) complexes and triorganotin(IV)-chloro adducts, can be extracted from the analysis of infrared spectra of the free and coordinated ligands. In fact, the bands attributable to ν_{OH} of the water molecules are present both in the free amoxicillin, ampicillin and methicillin and in all the hydrated complexes, also if as broad band, due probably to hydrogen bonded water molecules. The $\nu_{NH3}{}^+$ of the β-amino-*p*-hydroxyphenyl group, present in amoxicillin and ampicillin between 2,700 and 2,500 cm^{-1}, bd disappeared in the corresponding organotin(IV) complexes owing to the deprotonation of $NH_3{}^+$.

The bands attributable to ν_{NH}, to lactamic $\nu_{C=O}$, to amidic $\nu_{C=O}$, to ν_{CN}, to ν_{CS} have been all identified both in the free and coordinated antibiotics [78–82]. The stretching bands ν_{NH} were shifted at higher wavenumbers, whereas the lactamic and amidic

$\nu_{C=O}$ were shifted towards lower wavenumbers. The bands attributable to CN and CS stretchings, ν_{CN} and ν_{CS}, were practically found in the same position, in all of the complexes.

The above mentioned findings would suggest an involvement of lactamic C=O in coordinating tin(IV) [102–103], but would exclude any involvement of lactamic nitrogen and thiazolidinic sulfur in tin(IV) coordination. As far as the amidic NH and C = O stretchings are concerned, the shifts observed, in all the synthesized derivatives, in respect to the free ampicillin and methicillin ligands, are a consequence of intermolecular hydrogen bondings [104]. In conclusion, differences between free and coordinated antibiotics occurred both in the 4,500 to 3,100 and in the 1,650 to 1,300 cm^{-1} regions, apart from the presence down to 600 cm^{-1} of the characteristic absorptions of the organotin(IV) moieties [105–108].

There are three possible structures about the organotin, two symmetrical and one asymmetrical. The –O–Sn–O– structure can be tetrahedral where the carbonyl is not directly associated with the organotin moiety. This structure is referred to as the non-bridging structure. By comparison, the carbonyl oxygen can be directly associated with the tin resulting in the formation of an octahedral structure. This structure is referred to as a bridging structure since the carbonyl oxygen bridges onto the tin. In general, when the symmetrical –O–Sn–O– structure is present, the product takes on a distorted octahedral structure whereas asymmetrical structures occur as distorted tetrahedral structures when the asymmetric - O–Sn–N– structure is present. The difference between the bridging and nonbridging structures results in a difference in the location of the carbonyl stretching bands. Bridging structures show a strong asymmetric stretching band at about 1,580 cm^{-1} and a weaker band assigned to symmetric stretching near 1,400 cm^{-1}. Non-bridging structures exhibit a strong asymmetric stretching band near 1,650 cm^{-1} and a weaker symmetrical stretching band near 1,360 cm^{-1}.

Finally, ν_{asCOO^-} and ν_{sCOO^-} in the coordinated antibiotic ligands are likely to occur from 1,582 to 1,621 and 1,360 to 1,456 cm^{-1}, respectively, with $\Delta\nu$ |=ν_{asCOO^-} – ν_{sCOO^-}] ranging from 141 to 275 cm^{-1}, following the salification of the carboxylic group [109,110].

The $\Delta\nu$ values are increased over 200 cm^{-1} (210–275 cm^{-1}), both in the R_2SnClantibiotic·$n$$H_2O$ (R = Me, Bu, Ph; antibiotic = amoxicillin, n = 2; ampicillin, methicillin, n = 1; penicillinG, n = 0), and R_3Snantibiotic·H_2O (R = Me, Bu, ph; antibiotic = amoxicillin and ampicillin) suggesting a monodentate ester type coordination [109,110] of the carboxylate group towards the tin(IV) atom, whereas in R_3SnClantibioticNa·$n$$H_2O$, (R = Me, Bu, Ph; antibiotic = amoxicillin, n = 2; ampicillin, methicillin, n = 1; penicillinG, n = 0), $\Delta\nu$ ranged between 148 cm^{-1} in Bu_3SnClamNa·H_2O and 197 cm^{-1} in Me_3SnClpenGNa derivatives, in which carboxylate group seems not involved in coordination.

In conclusion, IR evidence indicates that tin(IV) reached five coordination in R_2SnClantibiotic·$n$$H_2O$ coordinating lactamic C=O and the oxygen atom of the ester type carboxylic group, R_3SnClantibioticNa·$n$$H_2O$ through the coordination of only the lactamic C=O.

In R_3Snantibiotic·H_2O, the antibiotic coordinates the triorganotin(IV) moieties through monoanionic bidentate bridging carboxylate groups.

Where cephemic antibiotic are concerned, the different ring arrangement favors a chelating behavior from carboxyl-carbonyl couple. The larger the bite angle, the more probable a skew octahedral arrangement for tin atom. For example, in diorganotin(IV) complexes, $Alk_2Sn(IV)OHceph{\cdot}H_2O$, complexation leads to a octahedron. But to attain the polymer, a coordination by aminic group of the side chain is required, thus differentiating from penemic series. Moreover, to complete the coordination sphere of tin an OH^- group is bound to the metal. Even in this case, such deductions came from IR findings. In the case of triorganotin(IV) derivatives, $Alk_3Sn(IV)ceph{\cdot}H_2O$, the situation is completely different: the conjugate effect of large bite angle and steric hindrance by three organic groups leads to polymeric complexes with tin in tbp environment, with carboxylic groups acting as bidentate ligands, bridging between organotin centers. This can be deduced from the absence of modifications in β-lactamic stretching absorption upon coordination [82].

The diorganotin(IV) ciprofloxacinate polymer is complicated because of the presence of an additional carbonyl, the ring ketone, assigned at about $1{,}623\,cm^{-1}$.

The strong peak at $1{,}708\,cm^{-1}$ in the ciprofloxacin spectrum, assigned to the carbonyl group of the carboxylic acid, is missing in the spectrum of the products. A new band is found in all of the polymer spectra at $1{,}578\,cm^{-1}$. This band is assigned to the asymmetric stretching for bridged carboxylic groups. A new band is also found at about $1{,}420\,cm^{-1}$ for all of the polymer products assigned to the

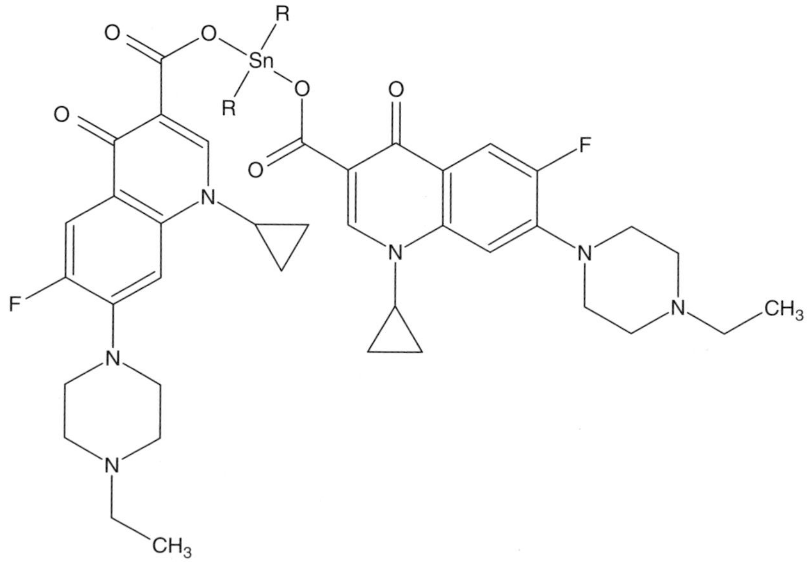

(13)

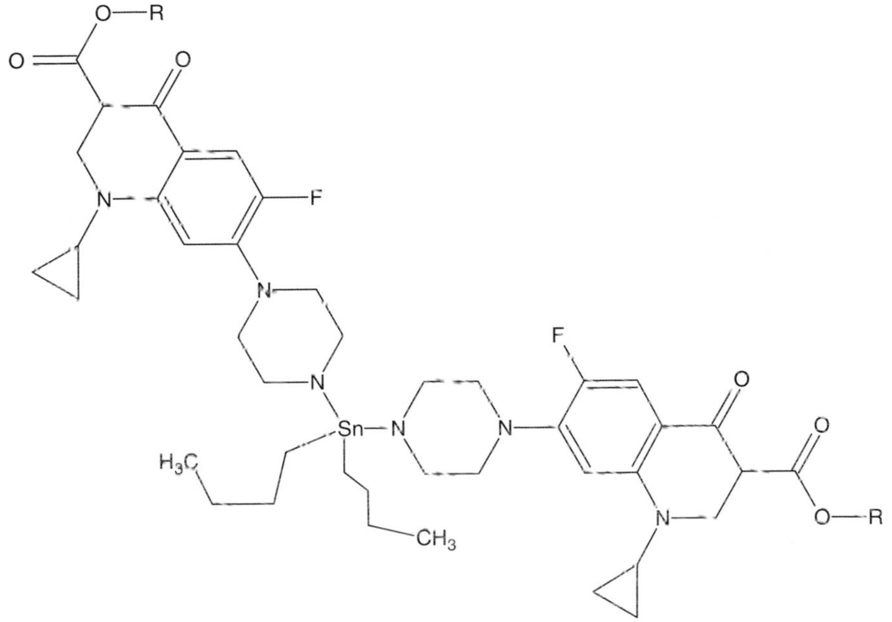

(14)

symmetric carbonyl stretching. The ketone carbonyl band is also present at about 1,620 cm^{-1} for all of the polymer products.

The infrared spectrum data are then consistent with the organotin moiety being present in a distorted octahedral structure in the symmetric –O–Sn–O– structure. The data are also consistent with the structural assignments given by the Mössbauer data.

8 Reaction Implications

The most reasonable way symmetrical structures are formed about the organotin moiety is that preferential addition of one of the Lewis bases, either the nitrogen or oxygen, to the organotin occurs with subsequent growth occurring from coupling of these units.

We believe that the initial growth step in the polymerization is the formation of the –O–Sn–O– based on the following reaction with antibiotics, in particular enrofloxacin, occurs rapidly and in good yield. Here, only the enrofloxacin-Sn-enrofloxacin compound can form (**13**).

By comparison, reaction with the ester of ciprofloxacin where only the amine reacts gives poor yield of the corresponding dimer (**14**).

Both products were formed employing reaction conditions similar to those employed for the polymer synthesis. As noted, the yield for the enrofloxacin was high whereas the product yield for the ciprofloxacin ester was low, consistent with the initial formation of the –O–Sn–O– product.

9 Chain Length

Average molecular weight was determined employing light scattering photometry. The three $R_2Sn(IV)$ciprofloxacinate polymers examined in this study have the following molecular weights and chain lengths: diethyltin(IV)ciprofloxacinate $3.7 \cdot 10^6$ daltons and 7,400 units; dibutyltin(IV)ciprofloxacinate $5.8 \cdot 10^6$ daltons and 10,300 units; and diphenyltin-ciprofloxacin $4.1 \cdot 10^6$ daltons and 7,800 units. Thus, the materials studied here are high polymers.

10 Importance of Present Products

The organotin-antibiotic products exhibit a wide range of biological activity. Against a large battery of bacteria, the polymers are able of inhibiting all tested bacteria including *Sacchyromyces cerevisiae*, *Staphylococcus aureus*, *Bacillus subtilis, Escherichia coli*, *Pseudomonas aeruginosa*, and *Alcaligenes faecalis*. Some of the polymers are also able to inhibit a number of DNA viruses such as *vaccinia* (smallpox), *herpes simplex* (herpes), and *varicella zoster* (chicken pox and shingles). It also inhibits the growth of a number of cell lines including L929, BS-C-1, Balb 3-T-3, vero cells, and 143 cells.

11 Conclusions

According to the information drawn from Mössbauer spectra, there are three major configurations around tin atoms for the organotin antibiotic polymers: one is trigonal bipyramidal, both for penem and cephem antibiotics, while a tetrahedral consisting of a N–Sn–N structure, and a distorted tetrahedral or a skew trapezoidal *trans*-R_2SnO_4 configuration consisting of a O–Sn–O structure for diorganotin ciprofloxacinate. All the structures are considered as symmetric configurations. The expected asymmetric structure, N–Sn–O for ciprofloxacinate, was not detected. One possible reason for the existence of only the symmetrical structures is the high reactivity of O–Sn bond and high stability of O–Sn–O structure due to the chelating of carboxyl group allowing the O–Sn–O structure to preferentially form leaving the N to react with remaining Sn atoms. As noted before, the structural results are consistent with the infrared spectral data for the compounds.

References

1. Goldanskii VI, Herber RH (eds). (1968) Chemical applications of Mössbauer spectroscopy. Academic Press, New York.
2. Greenwood NN, Gibb TC. (1971) Mössbauer spectroscopy. Chapman and Hall, London.
3. Gibb TC. (1976) Principles of Mössbauer spectroscopy. Chapman and Hall, London.
4. Herber RH (ed). (1984) Chemical Mössbauer spectroscopy. Plenum Press, New York.
5. Parish RV. (1984) Structure and bonding in tin compounds. In: Long GJ, (ed), Mössbauer spectroscopy applied to inorganic chemistry, Vol. 1. Plenum Press, New York, p. 527.
6. Harrison PG. (1989) Investigating tin compounds using spectroscopy. In: Harrison PG (ed), Chemistry of tin. Blackie, Glasgow, p 60
7. Parish RV. (1990) Mössbauer spectroscopy. In: NMR, NQR, EPR, and Mössbauer spectroscopy in inorganic chemistry. Ellis Horwood, New York, p 128.
8. Zuckerman JJ. (1984) Organotin-119m Mössbauer spectroscopy, In: Herber RH (ed), The first quarter century in chemical Mössbauer spectroscopy. Plenum Press, New York, p 267.
9. Stevens JG, Stevens VE. (1978) Mössbauer effect data index. A. Hilger, London and IFI-Plenum, New York. ; Stevens JG, Stevens VE. (1995) Mössbauer effect refs Data J., Vol. 1–18. Mössbauer Effect Data Center, Ashevill, North Carolina.
10. Cranshaw TE. (1984) Mössbauer spectrometers and calibration, In: Long GJ (ed), Mössbauer spectroscopy applied to inorganic chemistry, Vol. 1. Plenum Press, New York, p 27.
11. Dunlap BD, Kalvius GM. (1978) Theory of Isomer Shifts, In: Shenoy GK, Wagner FE, (eds), Mössbauer isomer shifts. North Holland, Amsterdam, p 15.
12. Schmiedgen R, Huber F, Preut H, et al.. (1994) Appl Organomet Chem. 8:397
13. Protsky AN, Bulychev BM, Soloveichik GL, Belsky VK. (1986) Inorg Chim Acta. 115:121.
14. Biddle BN, Gray JS, Crowe AJ. (1990) J Chem Soc Dalton Trans. 419.
15. Casas JS, García Martínez E, Sánchez González A, et al. (1995) J Organomet Chem. 493:107.
16. Kumar Das VG, Keong YC, Ng SW, Wei C, Mak TCW. (1986) J Organomet Chem. 311:289.
17. Cheng ZZ, Chen L, Xie Q, Wang HK, Zhang JK, Luo S, Huang J. (1987) Mössbauer Effects Refs Data J. 6:98, 116.
18. Dallaire MA, Brook AD, Bain CS, Frampton JF. (1993) Can J Chem. 71:1676.
19. Jurkschat K, Tzschach A, Weichmann H, Rajczy P, Mostafa MA, Korecz L, Burger K. (1991) Inorg Chim Acta 179:83.
20. Shenoy GK. (1984) Mössbauer-Effect Isomer Shifts, In: Long GJ, (ed), Mössbauer spectroscopy applied to inorganic chemistry, Vol.1. Plenum Press, New York, p.
21. Sosinski BA. (1984) Aspects of Organoiron Mössbauer Spectroscopy, In: Herber RH, (ed), Chemical Mössbauer spectroscopy. Plenum Press, New York, p 1.
22. James BD, Gioskos S, Chandra S, Magee RJ, Cashion JD. (1992) J Organomet Chem. 436:155.
23. Sandhu GK, Verma SP, Moore LS, Parish RV. (1986) J Organomet Chem. 315:309.
24. Barbieri A, Giuliani AM, Ruisi G, Silvestri A, Barbieri R, (1995) Zanorg Allg Chem. 621:89; Δ_a in the T range 77.3, 155 K.
25. Mundus-Glowacki B, Huber F, Preut H, Ruisi G, Barbieri R. (1992) Appl Organomet Chem. 6:83.
26. Sánchez-González A, Casas JS, Sordo J, Russo U, Lareo MI, Regueiro BJ. (1990) J Inorg Biochem. 39:227.
27. Sandhu GK, Boparoy NS. (1991) J Organomet Chem. 420:23.
28. Burger K, Nagy L, Buzás N, Vertes A, Mehner H. (1993) J Chem Soc Dalton Trans. 2499.
29. Lo KM, Selvaratnam S, Ng SW, Wei C, Kumar Das VG. (1992) J Organomet Chem. 430:149.
30. Molloy KC, Quill K, Cunningham S, McArdle P, Higgins T. (1989) J Chem Soc Dalton Trans. 267.
31. Smith PJ, Day RO, Chandrasekhar V, Holmes JM, Holmes RR. (1986) Inorg Chem. 25:2495.

32. Sandhu GK, Verma SP, Moore LS, Parish RV. (1987) J Organomet Chem. 321:15.
33. Cunningham D, McGinley J. (1992) J Chem Soc Dalton Trans. 1387.
34. Pettinari C, Lorenzotti A, Sclavi G, et al (1995) J Organomet Chem. 496:69.
35. Singh NK, Sharma U, Kulshreshtha SK. (1990) J Organomet Chem. 382:375.
36. Bonire JJ, Crowe AJ, Smith PJ. (1994) Appl Organomet Chem. 8:433.
37. Mallela SP, Tomic ST, Lee K, Sams JR, Aubke F. (1986) Inorg Chem. 25:2939.
38. Carini C, Pelizzi G, Tarasconi P, Pelizzi C, Molloy KC, Waterfield PC. (1989) J Chem Soc Dalton Trans. 289.
39. Gielen M, Bouhdid A, Kayser F, et al. (1995) Appl Organomet Chem. 9:251.
40. Spiering H. (1984) The Electric Field Gradient and the Quadrupole Interaction, In: Long GJ (ed), Mössbauer spectroscopy applied to inorganic chemistry, Vol. 1. Plenum Press, New York, p 77.
41. Kolk B. (1984) Studies of Dynamical Properties of Solids with the Mössbauer Effect, In: Horton GK, Maradudin AA, (eds), Dynamical properties of solids, Vol. 5. North Holland, Amsterdam, p 3.
42. Herber RH. (1984) Structure, Bonding, and the Mössbauer Lattice Temperature, In: Herber RH, (ed)), Chemical Mössbauer spectroscopy. Plenum Press, New York, p 199.
43. Bauminger ER, Nowick I. (1986) The Dynamics of Nuclei studied by Mössbauer Spectroscopy, In: Dickson DPE, Berry FJ, (eds), Mössbauer spectroscopy. Cambridge University Press, Cambridge, UK, p 219.
44. Goldanskii VI, Makarov EF. (1968) Fundamentals of Gamma-Resonance Spectroscopy, In: Goldanskii VI, Herber RH (eds), Chemical applications of Mössbauer spectroscopy. Academic Press, New York, p1.
45. Stöckler HA, Sano H. (1969) J Chem Phys. 50:3813.
46. Lang G. (1963) Nucl Instrum Methods. 24:425.
47. Sano H, Mekata Y. (1975) Chem Lett. 155.
48. Matsubara S, Katada M, Sato K, Motoyama I, Sano H. (1979) J de Physique 40:C2–363.
49. Barbieri R, Silvestri A, Barbieri A, Ruisi G, Huber F, Hager C–D. (1994) Gazz. Chim. Ital. 124:187.
50. Molloy KC, Quill K. (1985) J Chem Soc Dalton Trans. 1417.
51. Brooks JS, Clarkson RW, Williams JM. (1983) J Organomet Chem. 251:63.
52. Stöckler HA, Sano H, Herber RH. (1967) J Chem Phys. 47:1567.
53. Barbieri R, Cefalù R, Chandra SC, Herber RH. (1971) J Organomet Chem. 32:97.
54. Barbieri R, Silvestri A, Pellerito L, Gennaro A, Petrera M, Burriesci N. (1980) J Chem Soc Dalton Trans. 1983.
55. Barbieri R, Huber F, Pellerito L, Ruisi G, Silvestri A. (1998) ^{119}Sn Mössbauer Studies on Tin Compounds. Recent developments, In: Smith PJ, (ed), Chemistry of tin, Vol. 2. Blackie Academic & Professional, Glasgow, p 496.
56. Goldanskii VI. (1971) Vysokomolek. Soed Seriya A. 13:311.
57.Apperley D, Davies N, Harris R, Brimah A, Eller S, Fischer R. (1990) Organometallics. 9:2672.
58. deMello VJ, daS.Maia JR, deOliveira TT, Nagem TJ, Ardisson JD, deO.Patricio PS, deLima GM. (2004) Main Group Metal Chemistry. 27:309.
59. Deacon P, Devylder N, Hill M, Mahon M, Mollow K, Price G. (2003) J Organomet Chem. 687:45.
60. Barbieri R, Pellerito L, Silvestri A, Ruisi G, Noltes J. (1981) J Organometal Chem. 210:43.
61. Grigoriev E, Yashina N, Prischenko A, et al. (1995) Appl Organometal Chem. 9:11.
62. Molloy K, Blunden S, Hill R. (1988) J Chem Soc Dalton Trans Inorg Chem. 1259.
63. Ford B, Sams J. (1971) J Organomet Chem. 31:47.
64. Stapfer C, Herber R. (1973) J Organometal Chem. 56:175.
65. Lo K, Das V, Yip W, Mak T. (1991) J Organometal Chem. 412:21.
66. Barbieri R, Silvestri A, Filippeschi S, Magistrelli M, Huber F. (1990) Inorg Chim Acta. 177:141.
67. Danish M, Alt H, Badshah A, Ali S, Mazhar M, Islam N. (1995) J Organomet Chem. 486:51.

68. Nath M, Jairath R, Eng G, Song X, Kumar A. (2005) Spectrochimica Acta Part A.Mol Biomol Spectrom. 61:77.
69. Nath M, Jairath R, Eng G, Song X, Kumar A. (2005) Spectrochimica Acta Part A Mol Biomol Spectrom. 62:1179.
70. Nath M, Jairath R, Eng G, Song X, Kumar A. (2005) Spectrochimica Acta Part A Mol Biomol Spectrom. 61:3155.
71. Nath M, Jairath R, Eng G, Song X, Kumar A. (2005) J Organomet Chem. 690:134.
72. Bancroft GM, Platt RH. (1972) Adv Inorg Chem Radiochem. 15:59.
73. Collins RL, Travis JC. (1967) The electric Field Gradient Tensor, In: Gruverman IL, (ed), Mössbauer effect methodology, Vol 3. Plenum Press, New York, p 123
74. Bancroft GM, Kumar Das VG, Sham TK, Clarck MG. (1976) J. Chem Soc Dalton Trans. 643.
75. Barbieri R, Silvestri A, Huber F, Hager CD. (1981) Can J Spectr. 26:194.
76. Sham TK, Bancroft GM. (1975) Inorg Chem. 14:2281.
77. Ng SW, Wei C, Kumar Das VG, Mak TCW. (1987) J Organomet Chem. 334:295.
78. Pellerito L, Maggio F, Consiglio M, Pellerito A, Stocco GC, Grimaudo S. (1995) Appl Organomet Chem. 9:227.
79. Pellerito L, Maggio F, Fiore T, Pellerito A. (1996) Appl Organomet Chem. 10:393.
80. Maggio F, Pellerito A, Pellerito L, Grimaudo S, Mansueto C, Vitturi R. (1994) Appl Organomet Chem. 8:71.
81. Di Stefano R, Scopelliti M, Pellerito C, et al. (2002) J Inorg Biochem. 89:279.
82. Di Stefano R, Scopelliti M, Pellerito C, et al. (2004) J Inorg Biochem. 98:534.
83. McGrady MM, Tobias RS. (1964) Inorg Chem. 3:1960.
84. McGrady MM, Tobias RS. (1965) J Am Chem Soc. 87:1909.
85. Barbieri R, Faraglia G, Gustiniani M, Roncucci L. (1964) J Inorg Nucl Chem. 26:203.
86. Roncucci L, Faraglia G, Barbieri R. (1964) J Organomet Chem. 1:427.
87. Nelson WH, Martin DF. (1965) J Organomet Chem. 27:89.
88. Tanaka T, Komura M, Kawasaki Y, Okawara R. (1964) J Organomet Chem. 1:484.
89. Kawasaki K, Tanaka T, Okawara R. (1966) J Organomet Chem. 6:95.
90. Ramos VB, Tobias RS. (1973) Spectrochim Acta Part A. 29A:953.
91. Ramos VB, Tobias RS. (1974) Spectrochim Acta Part A. 30A:181.
92. Miller GA, Schlemper EO. (1973) Inorg Chem. 12:677.
93. Schlemper EO. (1967) Inorg Chem. 6:2012.
94. Nelson WH, Aroney MJ. (1973) Inorg Chem. 12:132.
95. Kepert DL. (1977) Inorg Chem. 12:1.
96. Brahma SK, Nelson WH. (1982) Inorg Chem. 21:4076.
97. Nelson WH, Howard WF Jr, Pecora R. (1982) Inorg Chem. 21:1483.
98. Howard WF Jr, Nelson WH. (1982) Inorg Chem. 21:2283.
99. Howard WF Jr, Crecely RW, Nelson WH. (1985) Inorg Chem. 24:2204.
100. Clark MG, Maddock AG,. Platt RH. (1972) J Chem Soc Dalton Trans. 281.
101. Pellerito L, Maggio F, Pellerito A. (1994) 35th Meeting of the Mössbauer spectroscopy discussion group, University of Nottingham, p 21.
102. Sodhi GS, Bajaj RK, Kaushik NK. (1984) Inorg Chim Acta. 92:L27.
103. Kamrah SG, Sodhi GS, Kaushik NK. (1985) Inorg Chim Acta. 107:29.
104. Dexter DD, Van der Veen JM, Perkin I. (1978) J Chem Soc. 185.
105. Beattie IR, McQuillan GP. (1963) J Chem Soc. 1519.
106. Taimsalu P, Wood JL. (1964) Spectochim Acta. 20:1045
107. Butcher FK, Gerrard W, Mooney EF, Gees RG, Willis HA. (1964) Spectrochim Acta. 20:51.
108. Whiffen DH. (1956) J Chem Soc. 1350.
109. Deacon GB, Phillips RJ. (1980) Coord Chem Rev. 33:227.
110. Deacon GB, Huber F, Phillips RJ. (1985) Inorg Chim Acta. 104:41.

Fundamentals of Fragmentation Matrix Assisted Laser Desorption/Ionization Mass Spectrometry

Charles E. Carraher Jr., Theodore S. Sabir, and Cara Lea Carraher

1 Introduction

In 1981 Barber and Liu and coworkers [1,2] independently introduced the concept of employing matrix-assisted desorption/ionization where the absorption of the matrix is chosen to coincide with the wavelength of the employed laser to assist in the volatilization of materials. In 1988 Tanaka, Hillenkamp and coworkers [3,4] employed the laser as the energy source giving birth to matrix-assisted laser/desorption mass spectroscopy (MALDI MS).

MALDI MS was developed for the analysis of nonvolative samples and was heralded as an exciting new MS technique for the identification of materials with special use in the identification of polymers. It has fulfilled this promise to only a limited extent. Whereas it has become a well-used and essential tool for biochemists in exploring mainly nucleic acids and proteins, it has been only sparsely employed by synthetic polymer chemists. Why has MALDI MS been largely neglected by most synthetic polymer chemists? The answer involves the lack of congruency between the requirements of MALDI MS and most synthetic polymers. This chapter describes a new approach that allows MALDI MS to be of greater use to the analysis of synthetic polymers and investigates why it has not been useful in the past to the analysis of such polymers. We have employed various forms of MALDI MS for about 5 years and have developed a technique that will be described here. We have previously reported some of our results in journals and national meetings but have not previously given the foundational data that allows this technique to be viable.

A.S. Abd-El-Aziz et al. (eds.), *Inorganic and Organometallic Macromolecules: Design and Applications.*

We have recently employed MALDI MS for the identification of a number of metal and nonmetal containing polymers [5–11]. The technique employed by is not straight-forward MALDI MS. Classical MALDI MS requires that the material be soluble in a suitable solvent. A "suitable solvent" means a solvent that is sufficiently volatile to allow it to be evaporated prior to the procedure. Further, such a solvent should dissolve both the polymer and the matrix material. Finally, an ideal solvent will allow a decent level of polymer solubility, preferably a solubility of several percentage and greater. For most synthetic polymers, these qualifications are only approximately attainted. Thus, traditional MALDI MS has not achieved its possible position as a general modern characterization tool for synthetic polymers. By comparison, MALDI MS is extremely useful for many biopolymers where the polymers are soluble in water. Even so, we have employed a variety of MALDI MS techniques that appear to be suitable for poorly or insoluble samples. Even though MALDI MS is a mild or gentle technique causing a minimum of fragmentation, we have observed fragmentation in polymer samples [5–12].

We have named our approach Fragmentation Matrix-Assisted Laser Desorption/Ionization mass spectrometry, or simply F MALDI MS, because it is the fragmentation fragments that are emphasized in the study. The technique should be applicable to any solid when the proper operating conditions are employed.

Although the technique is useful for nonmetal and metal-containing polymers, this chapter emphasizes its use with metal-containing polymers. Identification of polymer structure is an essential characterization imperative. For metal-containing polymers, this is more difficult for metal-containing polymers. The difficulty involves both increased complexity of possible structures and decreased (or lack of) solubility of the polymers. F MALDI MS allows for ready identification of the repeat unit and can allow determination of some low-molecular-weight range chains. This chapter describes our overall approach and presents data that justifies the use of this technique for the identification of polymer structure.

2 Experimental

Sample preparation varies but essentially we grind together our polymer with the matrix dissolved in a suitable solvent and employ this mixture for our analysis.

A HP MdlG2025A MALDI-TOF mass spectrophotometer was employed to obtain MS spectra in the range of 400 to 1,200 daltons. This MALDI MS is considered a low-mass range instrument and a low-end performance-wise instrument by today's standards. For the high-range studies, high-resolution electron impact positive ion matrix assisted laser desorption ionization time of flight (HR MALDI-TOF) mass spectrometry was carried out employing a Voyager-DE STR BioSpectrometer (Applied Biosystems; Foster City, CA). The standard settings were used with a linear mode of operation and an accelerating voltage of 25,000 volts; grid voltage 90% and an acquisition mass range of 2000 to 100,000. Two hundred shots were typically taken for each spectrum. This is considered a "state-of-the-art" instrument.

Several matrix materials were employed but here only results employing α-cyano-4-hydroxycinnamic acid are included.

3 Results and Discussion

3.1 Theory

Analysis is not straight-forward because most elements contain naturally occurring isotopes. This includes even simple elements such as hydrogen and carbon. As the number of atoms increases, spreading occurs [13,14] so that by the time there are 20 or more units of even a somewhat simple polymer such as nylon 66, the spread is significant. Even so, computer models show that whereas spreading continues as the number of units increases, the abundance of particular elements appears to approach the natural abundance of the particular elements allowing the identification of higher molecular weight ion fragments [14].

We began by running blanks that contained the matrix and solvent. Ion fragments were found that corresponded to the matrix and its decomposition. Corresponding MS containing the polymer were then run. These spectra did not contain ion fragments from the matrix and solvent consistent with the relative abundance of fragments from the polymer being greater than those generated by the matrix/solvent. Conditions were such that different matrix/solvent combinations worked better for different polymers so that some effort was taken to get a decent match. Typically, an accumulation of many laser "shots," from 100 to 500, were employed to accumulate a decent population of fragments. Because of spreading, it was somewhat surprising that preferred ion fragments were even distinguishable for fragments containing 200 or more units. Even so, ion fragments that were 3 or more times the background are found. It is these fragments that are identified.

For shorter chains, it is possible to identify intact chains including the identification of end-groups. For high polymer fragments, typically no fragmentation occurs other than that necessary to generate the particular fragmentation. We have identified fragments in a continuous sequence from dimmers and trimers to over 200 units long. The results allow the identification of materials that are oligomeric to polymeric and which are either insoluble or only partly soluble. At lower masses the technique also allows for identification of the repeat unit as well as natural isotopic abundances. Identification of natural isotopic abundances is important in verifying the assignments for units that contain elements such as tin and antimony that contain a number of isotopes in reasonable percentages.

As the mass of the fragment increases, more precise masses of each element have been used rather than simply rounding off masses to the nearest whole number. (All masses are given in amu or daltons.) Thus, C = 12.011 rather than simply 12, H = 1.00794 rather than 1, and so on. Also, precise masses of the metals

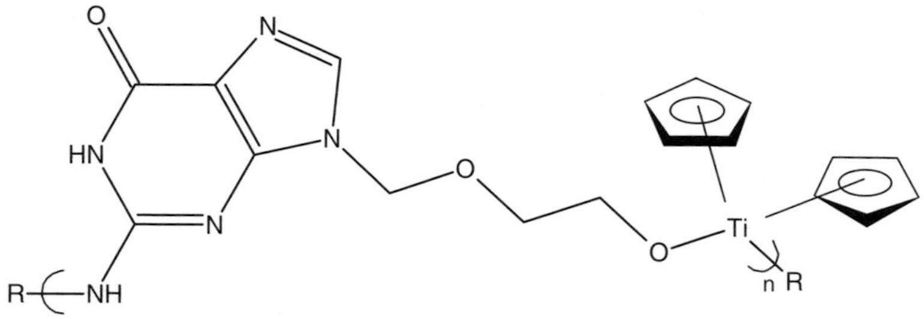

(1)

were employed so that Ti = 47.88 rather than 48. This gives the mass for the product of titanocene dichloride and acyclovir as 401.260 (**1**).

For the Cp_2Ti-Ac the "precise" value compared with the "rounded off" value for each unit was only 0.26 amu which was not too significant for the lower mass range but became increasingly significant as the number of repeat units increases. For 30 units, the difference was only 8 amu units (12,037.8 compared with 12,030). For 100 units this difference increases to 26 amu (40,126 compared to 40,100) or a difference of more than 1 oxygen or methylene moiety. The assignment of the 96,898 ion-fragment was 242 units plus a NH unit, minus the acyclovir moiety. The calculated value was within 1 amu of the ion fragment value. Using the rounded off value of 401, a value of 96,834 was calculated for this assigned structure with a difference in 64 amu that could correspond to a O–CH_2–CH_2–O or almost a cyclopentadiene (65 amu) group.

An alternate calculation involves assuming that the most abundant ion will be composed entirely of the most abundant isotope. For Zr this isotope is 90. Thus, rather than using for the repeat unit for the zirconocene dichloride-acyclovir product of 444.6038, a mass repeat unit value would be 443 or a difference of 1.6038 amu. Again, for short chains this is not too damaging, but for longer chains it would be misleading. Thus, for 30 units the values would be 1,338.1 compared with 13,290 or a difference of 48 amu that might correspond to O–CH_2–CH_2 for the present system. For the ion fragment at 40,488, assigned as 91 units plus O–CH_2, calculated mass is 40,492 or a difference of 4 amu, whereas using the most abundant isotope value the mass assigned would be 40,343 or a difference of 149 amu.

In reality, using computer programs, the isotopic contribution approaches an average, after 10 units, with that average corresponding to the atomic weight reported in periodic tables which is a summation of the various isotopic masses times the natural abundance.

Preliminary results are consistent with F MALDI MS being a new general use characterization tool for oligomers and polymers. Employing a low-mass-range MALDI MS it is reasonable to identify ion fragments to a mass of about 1,000 daltons and using a high range MALDI mass spectrometer it is reasonable to identify fragments to 100,000 daltons, the useful upper limit for the MALDI MS employed. Thus, the combination of the high-energy laser source and matrix allows for the creation and identification of much higher mass ion fragments than is possible for classical electron impact mass spectrometry, EI MS (which is typically below 1,000 daltons).

3.2 *Application*

3.2.1 Low-Range F MALDI MS

For the next few examples results employing a low-range MALDI MS instrument are given. This MALDI MS had an upper useful limit of about 1,100 daltons. We used F MALDI MS to identify fully-intact chains for some lower molecular weight products. For instance, the product of dioctyltin dichloride and norfloxacin (**2**), has a M_w of 4.2 × 10^3 M which corresponds to about 6 repeat units [5,6]. Table 13.1 contains the most abundant fragment clusters in the range of 600 to 1,050 daltons (all fragment weights are given in daltons). Three end groups are possible—NH, COCl and COOH. It is difficult to distinguish between HN–R as an end group and N–R derived from fragmentation of the chain creating N–R end groups because the instrument is not

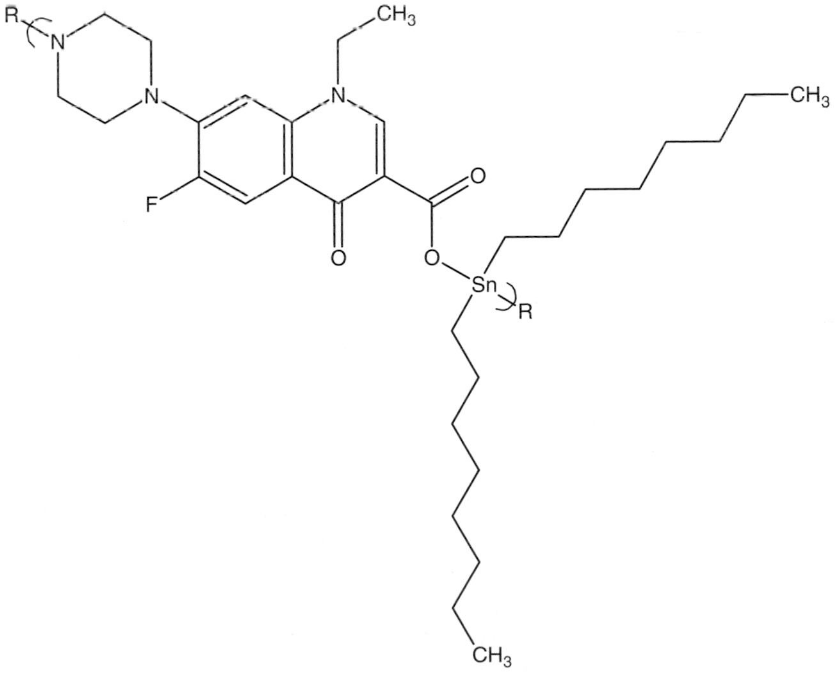

(2)

Table 13.1 Most abundant fragment clusters (600–1050 daltons) for the product of dioctyltin dichloride and norfloxacin where Oct = octyl

m/e	(Possible) Assignment
663	One Unit
699	One unit+Cl
718	Cl-Unit-Sn-OH minus 3 Oct
779	Unit-Sn minus 2Oct
869	pip-Unit-Sn minus 2Oct
1,024	Unit-Sn-OH
1,035	Unit-Sn-2OH

Table 13.2 Ion fragments in the 400 to 1000 amu range from the product of levodopa and adipoyl chloride. Here the letter "L" , is used to describe the levodopa portion whereas "A" represents the adipoyl moiety. The letter U represents one repeat unit and Amac represents the amino acid group, –CH(NH_2)COOH

m/e	(Tentative) Assignment	m/e	(Tentative) Assignment
417	ALA	544	2U-Amac
432	LAL-CH_2,Amac	550	2U-CO_2,NH_2
439	LAL-CO_2,O	566	2U-CO_2
446	LAL-CO_2,NH_2	583	2U-2O
455	LAL-CO_2	596	2U-CO
490	LAL-O	607	2U-NH_2
504	LAL	622	2U
526	2U-CH_2,Amac	795	3U-CO_2 Amac
883	3U-CO_2	899	3U-CO
		971	3U+CO_2

capable of distinguishing one dalton mass difference. In comparison, it is possible to identify COCl and COOH end groups. For this product, end groups are identified as being both COCl (at 699) and COOH (1,024 and 1,035). Intact chains containing one and one and a half units are found (at 699 and 1,035).

End-groups can also be identified for polymers with moderate chain lengths. For instance, for the reaction of organic acid dichlorides and L-dopa [10], there are three reasonable end-groups. These are R–CO–Cl, R–CO–OH, and R′–OH. The presence (or absence) of Cl end groups and the amount of these end groups, that is one or two, is ascertained by looking at the isotopic abundance of Cl-associated fragments. Table 13.2 contains representative data, for the product from levodopa and adipoyl chloride (M_w 6.3×10^3; DP = 20) (**3**). No chlorine end-groups are identified in the fragments. R–COOH end groups are found (for instance at 971).

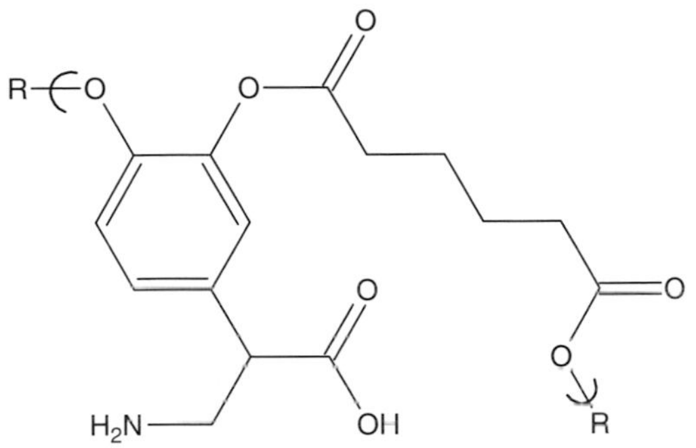

(3)

Table 13.3 Ion fragments in the 400 to 1000 amu range for the product of levodopa and isophthaloyl chloride

m/e	(Tentative) Assignment	m/e	(Tentative) Assignment
409	ILI-2CO	634	2U–O
432	ILI-CO	658	2U
459	ILI	702	2U+CO_2
522	LIL	899	3U-Amac
547	LIL+CO	987	3U
575	LIL+2CO		

Similar results were found for the product of levodopa and isophthaloyl chloride (Table 13.3; M_w 3.4 × 10^4; DP = 103). The site of bond breakage along the backbone of the chain appeared to be at the ester linkage, as expected. Another site for frequent bond breakage was the amino acid-containing moiety.

Thus, up to three units are easily identifiable employing a low-range instrument. Also, sites of preferred bond breakage are easily identifiable.

3.2.2 High-Range F MALDI MS

Next, the focus is on the use of a state-of-the-art MALDI MS. Figure 13.1 contains the F MALDI MS for the matrix employed in the present study, α-cyano-4-hydroxycinnamic acid (**4**), and solvent alone for the 100 to 100,000 daltons range. Figure 13.2 contains the MALDI MS spectra over the range of 100 to 1,000 daltons. The sole fragments found in Figures 13.1 and 13.2 were assigned to the matrix and its fragmentation and were found about 247 and below. The major frag-

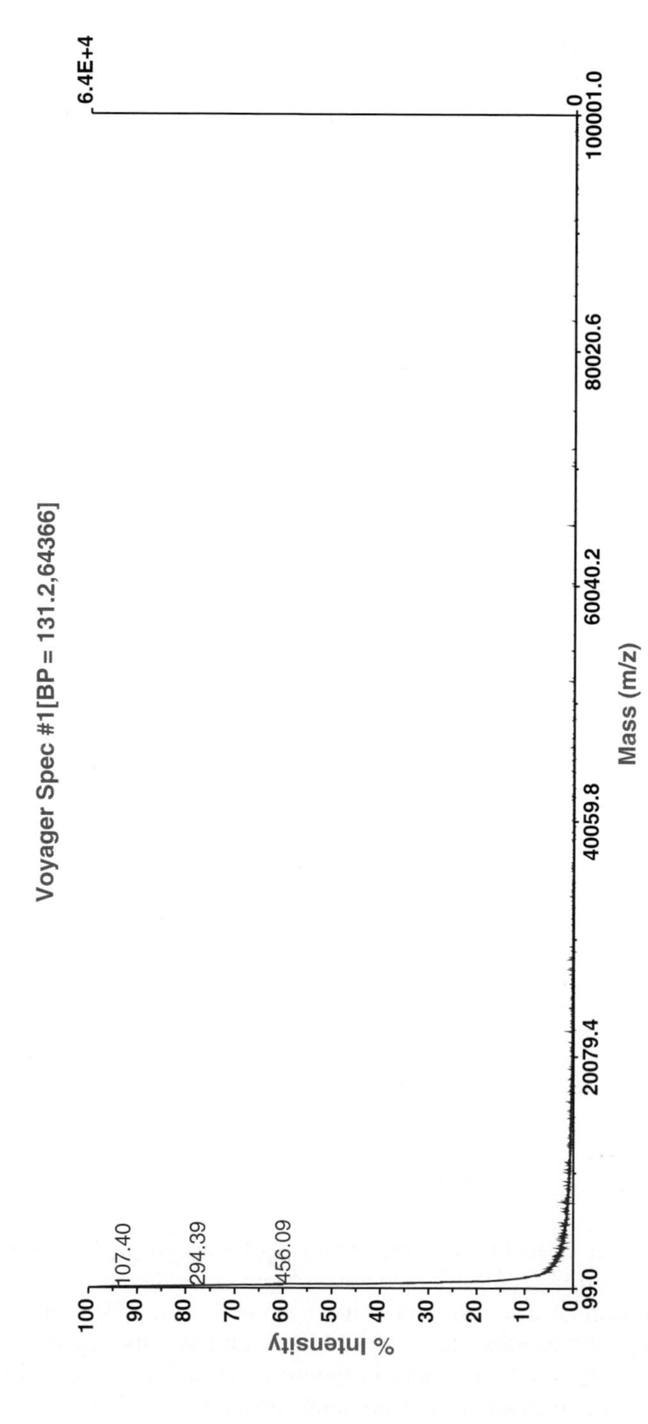

Figure 13.1 F MALDI MS of the matrix, α-cyano-4-hydroxycinnamic acid over the mass range of 100 to 100,000 daltons

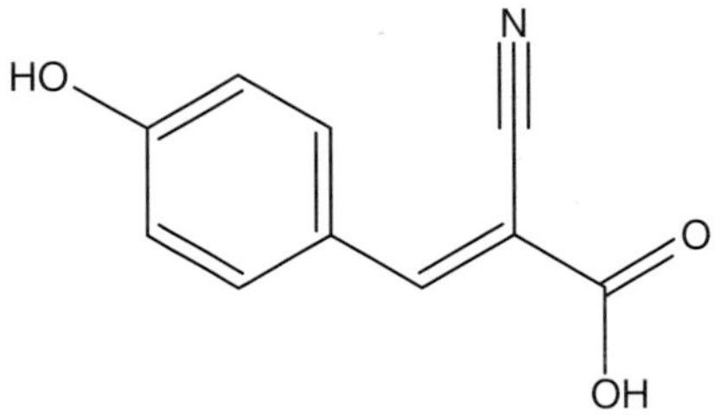

(4)

ments were found at 177 (matrix minus CO_2), 207 (matrix minus methyl), 225 (matrix), and 247 (matrix plus Na).

For comparison, Figure 13.3 contains a similar spectra except for the polymer derived from 1,1′-dicarboxylferrocene sulfonyl dichloride and acyclovir (M_w = 1.3×10^5; **5**). In Figures 13.1 and 13.2 no significant fragments were found above ~247 dalton, but for the acyclovir product ion fragments were found up to about 100,000 daltons that corresponded to fragments over 200 units long (Tables 13.4 and 13.5). Further, in the 100 to 1,000 daltons range, no ion fragments derived from the matrix were found. Ion fragments for the product of 1,1′-disulfonylferrocene and acyclovir in the 100 to 1,000 daltons range are given in Table 13.4. Assignments for the major fragments given in Figure 13.3 are given in Table 13.5.

Figure 13.4 contains an expansion of the about 3,000 to 4,000 dalton mass range for the matrix. Only background is found. By comparison, Figure 13.5 contains a similar expansion for the product of 1,1′-disulfonylferrocene and acyclovir. Numerous fragments are found for this expanded range beyond the background "noise" shown in Figure 13.4.

The agreements of the proposed structures and the masses are generally within ±3 daltons to 100,000 daltons or within 3×10^{-3}%.

F MALDI MS is also capable of identifying isotope abundances. Figure 13.6 contains the low-range F MALDI MS for the product of triphenylantimony and acyclovir ($M_w = 7.7 \times 10^6$) (**6**). Antimony has two natural occurring isotopes, Sb-121 at 57% and Sb-123 at 43%. Table 13.6 contains assignments for the fragments given in Figure 13.6.

The couplets have ratios of about 57/43 that correspond to the natural abundance of antimony isotopes (Table 13.7) consistent with the presence of one antimony atom. Spectra at higher masses were found that corresponded to the presence of 2, 3, and 4 antimony atoms within fragments that corresponded to approximately dimers, trimers, and tetramer-containing fragments (Table 13.8). In each case the agreement was reasonable and consistent with the number of antimony atoms assigned to the proposed structure.

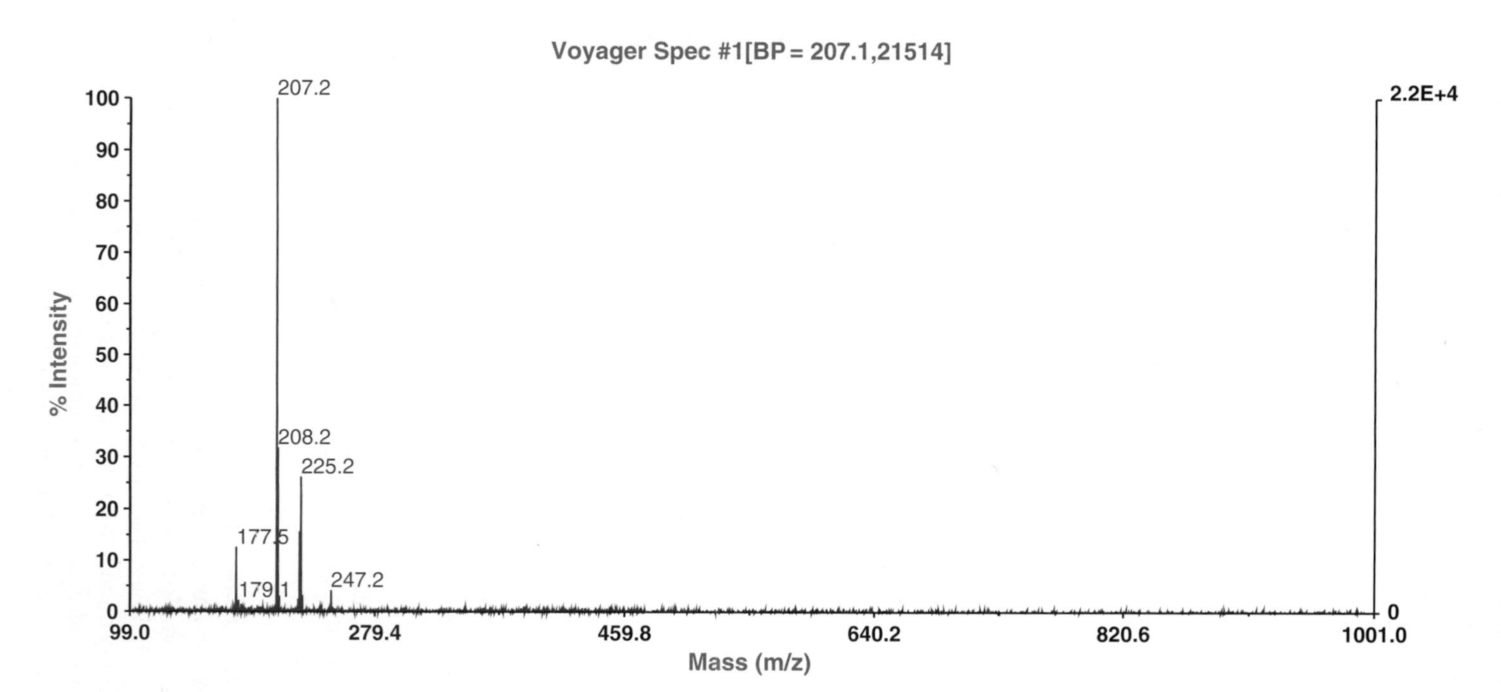

Figure 13.2 F MALDI MS for the matrix α-cyano-4-hydroxycinnamic acid for the mass range of 100-1,000 daltons

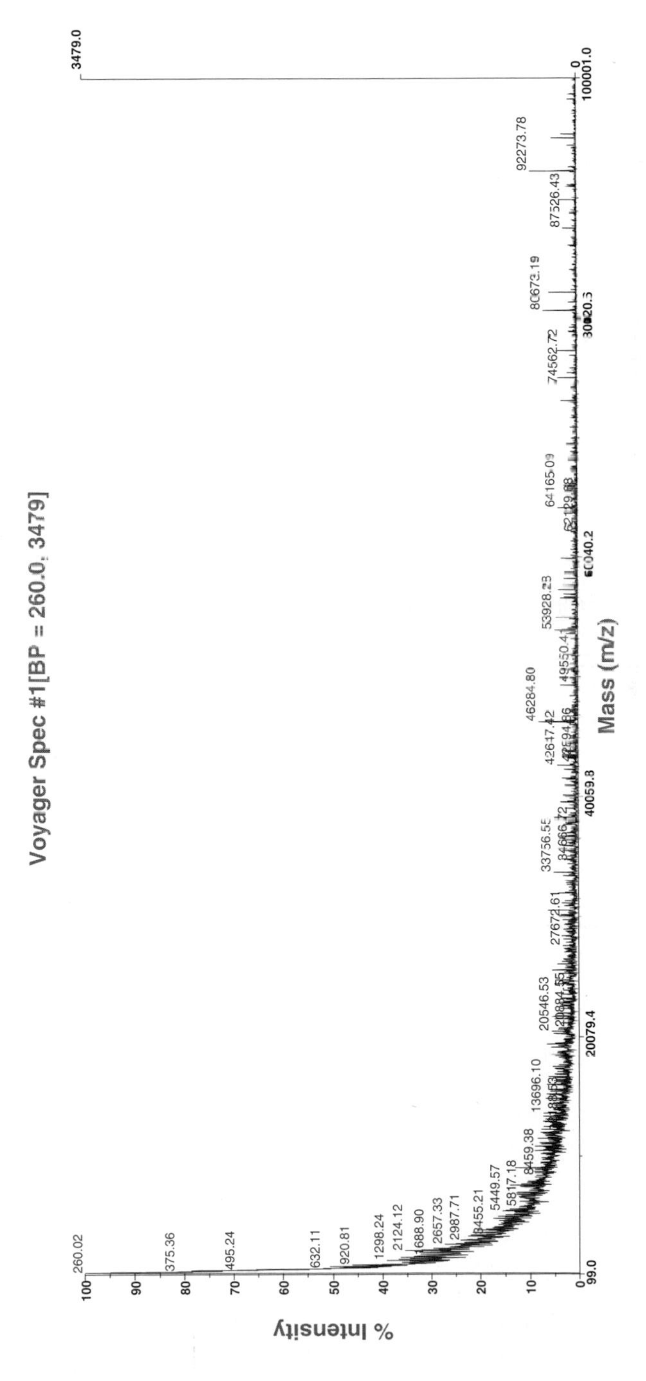

Figure 13.3 F MALDI of the product of 1,1′-dicarboxylferrocene sulfonyl dichloride and acyclovir over the mass range of 100 to 100,000 daltons

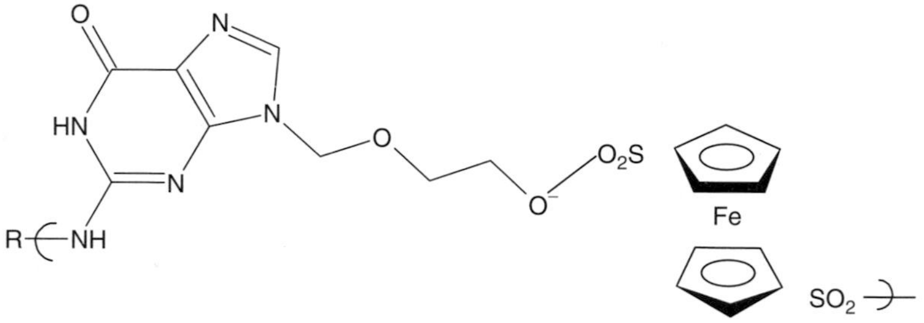

(5)

Table 13.4 Fragments from the product of 1,1′-disulfonylferrocene and acyclovir over the mass range of 100 to 1,000 daltons

m/e	Assignment	m/e	Assignment
113	SO2OCC	133	Pu
165	NHPuC	189	Fe
238	ASO2-OCC	405	U-Pu
421	U-SO2OCC	438	U-SO2OC
544	U	612	USO2O
644	U+SO2OCC	700	U+NPuC
756	U+A	882	U+FeNPuC-SO2
939	2U-Pu		

Table 13.5 Fragments derived from the product of 1,1′-disulfonylferrocene chloride and acyclovir in the range of 5,000 to 100,000 daltons (where Fe is the ferrocene moiety and Pu is the pyrimidine unit)

	(Tentative)		(Tenative)
m/e	Assignment	m/e	Assignment
5,450	17U+NPu	5817	19U-SO_2OCC
8,459	27U+OC	13696	44U-OCC
20,547	66U-OCCO	27673	89U-SO_2OCC
33,757	108U+OCC	42647	137U-SO_2OCCO
46,285	148U+SO_2O	49550	159U-SO_2O
53,928	173U-OCCOC	64165	206U-Pu
74,563	239U-OCC	80673	259U-Fe
87,526	281-Fe	92274	296U-SO_2OCCO

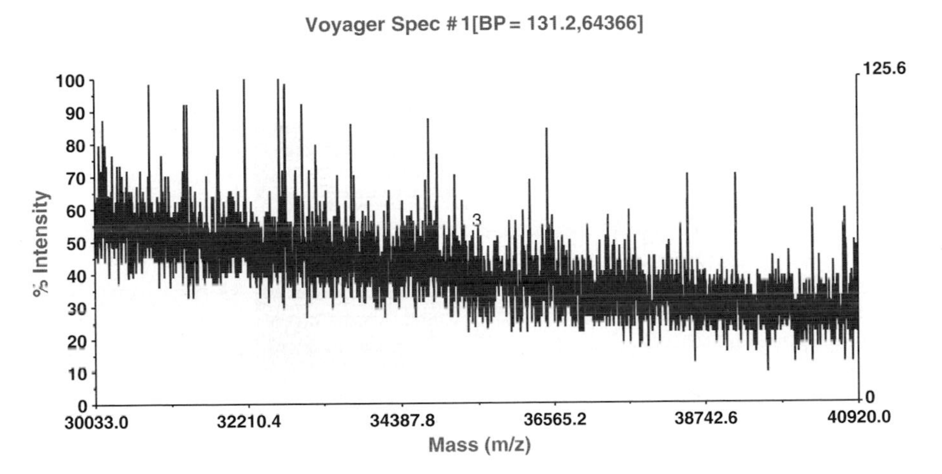

Figure 13.4 Expanded mass range of 3,000 to 4,000 daltons for the matrix derived from Figure 13.1

4 Further Considerations

In the examples reported above, analysis of higher mass fragments were relatively straight forward because the only bond scission found for the most abundant fragments were those occurring within the polymer backbone required to create the fragment. Additional bond breakage occurred in some instances, making such assignments more difficult, but still possible. For the product of dibutyltin dichloride and ticarcillin loss of units away from the polymer backbone occurs. The most abundant fragment clusters and assignments for fragments appear in Table 13.9. In the presentation of data several abbreviations are used in the assignment of fragment identities. The letter "T" is used to represent the ticarcillin moiety (minus two protons); Sn represents the organotin moiety, that is R_2Sn moiety; U represents one unit, 2U represents two units, and so on; AB represents the azabicyclo ring system (**7**), and SP is used to represent the thiophene moiety.

The bonds most broken are indicated below by arrows (**8**). Nonbackbone breakage is minimal with only the thiophene group lost on a regular basis.

Of interest is the lack of significant loss of the alkyl or aryl groups on tin. This is in contrast with other HR EI MS [5–12] and is consistent with the overall more "gentle" nature of the particular instrument used in this study. In comparison, we found that loss of organotin alkyl groups does occur with the low-range instruments employed by us (Table 13.1).

Another difference we found between the two instruments was the loss of the cyclopentadiene groups for the lower range instrument that did not occur with the high range instrument. Thus, results from the product of titanocene dichloride and norfloxacin (**9**) are given in Table 13.9 derived from the low-range instrument. Ion fragments consistent with up to 5 units were found. The most common breakage

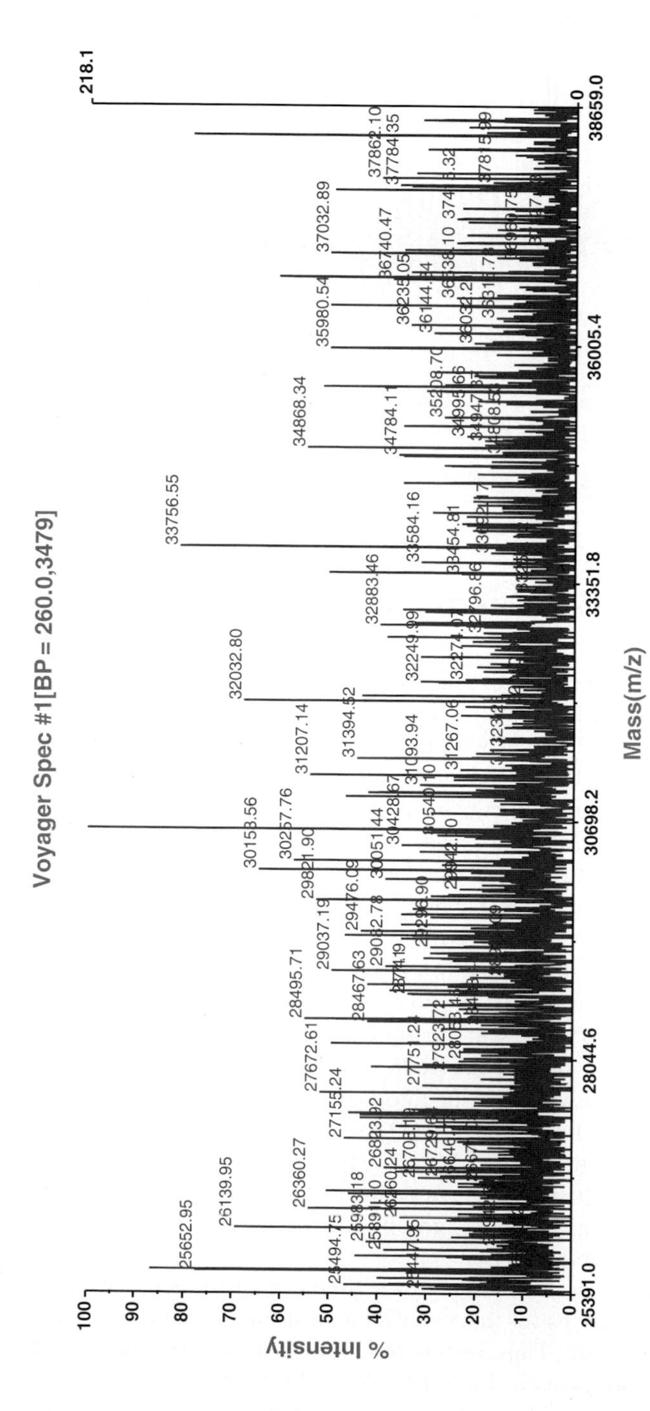

Figure 13.5 Expanded mass range for the product of 1,1′-disulfonylferrocene and acyclovir

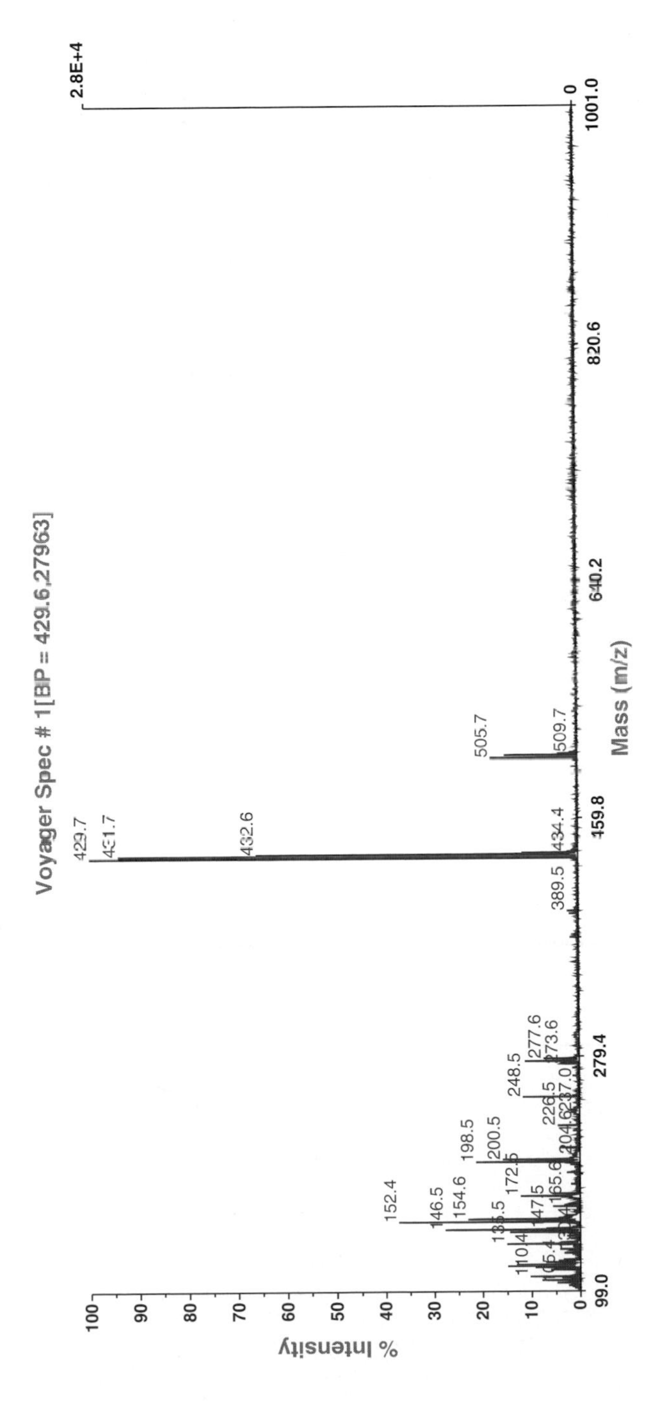

Figure 13.6 Long range F MALDI MS for the product of triphenylantimony dichloride and acyclovir

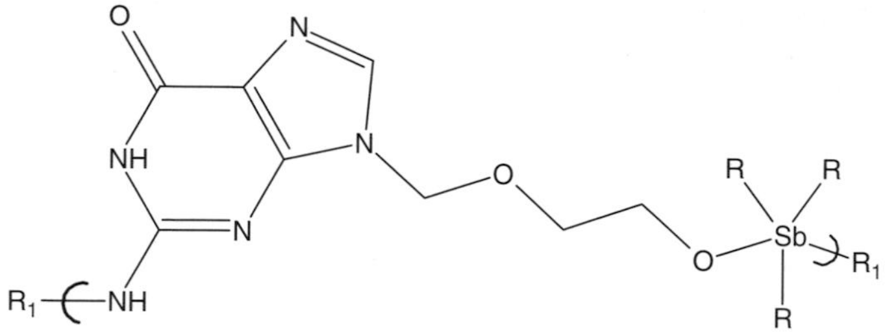

(6)

Table 13.6 Fragments derived from the product of triphenylantimony dichloride and acyclovir employing α-cyano-4-hydroxycinnamic acid as the matrix

m/e	Assignment	m/e	Assignment
136	Pu	152,154	SbOC
198,200	SbPh	248	Ac,Na
278,280	$SbPh_2$		
430,432	$Ph_3SbOCCOC$		
507,509	U–OCCOC		

Table 13.7 Isotope abundance ratios for various single antimony-containing fragments from Figure 13.6

m/e	121	123
Natural Abundance,%	57	43
SbPh	198	200
Found	57	43
$SbPh_2$	278	280
Found	57	43
Ph_3 SbOCCOC	430	432
Found	52	48
U-OCCOC	507	509
Found	55	45

Table 13.8 Isotope abundance ratio for fragments containing multiple antimony atoms for the product of triphenylantimony dichloride and acyclovir

Two antimony-containing fragments					
Calculated percentage	33	49	18		
2U-COCCO; m/e	1,078	1,080	1,082		
Found	36	46	18		
2U-NPuCO; m/e	1,323	1,325	1,327		
Found	35	48	17		
Three antimony-containing fragments					
Calculated percentage	19	42	31	8	
3U; m/e	1,718	1,720	1,722	1,724	
Found	15	47	30	8	
Four antimony-containing fragments					
Calculated percentage	11	32	36	18	3
4U-NPuOC	2,121	2,123	2,125	2,127	2,129
Found	11	31	38	16	4

Table 13.9 Most abundant ion fragment clusters for the product from dibutyltin dichloride and ticarcillin; >500 daltons

m/e	Structural Assignment	m/e	Structural Assignment
3,708	6U	1,760	3U-SPCH
3,625	6U-SP	1,618	3U-Sn
3,580	6U-SPCHCO	1,582	2U+T-CO
3,472	6U-Sn	1,555	2U+T-$CHCO_2$
3,256	6U-Sn,SP	1,407	2U+O_2CCHSPCO
3,189	5U+O_2CCHCONH	1,264	2U+O
3,133	5U+CO_2	1,236	2U
3,090	5U	1,153	2U-SP
2,968	5U-SPCHCO	1,012	U+Sn,O_2CCHSPCO
2,812	5U-Sn,CO_2	993	2U+$SPCHCO_2$-T
2,708	5U-T	883	U+Sn,CO
2,611	4U+$SPCHCO_2$	866	U+Sn,O
2,431	4U-CO_2	851	2U-T
2,240	4U-Sn	809	U+O_2CCHSPCONH
2,090	4U-T	704	U+O_2CCHCO
2,008	3U+O_2CCHSPCO	665	U+CO_2
1,854	3U	633	U+O
1,838	3U-O	529	U-SP
1,810	3U-CO_2		
1,771	3U-SP		

Table 13.9 Abundant ion fragments clusters (>600 Daltons) derived from the MALDI mass spectrometry of the product of titanocene dichloride and norfloxacin where Ti = Cp_2Ti, pip = piperazinyl, and Nor = norfloxacin moiety

m/e	(Tentative) Assignment	m/e	(Tenative) Assignment
642	One unit-Ti-Cl minus Cp	686	One unit-Ti-OH
707	One unit-Ti-Cl	861	Twounits-OHminusCp,pip
884	Two units-OH minus 2Cp	957	Two units-Cl minus Cp
1,004	Two units-OH	1,028	Two units-Cl
1,047	2Units-Ti-OH minus 2Cp	1,205	HO-Two units-Ti-OH
1,226	Two units-Nor minus pip	1,369	Threeunits-OH minus 2Cp
1,381	3Units-Ti minus pip,3Cp	1,403	Three units minus pip
1,547	Three units-Ti-OH minus 2Cp	1,614	3Units-Ti-pip minus 2Cp
1,690	Three units-Ti-OH	1,779	4Units minus 3Cp
1,832	4Units minus pip,Cp	2,051	4Units-Ti-pip minus 3Cp
2,250	Nor-4Units-Ti-Cl minus 4Cp	2,480	5U-Cp_2Ti
2,560	5U		

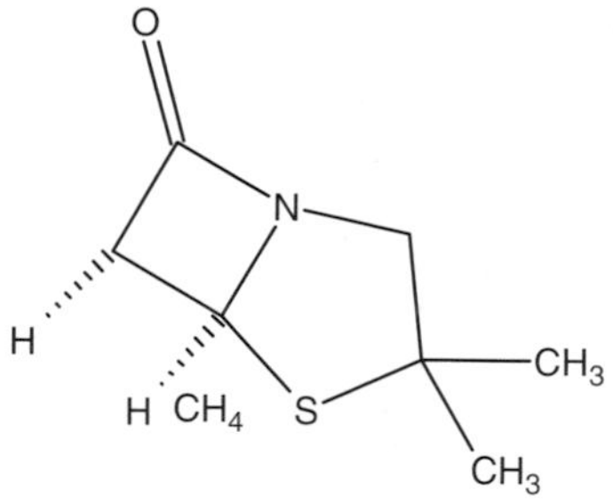

(7)

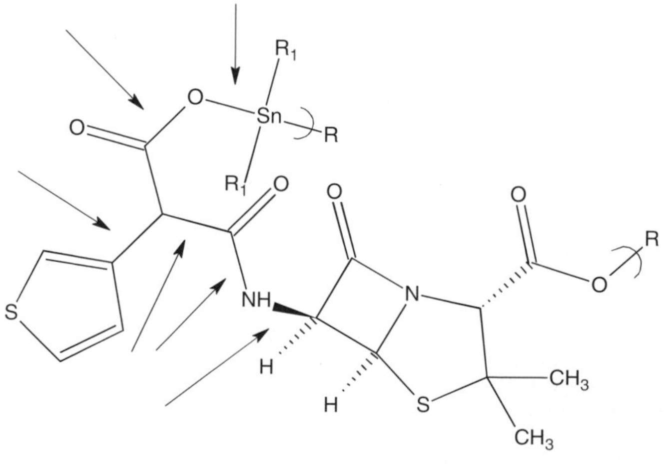

(8)

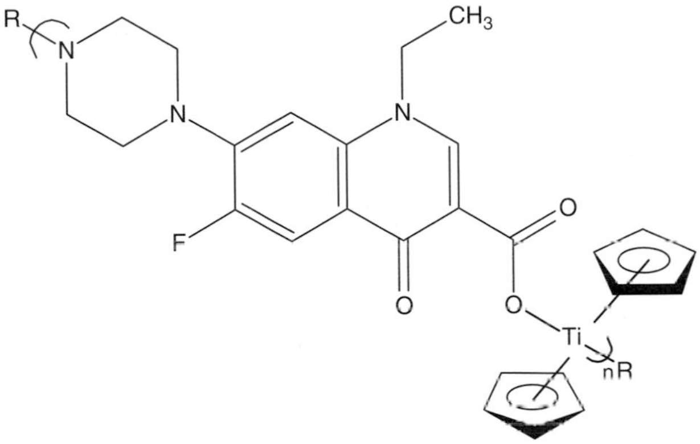

(9)

points were those expected, the Ti–O, Ti–pip, and breakage between the pip and the larger ring system of the norfloxacin. Furthermore, as with the HREI results for the Group IVB metallocene products, loss of Cp groups was common. This loss of Cp groups was a reoccurring theme for the present study and for studies involving the mass spectral analysis of other Group IVB products.

In comparison, F MALDI data obtained from the high-range instrument is shown in Table 13.10 for the product of titanocene dichloride and acyclovir (**1**). No loss of the cyclopentadiene groups occurs.

Thus, attention should be paid to the capabilities and functioning of the particular MALDI MS employed in making assignments.

Even with the high-mass ion fragments, the ion fragments were actually clumps or groups of ions reflecting the isotopic range for the particular chains. Only the ion fragment with the highest abundance was reported.

Thus, F MALDI MS is capable of giving end groups for low- to moderate-length chains, and for longer chains it is able to give fragments that can be assigned up to the 300 DP range. It also allows a determination of sites of least stability towards the employed laser. It is a potentially important tool for polymeric materials that do not lend themselves to the requirements to do classical MALDI MS.

Table 13.10 Ion fragments and assignments for the product of titanocene dichloride and acyclovir in the mass range of 2,000 to 100,000 amu (daltons)

m/e	(Proposed) Assignment	m/e	(Proposed) Assignment
2,476	6U+OCCOC	2,559	6U+Ac-OCCOC
2,615	6U+Ac-O	2,666	6U+Ti,OCCOC
2,738	7U-OCCOC	2,827	7U+O
2,940	7U+Ac-OCCOC,O	3,010	7U+Ac-O
3,052	7U+Ti,OCCO	3,104	7U+Ti,OCCOC,OC
3,197	8U-O	3,277	8U+OCCO
3,353	8U+Ac-OCCOC	3,447	8U+Ti,OCCO
3,516	8U+Ti,OCCOC,OCCO	3,654	9 U+OCC
3,836	9U+Ac	4,066	10U+OCCO
4,207	10U+Ac-OC	4,321	10U+Ti,OCCOC,OCCO
4,333	10U-OCCOC	4,388	11U-OC
4,414	11U	4,494	11U+OCCOC
5,273	13U+OCCO	5,347	13U+Ac-OCCOC,O
5,491	13U+Ti,OCCOC,O	5,598	14U-O
5,675	14U+OCCO	5,786	14U+Ac-OCCO
6,001	15U-O	6,020	15U
6,175	15U+Ac-OCCOC,OC	6,381	16U-OCC
6,587	16U+Ac-OCCO	6,732	16U+Ti,OCCOC,OCCO
6,808	17U-O	7,252	18U+OC
7,384	18U+Ac-OCCO	7,468	18U+Ti,OCCOC
7,836	19U+Ti,2O	7,975	20U-OCCO
8,009	20U-O	8,130	20U+Ac-OCCOC,OCC
8,212	20U+Ac-OC	8,525	21U+Ac-OCCOC,OCC
8,691	21U+Ti,OCCOC,O	9,186	23U-OCC
9,578	24U-OCC	9,633	24U
9,780	24U+Ac-OCCOC	9,804	24U+Ac-OCC
9,912	24U+Ti,OCCOC,OC	9,946	24U+Ti,OCCOC,OCCO

(continued)

Borazine Based Preceramic Polymers for Advanced BN Materials

Samuel Bernard, David Cornu, Sylvain Duperrier, Bérangère Toury, and Philippe Miele

1 Background

To properly explain the subject area of this chapter, it is important to specify that boron containing polymers can be classified into two different categories depending on the expected final uses or applications:

1. Functional polymers in which boron-containing groups permit to tailor the physical and/or chemical properties of the materials (e.g., flame retardants, neutron capture/sensor systems, reversible hydrogen storage, optoelectronic or biomedical applications)
2. Non-oxide preceramic polymers which are expected to yield, under convenient thermal and chemical conditions, boron-containing amorphous or crystallized ceramics including boron nitride (BN), boron carbide (B_4C), boron carbonitride (B–C–N), and boron silicon carbonitride Si–B–C–N.

The purpose of this chapter is to provide an overview of the chemistry, processing and application of boron-containing preceramic polymers in the BN system. The non-oxide precursor route, also called the Polymer Derived Ceramics (PDCs) route, represents a chemical approach based on the use of air- and/or moisture-sensitive (molecular or polymeric) precursors by means of standard Schlenk techniques and vacuum/argon lines. This precursor route allows the chemistry (e.g., elemental composition, compositional homogeneity and atomic architecture) of molecular precursors to be controlled and tailored in order to provide the ensuing preceramic polymers

A.S. Abd-El-Aziz et al. (eds.), *Inorganic and Organometallic Macromolecules: Design and Applications.*

with desirable composition, structure and thermal properties [1–2]. The preceramic polymers receive increased attention to prepare advanced nonoxide ceramics with a tailored chemical composition and a closely defined nanostructural organization by convenient thermal/chemical treatment (curing and thermolysis processes) under controlled atmospheres [3–7]. The main advantage of the PDCs route lies in the possibility to design preceramic polymers whose rheological properties allow the fabrication of near-net-shapes in a way not known from other techniques.

This chapter will be divided into three parts corresponding to the 3 main class of boron-containing preceramic polymers which are used for BN preparation: (1) polyborazylene and other polymers derived from borazine $H_3B_3N_3H_3$, (2) poly[*B*-aminoborazines] obtained from *B*-chloroborazines, and (3) polyborylborazines, a new class of BN polymer exhibiting increased rings-spacing. The potentialities of some preceramic polymers to generate non-oxide ceramics through the PDCs route will be illustrated within this chapter through some examples of realizations including BN powders, coatings, matrices and fibers. In a final part, some outlooks will be drawn and we will try to put into the light further scientific challenges in the field.

2 Boron-Containing Polymers

The review is in part based on our own work related to the poly(alkylaminoborazines)- and poly(borylborazines)-derived boron nitride fibers, but will also cover a comprehensive state of the art including the published literature in this field.

2.1 Boron-Containing Polymers Derived from Borazine

Polymers included in this category are generally called polyborazylenes. Synthetic route to these polymers have been reviewed [8]; thus, we provide an overview of typical synthesis procedures starting from borazine derivatives and leading to BN preceramic polymers.

Borazine (Figure 14.1), which was originally synthesized in 1926 by Stock et al. [9], finds considerable interest as molecular candidate for the preparation of boron nitride, beause it offers the advantages of a source of boron and nitrogen elements with the correct atomic ratio and geometry to yield polymers (i.e., polyborazylene, then boron nitride in a high ceramic yield) [8].

Its synthesis was recently reported by Wideman and Sneddon through three one-step procedures using various starting compounds including 2,4,6-trichloroborazine, metal borohydrides and ammonia-borane [10].

Because of the difficulties encountered for handling borazine due to its high reactivity and volatility at room temperature, very few efforts were devoted to the preparation of borazine-based preceramic polymers. Sneddon et al. have demonstrated that polymers derived from borazine, namely polyborazylenes, could be

Figure 14.1 Structure of the borazine

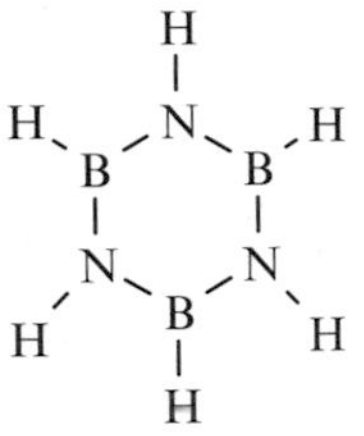

simply produced by heating the liquid borazine in vacuum at moderate temperature (T = 70 °C; 48 hours) through thermally-induced dehydrocoupling reactions of N–H and B–H units (Eq. 14.1) [11–13].

Vacuum/70° C (14.1)

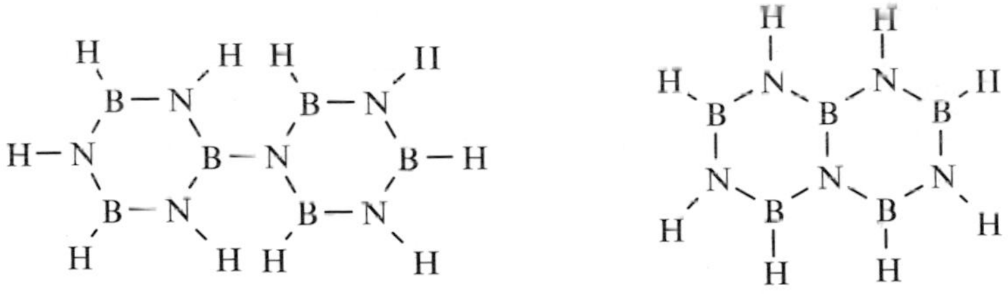

Figure 14.2 Biphenyl and naphtalen-type units formed during thermolysis of borazine

The empirical formula of the resulting thermoset polymer ($B_{3.0}N_{3.0}H_{3.0}$) suggests the formation of a complex structure made of linear and branched-chain by extra-ring B–N bonds (biphenyl-type unit) and fused-cyclic segments (naphtalenic-type unit) which can be related to those of polyphenylenes (Figure 14.2).

A more detailed structural study of the polyborazylene by ^{11}B and ^{15}N solid-state NMR allowed the identification of BHN_2 and BN_3 environments as well as NHB_2 and NB_3 sites suggesting that thermolysis occurs in part through ring-opening pathways [14]. This mechanism seems to lead to eight-membered ring structure (Figure 14.3) in addition to the expected six-membered borazine rings.

Typical polymer samples, which display an average molecular weight (M_w) of 7600 g mol^{-1}, are solids soluble in polar solvents providing an interesting source for coating applications in solution. In contrast, access to complex shapes of ceramics through melt-processing is limited as a result of the high chain entanglement of the

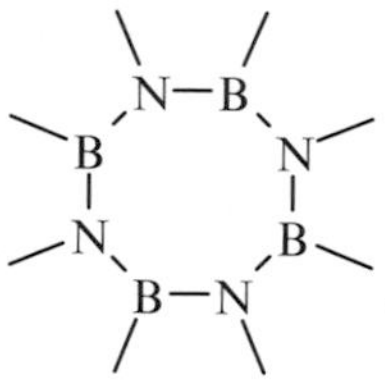

Figure 14.3 Basal eight-membered ring structure proposed for the polyborazylene

polymer and latent reactivity of B–H and N–H units at low temperature resulting in the occurrence of crosslinking reactions prior melting. Nevertheless, in accordance with (i) the absence of carbon, (ii) a structure close to that of *h*-BN, and (iii) the presence of low-weight leaving groups, this polymer provides one of the best routes to prepare *h*-BN powder at low temperature (1200 °C; argon or ammonia atmosphere) in a high-ceramic-yield (85–93%) through hydrogen release.

Economy and Kim [14] have proposed an alternative synthesis route to prepare more tractable polyborazylenes. Their objective was to develop a polymer with a low viscosity that allows the preparation of BN matrices by impregnation technique for carbon fiber-reinforced composite materials. In contrast to the synthesis procedure reported by Sneddon et al. [11–13], they have performed the thermolysis of the borazine in a nitrogen atmosphere at 70 °C for 40h. The as-obtained viscous polymer displays a different chemical composition ($B_{3.0}N_{3.6}H_{3.7}$), and could readily wet ceramic fibers. The green composite was first cured under applied pressure to 400 °C including a weight loss of 5%, and then pyrolyzed between 1200–1500 °C in a nitrogen atmosphere in a 83% ceramic yield. The lower ceramic yield compared with that described by Sneddon et al. [10–13] results from a lower chain branching in the polymer. Authors have showed that heating up to 1200 °C is very important to produce a stabilized BN matrice.

In another paper [15], these authors have prepared the first inorganic mesophase by low-temperature thermolysis of the borazine. Following the synthesis procedure reported by Sneddon et al. [11–13] (Eq. 14.1), the prediction made on the formation of biphenyl and naphtalen-type units and requirements for the formation of pitch mesophases [16], the authors have modified the thermolysis rate to maintain a certain mobility in as-formed molecules, and therefore obtain optically anisotropic phases during thermolysis of borazine. The formation of a liquid-crystalline phase during thermolysis provides an efficient way to produce a final BN material with a high degree of crystalline order by heating to 1800 °C.

Transition-metal-catalyzed reactions of borazine with a series of olefins is an alternative method of turning polyborazylenes into processable materials (Eq. 14.2) [17–20].

Some of these compounds lead to poly(*B*-alkenylborazines) (Eq. 14.2) through thermally-induced polymerizations at moderate temperature, leading to boron nitride-based ceramic in ammonia and nitrogen atmosphere. Depending on the reaction conditions, it is interesting to note that either insoluble or soluble polymers are produced. The latter is particularly valuable for producing either boron nitride-based films or coated ceramics with both high ceramic and chemical yields.

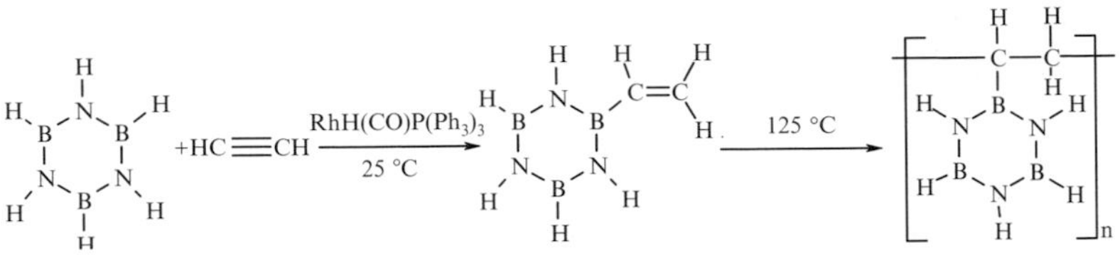

$$+HC\equiv CH \xrightarrow[25\ ^\circ C]{RhH(CO)P(Ph_3)_3} \quad \xrightarrow{125\ ^\circ C} \qquad (14.2)$$

As the preparation of ceramic fibers is one of the most demanding prospects, the preparation of dialkylamine-modified polyborazylenes was an interesting strategy of designing spinnable polyborazylenes [21–22]. This novel method of improving the processability of polyborazylene by its reaction with dialkylamine (Eq. 14.3) at 75 °C in vacuum was described by Sneddon et al. The amine-modified polyborazylene exhibits inter-rings B–N bonds with the pendent amine. These bonds are formed by dehydrocoupling reactions of the amine N–H units and borazine B–H units leading to the polyborazylene without destroying borazine rings.

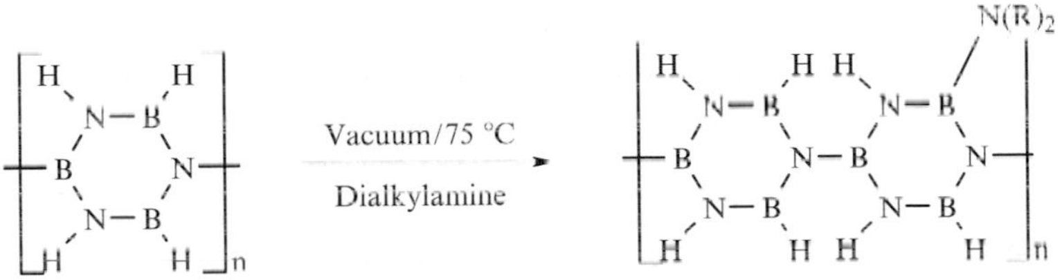

$$\xrightarrow[\text{Dialkylamine}]{\text{Vacuum/75 °C}} \qquad (14.3)$$

$R = C_2H_5, C_5H_{11}$

In contrast to polyborazylene which crosslinks prior melting, the dipentylamine-modified polyborazylenes are soluble in hydrocarbons and most interestingly, become fluid without pronounced crosslinking or decomposition upon heating from 75 to 95 °C. It was postulated that dipentylamino groups lower the glass transition temperature of polyborazylene, because they block the crosslinking progress during thermolysis and act as plasticizers to render polymer more tractable when heating at a sufficient temperature. Green fibers could be generated from the dipentylamine-containing polymer at temperatures ranging from 75 to 95 °C. As-spun fibers could then be cured in air before pyrolysis to 1000 °C in an ammonia atmosphere to obtain boron nitride fibers with 10 µm in diameter in a ceramic yield which is significantly reduced (~64 %) in comparison with that of polyborazylene (85–93%). Resulting boron nitride fibers display poor mechanical properties.

Hence, these works provide information on the feasibility of producing fibers starting from polyborazylenes. It was clearly demonstrated that the high degree of branching associated with the formation of fused-rings and extra-rings B–N bonds in the polyborazylene structure are not adapted to the spinning of such polymers. The addition of pendent groups improves the processability of polyborazylene, because they act as plasticizers to render the polymer more tractable, but the lack of flexible bonds in the maintained basal polyborazylene network did not allow the preparation of high performance fibers.

2.2 *Boron-Containing Polymers Derived from B-Chloroborazine*

A second variety of BN preceramic polymers (i.e., poly[*B*-aminoborazines]) were well described in the literature [8,23]. Synthetic routes to poly(*B*-aminoborazines) involve first nucleophilic reactions of the *B*-chloroborazine with appropriate linking reagents then deamination reactions of the as-obtained *B*-aminoborazine. The resulting polymers are composed of B_3N_3 rings mainly linked through nitrogen-bonded organic groups of the –NR-type. Such polymers can be synthesized starting either from aminoborane or *B*-chloroborazine [8,23–24]. Since the use of aminoborane is widely reported [24], we have provided an outline of typical methods which involve the use of *B*-chloroborazines as a source of poly(*B*-alkylaminoborazines).

Because borazine is a poorly handling compound that is difficult to work with, most of the researchers focused their work on the use of *B*-trichloroborazine (Figure 14.4).

This molecule is likely the ideal molecular precursor for the preparation of processable polymers and the subsequent conversion into *h*-BN complex forms.

Two chemical approaches, which involve nucleophilic reactions of the chlorine atoms attached to boron with linking reagents, are reported in the literature.

In a two-step synthesis procedure, the *B*-trichloroborazine reacts with ammonia or amine derivatives as linking reagents to form *B*-aminoborazines which are subsequently thermolyzed at moderate temperatures to generate poly[*B*-aminoborazine] [25–47]. Pioneering works in this chemistry have described the preparation of *B*-aminoborazine-type polymers through a two-step procedure [31,32,34], but no attention has been given to the materials as BN precursors.

In the 1970s, Tanigushi et al. reported the preparation of boron nitride fibers from a melt-spinnable poly[*B*-aminoborazine] (Eq. 14.4) [25]. However, little information has been provided and the lack of reproducibility in the melt-spinning process rendered the production of BN fibers difficult.

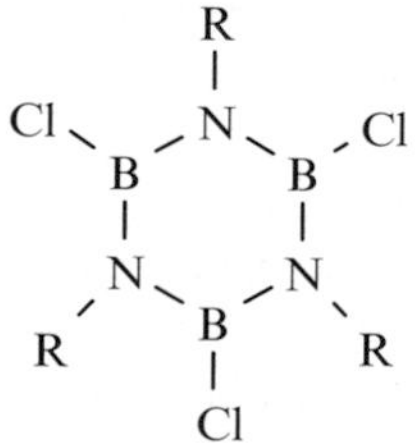

Figure 14.4 Structure of the *B*-trichloroborazine

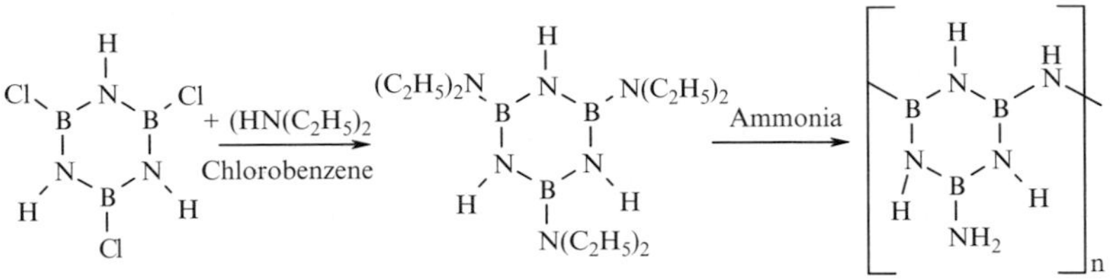

(14.4)

Following a similar approach, starting from *B*-trichloroborazine, Kimura et al. [26] as well as Paine et al. [27] reported the ammonolysis reaction of the as-obtained *B*-trialkylaminoborazine (R = C_2H_5[26], CH_3[27]) followed by its thermolysis to prepare untractable polymers composed of bridged units (–N(H)–) and fused rings. The former are formed by condensation reaction while the latter are formed by borazine ring-opening pathways. Poly[*B*-aminoborazine] is converted into boron nitride powder in 80% ceramic yield through a thermal treatment in a nitrogen atmosphere up to 1800 °C [26].

Using methylamine as linking reagent, Kimura et al. [28] reported the synthesis of poly[*B*-(trimethylamino)borazine] by reaction of methylamine with 2,4,6-trichloroborazine followed by self-condensation of the as obtained 2,4,6-(trimethylamino)borazine above 200 °C in a nitrogen atmosphere (Eq. 14.5). Infusible preceramic BN polymers are produced as a result of an important chain entanglement through $-N(CH_3)-$ bridges.

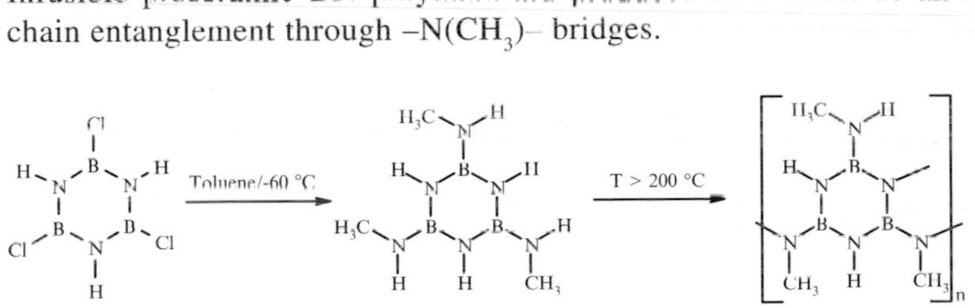

(14.5)

Kimura et al. described a melt-spinnable poly[*B*-(methylamino)borazine] by controlling the crosslinking progress during thermolysis through addition of laurylamine as a blocking agent for the crosslinking progress during thermolysis. Curing in air followed by pyrolysis in ammonia and nitrogen atmospheres generated BN fibers with suitable mechanical properties (Tensile strength, $\sigma = 0.98$ GPa; Young's modulus, $E = 78$ GPa).

Extending this work, we investigated the synthesis of poly(*B*-trimethylaminoborazines) obtained by thermolysis of the *B*-trimethylaminoborazine without blocking agents [29]. A series of melt-spinnable poly[*B*-trimethylaminoborazine] can be generated by self-condensation of the *B*-trimethylaminoborazine by applying different thermolysis temperatures in an argon atmosphere. It could be deduced from ^{15}N solid-state NMR and GC/MS experiments that the structure of a typical melt-spinnable poly[*B*-trimethylaminoborazine] ($[B_{3.0}N_{4.6}C_{2.1}H_{9.5}]_n$, $T_{synthesis} = 180$ °C) is composed of borazine rings, connected via a majority of $-N(CH_3)-$ bridges and direct inter-rings–N bonds forming a crosslinked network with a low proportion of terminal $-N(H)CH_3$ groups [30]. Bridging units were initially suggested by Lappert et al. [31] for the preparation of polymers in which borazinic rings are linked by amino groups –N(R)– (Eq. 14.6), whereas NB_3-containing motifs (i.e., extra-ring B–N bonds) were initially proposed by Gerrard et al. [32] and have been demonstrated by Paciorek et al. (Eq. 14.7) [33].

$$>B-N(R)-H + >B-N(R)-H \longrightarrow >B-N(R)-B< + RNH_2 \quad R = H, \text{alkyl} \tag{14.6}$$

$$>N-H + >B-N(R)-H \longrightarrow >B-N< + RNH_2 \quad R = H, \text{alkyl} \tag{14.7}$$

It has been shown that $-N(CH_3)-$ bridges provide flexibility, whereas $-N(H)CH_3$ end groups act as plasticizers to provide a tractable polymer upon heating at moderate temperature. As a proportion of bridging units and methylamino end groups decrease upon heating to the detriment of inter-rings B–N bonds, melt-spinnability is reduced for thermolysis above 195 °C [30].

Bridging –N(H)– units and/or $-NH_2$ ending units could be considered in a low proportion in such system. Therefore, ring-opening pathway most probably occurs during thermolysis as suggested by Toeniskoetter et al. [34] and confirmed by Paciorek et al. during the self-condensation of *B*-(trianilino)borazine [33]. Paciorek et al. have described a series of concurrent thermolysis mechanisms to obtain a poly[*B*-(trianilino)borazine] in which ring-opening pathways (Eq. 14.8) proceeded along the occurrence of above-described mechanisms (Eqs. 14.6 and 14.7).

(14.8)

Green fibers derived from melt-spinnable polymers have been cured in an ammonia atmosphere, then pyrolyzed in ammonia and nitrogen atmospheres to generate at 1800 °C boron nitride fibers with excellent mechanical properties. It was interesting to note that boron nitride fibers, with controlled mechanical properties and crystallinity, could be produced through simply changing the melt-spinning conditions (Table 14.1) [35].

A large variety of nitrogen-containing linking reagents, some of them basically being derivatives of ammonia, are available. Choosing special amines instead of methylamine provides an opportunity of major importance, because through the functionality of the amine—primary or secondary—the degree of crosslinking, and thus the viscosity and the ceramic yield of the resulting polymer, can be tuned. In addition, such works provide a comprehensive mechanistic study of the thermolysis

Table 14.1 Boron nitride fibers with controlled mechanical properties and crystallinity

BN Fibers	Diameter (μm)	Strength σ (GPa)	Modulus E (GPa)	Strain ε (%)	d_{002} (Å)	L_c (Å)	L_a (Å)	$d_{10/100}$ (Å)
A	7.8	1.5	105	1.30	3.45	31	86	2.13
B	8.3	1.5	130	1.10	3.44	42	110	2.13
C	9.8	1.0	235	0.43	3.37	100	280	2.14
D	10.5	2.0	280	0.55	3.38	109	280	2.15
E	9.3	1.6	315	0.46	3.36	120	310	2.15
F	7.4	1.4	400	0.31	3.35	117	357	2.16

process of various types of *B*-alkylaminoborazines. With this in mind, we developed a route for the synthesis and thermolysis of a series of *B*-trialkylaminoborazines using various nitrogen-based linking reagents [36].

In accordance with Gerrard's observations [32], if dialkylamino groups ($R = CH_3, C_2H_5$) are attached to the boron atoms in 2,4,6-tri(dialkylamino)borazine ($R = CH_3, C_2H_5$), only dialkylamine ($HN(CH_3)_2$, $HN(C_2H_5)_2$) is identified by GC/MS during self condensation. This is exclusively consistent with a mechanism of formation of direct inter-rings B–N bonds (Eq. 14.9) [36].

$$2\ \left[(R_2N)_3B_3N_3H_3\right] \xrightarrow{-\,R_2NH} \text{B–N linked borazine rings bearing } R_2N,\ H \text{ and } N(Et)_2 \text{ groups} \qquad (14.9)$$

$R = CH_3, C_2H_5$

Interestingly, a detailed picture of both the thermolysis mechanisms and the role of bonds which compose poly(*B*-alkylaminoborazines) on the degree of crosslinking and processability could be provided by synthesizing asymmetric *B*-alkylaminoborazine models. We developed a route through the two-step nucleophilic reaction of *B*-trichloroborazine with various amine derivatives yielding molecular compounds **1–3** [37]. Comparison was made with the symmetric *B*-trimethylaminoborazine (**4**) that provided melt-spinnable polymers as described above (Figure 14.5).

The self-condensation of each *B*-alkylaminoborazine in an argon atmosphere allows the establishment of bridging units as well as direct B–N bonds yielding the related poly[*B*-alkylaminoborazine], thus revealing competition between the different reaction mechanisms. The proportion of bridging units has been shown to decrease going from **4**- to **1**-derived polymers.

It has been confirmed that the presence of bridge-type bonds in such poly (*B*-alkylaminoborazines) confers flexibility and an increased melt-spinnability, thus leading to the conclusions that melt-spinnability increases from **1**- to **4**-derived polymers. In addition, the ceramic yield is reduced with the increased proportion of $–N(CH_3)_2$ in the *B*-(alkylamino)borazine leading to important shrinkages during the further polymer-to-ceramic conversion of green fibers derived therefrom. Hence, it is relevant to assume

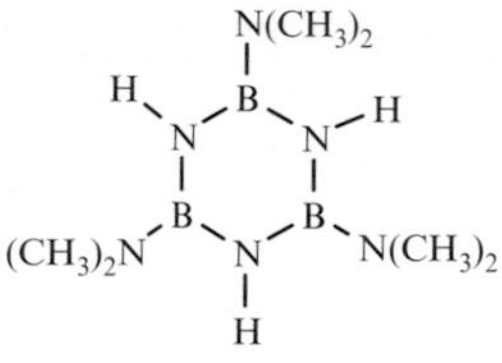

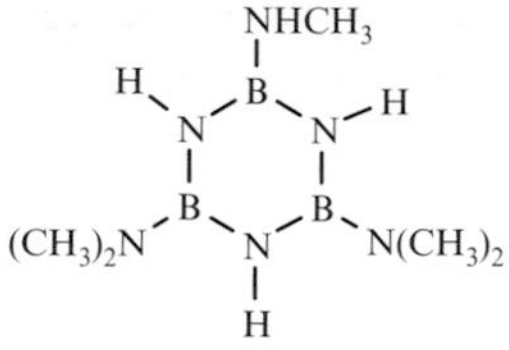

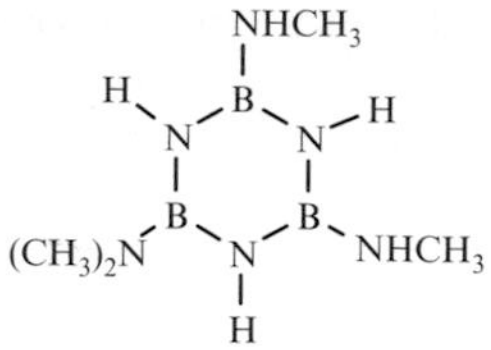

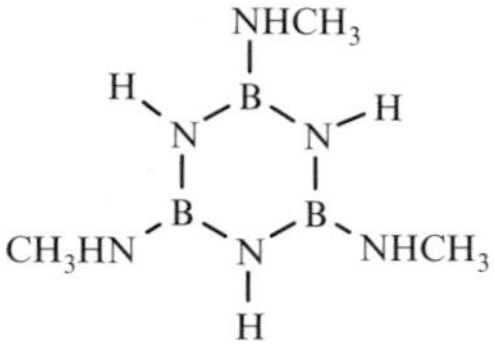

Figure 14.5 Structure of *B*-alkylaminoborazine derivatives

that both morphological variations are observed in the resulting fibers and mechanical properties are reduced in going from **4** to **1**-derived boron nitride fibers (Table 14.2).

The second approach which generates poly(*B*-aminoborazines) is based on the reaction of *B*-chloroborazine derivatives with linking reagents such as silazanes in an one-step synthesis procedure. The earliest successful efforts to prepare poly[*B*-aminoborazine] according to this one-step synthesis procedure are those made by Paine et al. [38] They have synthesized a large variety of polymers by reaction of *B*-chloroborazine derivatives with disilazanes.

Starting from the *B*-trichloroborazine, they have obtained highly 3D cross-linked gels by reaction with hexamethyldisilazane (Eq. 14.10).

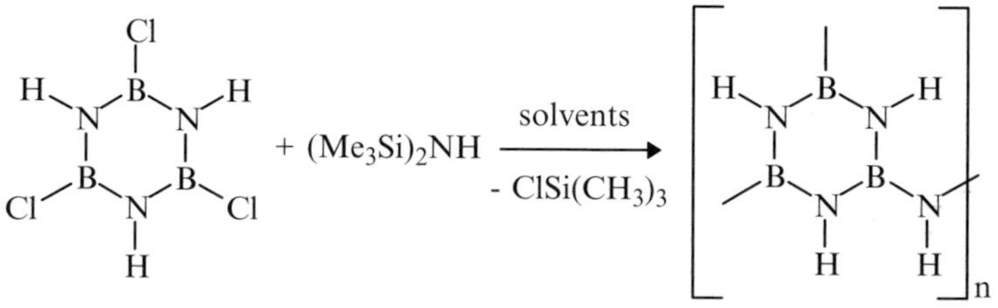

$$+ (Me_3Si)_2NH \xrightarrow[-\ ClSi(CH_3)_3]{\text{solvents}} \qquad (14.10)$$

In order to provide a detailed picture of the thermolysis mechanisms, model reactions with substituted *B*-chloroborazines and heptamethyldisilazane have been made [38–41]. Thermolysis mechanisms through bridging unit (–N(H)–, –N(CH_3)–) formation have been demonstrated. N(H)Si$(CH_3)_3$ ending groups are shown to remain after thermolysis. Pyrolysis of polymer at 900 °C under vacuum, then briefly up to 1200 °C in air results in the formation of *h*-BN. Using an alternative pyrolysis procedure in an ammonia atmosphere to 1200 °C, then 1600 °C in a nitrogen atmosphere, highly crystallized *h*-BN is generated.

Building-up these results through the preparation of highly microporous BN [42] and thin coating [43], they have devoted their study to the synthesis of linear

Table 14.2 Characteristics of typical melt-spinnable **1–4**-derived polyborazines and typical morphology and mechanical properties of BN fibers derived from **1–4**

	1-Derived polymer [37]	**2**-Derived polymer	**3**-Derived polymer	**1**-Derived polymer
Empirical formula	$[B_{3.0}N_{4.6}C_{2.1}H_{9.5}]_n$	$[B_{3.0}N_{4.6}C_{2.8}H_{10.2}]_n$	$[B_{3.0}N_{4.4}C_{2.8}H_{10.6}]_n$	$[B_{3.0}N_{3.9}C_{2.6}H_{9.9}]_n$
Spinning temperature (°C)	173	165	120	150
Ceramic yield (%)	55.5	52.4	51.3	*
Morphology				
Mechanical properties	σ = 1.6 GPa	1.4 GPa	0.5 GPa	*

* Not measured

poly(*B*-aminoborazine) using a blocking group attached to one boron atom in *B*-trichloroborazine (i.e., *N*-dimethyl *B*-dichloroborazine) (Eq. 14.11) [44].

$$\text{B-N(CH}_3)_2\text{-B,B'-dichloroborazine} + (Me_3Si)_2NR' \xrightarrow[-\,ClSi(CH_3)_3]{\text{hexane}} [\text{B-N(CH}_3)_2\text{ borazine}-N(R')-]_n \xrightarrow[\text{diethyl ether}]{H_3B{\bullet}THF} [\text{B-H borazine}-N(R')-]_n \qquad (14.11)$$

R' = H, CH_3

These insoluble gels are interesting candidates for the preparation of BN powders after pyrolysis at 1200 °C in ammonia and nitrogen atmospheres. Oxidation resistance of these ceramics are as high as 900 °C.

A similar approach has been reported by Paciorek et al. [45–46] Poly[*B*-aminoborazines] have been prepared through nucleophilic reactions of the chlorine atoms in the *B*-trichloroborazine (Eq. 14.12).

$$5\ B_3N_3H_3Cl_3 + 11((CH_3)_3Si)_2NH \xrightarrow[-35\ ^\circ C]{\text{Benzene/hexane}} [B_3N_3H_3(NH)]_n + 15ClSi(CH_3)_3 \qquad (14.12)$$

The as-synthesized polymer is soluble in pentane and hexane but does not melt upon heating preventing the preparation of complex shapes by melt-processing such as melt-spinning. However, its solubility offers potential for the preparation of BN matrices and coatings for ceramic fibers through pyrolysis up to 1000 °C in an ammonia atmosphere. A significant improvement of the oxidation resistance of carbon fibers has been observed with BN matrices.

It is further noted that a low-melting polymer could be generated from the reaction of *N*-trimethylsilyl *B*-trichloroborazine with excess of hexamethyldisilazane (Eq. 14.12) when the sample is exposed to liquid ammonia. A foamy white solid is obtained and could be hand drawn to form green fibers, but no information is provided on the mechanical properties of samples [47].

2.3 Polyborylborazine: A New Class of Boron-Containing Polymer Exhibiting Increased Ring-Spacing

In 1984, Wynne and Rice reported what would be an ideal polymeric precursor for fibers preparation [48]. They suggested that 6-membered rings (to avoid reversion reaction) linked through 2 to 5 atom-bridges (to decrease rigidity) represent the best compromise between ceramic yield and viscoelastic properties optimized for melt-shaping process, like melt-spinning operation (Figure 14.6).

Even if the rheology of inorganic polymers is much more complex than one can thought from this view, Wynne and Rice put into the light an exciting scientific challenge: how to improve the length of the bridge (spacer size) between the borazine rings into BN polymeric precursors? An answer was given 16 years ago by Clement and Proux [49], who established that transamination reactions can permit conversion of *B*-tri(dimethyl)-*N*-trimethylborazine into various polymers exhibiting larger ring-spacing compared to polyborazines derived from *B*-trialkylaminoborane or *B*-(trialkylamino)borazine (Eq. 14.13).

$$\xrightarrow[-\ 2nHN(CH_3)_2]{+\ H{-}X{-}H}$$

with X : $-NH-(CH_2)_6-NH-$; $-N(CH_3)-(CH_2)_6-N(CH_3)-$; piperazine ($-N\langle\ \rangle N-$) ; $-NH-C_6H_4-C_6H_4-NH-$

(14.13)

To our knowledge, these derivatives represent the first examples of borazine-derived polymers exhibiting several atom-bridges between the rings. Their main drawback is that they incorporate carbon atoms into their backbones, which drastically reduces their interest as BN precursors. Knowing this, we improved the Wynne and Rice model with an ideal BN polymeric precursor designed for melt-drawing fibers, namely polyborylborazine (Figure 14.7). This polymer structure

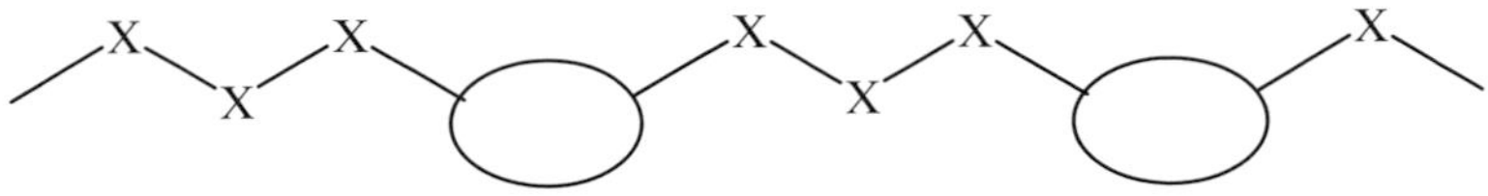

Figure 14.6 Idealyzed structure of a polymer designed for melt-processing

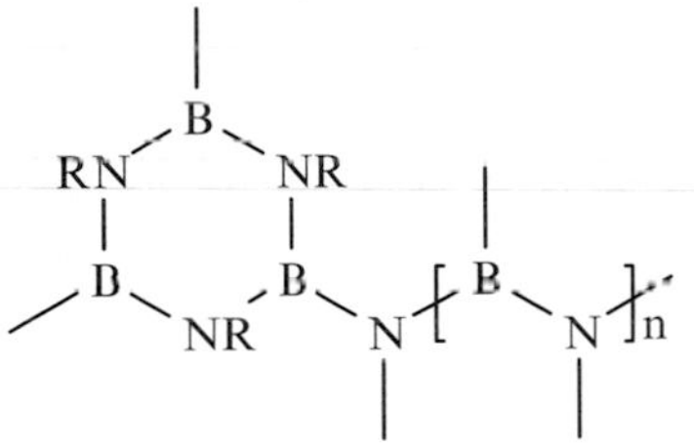

with n = 1,2

Figure 14.7 Model of an ideal BN polymeric precursor

can be considered as ideal since its "flexible" backbone corresponds to the sequence found in *h*-BN ceramics.

Two routes were envisaged for the preparation of such polymer: a "thermal way" consisting in the thermolysis of tailored molecular precursors, and by opposition a "low-temperature chemical way" which can be assimilated to a "metathesis route."

2.3.1 Thermal Route: From Borylborazine to Polyborylborazine

With knowledge of *B*-trialkylaminoborazine and polyborazine derived therefrom, an envisioned route to polyborylborazine was the thermolysis of tailored molecular precursors under convenient atmosphere. For that purpose, we decided to explore the chemistry of a new class of 2,4,6-trialkylaminoborazines in which the alkylamino –N(H)R groups bore by the B_3N_3 rings would incorporate boron atoms. As mentioned above, 2,4,6-trialkylaminoborazines, $(NHR)B_3N_3H_3$, can be easily prepared from the reaction of 2,4,6-trichloroborazine, $Cl_3B_3N_3H_3$, with a molar excess of the corresponding alkylamine, RNH_2. With the structure of Figure 14.6, there are two possibilities: (1) both trialkylaminoborane $(B(NHR)_3)$, and dialkylaminoborylamine $([(RHN)_2B]NH)$ can match the point and play the role of the amine in the reaction pathway. Considering that the chemistry and synthetic accessibility of *B*-trialkylaminoborane are by far much easier than that of the diborylamine derivatives, we focused our attention exclusively on trialkylaminoboranes.

In addition to our work, there are, to our knowledge, only two other examples of borylborazine derivatives: $R_3B_3N_3(BR_2)_3$ in which the nitrogen atoms of the ring bound to dialkylboryl groups [50], and *N*-trimethyl-*B,B'*-dimethyl-*B"*-[dimethylborylmethylamino]borazine [51] (Eq. 14.14). Neither compound is as convenient as the BN molecular precursor because of their structural lack of latent reactive groups, which affords subsequent polymerization or polycondensation during thermolysis.

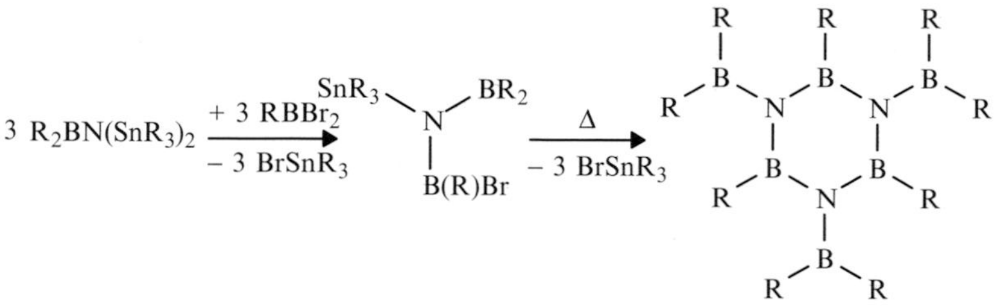

$$3\ R_2BN(SnR_3)_2 \xrightarrow[-\ 3\ BrSnR_3]{+\ 3\ RBBr_2} (R_3Sn)(BR_2)N\text{–}B(R)Br \xrightarrow[-\ 3\ BrSnR_3]{\Delta} [R_2B\text{–}N\text{–}B(R)]_3 \quad (14.14)$$

To begin, we examined the chemistry of borylborazine derivatives in B_3N_3 core surrounded by three aminoboryl groups. We found that this compound can be easily obtained from the reaction of $Cl_3B_3N_3H_3$ with symmetric or asymmetric trialkylaminoboranes, $(R(H)N)B(NRR')_2$ (with R= alkyl, R'=H or alkyl) in a 1 : 3 molar ratio, and in presence of a tertiary amine (e.g., Et_3N) to precipitate the corresponding amine hydrochloride (e.g., $Et_3N.HCl$) [52]. Equation 14.15 presents the general reaction scheme for preparing borylborazine derivatives. This reaction must be driven at low temperature (–10 °C) to avoid the possible polycondensation reactions of aminoborane [53].

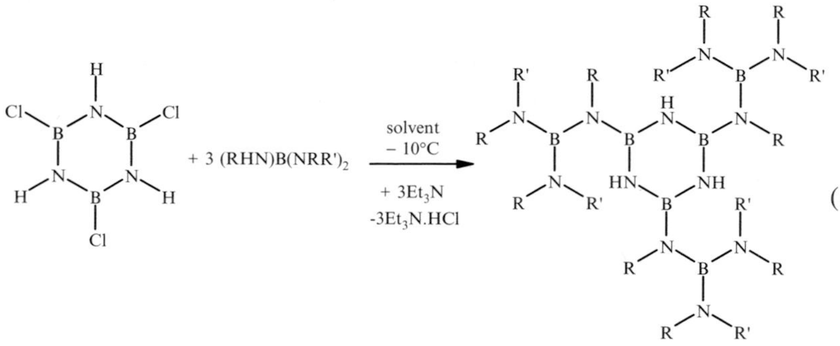

$$Cl_3B_3N_3H_3 + 3\ (RHN)B(NRR')_2 \xrightarrow[+\ 3Et_3N,\ -3Et_3N.HCl]{\text{solvent},\ -\ 10°C} [(NRR')_2B(NR)]_3B_3N_3H_3 \quad (14.15)$$

The borylborazine, which is collected as a viscous liquid or a solid depending on the nature of the alkyl group, can be prepared in quantitative yield as a result of the high reactivity of the B–Cl groups towards primary and secondary amines. Various borylborazines can be prepared by the synthetic pathway depicted above and Figure 14.8 shows the X-ray diffraction crystal structure of $[(NPr^i_2)_2B(NMe)]_3B_3N_3H_3$ obtained from $Cl_3B_3N_3H_3$ and the asymmetric alkylaminoborane $(Pr^i(H)N)B(NPr^i_2)_2$ [54]. It is interesting to note that in the solid state, the B_{ring}–$N_{exocyclic}$ bonds are in the plan defined by the B_3N_3 ring.

The next step is the thermal conversion of borylborazine into polyborylborazine. With regard to the continuous elimination of starting alkylamine as the main by-product during the thermal polycondensation of 2,4,6-trialkylaminoborazine, we expected the continuous elimination of starting *B*-alkylaminoborane during the thermal polycondensation of borylborazine. Therefore, the thermolysis had to be conducted *in vacuo* in order to continuously distillate the evolving $B(NHR)_3$ from the reaction mixture. This procedure elimiates the possibility of any parasitic polycondensation reaction from $B(NHR)_3$.

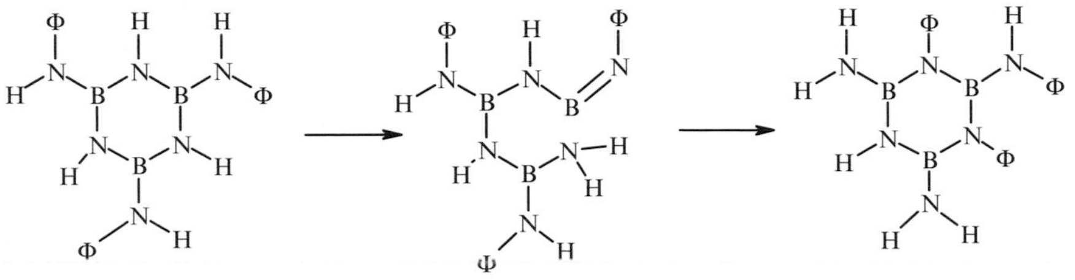

Figure 14.8 ORTEP drawing of $[(NPr^i_2)_2B(NMe)]_3B_3N_3H_3$

As we found, borylborazines behave as 2,4,6-trialkylaminoborazines and two competitive mechanisms occurred to link the rings during the thermolysis *in vacuo*: the formation of direct B–N bonds between two rings, and the formation of three-atom bridges, *–N–B–N–* bonds, between two rings (Eq. 14.16). Similar to 2,4,6-trialkylaminoborazines, the latter linkage mechanism is preponderant according to liquid and solid-state NMR investigations [55].

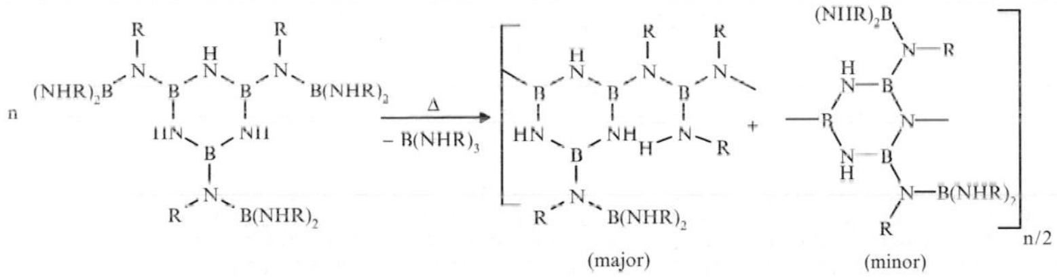

(14.16)

Figure 14.9 presents a typical ^{11}B NMR spectrum of a polyborylborazine obtained by the thermolysis *in vacuo* of 2,4,6-tri[(bismethylaminoboryl)methylamino]borazine, $\{[(NHMe)_2B](Me)N\}_3B_3N_3H_3$ [56].

This spectrum displays four signals attributed to the four kinds of boron site contained in the polymer, namely, from high field to low field, boron atoms from terminal aminoboryl groups, boron atoms from aminoboryl bridges, boron atoms from borazine rings bearing the bridges, and finally, boron atoms from borazine rings bonded directly to another ring.

As an illustration of the utilization of polyborylborazine as BN polymeric precursor, BN matrices incorporating carbon or nicalon micrometric fibers and oxidation-protective coatings on graphite substrates were prepared [57]. Moreover, polyborylborazine displaying adequate melt-viscosity for spinning have been prepared from 2,4,6-$[(NHPr^i)_2B(NPr^i)]_3B_3N_3H_3$ [57]. Indeed, this polymer allowed us to obtain kilometers of continuous BN fibers of 10.5 μm diameter (Figure 14.10) [56].

Despite the low ceramic yield of this polymer, related to the presence of heavy isopropyl groups, these fibers exhibited interesting mechanical performances with a tensile strength of 1.1 GPa and a Young' modulus of 170 GPa [56].

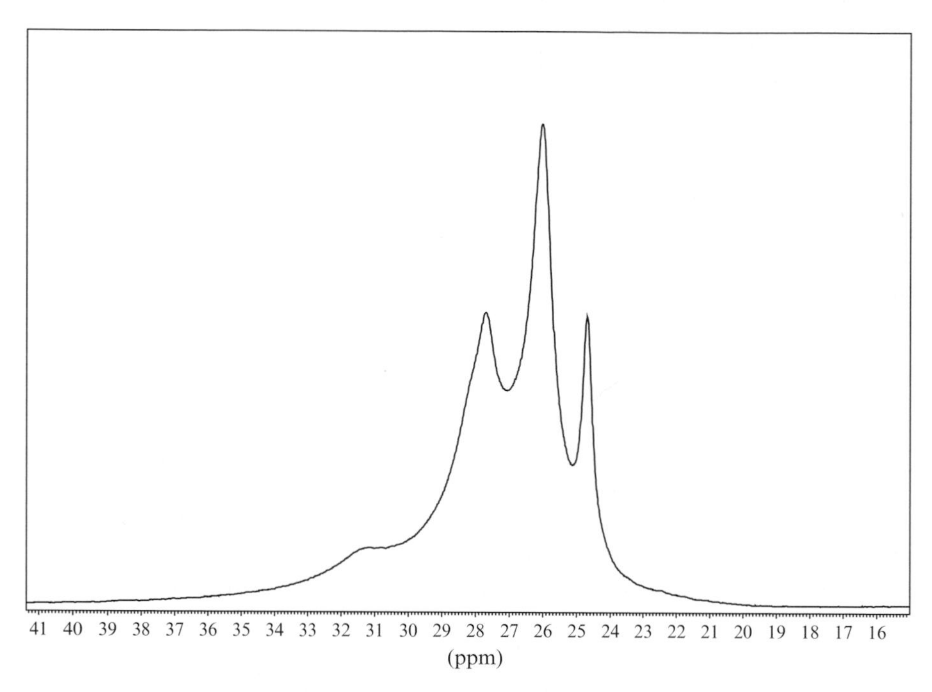

Figure 14.9 Typical ^{11}B NMR spectrum of a polyborylborazine

Figure 14.10 SEM images of BN fibers

On this basis, it was interesting to examine whether the mechanical performances of the ensuing BN fibers could be increased by an improvement in ceramic yield of the polymeric precursor. To this end, 2,4,6-$[(NHMe)_2B(NMe)]_3B_3N_3H_3$ was thermolysed in a vacuum at up to 150 °C [56–57]. It is important to note that, given the high reactivity of the $-B(NHMe)_2$ groups, the starting molecular precursor could be isolated as itself at room temperature and already contained a panel of its oligomers (obtained via polycondensation reaction) (*see* Figure 14.10). The kinetic of polycondensation is by far faster than that of 2,4,6-$[(NHPr^i)_2B(NPr^i)]_3B_3N_3H_3$ as a result of the high reactivity of the $-B(NHMe)_2$ groups. Therefore, from a technical viewpoint, this resulted in difficulty of controlling the polycondensation degree (or T_g value). The problem lies with the preparation of polyborylborazines exhibiting visco-elastic behaviors, which are not favorable for melt-spinning. This may be the result of a more crosslinked structure and/or a higher scattering of the molecular masses within the samples.

2.3.2 Metathesis Route to Polyborylborazine

As previously mentioned, the main drawbacks of the thermal route to polyborylborazine are (i) the presence of both direct intercyclic bonds and three-atom bridges between the rings and (ii) a difficulty to control the polycondensation rate when working with high-ceramic yield but also high-reactive dimethylaminoboryl groups. One solution we investigated was a metathesis route based on the room temperature reaction of *B*-chloroborazine with trialkylaminoborane [58,59]. In order to reduce the crosslinking in the polymer structure and therefore potentially increase its melt-shaping ability, we started with 2-dialkylamino-*4,6*-dichloroborazine instead of *B*-trichloroborazine. The reaction was also conducted in presence of a tertiary amine (e.g., Et_3N) to precipitate the corresponding hydrochloride (e.g., $Et_3N.HCl$). The dialkylamino group bound to the B_3N_3 ring was not reactive toward $B(NHR)_3$ and was retained during the room temperature reaction. Moreover, it can be considered as a latent crosslinking agent because it reacts at high temperatures with ammonia to afford reticulation (e.g., shape-keeping after the melt-processing step). After reaction, the polyborylborazine, which remained soluble during the whole process, was easily obtained through filtration and solvent-elimination steps.

The main advantage of this technique is that polyborylborazine displayed only three-atom bridges, *–N–B–N–* between the rings. As an illustration, a polyborylborazine was prepared at room temperature from $(Me_2N)Cl_2B_3N_3H_3$ and $B(NHMe)_3$ in a 1:1 molar ratio, and in presence of triethylamine (Eq. 14.17).

$$n\,(Me_2N)Cl_2B_3N_3H_3 + (n+1)\,MeHN\text{–}B(NHMe)\text{–}NHMe \xrightarrow[-\,2nEt_3N.HCl]{+\,2nEt_3N} \text{polyborylborazine} \qquad (14.17)$$

The polymer characterization clearly supports the proposed structure. For instance, Figure 14.11 presents a typical ^{15}N CP MAS solid state NMR spectrum

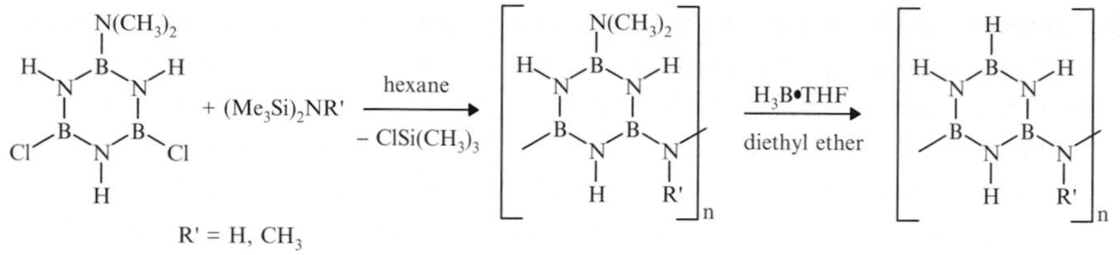

$N(CH_3)_2$-substituted chloroborazine + $(Me_3Si)_2NR'$ $\xrightarrow[-\ ClSi(CH_3)_3]{hexane}$ $[\ldots]_n$ $\xrightarrow[diethyl\ ether]{H_3B \bullet THF}$ $[\ldots]_n$

R' = H, CH_3

Figure 14.11 Typical solid state ^{15}N CP MAS spectrum

Figure 14.12 SEM images of BN fibers derived from polyborylborazines

showing four peaks after fitting [55]. They can be attributed to the four kinds of nitrogen sites found within the polymer, from low to high chemical shift values: B*N*(H)Me, B*N*Me_2, B_2*N*Me and B_2*N*H.

As expected, this polymer was suitable for a spinning process and continuous green fibers were melt-spun [56,58–59]. Figure 14.12 shows SEM pictures of the 10 µm diameter BN fibers obtained after a convenient thermal and chemical treatment up to 1800 °C. These fibers were striated lengthwise without surface defect. The circular cross-section is retained and numerous randomly distributed flakes are visible. The presence of these defects can be related to the removal of the carbon species and the volatilization of low-molecular-weight species upon heating. Despite these defects, the BN fibers show promising mechanical properties with a tensile strength above 1.0 GPa and an average Young's modulus value of 200 GPa.

3 Future Considerations

In several advanced sectors electronic, opto-electronic, optic, transport, and aeronautic and aerospace technology, the synthesis of new materials displaying wide sets of functional or structural properties is a challenge which can be tackled via the chemistry of molecular and polymeric precursors, with numerous advantages. For instance, this chapter has demonstrated that the pyrolysis of preceramic polymers allows for fabrication of ceramic fibers without use of pre-prepared ceramic powders. Several borazine-based polymeric precursors of boron nitride have been described and compared in terms of their ability to generate BN fibers. Our results led to a strategy for obtaining meltable and spinnable polyborazines based on the functionalization of the boron atoms of borazinic derivatives. In our opinion, the borylborazine route, which permits the elaboration of linear polymers with –N–B–N– flexible bridges provided by the utilization of a "two-points" polymerization, is an excellent example of what the mastery of the chemistry of the precursors can provide. It was also shown that the ceramic yield of a precursor, which is classicaly considered the key point in materials science, must be counterbalanced with its chemical structure when the shaping of the ceramic needs to be controlled.

What is clear is that considerable efforts should be devoted to the development and finalizing of tailored molecular and polymeric precursors for preparation of boron nitride fibers displaying mechanical properties required for integration in composite materials as reinforcing agents. Furthermore, the design of new precursors should be developed by linking novel substituents to the borazinic monomers in order to introduce new types of linkages in derived polymers. The development of novel bonding, along with different energies, should play a crucial role in synthesizing linear polymers with a controlled architecture.

Improvements in the control of polymerization reactions is essential and studies should be directed toward measurement of the molecular masses—M_n and M_w—of inorganic polymers through use of suitable Steric Exclusion Chromatrophy techniques. The air- and moisture-sensitivity of these boron-based polymers make this a difficult undertaking, and collaboration will be required throughout the scientific community. With regard to fibers that imply a melt-spinning step, methods determining the rheological parameters of such polymers—such as viscosity as a function of processing temperature—should be investigated. Advances in this area will also require the skills and knowledge of those studying chemistry, materials science, rheology, and modeling.

References

1. Wynne KJ, Rice RW. (1984) Ann Rev Mater Sci. 14:297.
2. Blum YD, Schwartz KB, Laine RM. (1989) J Mater Sci. 24:1707.
3. Chantrell PG, Popper EP. (1964) In Popper (ed), : Special ceramics, Vol. 4. Academic Press, New York, , p 87.
4. Peuckert M, Vaahs T, Brück M. (1990) Adv Mater. 2:398.

5. Bill J, Aldinger F. (1995) Adv Mater. 7:775.
6. Segal D. (1997) J Mater Chem. 7:1297.
7. Greil P. (2000) Adv Engineer Mater. 2:339.
8. Paine RT, Sneddon LG. (1994) Chemtech. 29.
9. Stock A, Pohland E. (1926) Chem Ber. 59:2215.
10. Wideman T, Sneddon LG. (1995) Inorg Chem. 34:1002.
11. Fazen PJ, Remsen EE, Sneddon LG. (1991) Polym Preprints. 32:544.
12. Sneddon LG, Mirabelli MGL, Lynch AT, Fazen PJ, Su K, Beck JS. (1991) Pure Appl Chem. 63:407.
13. Fazen PJ, R\emsen EE, Beck JS, Carroll PJ, McGhie AR, Sneddon LG. (1995) Chem Mater. 7:1942.
14. Gervais C, Maquet J, Babonneau F, Duriez C, Framery E, Vaultier M, Florian P, Massiot D, (2001) Chem Mater. 13:1700.
15. Kim D-P. (1993) J Economy Chem Mater. 5:1216.
16. Kim D-P. (1994) J Economy Chem Mater. 6:395.
17. Lynch AT, Sneddon LG. (1987) J Am Chem Soc. 109:5867.
18. Lynch AT, Sneddon LG. (1989) J Am Chem Soc. 111:6201.
19. Su K, Remsen EE, Thompson HM, Sneddon LG. (1991) Macromolecules, 24:3760.
20. Fazen PJ, Sneddon LG. (1994) Organometallics. 13:2867.
21. Wideman T, Sneddon LG. (1996) Chem Mater. 8:3.
22. Wideman T, Remsen EE, Cortez E, Chlanda VL, Sneddon LG. (1998) Chem Mater. 10: 412.
23. Paine RT, Narula CK. (1990) Chem Rev. 90:73.
24. Maya L. (1996) Appl Organomet Chem. 10:175.
25. Tanigushi L, Harada K, Maeda T. (1976) Chem Abstr. 85:96582v.
26. Kimura Y, Kubo Y, Hayashi N. (1992) J Inorg Organomet Polym. 2:231.
27. Narula CK, Schaeffer R, Datye A, Paine RT. (1989) Inorg Chem. 28:4053.
28. Kimura Y, Kubo Y, Hayashi N. (1994) Comput Sci Technol 51:173.
29. Miele P, Toury B, Bernard S, Cornu D, Ayadi K, Roussean L, Beauhaire G, (2004) US Patent No. 2004044162.
30. Duperrier S, Gervais C, Bernard S, Cornu D, Babonneau F, Balan C, Miele P. (2007) Macromiol 40:1018
31. Lappert MF. (1959) Proc Chem Soc. 59.
32. Gerrard W, Hudson HR, Mooney EF. (1962) J Chem Soc. 113.
33. Paciorek KJL, Harris DH, Kratzer RH. (1986) J Polym Sci. 24:173.
34. Toeniskoetter RH, Hall FR. (1963) Inor Chem. 2:29.
35. Bernard S, Chassagneux F, Berthet MP, Cornu D, Miele P. (2005) J Am Ceram Soc. 88: 1607.
36. Toury B, Miele P, Cornu D, Vincent H, Bouix J. (2002) Adv Funct Mater. 12:228.
37. Toury B, Bernard S, Cornu D, Chassagneux F, Létoffé J-M, Miele P. (2003) J Mater Chem. 13: 274.
38. Narula CK, Schaeffer R, Paine RT. (1987) J Am Chem Soc. 109:5556.
39. Narula CK, Paine RT, Schaeffer R. (1988) Inorganic and Organometallic Polymers. In:Zeldin M, Wynne KJ, Allcock HR, (eds), Macromolecules containing silicon, phosphorus, and other onorganic elements. ACS Symposium series 360.
40. Narula CK, Paine RT, Schaeffer R. (1986) Mater Res Soc Symp Proc. 73:383.
41. Narula CK, Lindquist DA, Fan M-M, Borek T.T, Duesler Z.N., Datye A.K., Schaeffer R, Paine R.T., (1990) Chem Mater. 2:377.
42. Borek TT, Ackerman W, Hua DW, Paine RT, Smith DM. (1991) Langmuir. 7:2844.
43. Rye RR, Borek TT, Lindquist DA, Paine RT. (1990) J Am Ceram Soc. 73:1409.
44. Narula CK, Schaeffer R, Datye AK, Borek TT, Rapko BM, Paine RT. (1990) Chem Mater. 2:384.
45. Paciorek KJL, Masuda SR, Kratzer RH. (1991) Chem Mater. 3:88.
46. Paciorek KJL, Kratzer RH. (1992) Eur J Solid State Inorg Chem. 29:101.
47. Paciorek KJL, Kratzer RH. (1988) Ceram Eng Sci Proc. 9:993.

48. Wynne KJ, Rice RW. (1984) Ann Rev Mater Sci. 14:297.
49. Clement R, Proux Y. (1969) Bull. Soc Chim Fr. 2:558.
50. Noth H, Wrackmeyer B. (1978) Nuclear magnetic resonance spectroscopy of boron coumpounds. Springer Verlag, New-York.
51. Nöth H, Otto P, Storch W. (1985) Chem Ber. 118:3020.
52. Cornu D, Miele P, Bonnetot B, Guenot P, Mongeot H, Bouix J. (1998) Main Group Met Chem. 21:301.
53. Mongeot H, Guilhon F, Miele P, Cornu D, Bonnetot B. (1997) J Solid State Chem. 133:164.
54. Toury B, Cornu D, Lecocq S, Miele P. (2003)Appl Organomet Chem. 17:68.
55. Toury B, Gervais C, Dibandgo P, Cornu D, Miele P, Babonneau F. (2004) Appl Organomet Chem. 18:227.
56. Miele P, Toury B, Cornu D, Bernard S. (2005) J Organomet Chem. 690:2809.
57. Cornu D, Miele P, Toury B, Bonnetot B, Mongeot H, Bouix J, (1999) J Mater Chem. 9:2605.
58. Toury B, Miele P. (2004) J Mater Chem. 14:2609
59. Toury B, Cornu D, Chassagneux F, Miele P. (2005) J Europ Ceram Soc. 25:137.

Recent Advances in High-Temperature Network Polymers of Carboranylenesiloxanes and Silarylene-Siloxanes

Manoj K. Kolel-Veetil and Teddy M. Keller

1 Introduction

Since the discovery of siloxanes by Kipping and coworkers [1], classical polysiloxanes $-[(R)(R')SiO]_n-$, or silicones, have been extensively studied and some of them commercialized as early as the 1940s. Their broad range of unique properties has facilitated their application in such diverse fields as aeronautics, biomedical, cosmetics, surface water-proofing, sealants, and unmolding agents. The most impressive properties of silicones are their very low glass transition temperatures (T_g) resulting from the flexibility within their backbones and their low surface tension. These two properties account for their wide range of applications.

Silicones also exhibit impressive thermal stability, the origins of which can be appreciated from their basic structural characteristic, the siloxane bond –Si–O–. It is a partially ionic bond that also possesses some characteristics of a double bond. The former property result from the relatively large difference in the electronegativities between silicon and oxygen, determined by Pauling [2] as 1.8 and 3.5, respectively. This difference results in an estimated 37 to 51% ionic character of the Si–O bond [2,3]. The latter characteristic, however, is associated with the partial overlap of the vacant low-energy silicon *d* orbitals with the *p* orbitals of oxygen,

A.S. Abd-El-Aziz et al. (eds.), *Inorganic and Organometallic Macromolecules: Design and Applications.*

which occurs because of the relatively large difference in sizes between these two atoms, thus enabling oxygen to back-donate its lone electron pairs and create a $d\pi$-$p\pi$ bond in addition to the normal σ bond [4].

These two fundamental characteristics of the siloxane bond are, in essence, the reasons for its pronounced thermal stability and specific chemical behavior. However, under certain conditions such as in an acid or a base medium, or at high temperature, a silicone backbone consisting of polarizable siloxyl units is susceptible to degradation by ionic reactions. As a result of such reactions, the backbone is observed to depolymerize by chain scission of some of the –Si–O–Si moieties through a six-centered mechanism resulting in the formation of the thermodynamically more stable low-molecular-weight cyclic products and shorter linear chains (Figure 15.1) [5]. This intramolecular cycloreversion or depolymerization of silicones has been observed to occur from as few as four consecutive Si–O bonds [6]. To alleviate this problem of depolymerization, several groups of hybrid silicones were synthesized as early as the 1950s, and were generated through inclusion of large and bulky groups such as alkyl, aryl, alkyl, aryl, carboranyl, and fluoroalkyl in the backbone of silicone precursors [7]. It was theorized that the presence of such large rigid groups would provide pronounced resistance to depolymerization by preventing formation of cyclic siloxane volatile products, a result of making the reaction considerably more difficult.

Among the hybrid silicones that were developed, the most extensively researched were the silarylene-siloxanes and the carboranylenesiloxanes (Figure 15.1) [7]. In each of these silicone groups a percentage of oxygen atoms in the parent backbone are substituted with aromatic or carboranyl groups, respectively. In the case of carboranes, several research groups in the mid-1960s began investigating the effect that incorporation of closodicarbaborane units into chains and networks of almost any known type of polymers would have [8]. It was soon recognized that the carborane units attached as pendant groups to the otherwise organic or inorganic backbone of the polymer produced no practical advantage [9]. However, if the carborane became part of the polymeric chain (e.g., regularly interspersed *among* repeating –Si(R)$_2$O-units) they induced an appreciable thermal stability in the product [10]. In polymeric systems, it was observed that the *meta* and *para* congeners of the closodicarbaborane

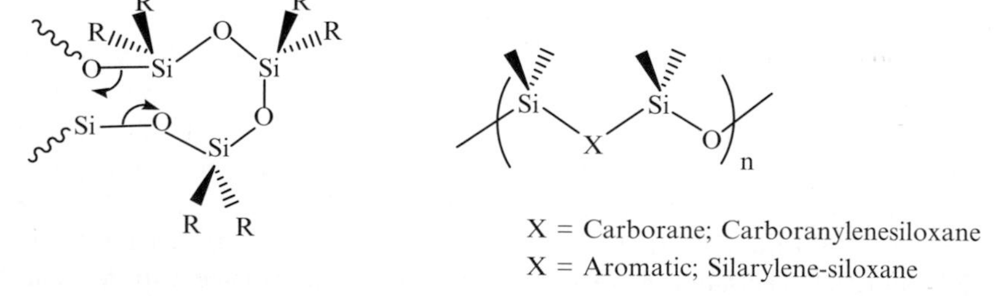

Figure 15.1 Schematic representation of (**a**) a six-centered mechanism for depolymerization of siloxanes and (**b**) hybrid siloxanes

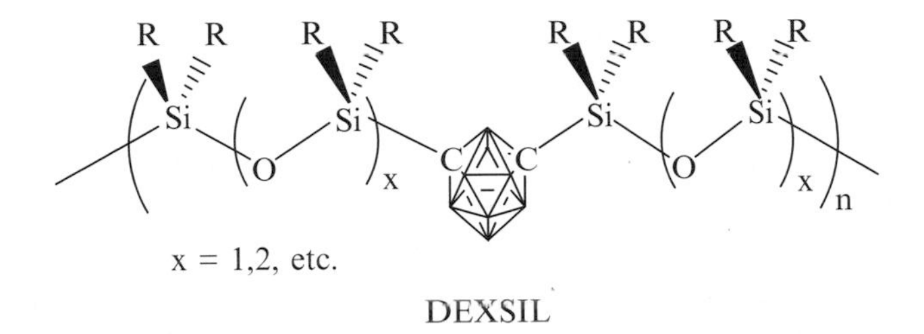

Figure 15.2 DEXSIL polymer

with the formula, $C_2B_{10}H_{12}$, imparted the best properties relative to other carborane molecules. The *ortho* congener was observed to form stable monomeric, or at best dimeric, cyclic species that were of no use in polymer formation [11]. The best group of thermally stable high-molecular-weight hybrid carboranylenesiloxane polymers were obtained from the family of poly-*m*-carboranylenesiloxane, which became renowned under the trade name DEXSIL (Figure 15.2) [10,12].

Around the same time, several groups attempted to incorporate aromatic units into organic and inorganic polymers [13]. As was the case with carboranes, it was observed that there was an appreciable improvement in the thermal stability of silicones only when the aromatic groups were incorporated into the backbone and not as pendant groups on the silicones. The incorporation of aromatic groups into the backbones of silicones was achieved through condensation reactions of disilanol monomers with various reactive silane monomers. Four principal methods were developed: (1) the chlorosilane route [14], (2) the aminosilane route [15], (3) the acetoxysilane route [16], and (4) the ureidosilane route [17], based on the reactive functionality on the siloxane monomer that reacted with the alcohol functionality on the siloxane reactant to yield the hybrid silicone.

The improved performance of these two groups of hybrid silicones was the direct consequence of the carborane or aromatic-imparted strengthening of the Si–O bonds and their described impedance of the depolymerization reaction. Additionally, in the case of the carboranylenesiloxanes, the thermo-oxidative stability of the boron-containing carboranes [18] also contributed to the polymer's improved stability. However, it was soon recognized that forming thermally robust siloxane polymers through inclusion of large and rigid molecules between the flexible siloxane segments resulted in overall reductions in polymer elasticity [7]. Hence, even in its early stages, research and development of hybrid silicones was geared toward preparing structures which took advantage of the high-temperature stabilization while simultaneously maintaining the low temperature properties of polysiloxanes. The generation of networked or vulcanized versions of the developed hybrid silicones was pursued to produce polymers of improved mechanical and thermal properties that also possessed enhanced processablity. The vulcanization of polysiloxanes was traditionally achieved through either an "activated cure," in which various crosslinking agents were utilized together with pressure and heat, or through the "room-temperature vulcanization" (RTV) processes in which no external energy was necessary.

Most of the activated cure methods were known to make use of free radical initiators such as organic peroxides to cause crosslinking reactions. The RTV cures were achieved either by a condensation reaction in which reactive polymer end-groups such as silanol or acetoxysilane units were combined in the presence of a catalyst, or by platinum-catalyzed hydrosilation reactions of pendant groups such as vinyl or allyl [7]. Similar routes were also explored for the vulcanization of carboranylenesiloxanes and silarylene-siloxanes. The majority of the reports on the vulcanization of carboranylenesiloxanes polymers and the properties of their vulcanizates were concerned with the *m*-$CB_{10}H_{10}C$–containing polymers. Several different unsaturated groups were examined as potential reactive sites for the vulcanization of carboranylenesiloxane polymers. The reported groups included organic unsaturations such as vinyl, allyl, and vinyl-*o*-carboranyl units [19]. It was discovered that the best combination of properties was obtained, particularly before heat aging, from the vulcanization of the vinyl-*o*-carboranyl groups. However, this approach was later abandoned because of the difficulties in preparing such compounds. Vulcanization of carboranylenesiloxanes was also attempted through reactions with organic peroxides at high temperatures and pressures. One example is the production of a network polymer through polymerization of the vinyl groups of a pendant vinyl-containing carboranylenesiloxane by organic peroxides in air at 315 °C for 300 hours [20]. This polymerization could also be accelerated under pressure, resulting in completion in about 2 hours. Room temperature vulcanization of *m*-$CB_{10}H_{10}C$ carboranylenesiloxane elastomers has also been attempted and reported based on either: (1) platinum-catalyzed silylhydride-vinylsilane reaction or (2) the reaction of tris(acetoxy)silane with carboranylsilanol terminated prepolymers. However, neither of these methods was developed to the level of practical use [21].

The vulcanization of silphenylene-siloxane polymers was reported at room temperature either in solution or in bulk with silicate esters [22]. For crosslinking in solution, toluene was used as solvent, and the additives used for curing consisted of 20 parts of partially hydrolyzed ethyl silicate and 2.5 parts of dibutyltin diacetate to 100 parts of the polymer. In another report, the crosslinking of silarylene-siloxane polymers, which contained *p*-phenylene and *p,p′*-diphenyl ether units in their silarylene groups, and various amounts of vinylsiloxane groups with pendant vinyl groups were performed using dicumyl peroxide as the crosslinking reagent because of its low reactivity to methyl side groups and its high reactivity toward vinyl groups [23]. These polymers were designed to permit controlled vulcanization for the preparation of elastomers with specific crosslinking densities so as to enable an investigation of their mechanical properties as functions of different crosslink densities. A report containing two vulcanization reactions of vinyl-containing silarylene-siloxane polymers using in one case 2,5-dimethyl-2,5-di (*t*-butyl-peroxy)hexane and in the other dicumyl peroxide as the crosslinking agent is also available [24]. In these vulcanizations, compounded gum-stocks were first milled and then vulcanized under unspecified pressure for 20 minutes at 170 °C, followed by another cycle of 16 hours at 232 °C [24].

Recently, however, research into hybrid silicones of silarylene-siloxanes and carboranylenesiloxanes has been focused on development of hybrids with crosslinkable groups in the backbone of the polymer. Keller and coworkers [25–36] modified the chemistry of the carboranylenesiloxane system to incorporate unsaturation within the backbone, or at the terminal sites, as a source for the polymerization to a thermoset or network. The developed precursors contained either crosslinkable diacetylene groups (in oligomers) [25] or RTV organic groups such as vinyl or ethynyl group (in monomers) which enable their conversion into extended network polymers [30]. The network polymer in each case was generated by the thermal polymerization of the diacetylene group or by the hydrosilation reaction of vinyl or ethynyl units, respectively. The initial examples of the thermally- and hydrosilatively-formed carboranylenesiloxane and silarylene-siloxane network polymers were observed to possess plastic characteristics [25,30]. More recent modifications in the synthetic strategies for production of such materials have resulted in the development of elastomeric versions of network polymers for both of the systems. This chapter briefly summarize these developments.

2 Research and Discussion

2.1 *Theory*

2.1.1 Solid-State Polymerization of Diacetylene Groups

Even though the polymerization of diacetylenes in the hybrid silicones described in this review per se is not conducted in their solid states, an understanding of the solid-state polymerization of diacetylenes is deemed useful for a greater appreciation of the crosslinking processes occurring in these hybrid silicones. The diacetylene solid-state polymerization has been studied extensively by Wegner [37]. The polymerization is known to produce polydiacetylenes with an extensive π-conjugation in the backbone [38]. The available modes of polymerization for diacetylenes are the 1,2-addition, 1,4-addition, and cyclic trimerization reactions [39]. From his studies in the 1970s, Wegner concluded that the most prevalent mechanism in the diacetylene solid-state polymerization is the 1,4-addition mechanism (Figure 15.3). The polymerization is usually carried out by subjecting the monomer crystals to thermal annealing, ultraviolet (UV) radiation, or by high-energy irradiation [37,38,40]. A detailed description of diacetylene solid-state polymerization theory is beyond the scope of this chapter. In general, the theory includes a least-motion criterion which predicts the solid-state reactivity of different diacetylene phases and the most likely polymerization mode for a particular phase from a two-parameter description of the relationship between neighboring monomer molecules [41]. The theory also considers the phase stability criteria for characteristic free energy diagrams that explain the phase behaviors that have been observed

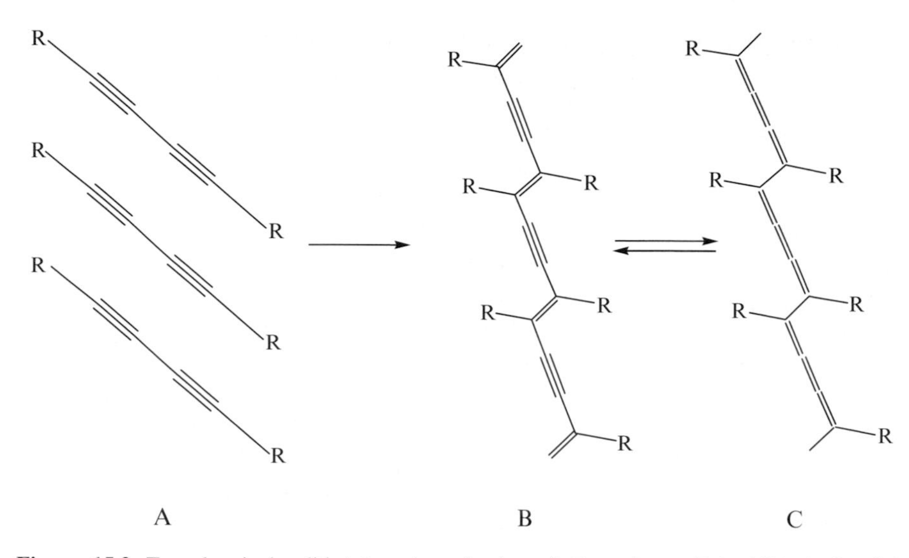

Figure 15.3 Topochemical solid-state polymerization of diacetylenes (1,4-addition). Parallel diacetylene molecules (A) reacting to form the *trans, trans* polymer, which can be represented by the alternate monomeric structures (B) and (C)

for various polymerizing diacetylenes. It also takes into account the reaction uniqueness criteria as they relate to monomer site point group symmetry, symmetry relations between mutually reacting monomer molecules, and dimensional changes during polymerization. It has been observed that during a diacetylene solid-state polymerization reaction, as the reaction proceeds, a homogeneous solution of the polymer chains is formed in the solid monomer matrix. The polymers are usually brightly colored and are generally insoluble in common organic solvents in which the monomers are soluble [42]. The difference in the degree of solubility of the monomer and the polymer in common solvents has been utilized to extract the unreacted monomer from the partially polymerized samples and to obtain a measure of the monomer to polymer conversion [40].

2.1.2 Hydrosilation Reaction

The addition of Si–H bonds to organic unsaturations such as olefins, acetylenes, and ketones is known as the hydrosilation reaction. A majority of the hydrosilation reactions are catalyzed by soluble transition-metal complexes and are known as homogeneous hydrosilation reactions [43]. Catalytic hydrosilation reactions are known to be very complex reactions. However, a few generalizations have been drawn about the mechanism of such reactions (Figure 15.4). Homogeneous olefin hydrosilation are presumed to start with an obligatory oxidative addition of the Si–H bond in a *cis* fashion to the catalytic metal [44]. This process is followed

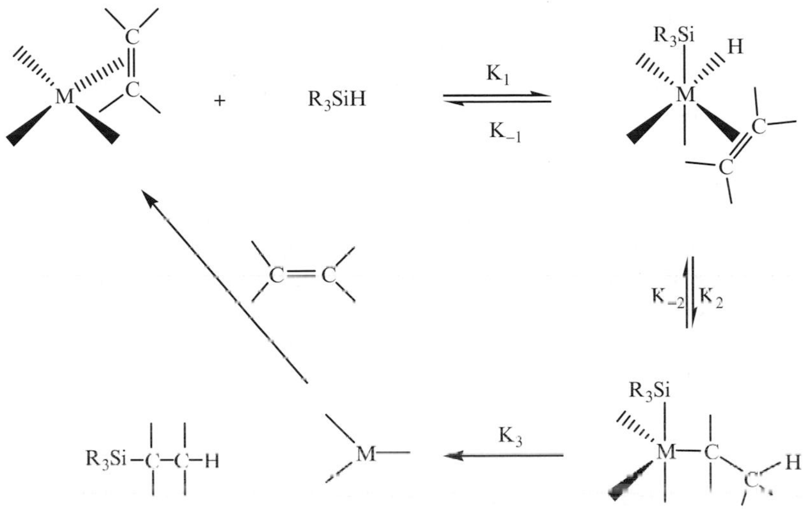

Figure 15.4 A generalized mechanism for catalytic hydrosilation of olefins

by a hydride migratory insertion that is followed by an irreversible reductive elimination of the final product. The oxidative addition and migratory insertion steps are known to be reversible in nature. Heterogeneously catalyzed hydrosilation reactions are also known [45]. Heterogeneous catalysts such as the Speier's catalyst and the Karstedt catalyst have been known to operate by similar heterogeneous mechanisms wherein the formation of a fine catalytically active colloidal Pt particles is believed to occur during the catalyst initiation step [46]. Hydrosilation reactions are important in the manufacture of silicones [47].

2.1.3 Carboranes

Carboranes are compounds of boron, carbon, and hydrogen in which both boron and carbon atoms are incorporated into three-dimensional, polyhedral skeletons of the general formula $C_pB_qH_{p+q}$ [48,49]. All known members of this family could be classified into three major groups: the closo-, nido-, and arachno-carboranes, where the first represent the $C_{0-2}B_qH_{q+2}$ compounds which are built of closed polyhedral cages, whereas the latter two include those carboranes in which the boron-carbon polyhedra resemble the shapes of nest skeletons. The closed-shell structures of the closo-dicarbaboranes are known to contribute to their astonishing chemical inertness, especially to acids. Thus, icosahedral carboranes are not attacked by hot sulfuric acid and are essentially inert to the conventional mixture of nitric and sulfuric acid that smoothly nitrates most aromatic organic species. Apparently the most stable closo-dicarbaboranes are the three members of the icosahedral family, the 1,2-, 1,7-, and 1,12-dicarba-closo-dodecaboranes better known by the more popular names *o*-, *m*-, and *p*-carborane, respectively.

A weak point of carboranes is their susceptibility to alkaline degradation. This weakness applies practically only to the species with adjacent carbon atoms. Thus, *o*-carborane is quantitatively converted to the nido-$[7,8\text{-}C_2B_9H_{12}]^-$ ion within a few hours by sodium methoxide in ethanol or by sodium (or potassium) hydroxide in ethanol [50]. With *m*-carborane the same type of degradation proceeds at least 100 times more slowly and a temperature of at least 70 °C is required for any practical effect [51]; the first observed degradation of the para isomer could only be forced using a 30% solution of potassium hydroxide in propylene glycol at 180 °C [52].

Carboranes, in general, and icosahedral carboranes, in particular, in their neutral and anionic forms are also known for their exceptional characteristics such as low nucleophilicity, high hydrophobicity, and their electron-withdrawing properties having highly polarizable σ-aromatic character [48,49].

2.2 *Carboranylenesiloxane Polymers Containing Thermally Crosslinkable or Vulcanizable Diacetylene Groups*

Since Wegner's studies on solid-state polymerization of diacetylenes, organosilicon polymers with bridging diacetylene units were being synthesized and reported by 1990 [53]. The interest in these polymers containing unsaturated organic units bridged by substituted silicon atoms primarily stemmed from their conductivity and nonlinear properties. The initial reported examples consisted of only silane or silylene polymers, and not siloxane polymer. The first example of the incorporation of a diacetylene group into a siloxane polymer was reported in 1994 by Henderson and Keller [25]. For the incorporation of the diacetylene group, the dilithia-diacetylide formation route used by Barton et al [53] was utilized to generate the reactive lithiated unsaturated organic unit which was subsequently reacted by a condensation reaction with a halogenated carboranylenesiloxane to produce the poly (carborane-siloxane-acetylene) polymers. In the synthesis which was a one-pot, two step reaction, hexachlorobutadiene was cleanly converted to dilithiobutadiyne through reaction with 4 equiv of *n*-BuLi in tetrahydrofuran (THF) and was subsequently reacted with an equi molar amount of the precursor carboranylenesiloxane, the Dexsil monomer, 1,7-bis(chlorotetramethyldisiloxy)-*m*-carborane, to generate the linear polymer **1** in high yield (Figure 15.5). The siloxane unit in the carboranylenesiloxane was the disiloxyl unit. The GPC analysis of the polymer **1** which is soluble in most common organic solvents had indicated the presence of low-molecular-weight species (M_w = 400) as well as higher molecular weight polymers (M_w = 4,900, M_n = 2,400) averaging to about 10 repeat units. Heating of **1** at 150 °C under reduced pressure resulted in the removal of lower molecular weight polymeric species, leaving the product in a 92 to 95% overall yield. The FTIR spectrum of **1** (Figure 15.6; *top*) exhibited prominent absorptions at 2,963 (C–H), 2,600 (B–H), 2,175 (C≡C), 1,260 (C–Si), and 1,081 (Si–O) cm^{-1}. The absorption of the internal diacetylenes at 2,175 cm^{-1} was found to be as intense as the other

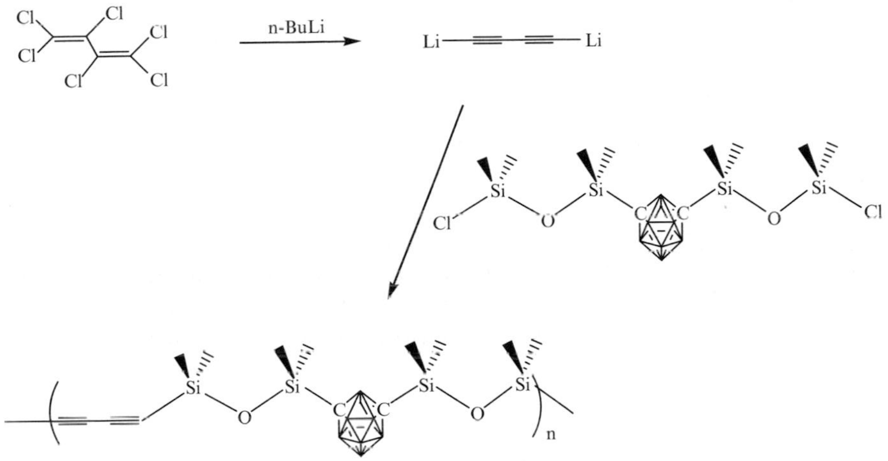

Figure 15.5 Synthesis of the poly(*m*-carborane-disiloxane-diacetylene) (**1**) reported by Henderson et al.

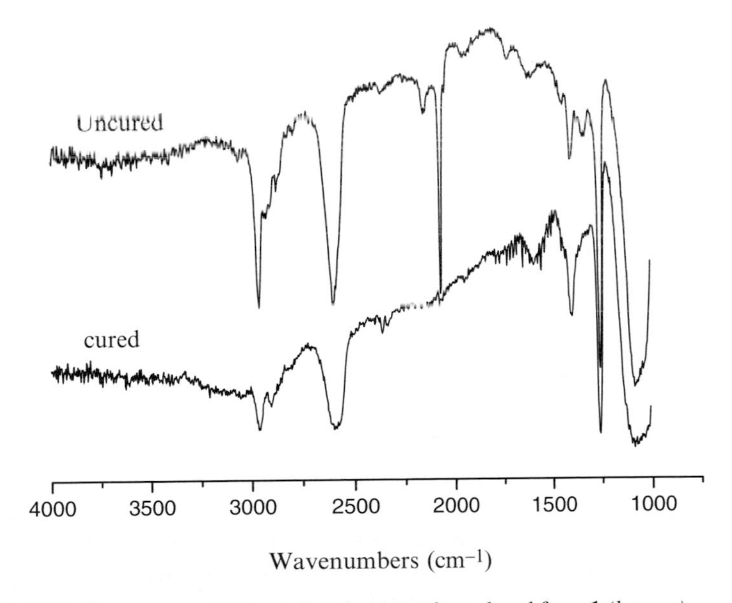

Figure 15.6 FT-IR spectrum of **1** *(top)* and the network produced from **1** *(bottom)*

vibrational absorptions of **1** even though internal alkynes are often known to possess very weak or nonexistent transitions [37].

The liquid linear polymer **1** was found to readily convert into a thermoset under thermal conditions by the crosslinking of diacetylene units as observed by the disappearance of the acetylenic absorption at 2,175 cm^{-1} and the appearance of a new, weak peak centered at 1,600 (C=C) cm^{-1} indicative of a 1,4-addition reaction

(Figure 15.6; *bottom*) [37]. DSC analysis revealed that the crosslinking of the diacetylenes commenced at 250 °C and peaked at 341 °C (Figure 15.11a). The completely crosslinked product was observed to be plastic in nature and was observed not to exhibit a glass transition temperature.

Pyrolysis of the crosslinked product to 1,000 °C in argon was reported to yield a black solid ceramic material in 85% yield. However, pyrolysis of the crosslinked product to 1,000 °C in air was observed to yield a ceramic with a yield of 92% presumably resulting from an uptake in weight on oxidation of the borons on the outer surface to B_2O_3 yielding a protective outer layer. Further characterizations of the ceramic products using scanning electron microscopy (SEM), X-ray photoelectron spectroscopy, Raman and scanning auger microprobe scattering studies were reported [54]. Samples heated under inert conditions to between 400 °C and 900 °C were observed to display little or reduced surface inorganic segregation, respectively, compared with material oxidized at 500 °C. The surface and the bulk of samples heated to 400 °C in argon consisted primarily of carbon, with lesser amounts of oxygen, boron and silicon. The microstructures and surface layers formed by oxidation at 500 °C and responsible for the resistance of the char to subsequent high-temperature oxidation were observed to consist of an antioxidative boron oxide and silicon oxide bilayer developed after oxidation at 500 °C. The SiO_2 layer was suggested to act as an oxygen barrier, whereas boron oxide was believed to be able to fill cracks formed in the silica layer due to thermal expansion mismatch with the underlying bulk. Prolonged oxidation exposure at 500 °C was observed to result in the volatilization of the boron oxide leaving a SiO_2 layer on the outer surface.

Soon after the report of Henderson et al., Son et al. reported the high-yield synthesis of similar diacetylene polymers containing only siloxane groups in 1995 (Figure 15.7) [26]. The disiloxane **2a** and trisiloxane **2b** variants of the polymers were synthesized following the convenient dilithiadiacetylide formation route. The high yields of the synthesis had represented an improvement over the synthesis of the disiloxane-diacetylene polymer reported by Parnell et al. [55] that was obtained via the oxidative coupling of 1,3-diethynyltetramethyldisiloxane, which had yielded only low molecular weight products in addition to some insoluble material. GPC analysis of both polymers indicated broad molecular weight distributions, with peak maxima occurring at ~10,000 (relative to polystyrene). This report is significant for its clear demonstration of the superiority of the dilithiadiacetylide formation route in incorporating diacetylene units in siloxane polymers. In addition, it also serves as a bench mark for comparing diacetylene polymers of carboranylenesiloxanes

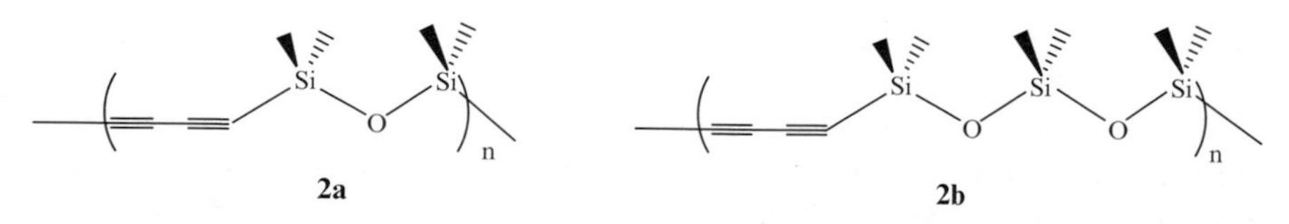

Figure 15.7 The poly(diacetylene-disiloxane) and poly(diacetylene-trisiloxane) systems of Son et al.

with their parent siloxane versions. The disiloxane and the trisiloxane polymers were studied by DSC analysis and were found to undergo vulcanization by thermal curing reactions of their diacetylene groups with the exotherms peaking at 289 and 315 °C for the disiloxane and trisiloxane polymers, respectively, to yield hard, void-free thermosets with T_g values between 144 and 170 °C. This was in contrast to the non existence of any glass transitions in the thermosets formed on curing of the diacetylene-containing carboranylenesiloxane polymers up to 450 °C [25].

The dilithiadiacetylide formation route was used subsequently by Sundar et al. in their reported synthesis of linear boron-silicon-diacetylene copolymers that explored an alternate boron source, namely, phenylboron dichloride (PBD) for carborane in the synthesis of diacetylene polymers of boron-containing siloxanes (Figure 15.8) [27]. The compositions of the boron-silicon-diacetylene copolymers **3a–d** are summarized in Figure 15.8. The mole percent ratios of silane/siloxane (dimethylsilane [DMS], tetramethyl disiloxane [TMDS], and hexamethyltrisiloxane [HMTS] to boron [PBD]) were reported as 90:10 and 80:20. They reported that only low-molecular-weight materials were formed when other siloxanes to boron ratios (such as 40:60 and 50:50) were used in the synthesis. The isolated copolymers **3a–c** were reported to be solids at room temperature whereas **3d** containing the trisiloxane group was reported to be a viscous liquid. Molecular weight determination using polystyrene as standard was performed only on the trisiloxane copolymer **3d** and was found to be in the 2,300 g/mol range because the copolymers **3a–c** were not completely soluble in THF. On thermally treating the four polymers to 1,000 °C in argon, char yields of 77.4, 75.0, 72.1, and 44.4% were obtained for **3a–d**, respectively. The char yields were reported to decrease with increasing siloxane spacer group length; DMS (**3a**: 77.4 wt%)>TMDS (**3b** and **3c**: 75.0 and 72.1 wt%)>HMTS (**3d**: 44.4 wt%). All of the polymers produced thermosets on crosslinking reactions of their diacetylene units under an inert atmosphere at 400 °C. However, no discussion on the nature of the thermosets was reported. They reported that a DSC analysis had revealed that the exotherms for the

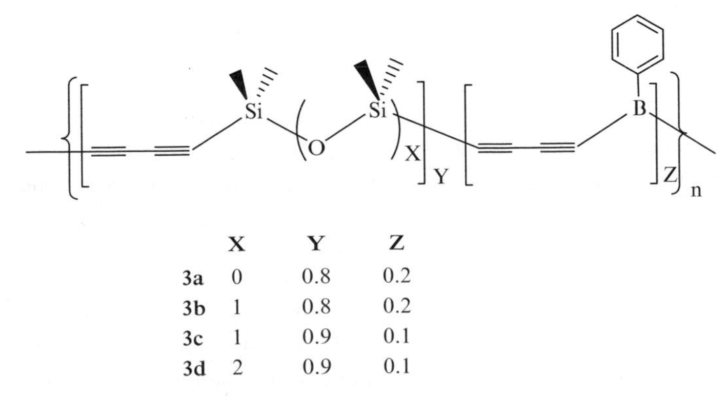

	X	Y	Z
3a	0	0.8	0.2
3b	1	0.8	0.2
3c	1	0.9	0.1
3d	2	0.9	0.1

Figure 15.8 The general structure of the (boron-silicon-diacetylene) copolymers of Sundar et al. *(top)* and the compositions of the representative polymers *(bottom)*

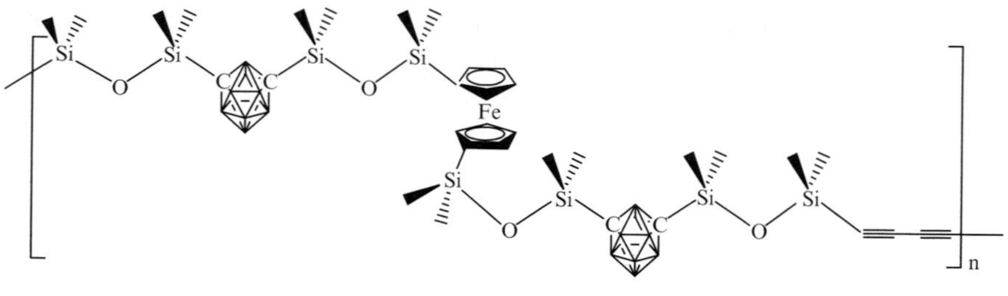

Figure 15.9 The linear (ferrocenyl-carboranylenesiloxyl-diacetylene) polymers of Houser et al.

crosslinking of the diacetylenes in **3a–d** had onsets ranging from 200 to 225 °C and peak maxima ranging from 285 to 300 °C.

Houser et al. reported the synthesis of linear ferrocenyl-carboranylenesiloxyl-diacetylene polymers in 1998, which are diacetylene-containing carboranylenesiloxanes wherein some of the carborane groups have been replaced by bridging organometallic ferrocenyl groups **4** (Figure 15.9) [29]. The polymers were produced by the reaction of dilithiobutadiyne with 2 equivalents of 1,7-bis(chlorotetramethyldisiloxyl)-*m*-carborane (Dexsil monomer) followed by treatment with dilithioferrocene-tmeda (1 equiv). The GPC measurements of this material were reported to yield a molecular weight of approximately 10,000 relative to polystyrene thereby accounting for the presence of ~10 repeat units in the polymer. Heat treatment of the polymer to 350 °C under an inert atmosphere was observed to result in the formation of a black, elastomeric thermoset with 98% weight retention. The curing of the diacetylenes exhibited an exotherm peaking at 280 °C in the product's DSC thermogram. It was reported that a small amount of shrinkage occurred during the formation of the thermoset. The elastomeric nature of the thermoset was in sharp contrast to the hard and tough thermosets obtained from siloxyl-diacetylene polymers such as poly (tetramethyldisiloxyl-diacetylene) reported by Son et al. [26] or the poly(carborane-tetramethyldisiloxyl-diacetylene) reported by Henderson et al. [25]. Thus, it represented the first example of a diacetylene-containing siloxane system which yielded an elastomeric networked thermoset upon curing. Further heating of the thermoset to 1,000 °C under N_2 afforded a hard, black ceramic in 78% ceramic yield.

In pursuit of elastomeric networks of the diacetylene-containing carboranylenesiloxane polymers, Kolel-Veetil et al. in 2003, reported the effects of the concentration dilution of the diacetylene units in the parent poly(carborane-disiloxane-diacetylene)s on the plasticities of the corresponding networks [34]. In the parent polymer, the concentration/ratio of the carborane, disiloxane and diacetylene moieties in its repeating unit was 1:2:1. The network polymer/thermoset obtained from this polymer had exhibited plastic characteristics at ambient temperature and had not displayed any discernible glass transitions in the product [25]. The study was based on the premise that the plasticity of the cured network of the parent system was a result of the high concentration of the crosslinkable diacetylenic groups in the parent polymer which had created an extensively crosslinked network on curing in which the resultant freedom of the –Si–O–Si– backbone flexibility was greatly reduced. It was, hence, postulated that a reduction in the concentration of the diacetylenic groups in the backbone of the precursor linear polymer **1** would afford an elastomeric networked material upon curing. Therefore, the elastomeric properties of networks formed from a linear poly(carborane-siloxane-acetylene)s should be controllable by varying the concentration of the diacetylene unit in the precursor polymer. In line with this postulation, the report included the synthesis and thermal characterization of three linear hybrid poly(carborane-disiloxane-diacetylene) systems, **5a–c**, that differed in the carborane:siloxane:acetylene ratios (**5a** ~ [2:3:1], **5b** ~ [4:5:1], and **5c** ~ [9:10:1]). A set each of blocky and alternating polymers were synthesized having the same carborane:siloxane:acetylene ratios (Figure 15.10). During the synthesis of each blocky polymer, two oligomeric intermediates, one that was a lithium end-capped carboranylenesiloxane and the other which was a chlorine terminated diacetylene-siloxane intermediate, were prepared individually and combined to afford the final viscous brown poly(carborane-disiloxane-diacetylene)s in high yield (85–95%) (Figure 15.10; *right*). Similar alternating polymers were also prepared by the reaction of the lithiated diacetylene with the respective dichlorodisiloxane-capped *m*-carborane (Figure 15.10; *left*).

The linear polymers were readily formed by both methods. A salient feature in both of the syntheses was the in situ production of a desired carboranylenesiloxane oligomer by the reaction of a lithiated carborane and a chlorinated siloxane species rather than the use of the commercial Dexsil monomer as in the case of Henderson et al. [25] and Houser et al. [29]. This has afforded an enormous degree of variation in the compositions of diacetylene-containing carboranylenesiloxane polymers that is not available by the conventional route of Henderson et al. [25]. The effect of the reduction in the diacetylene concentration was apparent in the positions and the intensities of the DSC peak maxima of the crosslinking endotherms of **5a–c**. Thus, it was observed that as the concentration of the diacetylene units in the polymer

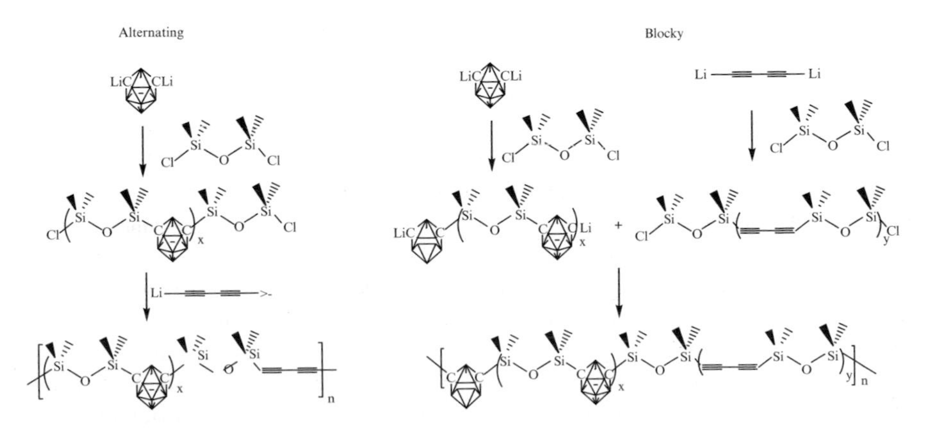

Figure 15.10 The synthetic schemes for the diacetylene-diluted (**a**) alternating *(left)* and (**b**) blocky poly (*m*-carborane-disiloxane-diacetylene)s reported by Kolel-Veetil et al.

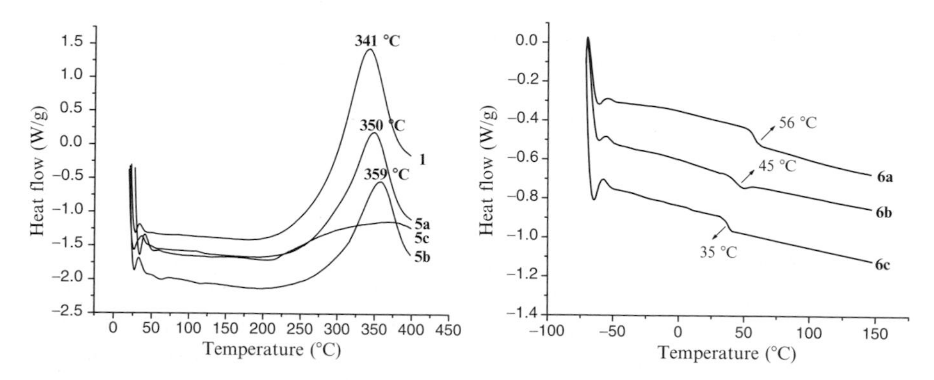

Figure 15.11 DSC thermograms of (**a**) the alternating poly(carborane-disiloxane-diacetylene) **5a** (3:2:1), **5b** (5:4:1), **5c** (10:9:1), and 1(2:1:1) *(left)* and (**b**) DCS thermograms of glass transitions of the crosslinked networks **6a**, **6b**, and **6c** produced from **5a**, **5b**, and **5c**, respectively *(right)*

decreased from **5a** to **5c**, the peak maximum of the corresponding endotherm shifted to higher temperature and also decreased in its intensity (Figure 15.11; *left*) indicating the requirement of a greater amount of heat and a longer duration for curing of the diacetylene units in the diacetylene-diluted systems. As expected the peak maximum and the intensity of the DSC exotherm of the parent poly(carborane-disiloxane-diacetylene)s, **1**, appears at the lowest temperature and as the most intense exotherm in a series of comparative DSC exotherms of **1** and **5a–c** (Figure 15.11; *left*). The crosslinked networks **6a–c** produced from the polymers **5a–c**, respectively, were reported to possess T_g values of 56 °C, 45 °C and 35 °C (Figure 15.11; *right*). Even though the T_g values of these networks were observed to decrease proportionally with an increase in diacetylene dilution in the precursor polymer, they were still observed to be predominantly plastic in nature at ambient temperature. However, in contrast, the crosslinked network obtained from **1** has been reported not to exhibit any glass transitions [25]. Importantly, this report very clearly demonstrated that the elasticity of the networks obtained from poly(carborane-disiloxane-diacetylene)s networks was controllable by the control of the concentration of the crosslinking groups in carboranylenesiloxanes.

Based on the observations with the diacetylene-diluted carboranylenesiloxanes containing the trisiloxyl groups, Kolel-Veetil et al., in 2004, synthesized and reported similar polycarboranylenesiloxane copolymers containing the trisiloxyl groups [35]. The substitution of the siloxane moiety was made in order to harvest the beneficial effect that a more flexible siloxane (trisiloxyl vs. disiloxyl) was expected to impart on the elasticity of the generated networks. Using the two synthetic methods described in Figure 15.10, they reported the synthesis of a set each of alternating and block copolymers with the same trisiloxane:carborane:diacetylene molar ratios of 10:9:1, 5:4:1, and 3:2:1, respectively. Both sets of copolymers were observed to yield networks upon thermal curing by the 1,4-addition reactions of the diacetylene units. Even though the two methods were observed to produce materials of comparable molecular weights and overall compositions, the variation in the copolymer sequence was discovered to cause noticeable differences in the calorimetric glass transitions for the copolymers in which the diacetylene group was most concentrated (i.e., the 3:2:1 and 5:4:1 copolymers) (Table 15.1). For a given reactant ratio, the alternating copolymers exhibited slightly higher molecular weights than the block copolymers.

The glass transitions (T_g) of the cured networks were determined from the DSC thermograms for both sets of polymers (Figure 15.12). Major glass transitions for both sets of polymers were found to occur below 0 °C (Table 15.1). Minor transitions also appeared for the alternating copolymers between 13 and 19 °C and at 46 °C, and for the block copolymers around 18 °C and between 60 and 65 °C. In general, as the diacetylene concentration increased in a set, yielding an enhancement in crosslink density, the T_g was observed to increase proportionally. The T_g values were found to be well defined for the two copolymers with lowest concentration of diacetylene groups (**7A**$_{10:9}$ and **8B**$_{10:9}$). In fact, these glass transitions (−46 °C for **7A**$_{10:9}$ and −49 °C for **8B**$_{10:9}$) were similar to the T_g of −50 °C reported for the uncrosslinked polycarboranylenesiloxane containing *m*-carborane and hexamethyltrisiloxane units [56]. For the alternating copolymers with higher diacetylene concentrations (**7A**$_{5:4}$ and **7A**$_{3:2}$), the glass transition remained relatively well defined. However, the block copolymers with higher diacetylene concentrations (**8B**$_{5:4}$ and **8B**$_{3:2}$)

Table 15.1 GPC molecular weights and thermal properties of the cured alternating ($7A_{3:2}$, $7A_{5:4}$ and $7A_{10:9}$) and block ($8B_{3:2}$, $8B_{5:4}$ and $8B_{10:9}$) polymers

Copolymer	M_n(Kg/mol)	M_w(kg/mol)	T_g(°C)	Char yield (%)
7A$_{3:2}$	4.3	6.6	−30	70
7A$_{5:4}$	3.3	5.4	−39	65
7A$_{10.9}$	5.2	8.6	−46	46
8B$_{3:2}$	3.4	6.2	−27	74
8B$_{5:4}$	2.9	4.3	−34	70
8B$_{10:9}$	4.1	6.7	−49	52

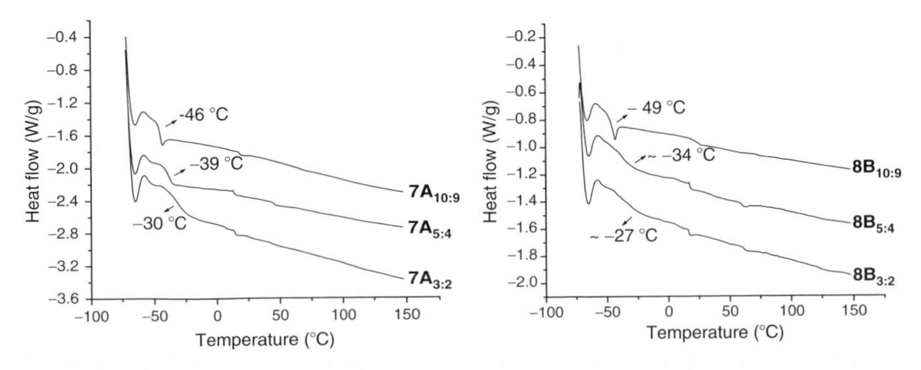

Figure 15.12 DSC thermograms of the glass transitions of the networks formed from (**a**) alternating and (**b**) blocky poly(*m*-carborane-trisiloxane-diacetylene)s

exhibited broadened glass transitions that reflect a greater distribution of segment lengths between cross links as compared with the samples with sharper glass transitions. The thermal stabilities of the crosslinked networks were analyzed gravimetrically by heating to 1,000 °C in N_2 (Table 15.1). For a given siloxane:carborane: diacetylene ratio, the char yields were about 4 to 6% higher for the block copolymers as compared to the alternating copolymers. Thus, this study demonstrated that a combination of diacetylene-dilution and the substitution of disiloxane by trisiloxane moieties in the parent diacetylene-containing carboranylenesiloxane **1** lead to the production of polymers that yield elastomeric networks upon curing.

2.3 Silarylene-Siloxane Polymers Containing Thermally Crosslinkable or Vulcanizable Diacetylene Groups

The first examples of silarylene-siloxanes polymers containing crosslinkable diacetylene groups were reported by Sundar et al. in 1997 (Figure 15.13) [28]. They reported the synthesis of a series of inorganic-organic linear diacetylenic hybrid polymers (**9a–e**) that were prepared by the polycondensation reaction of 1, 4-dilithiobutadiyne with 1,4-bis(dimethylchlorosilyl) benzene and/or 1,7-bis(tetra methylchlorodisiloxane)-*m*-carborane. These polymers in fact are a hybrid of the carboranylenesiloxane and silarylene-siloxane polymers. The polymers exhibited solubility in common organic solvents and were viscous liquids or low melting solids at room temperature. Broad prominent exotherms, attributed to the reaction of the diacetylenic units, were observed by DSC in the 306 °C to 354 °C temperature range. As in the case with the diacetylene-containing carboranylenesiloxane systems, the crosslinking of the diacetylenes units was observed to proceed by a 1,4-addition mechanism. The networks produced from all of the polymers were found to be thermosets. No molecular weight information was reported for the systems. When **9a–e** were analyzed by TGA to 1,000 °C under nitrogen, weight retentions between 79 and 86% were obtained. When heated to 1,000 °C, the polymers **9a–e** exhibited char yields of 83, 86, 80, 73, and 71%, respectively. The high char yield values reflected the thermal stabilizing effects of the aromatic unit

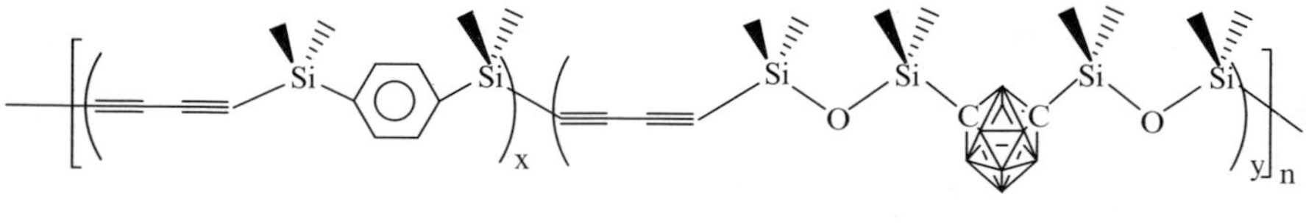

a: x = 1, y = 0; **b**: x = 3, y = 1; **c**: x = 1, y = 1; **d**: x = 1, y = 3; **e** = x = 0, y = 1

Figure 15.13 The hybrid silarylene-siloxane/carboranylenesiloxane reported by Sundar et al.

in the polymeric backone. When heated to 1,000 °C in a flow of air, the chars obtained from **9b–e** were reported to exhibit no weight losses and to actually attain a slight weight gain (1–4%) associated with the oxidation of boron to B_2O_3 and of silicon to SiO_2.

In 2002, Homrighausen et al. reported the synthesis of three groups of silarylene-siloxane polymers containing diacetylene crosslinking groups. The first of the reports involved the synthesis and characterization of a silarylene-siloxane-diacetylene polymer that converted to a thermosetting plastic on curing [31]. The linear polymer **10** was prepared via polycondensation of 1,4-bis(dimethylaminodimethylsilyl)-butadiyne with 1,4 bis(hydroxydimethylsilyl)benzene. The synthesis was an adapted version of the aminosilane-deficient method (Figure 15.14) [15]. It was determined that during the formation of linear polymer **10** some cleavage of the polymer occurred at the alkynyl carbon-silicon bond via dimethylamine reaction. This cleavage reaction was reported to disrupt the alternating nature of the polymer structure and to hinder the formation of a truly high molecular weight polymer. The weight average molecular weight for **10** was determined to be about 10,000 g/mol. Conversion to a thermoset was observed to occur through the crosslinking reactions

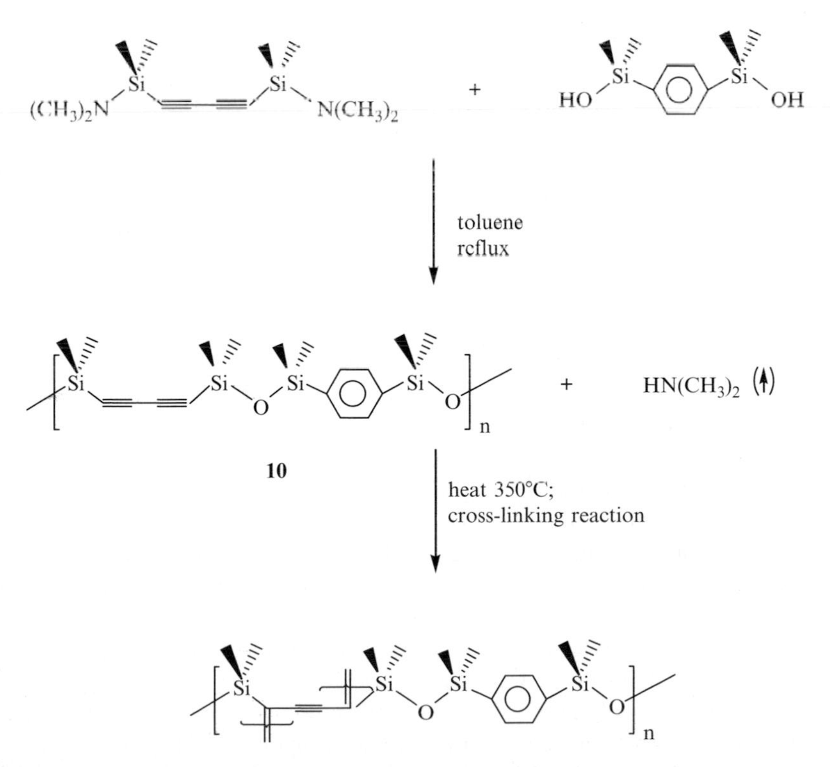

Figure 15.14 Synthesis of the thermosetting silarylene-siloxane-diacetylene polymer reported by Homrighausen et al.

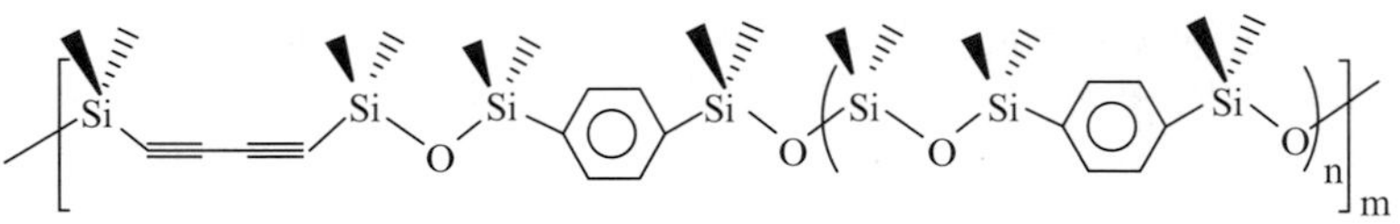

11a-c

11a n=1; **11b** n=2; **11c** n=3

Figure 15.15 Structure of the elastomeric diacetylene-diluted silarylene-siloxane-diacetylene polymer reported by Homrighausen et al.

of the diacetylene groups above 300 °C with the exotherm peaking at 336 °C in its DCS thermogram. The cross-linked polymer was observed not to display any glass transitions on evaluation.

The second set of networks of silarylene-disiloxane-diacetylene polymers **11a–c** (Figure 15.15) reported by Homrighausen et al. contained a series of linear silarylene-siloxane-diacetylene polymers with a varying diacetylene content synthesized, again, by an adapted version of the aminosilane-deficient method [32]. The linear polymers were prepared via polycondensation of 1,4-bis(dimethylamino dimethylsilyl)butadiyne [$(CH_3)_2N–Si(CH_3)_2–C≡C–C≡C–(CH_3)_2Si–N(CH_3)_2$] with a series of disilanol prepolymers. The low-molecular-weight silarylene-siloxane prepolymers terminated by hydroxyl groups were synthesized via solution condensation of an excess amount of 1,4-bis(hydroxydimethylsilyl)benzene with bis(dime thylamino)dimethylsilane. The length of the prepolymer was varied by adjusting the molar ratio of 1,4-bis(hydroxydimethylsilyl)benzene to bis(dimethylamino) dimethylsilane in the prepolymer synthesis. By altering the prepolymer length, the distance between the diacetylene units in the linear polymer was varied. Thus, the crosslinking density of the thermally prepared elastomers was controllable. The weight average molecular weight of **11b** was determined to be 10,000 g/mol by size exclusion chromatography. Heat treatment of silarylene-siloxane-diacetylene linear polymers **11a–c** was reported to result in a reaction between diacetylene units producing crosslinked (networked) polymers. The networked polymers were elastomeric materials that were completely insoluble in typical organic solvents. Thermal treatment of polymers **11a–c** to temperatures at or above where the crosslinking reaction occurred was reported to lead to the formation of soft, flexible, and void-free elastomeric materials. Thus, this second set of diacetylene-containing silarylene-siloxane polymers reported by Homrighausen et al. [32] validated the effectiveness of diluting the concentration of diacetylene units as a means for introducing elasticity in an otherwise plastic network.

The third set of silarylene-siloxane-diacetylene polymers that were reported by Homrighausen et al. also produced elastomeric networks on curing (Figure 15.16) [33]. The precursor polymers were produced by aminosilane–disilanol polycondensation reactions between a series of oligomeric, hydroxy-terminated silarylene–siloxane prepolymers and 1,4-bis(dimethylaminodimethylsilyl)butadiyne [$(CH_3)_2N–Si(CH_3)_2–C≡C–C≡C–(CH_3)_2Si–N(CH_3)_2$]. The oligomeric, hydroxy-terminated

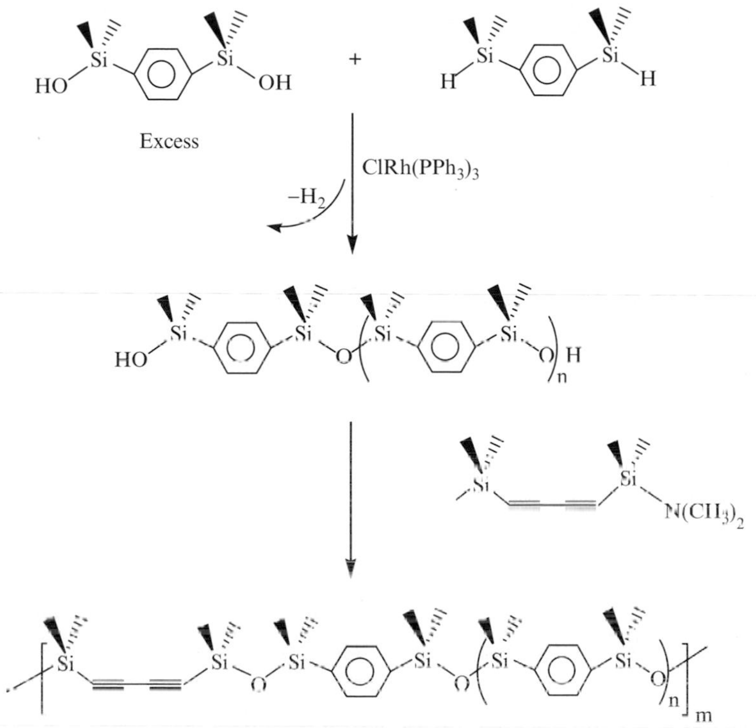

12a n=2, **12b** n = 4, **12c** n = 6, **12d** n = 8

Figure 15.16 Structure of the elastomeric diacetylene-diluted silarylene-siloxane-diacetylene polymer reported by Homrighausen et al. wherein Wilkinson's catalyst was used

silarylene–siloxane prepolymers of various lengths were prepared via dehydrogenative coupling between 1,4-bis(dimethylsilyl) benzene $[H(CH_3)_2\ SiC_6H_4\ Si(CH_3)_2H]$ and excess 1,4-bis(hydroxydimethylsilyl)benzene $[HO(CH_3)_2SiC_6H_4Si(CH_3)_2OH]$ in the presence of a catalytic amount of Wilkinson's catalyst $[(Ph_3P)_3RhCl]$. A number-average molecular weight ranging from 10,000 to 15,000 g/mol was obtained for **12a–d** with a polystyrene standard. The heat treatment of silarylene–disiloxane–diacetylene linear polymers **12a–d** resulted in a reaction between diacetylene units producing the crosslinked (networked) polymers. The networked polymers were reported to be completely insoluble in typical organic solvents. By varying the length of the hydroxy-terminated prepolymer in the synthesis of **12**, they were able to control the crosslinking density as demonstrated in the reports by Kolel-Veetil et al. [34,35]. Polymeric materials could, therefore, be obtained that produced slightly (elastomeric) to highly (plastic) crosslinked systems. The glass-transition temperatures for corresponding crosslinked polymers obtained from **12a–d** were determined by DSC as −29, −22, −29, and −29 °C, respectively.

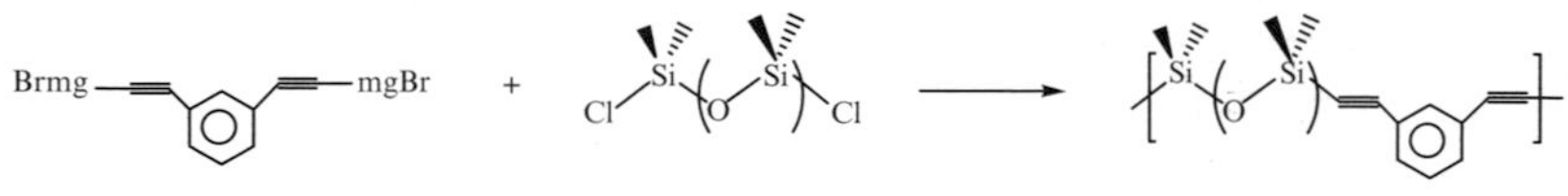

Brmg mgBr + Cl Si O Si Cl → Si O Si 13

Figure 15.17 Reported synthesis of a poly (siloxylene-ethylene-phenylene-ethylene)

Recently, Grignard reagents have been used to prepare hybrid silicones that contain phenyl and unsaturated organic groups, however, not necessarily as a diacetylene group [57,58]. The synthesis of poly (siloxylene-ethylene-phenylene-ethylene)s **13** has been reported by a reaction of a bischlorosiloxane with the bis-magnesium derivative of a diethynyl compound according to the following scheme (Figure 15.17). These compounds have been reported to be useful for composites with good heat resistance.

2.4 Hybrid Siloxane Network Polymers From Hydrosilation Reactions of Siloxane and Carboranylenesiloxane Monomers

Hydrosilation reactions are an important, if not the most important, class of reactions utilized in the manufacture of silicones [47]. Such reactions may also be conceived in the synthesis of elastomeric carboranylenesiloxanes and silarylene-siloxanes. In 1998, Houser et al. reported the syntheses of the monomeric carboranylenesiloxane 1,7-bis(vinyltetramethyldisiloxyl)-*m*-carborane **14**, and its hydrosilation reactions with the polymeric crosslinker, poly(methylhydrosiloxane) **15** (Figure 15.18) [30]. The reactions were catalyzed by the Speier's catalyst, H_2PtCl_6. The amount of the crosslinker was varied with regard to the amount of the monomer to determine the ratio of the reactants that imparted the highest thermal stability to the product polymer. This was done on the premise that the thermal stability of a polymeric material depended partly on its crosslinking density [59]. Three samples of **16** were prepared with monomer to polymer ratios of 18.2, 9.3, and 4.8. The values corresponded to a vinyl to Si–H group ratios of 1.04, 0.53, and 0.27. All of the products were found to be hard and colorless materials. None of the ratios was observed to yield an elastomeric product. Even though the reactions were performed with neat reagents, the formations of the products were reported to take a day to several days at room temperature.

In the interest of producing elastomeric network polymers, Kolel-Veetil et al. reported the modification of the preceding hydrosilation reaction system. They reported the Karstedt catalyst-catalyzed ambient-condition hydrosilation reactions of a monomeric vinyl **14** or ethynyl-containing **17** carboranylenesiloxane with three different monomeric branched siloxane crosslinkers in hexane (Figure 15.19) [36]. The reactions involving the vinylcarboranylenesiloxane were reported to

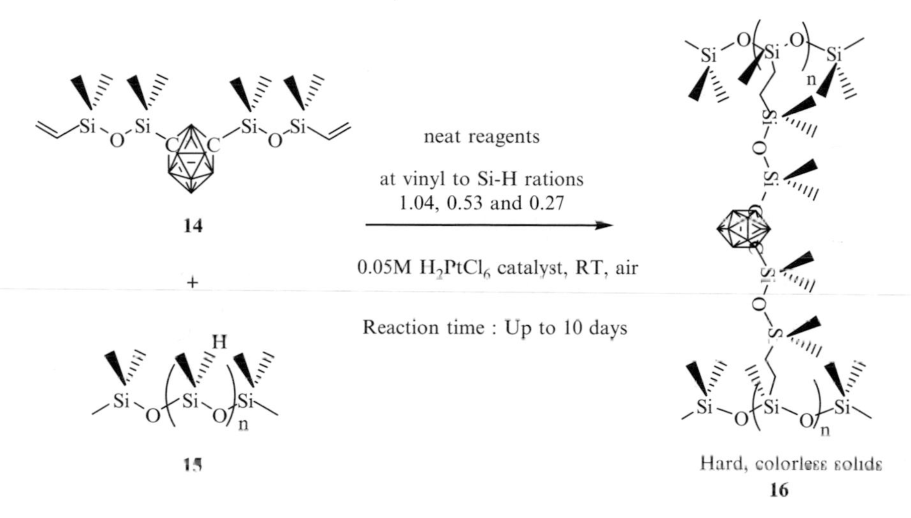

Figure 15.18 Hydrosilation reactions producing hard, colorless network plastics reported by Houser et al.

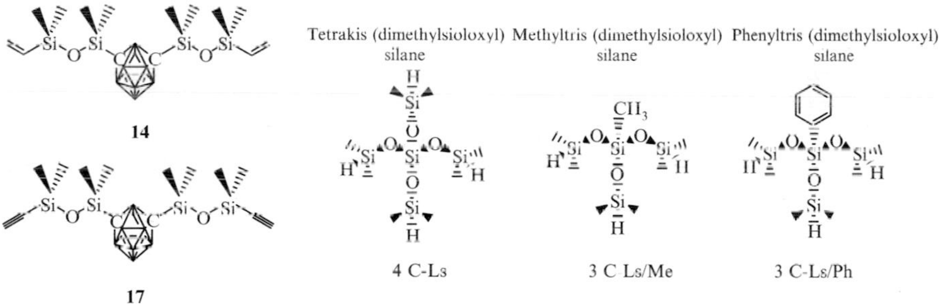

14

17

Tetrakis (dimethylsioloxyl) silane

Methyltris (dimethylsioloxyl) silane

Phenyltris (dimethylsioloxyl) silane

4 C-Ls

3 C-Ls/Me

3 C-Ls/Ph

Figure 15.19 The carboranylenesiloxane monomers and the branched siloxane crosslinkers used in the hydrosilation reactions reported by Kolel-Veetil et al.

produce a set of completely hydrosilated network polymers **18**. In the case of the ethynyl monomer, the reactions were carried out at two different ratios yielding a partially (**19**) and a completely saturated (**20**) set of network polymers. The logic behind their choice of the Karstedt catalyst was that it is more active than the Speier's catalyst for heterogeneous hydrosilation reactions due to its ability to form finer colloidal Pt particles during the catalyst initiation step [60]. The ambient reactions performed in hexane were observed to proceed rapidly to generate elastomeric network polymers in contrast to the slow formation of hard, colorless solids in the report by Houser et al. [30]. The flexible and transparent films of the saturated elastomeric network polymers from the vinyl monomer had T_g values below –35 °C while the T_g values of the films formed from the ethynyl monomer were below

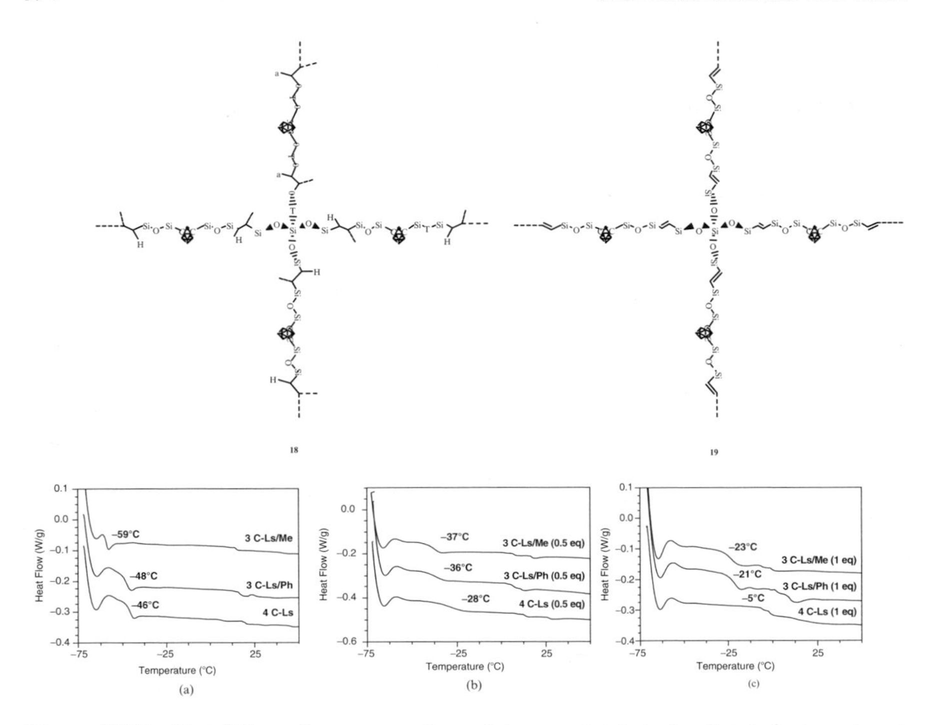

Figure 15.20 *(Top)* Schematic representations of the completely hydrosilated elastomeric network from **14** + 4 C–Ls (**18**) *(left)* and the partially hydrosilated elastomeric network from **17** + 4 C–Ls (**19**) *(right)*. *(Bottom)* DSC thermograms depicting the completely hydrosilated networks from **14** + 4 C–Ls (**18**) (**a**), and the partially and completely hydrosilated elastomeric network (**19**) (**b**) and (**20**) (**c**), respectively, from **17** + 4 C–Ls reported by Kolel-Veetil et al.

0 °C (Figure 15.20). The elastomeric polymeric networks from **14** and **17** were found to have degradation temperatures ranging from 500 to 550 °C.

3 Further Considerations

3.1 Observations

Recent developments in the area of the network polymers of carboranylenesiloxanes and silarylene-siloxanes have resulted in the generation of new synthetic routes and in the manipulation of the existing synthetic routes to yield elastomeric versions of the materials in both groups. In the synthesis of hybrid siloxanes of both groups of compounds, the utilization of diacetylene groups as crosslinking units has afforded some exceptional new materials as is evident from the examples reviewed in this chapter. Even though the evaluation of the elastic properties of the generated materials in these studies has been restricted to the determination of glass transition temperatures of the crosslinked networks, it has proven informative to compare such values with similar values of parent carboranylenesiloxane polymers. A great deal of information about a material can be obtained from its glass transition temperature. For example, there have been numerous efforts to understand the nature of glass transitions in materials and to relate T_g values to some easily measurable quantities of materials. Tanaka has proposed a relationship between T_g and the mean coordination number [61]. Gibbs and Di Marzio have developed a second-order phase transition model and obtained an empirical relationship (GDM equation) between the transition temperature and the density of cross linking agents inserted inside a system of molecular chains [62].

In the systems reported in this review, the manifestation of the elastomeric properties can be seen to be a synergistic effect of the crosslinking density, and the molecular weight and also the molecular structure of the constituent groups in the materials that are crosslinked. The elastomeric properties of the crosslinked networks from the diacetylene-containing carboranylenesiloxanes and silarylene-siloxanes with similar molecular weights are found to markedly depend on the nature of the constituent groups in the polymers and the concentration of the crosslinking diacetylene units in the polymer backbone. For example, even though the molecular weights of the poly (disiloxane-diacetylene) or poly(trisiloxane-diacetylene) [26], poly(ferrocenyl-siloxyl-diacetylene) [29], and the diacetylene-diluted poly(*m*-carborane-trisiloxane-diacetylene) [35] are reported to be 10,000, 10,000, and 4,300 to 8,600 g/mol respectively, the networks produced from the latter two are elastomeric in nature while the one produced from poly(disiloxane-diacetylene) or poly(trisiloxane-diacetylene) is a plastic with a T_g value of 144 or 170 °C, respectively. The T_g values can be shifted to larger values by exposing the plastics to high temperatures during postcure. Thus, the former two polymers with similar molecular weights have very different elastic properties. This marked difference may be attributed to the excessive degree of crosslinking in poly-(disiloxane-diacetylene) or poly(trisiloxane-diacetylene) due to the presence of a greater concentration of the diacetylene units in its repeating unit when compared to the other two systems.

Similar excessive degree of crosslinking of diacetylene (or 1,3-butadiyne) units in the silsequioxane network, $[O_{1.5}SiC \equiv CC \equiv CSiO_{1.5}]_n$ (**21**) has been exploited by Corriu et al. for the preparation of silica by the pyrolysis of such systems in air between 400 and 550 °C (Figure 15.21) [63]. The initial gel with bridging diyne units was discovered to have thermally crosslinked through ene-yne formation to form a rigid thermoset that on subsequent pyrolysis was reported to yield silica by oxidative decomposition. In the case of the diacetylene-diluted poly(carborane-trisiloxane-diacetylene) possessing the lowest range of molecular weights (4,300–8,000 g/mol), the presence of a decreased amount of crosslinking diacetylene units is believed to have contributed to the elasticity of their cured networks. In similar diacetylene-diluted poly(carborane-disiloxane-diacetylene), the T_g values of the crosslinked networks are found to be in the 30 to 50 °C range possibly resulting from a net increase in the crosslinking density arising from the shorter disiloxyl

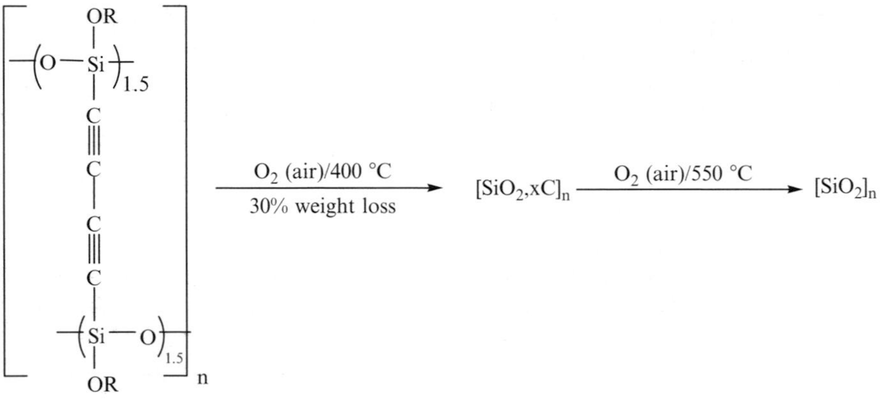

$$\left[-\!\left(O-\underset{\substack{|\\C\equiv C-C\equiv C\\|}}{\overset{OR}{Si}}\right)_{1.5} \cdots \left(\underset{OR}{Si}-O\right)_{1.5} \right]_n \xrightarrow[\text{30\% weight loss}]{O_2\text{ (air)/400 °C}} [SiO_2,xC]_n \xrightarrow{O_2\text{ (air)/550 °C}} [SiO_2]_n$$

R = H or Me

21

Figure 15.21 The synthesis of silica reported by Corriu et al. from an extensively crosslinked siloxane system by the crosslinking of diacetylene units

groups [34]. This effect that the diacetylene-dilution has on the elastomeric properties of the resultant networks is also evident in the networks produced from diacetylene-containing silarylene-siloxanes. Although the three groups of diacetylene-containing silarylene-siloxane polymers **10**, **11**, and **12** reported by Homrighausen et al. have molecular weights of 10,000, 10,000, and 10,000 to 150,000 g/mol, respectively, only the networks produced from **11** and **12** are seen to be elastomeric whereas **10** is observed to yield a thermosetting plastic due to an excessive concentration of crosslinking diacetylene groups.

In general, a study of the glass transition temperatures of the crosslinked networks of the recently developed carboranylenesiloxanes reveal that the T_g values decrease gradually in tandem with a reduction in crosslinking density in the precursor poly(carborane-siloxyl-diacetylene) oligomers. Although this trend in the decrease in the T_g values renders the disiloxyl systems progressively less plastic, a more dramatic decrease in the T_g values in the case of the networks of similar trisiloxyl systems render them elastomeric at ambient temperature. Such generalization can also be made in the case of the networks produced from the poly(silarylene-siloxane-diacetylene)s. The improvement in the elastic properties of crosslinked networks on crosslinking density-dilution is also very well demonstrated in the work by Kolel-Veetil et al. on the hydrosilation-cured networks [36]. However, in all of these systems, one of the disadvantages in reducing the crosslinking density is in the lowering of the thermal and thermo-oxidative stabilities of the less crosslinked systems in comparison to the parent systems. Thus, the most optimum elastomeric and thermal properties can be attained by a judicious manipulation of these two competing factors.

The effect that the nature of the constituent groups in a developed hybrid siloxane system has on its molecular weight is also worth evaluating. This is important as it has a direct bearing on the high-temperature thermo-oxidative stability of the polymer. It has been well established that hybrid siloxanes with ultra high-molecular-weights gain weight on heating in air compared with their low-molecular-weight counterparts [56]. This results from two competing oxidation processes, one of which causes crosslinking and the other fragmentation and volatilization. The relative rates of these two processes may change with changing polymer molecular weight because lower molecular weight polymer can produce volatile fragments more readily [56]. On perusing the examples in this chapter, it is apparent that the introduction of the crosslinking diacetylene groups in carboranylenesiloxanes substantially reduces the molecular weights of the resultant polymers when compared to the parent systems. For example, the molecular weights (either M_w or M_n) of the diacetylene-containing carboranylenesiloxanes are seldom seen to exceed 10,000 g/mol while many examples of the parent polymers with ultra high molecular weight (> 500,000) are available in the literature [7].

The effect that the steric nature of the group that replaces a diacetylene unit in a diacetylene-containing hybrid siloxane has on the product's molecular weight is evident when similar diacetylene-containing siloxanes, carboranylenesiloxanes, and silarylene siloxanes are compared. As a matter of clarification, it is to be noted that the ensuing observation is based completely on steric factors and has entirely ignored the differences in the chemical reactivity of the groups. In the three examples of **2a**, **10**, and **1**, whereas the replacement of a diacetylene group in **2a** by a 1,4-phenylene group does not seem to have had an effect on the molecular weight of the resultant **10** (**2a** and **10** have similar M_w s of ~10,000), the replacement of the diacetylene group by a *m*-carborane group seem to have reduced the molecular weight of the resultant **1** dramatically (**1** has a M_w of ~4,900). This could be attributed to the differences in the proximities of the respective sites of attachment of the two groups 1,4-*p*-phenylene and *m*-carborane in the polymer chain. Thus, the closeness of the two carbons sites (that are *meta* in positions) that involve in the polymerization in a *m*-carborane in comparison to the two carbon sites (that are *para* in position) in 1,4-*p*-phenylene group possibly makes it harder for the polymerization to proceed to a greater extent. It is well known in carboranylenesiloxane chemistry that progressively higher molecular weight materials could be obtained from carboranes with carbon sites that are farther apart, thus making *o*-carborane practically useless in producing polymers of any appreciable molecular weights [11]. Similar observations can also be made in connection with the linear boron-silicon-diacetylene copolymers (**3a–d**) reported by Sundar et al. When the mole percent ratios of silane/siloxane (DMS, TMDS, and HMTS) to boron (phenylborondichloride) were 90:10 (**3c** and **3d**) and 80:20 (**3a** and **3b**), respectively, the molecular weights of the polymers obtained were reasonably high. However, only low-molecular-weight materials were formed when other siloxanes to boron ratios (such as 40:60 and 50:50) were used in the synthesis. Again, this likely results from the increased proximity of the reactive carbon sites in 1,3-phenylborondichloride causing a decrease in the polymerization reaction in polymers containing a larger ratio of phenylborondichloride. However, the preceding arguments may be at odds with the observed differences in molecular weights of **2a** or **1** and the known carboranylenesiloxane (Dexsil) polymers. While ultra high molecular weight (>500,000) carboranylenesiloxane (Dexsil) polymers have been synthesized, the M_w of **2a** or **1** is only 10,000 or 4,900 g/mol, respectively. The lack of degree of polymerization of the diacetylene-substituted variants in contrast to the parent polymer that contain *m*-carborane alone and no diacetylene groups may have to do more with the differences in the chemical reactivity of the diacetylene and *m*-carborane groups than simply on the differences in their steric nature. A more in depth study of these features will be rather illuminating.

With the diacetylene-containing carboranylenesiloxane, it will be interesting to see what effects the crosslinking means will have on the elastomeric properties. As described in the section on the theory of the solid-state diacetylene polymerization, in addition to thermal annealing, UV radiation or high energy irradiation can effect the polymerization of diacetylenes [37,38,40]. In fact, Boileau et al. reported that UV-initiated crosslinking of diacetylene units contained in the main chain of a polysiloxane produced by the oxidative coupling of telechelic [2-(propyn-1) oxyethyl]oxypropyl-terminated polydimethylsiloxanes by a copper catalyst [64]. Even though it is unlikely that the use of UV radiation in place of thermal annealing as the means of crosslinking will cause any change in the mechanism of diacetylene polymerization, it should be interesting to compare the networks formed by both means.

In the case of all of the developed hybrid silarylene-siloxane and carboranylenesiloxane polymers, an evaluation of the nature of different phases containing aromatic or carborane groups will be illuminating to understand what contribution does each distinct phase has on the elasticity and other material properties of the generated networks. Other in-depth mechanical studies are also in order to better evaluate the high temperature elasticity of the materials.

3.2 Applications

3.2.1 High-Temperature and Miscellaneous

The interest in high-temperature elastomeric materials stems from the high demand for such materials for application in advanced technologies, particularly the aerospace, defense, and computer industries. Such materials are expected to have long-term thermal, thermo-oxidative and hydrolytic stability at and above 300 to 350 °C and to also have the ability to maintain pronounced flexibility to well below ambient temperature. The elastomeric crosslinked network polymers of carboranylenesiloxanes and silarylene-siloxanes described in this chapter possess similar properties and are therefore ideal candidates for a wide variety of engineering applications under unusual service conditions. In addition, these network polymers also have the utility in other applications that may or may not directly relate to their elastomeric properties and are available as a result of the presence of the introduced crosslinking groups such as diacetylene and ethynyl. Such applications are currently being evaluated.

One such application is in the thermo-oxidative protection of high-performance organic fibers. Keller has reported the use of the poly(carborane-disiloxane-diacetylene) polymer in the protection of carbon fibers [65]. Carbon fibers coated with poly (carborane–siloxane–acetylene) are reported to form a protective barrier against oxidation at elevated temperatures. When used as a matrix material (ceramic), the polymer was found to protect the carbon fibers from oxidative breakdown. The use of linear carborane-siloxane-acetylenic polymers as precursor materials for thermosets and ceramics for composite applications between 500 and 1,500 °C, respectively, in an oxidizing environment was also described. These linear polymers have the advantage of being extremely easy to process and to be converted into thermosets or ceramics because they are either liquids at room temperature or low melting solids. The elastomeric versions of the poly(carborane-siloxane-acetylene) and of the hydrosilated carboranylenesiloxanes should be even more convenient for use as protective coatings for high-performance fibers because of their ease of application. Kolel-Veetil et al. recently described the protection of high-performance organic fibers Zylon (PBO, poly (*p*-phenylenebenzobisoxazole), Kevlar, and carbon fibers by such polymers [66]. As these crosslinked networks are stable in air above the degradation temperatures of the organic fibres (450–700 °C), they prevent the catastrophic degradative oxidation of the high-performance fibers when applied as coatings.

The diacetylene-containing carboranylenesiloxanes have been evaluated to be exceptional high-temperature dielectric insulators in preliminary investigations [67]. The polymers have also shown impressive high-temperature adhesive characteristics during such investigations [67]. When obtained in their clear form, these materials can also conceivably function as mechanochromic sensors because of the crosslinking in the diacetylene units [68]. The elastomeric networks of hybrid silicones should also be able to function as high-temperature gas separation membranes [69].

In the case of the transparent hydrosilated networks of carboranylenesiloxanes obtained by hydrosilation reactions, the possibility exists for the use of the networks in coating and sensor applications. Additionally, with the partially hydrosilated network systems, the opportunity does exist for the reaction and sensing of molecules at the organic unsaturation sites that are uniformly dispersed in the hydrosilated membrane.

In the case of all of the carboranylenesiloxane network polymers produced either by thermal polymerization or by hydrosilation means, the opportunity exists for their use as neutron absorption materials that may find applications in nuclear reactor shields. The presence of boron atoms in these polymeric materials that are known to be exceptional neutron absorbing species by the well known boron neutron capture (BNC) reaction enhances the importance of them in nuclear applications [70].

3.2.2 Production of Ceramic Nanomaterials

The pyrolysis of various metallic derivatives of diacetylene-containing carboranylenesiloxanes, which has metal groups bound to the diacetylene fragment or being included in the polymer backbone, has been reported to afford an impressive group of ceramic nanomaterials with diverse conducting properties [71]. The derivatization of a diacetylene-containing carboranylenesiloxane was achieved by the complexation of the triple bonds in the diacetylene moiety with a chosen organometallic moiety. Depending on the ratio of the carboranylenesiloxane and organometallic moiety reactants, uncomplexed **22a**, partially complexed **22b**, and completely complexed **22c** carboranylenesiloxane could be obtained in various ratios from the reaction (Figure 15.22). Prior to conversion into the ceramic, the metallic derivatives are converted to a thermoset through the crosslinking reactions of **22a** and **22b**. The thermoset, thus formed, was pyrolyzed to desired temperatures to produce the ceramic nanomaterials with diverse magnetic and conducting properties.

Using this strategy, Kolel Veetil et al. recently reported the production of a superconducting mixture of β-Mo_2C and carbon nanotubes in an amorphous mixture of silicon and boron compounds by the pyrolysis of the metallic product obtained from $Cp_2Mo_2(CO)_6$ and a diacetylene-containing carboranylenesiloxane [71]. The equimolar reaction of $Cp_2Mo_2(CO)_6$ with the carboranylenesiloxane **1** in THF was found to result in the displacement of two of its carbonyl ligands by each triple bond in the diacetylene unit to yield a π-bonded complex. FT-IR and ^{13}C NMR spectral evidences corroborated the formation of metallic derivatives such as **22b** and **22c** and the retention of some unreached **22a**. During the thermoset formation, the successive losses of the labile carbonyl and cyclopentadienyl

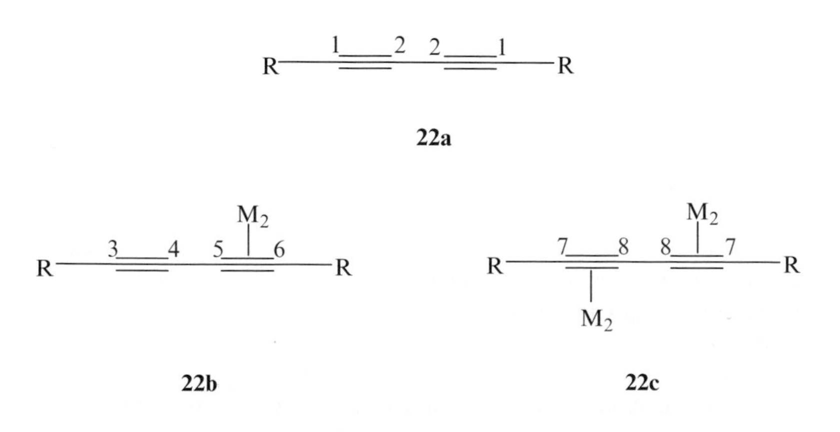

Figure 15.22 Schematic representations of the uncomplexed (**22a**), partially complexed (**22b**), and completely complexed (**22c**) metallic derivatives of a diacetylene-containing carboranylenesiloxane

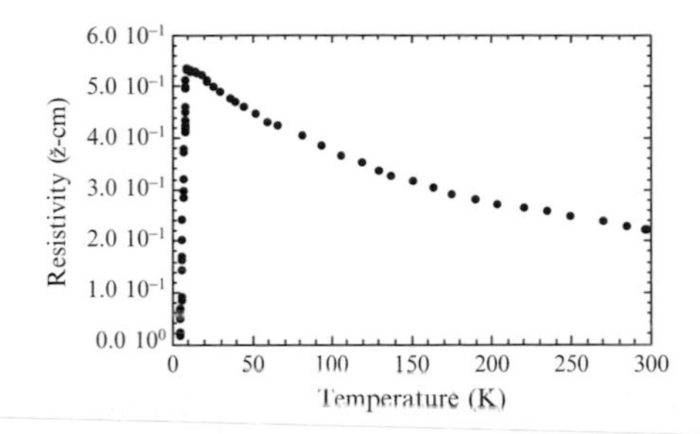

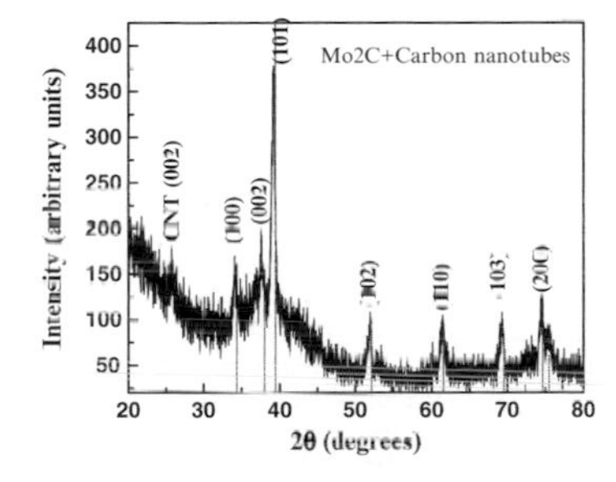

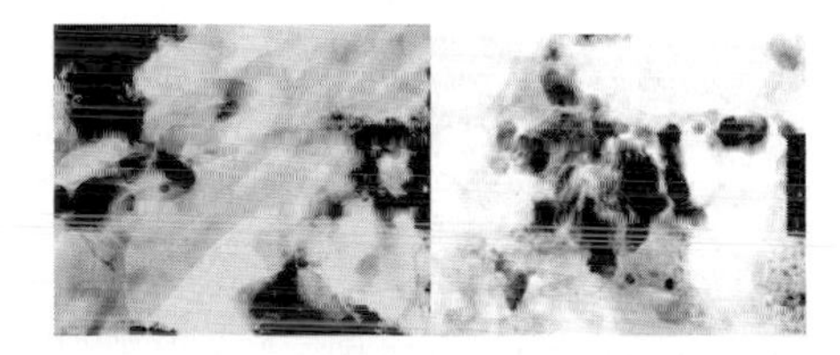

Figure 15.23 The resistivity plot, the XRD spectrum and TEM micrographs of the pyrolysis of the $Cp_2Mo_2(CO)_4$ complex of **1**

ligands were observed from the mixture around 222 and 267 °C, respectively. During the pyrolysis of the thermoset, the loss of silicon-bound methyl groups was also observed around 500 °C. On completion of the pyrolysis at 1,000 °C, concomitant formations of nanoparticles of β-Mo_2C and carbon nanotubes were observed as verified by X-ray diffraction and TEM studies (Figure 15.23). The nanoparticles were obtained in an amorphous matrix of silicon and boron compounds. Resistivity studies of the mixture revealed that it was superconducting in nature with a critical temperature (T_c) of 8 K (Figure 15.23). This represents only the third example of a formation of carbon nanotubes catalyzed by the metal Mo alone [72].

Nanomaterials of various transition metal borides, carbides and silicides with diverse magnetic and conducting properties have been obtained from **1** and other diacetylene-containing carboranylenesiloxanes using the same strategy as used in the β-Mo_2C formation. Some of the nanomaterials produced include CoB, FeCoB, CuB_2, and Fe_5Si_3.

In summary, recent advances in the area of network polymers of carboranylenesiloxane and silarylene-siloxanes have produced exceptional high-temperature elastomeric materials as a result of the incorporation of new methodologies in their syntheses. In addition to their high-temperature properties, the presence of various diverse constituents in the materials has opened avenues for the utilizations of these materials in myriad applications.

Acknowledgments The authors wish to thank the Office of Naval research for the financial support of this project.

References

1. (a) Kipping FS, Loyd LL. (1901) J Chem Soc. 79:449. (b) Kipping FS. (1904) Proc Chem Soc. 20:15. (c) Kipping FS. (1905) Proc Chem Soc. 21:65.
2. Pauling L. (1960) The Nature of the Chemical Bond, 3rd ed. Cornell University Press, Ithaca, New York.
3. Hannay NB, Smyth CP. (1964) J Am Chem Soc. 68:171.
4. Stone FGA, Seyferth D. (1955) J Inorg Nucl Chem. 1:112.
5. Thomas TH, Kendrick TC. (1969) J Polym Sci. A7:537.
6. Grassie N, MacFarlane IG. (1978) Eur Polym J. 14:875.
7. Dvornic PR, Lenz RW. (1990) High Temperature Siloxane Elastomers. Huthig & Wepf Verlag Basel, Heidelberg/New York.
8. Williams RE. (1972) J Pure Appl Chem. 29:569.
9. Mayes N, Green J, Cohen MS. (1967) J Polym Sci. 5:365.
10. Papetti S, Schaeffer BB, Grey AP, Heying TL. (1966) J Polym Sci 4:1623.
11. Grimes RN. (1970) Carborane polymers. In: Carboranes. Academic Press, New York, p 181.
12. (a) Mayes N, Greene J, Cohen MS. (1967) J Polym Sci 5:365. (b) Peters EN. (1979) J Macromol Sci Rev Macromol Chem. C17:173.
13. (a) Sveda, M. (1951) United States Patent Nos. 2,561,429 and 2,562,000. (b) Hyde JF. (1961) French Patent No. 1,263,448. (c) Klebanski AL, Fomina LP, Dolgoplosk SB. (1962) Zh Vses Khm Obsch 7:594. (d) Merker RL, Scott ML. (1964) J Polym Sci 2:15.
14. Wu TC. (1967) United States Patent No. 3,325,530.
15. Pike RM. (1961) J Polym Sci. 50:151.
16. Rosenberg H, Nahlovsky BD. (1978) Polymer Preprints.19(2):625.
17. Lenz RW, Dvornic PR. (1978) Paper presented at the 176th ACS Meeting, Miami Beach, FL, September 11–17.
18. (a) Reed CA. (1998) Acc Chem Res. 31:133. (b) Plesek J. (1992) Chem Rev. 92:269.
19. Schroeder H, Shaffling OG, Larchar TB, Frulla FF, Heying TL. (1966) Rubber Chem Tech. 39:1184.
20. Hedaya E, Kawakami JH, Kopf PW, et al. (1977) J Polym Sci Polym Chem Ed., 15 :2229.
21. Peters EN, Beard CD, Bohan JJ, et al. (1975) Technical Report CRL-T 818, Union Carbide Chemicals and Plastics Laboratory, Bound Brook, New Jersey and Tarrytown technical Center, Tarrytown, New York.
22. Berridge CA. (1958) United States Patent No. 2,843,555.
23. Livingston ME, Dvornic PR, Lenz RW. (1982) J Appl Polym Sci. 27 :3239.
24. Hundley NH, Patterson WJ. (1985) NASA Technical Paper 2476, G. C. Marshall Space Flight Center, Alabama.
25. (a) Henderson LJ, Keller TM. (1993) Polym Prepr. 34(1) :345. (b) Henderson LJ, Keller TM. (1994) Macromolecules. 27(6) :1660.
26. Son DY, Keller TM. (1995) Macromolecules. 28:399.
27. Sundar RA, Keller TM. (1996) Macromolecules. 29:3647.
28. Sundar RA, Keller TM. (1997) J Polym Sci Part A: Polym Chem. 35:2387.
29. Houser EJ, Keller TM. (1998) Macromolecules. 31:4038.
30. Houser EJ, Keller TM. (1998) J Polym Sci Part A: Polym Chem. 36:1969.
31. Homrighausen CL, Keller TM. (2002) Polymer. 43:2619.
32. Homrighausen CL, Keller TM. (2002) J Polym Sci Part A: Polym Chem. 40:88.
33. Homrighausen CL, Keller TM. (2002) J Polym Sci Part A: Polym Chem. 40:1334.
34. Kolel-Veetil MK, Keller TM. (2003) J Mater Chem. 13(7):1652.

35. Kolel-Veetil MK, Beckham HW, Keller TM. (2004) Chem Mater. 16:3162.
36. Kolel-Veetil MK, Keller TM. (2006) JPolym Sci Part A: Polym Chem. 44:147.
37. (a)Wegner G. (1974) Z. Naturforsch. B. 24:824. (b) Wegner G. (1970) Makromol Chem. 134:219. (c) Wegner G. (1971) Makromol Chem. 145:85.
38. Cantow HJ, (ed). (1984) Adv Polym Sci. 63:1.
39. (a) Newkirk AE, Hay AS, McDonald RS. (1964) J Polym Sci Part A, 2:2217. (b) Neenan TS, Callstrom MR, Scarmoutzos LM, Stewart KR, Whitesides GM. (1988) Macromolecules. 21:3528. (c) Coriu RJP, Guerin C, Henner B, Jean A, Mutin H. (1990) J Organomet Chem. 396:C35.
40. Chance RR, Patel GN. (1978) J Polym Sci Polym Phys Ed., 16:859.
41. Baughman RH. (1974) J Polym Sci Polym Phys Ed. 12:1511.
42. Bloor D, Change RR. (1985) Polydiacetylenes/:NATO ASI Ser. E 1.
43. Spier JL. (1978) Adv Organomet Chem. 17:407.
44. Chalk AJ, Harrod JF. (1965) J Am Chem Soc. 87:16.
45. Lukevics E, Belyakova ZV, Pomerantseva MG, Voronkov MG. (1977) In: Seyferth D, (ed). Journal of Organometallic Chemistry Library, vol. 5. Elsevier, Amsterdam.
46. (a) Crabtree RH, Mellea MF, Mihelcic JM, Quirk JM. (1982) J Am Chem Soc. 104:107. (b) Anton DR, Crabtree RH. (1983) Organometallics. 2.855.
47. Noll W. (1968) Chemistry and technology of slicones. Academic Press, New York.
48. Williams RE. (1970) Carboranes In: Progress in boron chemistry, vol. 2. Pergamon Press, New York.
49. Dunks GB, Hawthorne MF. (1973) Acc Chem Res. 6:124.
50. Wiesboek RA, Hawthorne MF. (1964) J Am Chem Soc. 86:1642.
51. Garrett PM, Tuble FN, Hawthorne MF. (1964) J Am Chem Soc. 86:5016.
52. Plesek J, Hermanek S. (1973) Chem Ind. (London), 381.
53. (a) Corriu RJP, Guerin C, Henner B, Kuhlmann TH, Jean A. French Patent No. 89, 05567. (b) Ijadi-Magshoodi S, Barton TJ. (1990) Macromolecules. 23:4485. (c) Barton TJ, Ijadi-Maghsoodi S, Pang Y. (1990) J Polym Sci Part A: Polym Chem. 28:955.
54. Pehrsson PE, Henderson LJ, Keller TM. (1996) Surf Interface Anal. 24:145.
55. Parnell DR, Macaione DP. (1973) J Polym Sci Polym Chem Ed. 11:1107.
56. Stewart DD, Peters EN, Beard CD, et al. (1979) J J Bohan Macromolecules. 12:373.
57. Buvat P, Jousse F, Nony F, Gerard JF. (2003) WO 2003076516 (CEA France); CA (2003) 139:246474.
58. Nony F. (2003), Ph D Thesis, INSA Lyon, November 14, 2003.
59. (a) Boury B, Corriu RJP, Douglas WE. (1991) Chem Mater. 3:487. (b) Allcock HR, McDonell GS, Riding GH. (1990) I Manners. Chem Mater. 2:425.
60. Lewis LN, Lewis NJ. (1986) J Am Chem Soc. 108:7228.
61. Tanaka S. (1984) Solid State Communications. 54:867.
62. (a) Adam G, Gibbs JH. (1965) J Chem Phys. 43:139. (b) Gibbs JH, Di Marzio EA. (1958) J Chem Phys. 28:373.
63. Corriu RJP, Moreau JJE, Thepot P, Man MWC. (1996) Chem Mater. 8:100.
64. Boileau S, Bouteiller L, Kowalewska A. (2004) Polym International. 53:191.
65. Keller TM. (2002) Carbon. 40(3):225.
66. Kolel-Veetil MK, Keller TM. (2005) Polym Mater Sci Eng. 92:188.
67. Kolel-Veetil MK, Keller TM. Unpublished results.
68. (a) Nallicheri RA, Rubner MF. (1991) Polymer Reprints 24(2):517. (b) Tsukruk VV, Reneker DH. (1993) Polymer Preprints, 34(2):312. (c) Tsukruk VV, Reneker DH. (1993) Ukranian Polymer Journal. 2(2):98.
69. Baker RW. Membrane technology and applications, Wiley, New York.
70. Locher GL. (1936) Am J Roentgenol Radium Ther. 36:1.
71. Kolel-Veetil MK, Qadri SB, Osofsky M, Keller TM. (2005) Chem. Mater 17(24):6101.
72. (a) Dai H, Rinzler AG, Nikolaev P, Thess A, Colbert DT, Smalley RE. (1996) Chem Phy. Lett. 260:471–475. (b) Motiei M, Calderon-Moreno J, Gedanken A (2002) Chem Phys Lett. 357:267–271.

Antimony-Containing Polymers

Charles E. Carraher Jr.

1 Introduction

This chapter reviews efforts to include organoantimony into polymers. This is not an exhaustive review but will contain most of the reports. Polymers containing organoantimony moieties have been reported to exhibit some potentially useful properties including an unusually low solution viscosity relative to molecular weight, antimicrobial activity, semiconducting properties, and flammability resistance. While antimony is typically found in both the plus 3 and plus 5 state, essentially all polymers are of the plus 5 state.

In 1975 we reported the initial synthesis of antimony polyesters and polyoxamines. The work was an extent of our efforts with other metal-containing electron deficient metal sites. Reactants typically employed in the formation of ogranoantimony polymers are triphenylantimony dichloride and dibromide, along with trimethylantimony dichloride. These materials are commercially available and as such are primary candidates for use as conveyers of the organoantimony moiety into polymers. These materials act similarly to regular organic acid chlorides with respect to hydrolysis by water and addition to typical mono-Lewis bases. Thus, the synthesis of organoantimony-containing polymers is a somewhat straightforward extension of our other work. The organoantimony polymers synthesized by our group emloyed the classical aqueous interfacial system where the organoantimony dihalide is contained within an organic solvent and the Lewis base is present in an aqueous phase. Reactions were generally rapid occurring within 30 seconds or less.

A.S. Abd-El-Aziz et al. (eds.), *Inorganic and Organometallic Macromolecules: Design and Applications.*

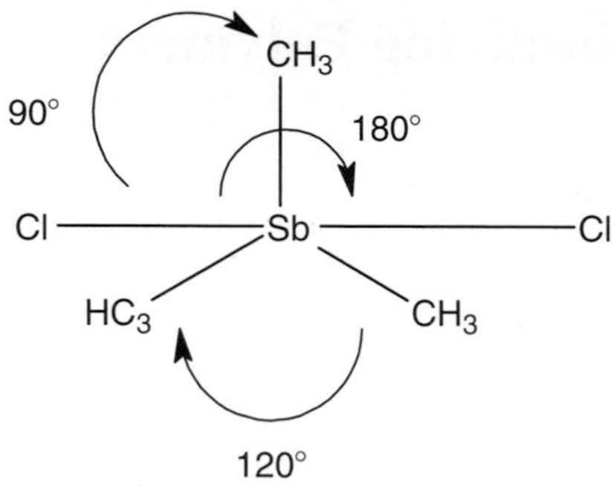

(1)

(1)

Organoantimony (V) dihalides exist in a triangular bipyramidal geometry with the typically bond angles as shown below for trimethylantimony dichloride (**1**). The two halides reside in the axial positions whereas the organic moieties are present on the equalitorial plane.

2 Polyesters

Unlike reactions with organic acid chlorides in the presence of water where hydrolysis proceeds faster than addition to the organic Lewis base, hydrolysis of organoantimony dihalides occurs more slowly than addition to typical Lewis bases as salts of carboxylic acids, allowing for the formation of esters. Thus, reaction of organoantimony dihalides with salts of monocarboxylic acids yields the diesters (**2**). [1,2]

We employed a simply extension of this employing salts of dicarboxylic acids to form polymers (**3**) [3–7].

A number of dicarboxylic acids were employed including, a series of *p*-phenylene dicarboxylic acids such as dimethylterephthalic, dimethylterphthalic acid, bromoterephthalic acid, nitroterephthalic acid, terephthalic acid, and dichloroterephthalic acid. The best yield of about 50% was achieved from terephthalic acid, consistent with steric concerns being important. A number of aliphatic diacids were employed including ferrocene-1,1′-dicarboxylic acid, fumaric acid, maleic acid, and oxalic acid.

Along with the ferrocene polymers, the analogous cobalticinium products were synthesized (**4**) [6,7]. If the hexafluorophosphate-containing monomer is dissolved in sodium hydroxide and immediately used a good yield is found on the order of 90%. But, if the solution is allowed to “age” for several hours, exchange occurs and poorer yields, on the order of 30%, are found.

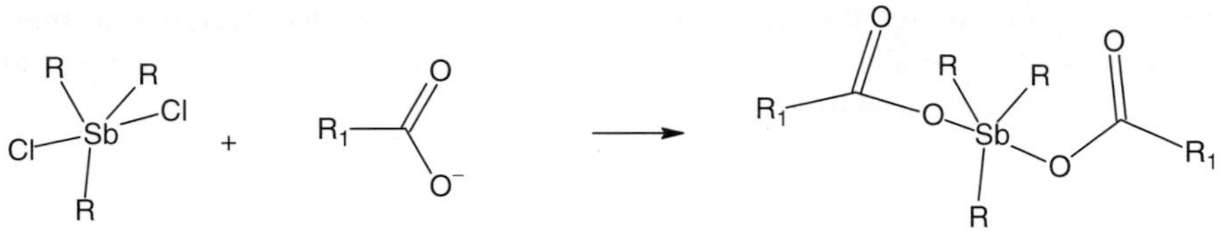

(2)

(2)

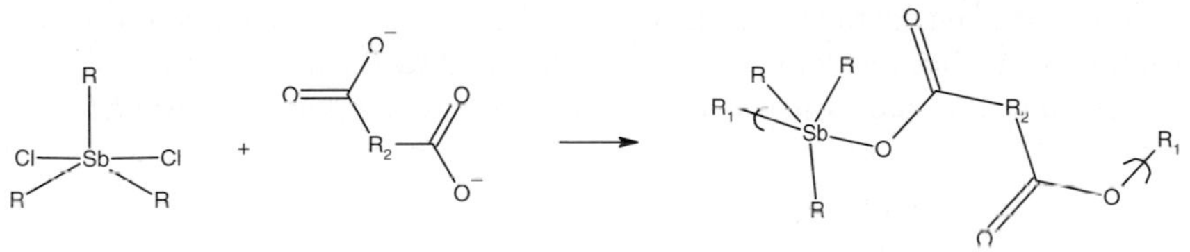

(3)

(3)

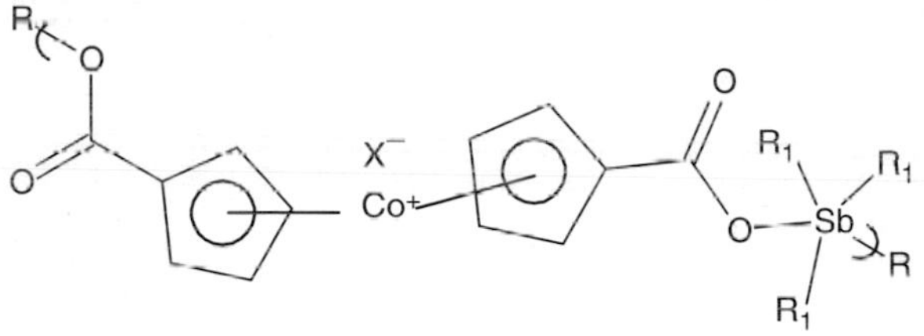

(4)

(4)

Whereas the hexafluorophosphate anion was typically employed, products containing a number of other anions were synthesized. Anions included the bromide and nitrate ions. The use of these other anions resulted in greatly reduced yields on the order of less than 10%. It is believed that the hexafluorophosphate moiety is closely associated with that of cobalticinium and, because of size, makes it more difficulty for water to approach. The water likely promotes hydrolysis of the polymer and possibly of the organoantimony chloride, either during reaction or as a growing end group.

Most of the organoantimony polyesters are near semiconductors with resistivity values generally within the range of 10^{10} to 10^{14} ohm cm [7]. Resistivity decreases slightly as pressure is applied consistent with most organic semiconductors. Resistivity increases with applied voltage. There was no clear trend for organic semiconductivity. In comparison, the cobalticinium were superior conductors with resistivity values in the range of 10^5 to 10^7 ohm cm well within the semiconductor range and approaching the near conductor range. Values were similar as the anion was varied. Resistivity slightly decreased as voltage increased, reaching a minimum and then generally increasing. Resistivity remained approximately constant as the pressure was increased.

3 Polyoxamines

Harrison and Zuckerman [8] reported the formation of monomeric oximes from reaction of triphenylantimony diethers (**5**). The analogous condensation of organoantimony dihalides with oximes was not reported by them.

Again, the shift to dioximes permitted the synthesis of polymers (**6**) [9,10].

One of the reasons for the synthesis of new polymers is their evaluation in light to general polymer theories. One such product is pictured below. Typically, polymers give a relatively large drag in dilute solutions because they reside in a number of flow planes. Some of these materials gave unusually low viscosities in comparison with their molecular weight. Three factors were identified as potential contributions to this behavior. First, many of these polymers are rigid-rod like. One such polymers structure is noted below (**7**).

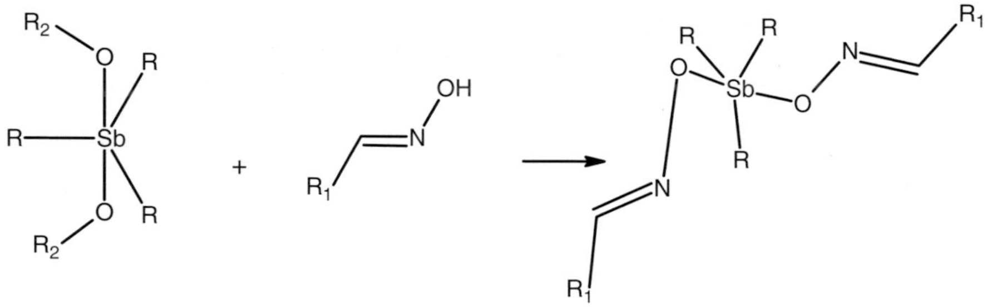

(5)

(5)

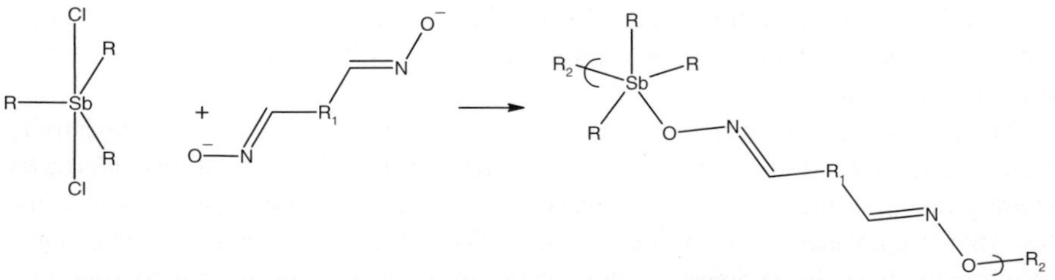

(6)

(6)

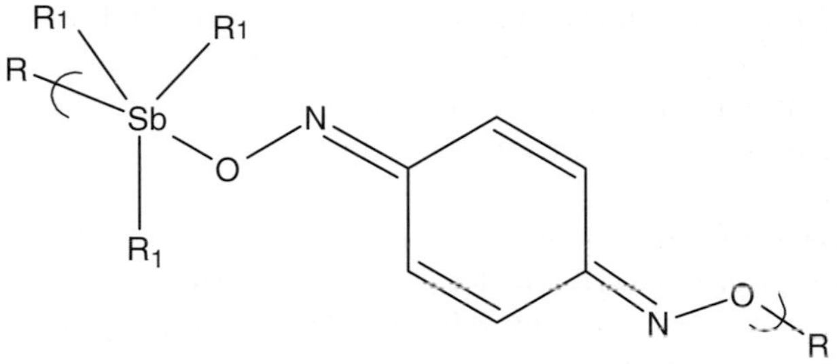

(7)

(7)

Resonance structures can be drawn that show that the polymer backbone from triphenylantimony dichloride and *p*-benzoquinone dioxime is quite stiff. The product had a low viscosity (6 mL/g) but a molecular weight of about 7×10^5 with a LVN/weight average molecular weight of 0.09×10^{-4}. In comparison, polystyrene has a LVN/M_w of 8×10^{-4} or about 100 times the antimony product. Polyethylene has a LVN/M_w value of 35×10^{-4}, or a value about 400 times that of the antimony polymer. Because of this stiffness, it is possible that the stiff chain is caught between flow planes rather than residing within a number of flow planes as modest flow occurs. This would result in a greatly decreased viscosity. Second, errors exist because of the possibility that such polymers fluorescence. Thus, this same polymer showed a M_w of 1.5×10^8 before correction for fluorescence. After correction, the M_w was reduced to 7×10^5. Third, anomalous scatter may result from color or absorption. This was tested for the antimony polymer and not found to be a factor.

Theoretically, the effect of stiffness can be understood in terms of the Flory and Fox description of *K*, from the Mark-Houwink relationship shown in Eq. 16.1 [11]:

$$\text{LVN} = KM^{a} \tag{16.1}$$

In Eq. 16.2 , *K* is described as follows [11]:

$$K = \phi(r_o^2/M) \tag{16.2}$$

where r_o is the end-to-end distance for the chain with a molecular weight *M* with a semiconstant referred to as the Universal Light Scattering Constant, ϕ.

Less flexible chains offer larger end-to-end distances per molecular weight unit, r_o^2/M, resulting in a larger *K* value. The *K* value for the antimony polyoxime is 95×10^{-3} in comparison to a value of *K* being 9×10^{-3} for polystyrene. Another factor that may contribute to the polymer residing in fewer flow planes is the poor solubility of the antimony polyoxime. Dimethly sulfoxide (DMSO), the solvent for which values are given above, is a poor solvent for the antimony polyoxime giving essentially no slope in the plot of intrinsic viscosity as a function of concentration. Sulfolane is a better solvent and a LVN of 39 ml/g is found more in line with the observed molecular weight.

This behavior for both new polymers produced and values for more common polymers have been reviewed and the behavior appears to be general for rigid poorly soluble polymers and is not related to the presence of metal or metal-like atoms in the polymer backbone.

4 Polyamines

Doak and Freedman [12] synthesized the monomeric organoantimony amines (**8**) from reaction of organoantimoney dihalides with monoamines. The analogous polyamines (**9**) were synthesized employing the diamine [13–15].

The products are oligomeric to high polymers with DP values ranging from 10 to 1,000. Yields ranged from about 10 to over 90%. A wide range of diamines were employed including simple aliphatic diamines such as 1,6-diaminohexane, simple aromatic diamines such as *p*-phenylenediamine, to more complex diamines such as adenine, 2,6-diamino-8-purinol, 4,4′-diaminodiphenylsulfon, Zineb, and 2,4-diamino-5(3,4-dimethoxybenzil)pyrimidine.

The products were tested for their ability to inhibit a wide range of bacteria, fungi, and yeast [13]. The ability to inhibit ranged from essentially no inhibition against any test organism for the product from adenine and triphenylantimony dichloride to inhibition of all ten test organisms for the 4,6-diamino-2-methyl-5-nitrosopyrimidine and triphenylantimony dichloride product.

The products were also tested for their ability to inhibit a number of cancer cell lines including BHK-21, L929, and HeLa cells [13]. The products generally exhibited decent cell inhibition with 50% growth inhibition values in the range of 5 μg/mL for the HeLa cells.

A number of other triphenylantimony dichloride-derived products were recently tested for their ability to inhibit Balb 3T3 cells as a measure of their ability as potential anticancer drugs [16]. This included those derived from cobalticinium-1,1′-dicarboxylic acid nitrate and thiopyrimidine. These materials showed GI50 values of about 10 micrograms/m.

$R_3SbCl_2 + R_1-NH_2 \longrightarrow R_3Sb(NH-R_1)_2$

(**9**)

(**8**)

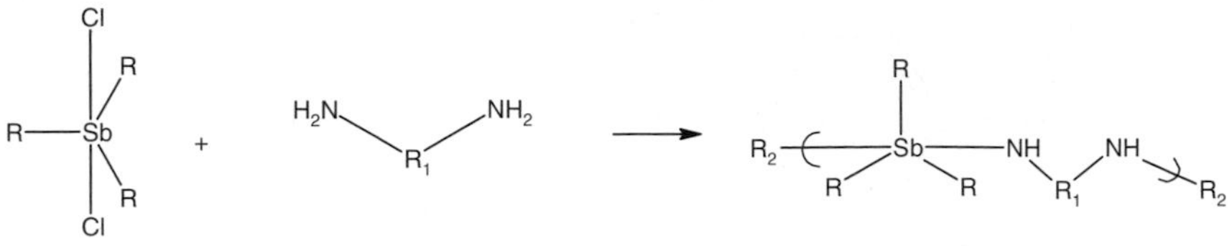

(9)

(9)

The effect of use of phase transfer agents and crown ethers was studied [15]. Comparison was made with the use of sodium hydroxide alone as the added base. Results were varied, with some added catalysts resulting in an increase in product yield and chain length and others giving a lower yield and chain length, and still others showing no change in product yield and chain length. For instance, the use of tetrahenyl phosphonium iodide showed an increase from 35 to 45% in product yield and a threefold increase in molecular weight for the reaction of 2,6-diamino-8-purinol and triphenylantimony dichloride. In comparison, the same phase transfer catalyst showed a decrease in yield from 38 to 28% and a fourfold decrease in chain length for the product of 2,5-dichloro-*p*-phenylenediamine and triphenylantimony dichloride in comparison to the simple sodium hydroxide system.

5 Polyethers

Mati and coworkers [17–21] synthesized a number of polyethers using a novel nitrate displacement polymerization. The structures of these materials is given below (**10–12**). This is part of an extensive study that includes evaluation of solubility parameters, biological characteristics, thermal properties, density, crystallinity, mechanical properties, and flame retarding ability. In fact, one of the most common uses for antimony oxides and organoantimony compounds is as flame retardants. The following is a description of some of these results.

Mechanical properties and hardness for NR and CR rubber impregnated by these antimony polymers were found to increase mildly. The flame retardancy of various rubber vulcanizates containing the antimony polymers increases based on the limiting oxygen index so that these materials may be nonleaching (nonmobile) antiflame additive materials. The polymers show decent activity against *Escherichia coli* bacteria and *Saccharomyces cerevisiae* yeast. They were also found to offer decent antifouling activity. The products are semiconductors with bulk resitivity values in the range of 10^8 to 10^9 Ω cm. The surface conductivities are also consistent with them being semiconductors.

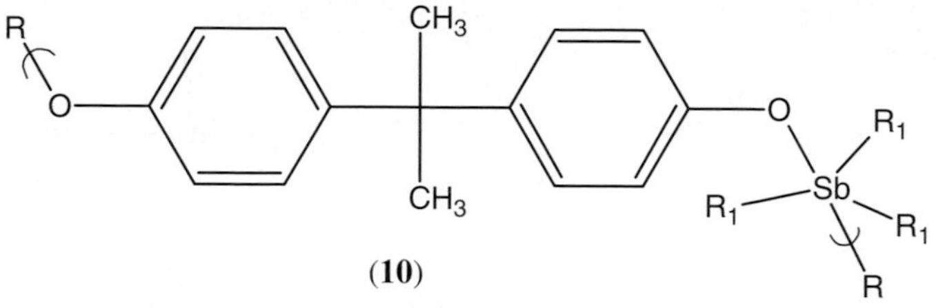

(**10**)

(10)

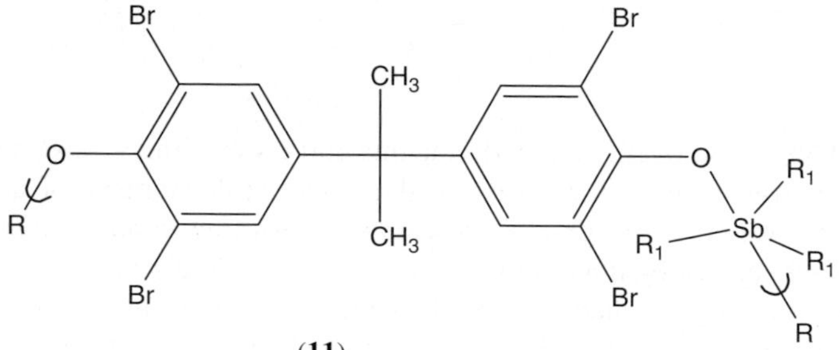

(**11**)

(11)

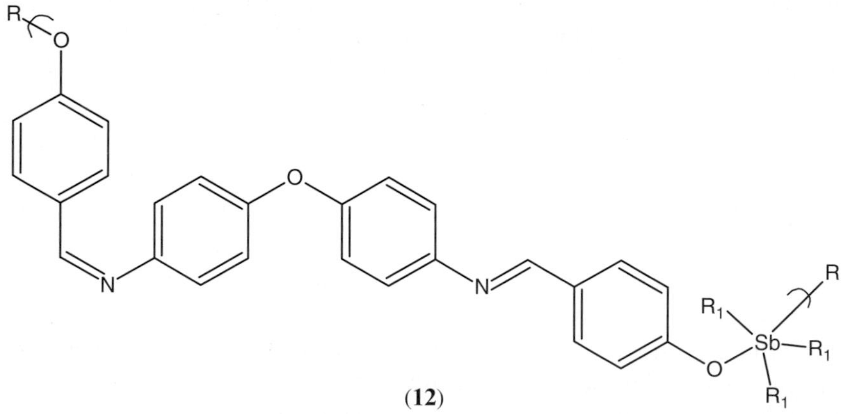

(**12**)

(12)

6 Incorporation of Known Drugs

Recently a number of organoantimony-containing polymers have been synthesized employing incorporation of known drugs. Incorporation of cephalexin was achieved [22,23] producing mixed amine-ester polymers (**13**).

These products were tested using Balb 3T3 cells [23]. The triphenylantimony derivative showed a GI50 of about 2 μg/ml whereas the product of trimethylantimony dibromide showed a GI50 of about 10 μg /ml.

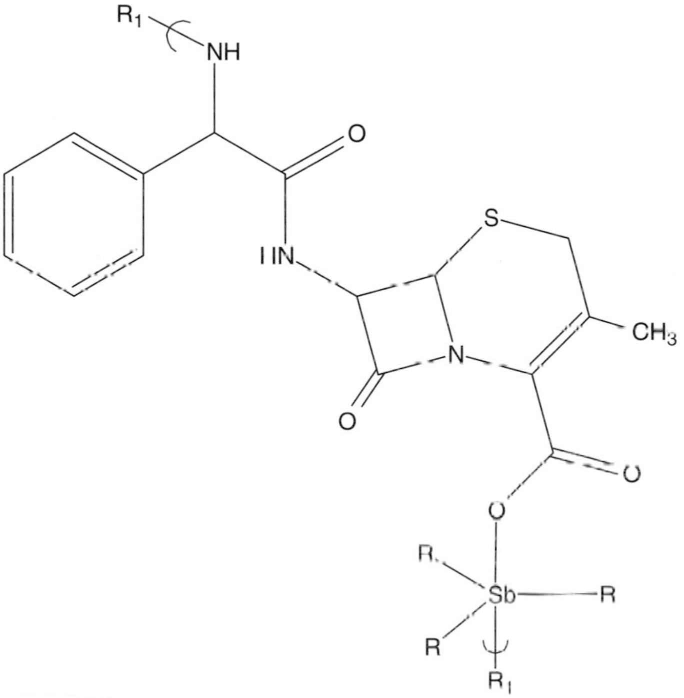

(**13**)

(**13**)

Ticarcillin was recently incorporated into similar products. As in the case of reactions utilized to produce the polyesters, ticarcillin possesses two acid groups that are neutralized by addition of sodium hydroxide producing the product below (**14**).

Reaction with organoantimony dihalides yielded products with the repeat unit shown below (**15**) [24].

Infrared spectra were consistent with the presence of both reactants and the formation of a new Sb bond. For the product of triphenylantimony dichloride and ticarcillin a new band about 1204 cm^{-1} (all IR bands given in cm^{-1}) is assigned to the Sb–O stretch. The upward shift in the salt carbonyl from about 1598 to 1645 was consistent with Sb–O–CO–R bond formation. A new band around 1280 was assigned to the Sb–O bond formation. An Sb–Ph stretch was found at 1440, smaller than that found for triphenylantimoney dichloride itself at 1434. Other phenyl-associated bands were found at 720 and 680, near those for triphenylantimony dichloride itself with bands at 732 and 686. The broad band centered at 3432 was assigned to NH stretching. The bands about 3059 were assigned to the C–H aromatic stretch present in the antimony moiety whereas those centering above this result from the C–H stretch in the ticarcillin. A series of bands from about 2990 to 2920 were derived from the C–H aliphatic found in ticarcillin.

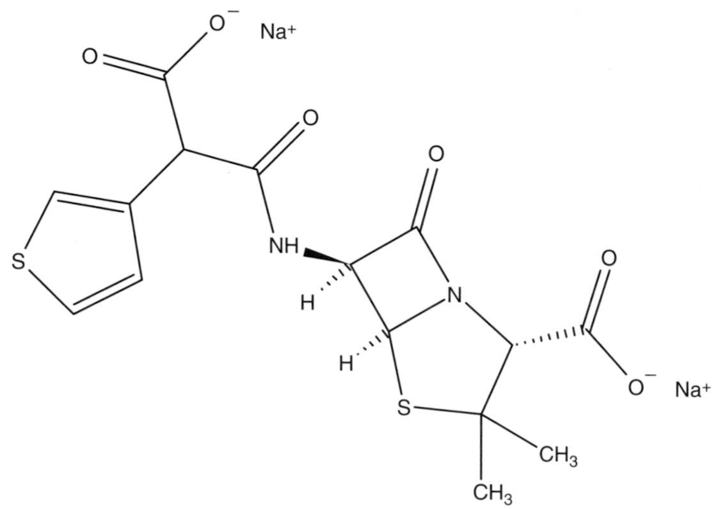

(**14**)

(**14**)

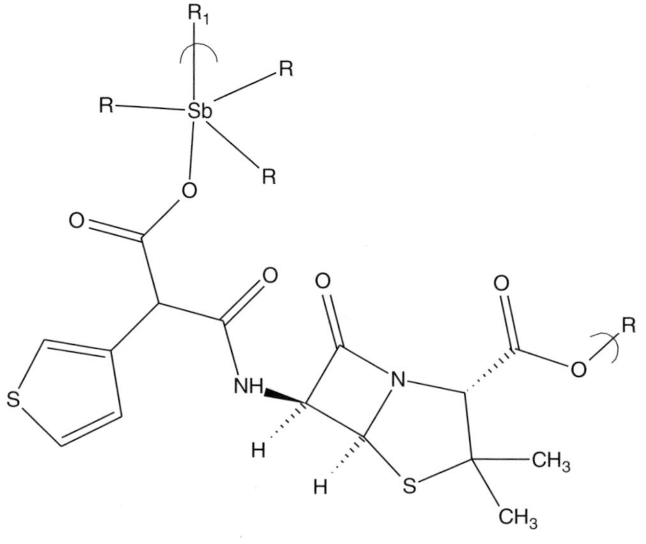

(**15**)

(**15**)

The antimony product was soluble in DMSO and partially soluble in HMPA, 1MPO, DMF, and sulfolane. The antimony product melts at about 156 °C becoming black at 222 °C with the coloration consistent with degradatoin of the product. Onset of coloration precedes the final darkening to black so that some thermal-induced degradation begins at a lower temperature.

Antimony has two natural-occurring isotopes, ^{121}Sb at 57% and ^{123}Sb at 43%. For a unit containing one antimony, the percentage of the 121-associated ion fragment should be 57% compared with 47% for the 123-containing unit. Table 16.1 contains some of the matches for the one-antimony ion fragments. The percentages are in rough agreement with what is expected.

Table 16.2 contains results for two, three and four units.

The presence of multiple antimony atoms in ion fragments causes a broadening of ion frequencies as seen in Tables 16.1 and 16.2. This broadening of ion frequencies, due to at least distributions of isotope abundances, causes a clustering of associated ion fragments. In Table 16.3, we note the maximum ion fragment mass number assuming that each noted ion location is really the most abundant ion

Table 16.1 Ion fragment abundances for one antimony-containing ion fragments

M/e	Assignment	%121-Containing Unit	% 123-Containing Unit
740,742	U	52	48
767,769	U+2O	53	47
777,779	U+CO_2	53	47
825,827	U+2CO_2	54	46
890,892	U+CO_2CHSP	55	45

Table 16.2 Isotope abundance results for multiple antimony-containing chains

2U+OH					
M/e	1,484	1,486	1,488		
Calculated (%)	32.5	49.0	18.5		
Found (%)	33.7	48.9	17.4		
3U+OH					
m/e	2,220	2,222	2,224	2,2226	
Calculated (%)	18.5	41.9	31.6	8.0	
Found (%)	20.8	39.6	31.3	8.3	
3U+triphenylantimoney and 2OH					
m/e	2,583	2,585	2587	2,589	2,591
Calculated (%)	10.6	31.9	36.0	18.1	3.4
Found ()%	12.5	35.6	38.1	9.4	4.4
4U+OH					
m/e	2,949	2,951	2,953	2,955	2,957
Calculated (%)	10.6	31.9	36.0	18.1	3.4
Found (%)	4.6	35.7	44.4	11.4	3.9

Table 16.3 Most abundant ion fragment clusters for the product from triphenylantimony dichloride and ticarcillin; >2000 daltons

m/e	Structural assignment	m/e	Structural assignment
2,006	3U-CO_2AB	3,874	5U+CO_2AB
2,093	3U+CO_2AB	4,227	5U+SbO_2CAB
2,222	3U+OH*	4,903	7U-O_2ABNHCO
2,404	3U+O_2CAB	5,432	7U-Sb,2OH*
2,583	7U+T*	6,475	9U-O_2CCHTP
2,632	3U+Sb,2Cl*	6,997	9U+T*
2,772	4U-O_2CCHTPCO	7,735	10U+Sb,2OH*
2,985	4U+CO_2	8,299	11U+T-O_2CCHTPCO
3,322	4U+T*	8,690	12U-O_2CCHTP
3,378	4U+Sb,O_2CHCO	8,988	12U+O_2CCHTPCO
3,631	5U-CO_2		

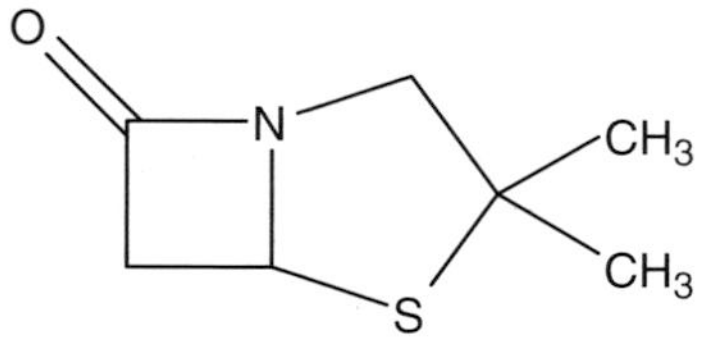

(16)

fragment location for that cluster of ion fragments. Furthermore, only the most abundant ion fragments are identified. Numerous secondary (less intense) ion fragments are present. In the presentation of data, several abbreviations will be used in the assignment of ion fragment identities. The letter "T" is used to represent the ticarcillin moiety (minus two protons); Sb represents the triphenylantimony moiety, that is Ph_3Sb moiety; U represents one unit, 2U represents two units, and so on; AB represents the azabicyclo ring system, shown in **16**, and SP will be used to represent the thiophene moiety. All ion fragment masses are given in m/e = 1, that is in daltons.

Table 16.3 contains results for the triphenylantimony polymer.

Some full units are found designated by a "*". Although there are some ion fragments that contain Cl end-groups, the bulk of the ion fragments do not. The presence of only a few Cl end-groups is consistent with the high tendency for the Sb-Cl to undergo hydrolysis.

Products have also been synthesized incorporating the known antiviral acyclovir as part of a project to synthesize third generation antiviral agents [25]. The products are of structure shown below in **17**.

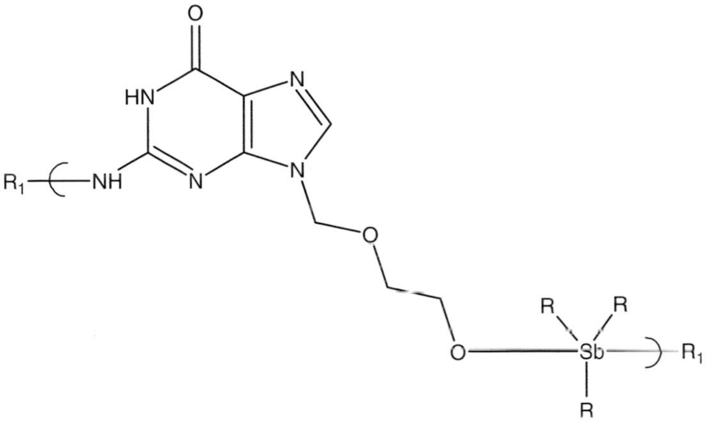

(17)

(17)

7 Other

Gudyno and coworkers [26] reported the used of cation exchanges based on antimony polymers. The exchange resins were able to remove strontium.

Belinskaya and coworkers [27] reported a similar study looking at the sorption of cadmium by inorganic cation exchangers based on antimony polymers. These products were also described as being of possible therapeutic use. Thus, rats were given doses of $CdCl_2$ and then given 100 mg of the silicoantimonic or phosphonoantimonic polymer resulting in 2 to 67% decrease in uptake of the cadmium by the liver and kidney in comparison to a control group of rats.

A patent by Predvoditelev and coworkers [28] describes the synthesis of antimony-containing polymers from reaction of trialky antimonites with such hydroxyl compounds as cellulose, starch, and poly(vinyl alcohol). The products presumably contain Sb–O units and as such are antimony ethers.

Feng and coworkers [29] reported on the synthesis and fluorescence properties of ternary complexes of polymer-bound organometal moieties including triphenylantimony. These complexes also contain the rare earth metal ions Eu, Sm, and Tb and thenoyltrifluoroacetone. The general structure is given below (**18**) where R = phenyl.

The fluorescence properties were investigated. The Eu^{+3} product was the best, giving fluorescence lifetimes between 0.35 and 0.47 MS. The relative intensity compared with a simple binary complex for the antimony product was over 150. The highest relative intensity was for the analogous arsenic product with a value of 300. The most intense emission peak occurred ay about 614 nm and corresponded

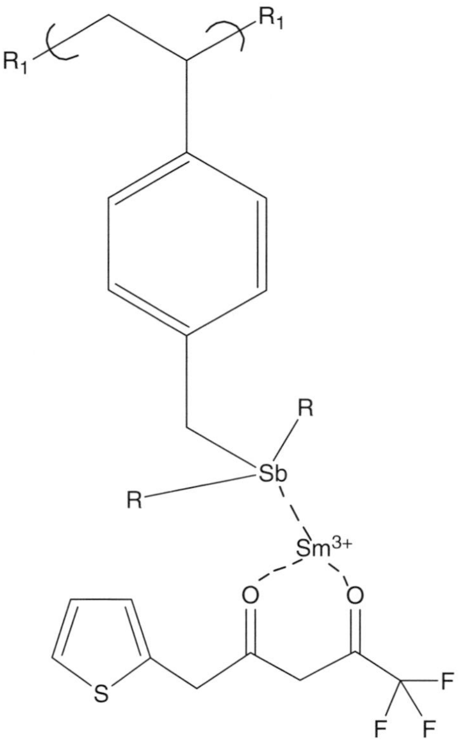

(**18**)

(18)

to the $^5D_o \rightarrow {}^7F_2$ transition of the Eu^{+3} (electric dipole transition that is sensitive to the crystalline field). Other peaks occurred at 592 and 648 nm and corresponded to the $^5D_1 \rightarrow {}^7F_j$ transition, which is insensitive to the crystalline field. Thus, the formation of polymeric ternary complexes enhances the fluorescence intensity of the rare earth ions tested. The precise results varied with the rare group and main group ions as well as the particular polymer employed.

References

1. Razuvaer G, Zinovera T, Brilkina T. (1969) Izv Akad Nauk SSSR. **188:**830.
2. Doak G, Freedman L. (1970) **Organometallic compounds of arsenic, antimony, and bismuth**. Wiley, New York, p 330.
3. Carraher C, Blaxall H. (1975) Polymer P. **16:**261.
4. Carraher C, Blaxall H. (1979) Angew Makromol Chemie. **83** 37.
5. Carraher C, Venable W, Blaxall H, Sheats J. (1980) J Macromol Sci.-Chem. **A14:**571.

6. Sheats J, Blaxall H, Carraher C. (1975) Polymer P. **16:**655.
7. Carraher C, Blaxall H, Schroeder J, Venable W. (1978) Org Coat Plast Chem. **39:**549.
8. Harrison P, Zuckerman J. (1970) Inorg Nucl Chem Let. **6:**5.
9. Carraher C, Hedlund L. (1975) Polymer P. **16:**264.
10. Carraher C, Hedlund L. (1980) J Macromol Sci.-Chem. **A14:**713.
11. Carraher C. (2003) **Polymer Chemistry**, Sixth Ed., Dekker, NY.
12. Doak G, Freedman LD. (1970) Org Chem Revs. **6:**574.
13. Carraher C, Naas M, Giron D, Cerutis DR. (1983) J Macromol Sci.-Chem. **A19:**1101.
14. Carraher C, Naas M. (1982) Polym P. **23:**158.
15. Carraher C, Naas M. (1984) In: Mathias L, Carraher C (eds), **Crown ethers and phase transfer catalysis in polymer science**. Plenum, New York.
16. Siegmann-Louda D, Carraher C, Pflueger F, Nagy D, Ross J. (2002) In: Carraher C, Swift G, (eds), **Functional condensation polymers**. Kluwer, New York.
17. Karak N, Maiti S. (1997) J Polym Mater. **14:**71.
18. Karak N, Maiti S. (1999) Angew Makromol Chemie. **265:**5.
19. Karak N, Maiti S. (1998) J Appl Polym Sci. **98:**927.
20. Karak N, Maiti S, Das S, Dey S. (2003) J Polym Mater. **20:**237.
21. Karak N, Maiti S, Sanigrahi S, Chowdhary R. (1998) Ind J Chem Tech. **5:**217.
22. Carraher C, McBride G, Zhao A. (2003) Polym Mater Sci Eng. **88:**424.
23. Siegmann-Louda D, Carraher C, Quinones Q, McBride G. (2003) Polym Mater Sci. Eng., **88** 390.
24. Carraher C, Morie K. (2005) Polym. Mater. Sci. Eng. **93:**387.
25. Sabir T, Carraher C. Unpublished results.
26. Gudyno T, Nikashina V, Belinskaya F. (1980) Neogran Ionoobmen Mater Leningrad. **2** 103.
27. Militsina E, Gigorova N, Runeva T, Borisov V, Seletskaya L, (1979) Ion Obmen Ionoobmetriya. **2:**47.
28. Predvoditelev D, Buyanova V, Konkin A. (1968) USSR Patent No. 237865.
29. Feng H, Jian S, Wang Y, Lei Z, Wang R. (1998) J Appl Polym Sci. **68:**1605.

Bacterial Inhibition by Organotin-Containing Polymers

Charles E. Carraher Jr., Yoshinobu Naoshima, Kazutaka Nagao, Yoshihiro Mori, Anna Zhao, Girish Barot, and Amitabh Battin

1 Bacteria Basics

Bacteria are small (about 0.0003 to 0.002 mm (0.00001 to 0.00008 in.) one-celled organisms living throughout our earth. A single cup of soil may contain more than 10 billion bacteria.

Most bacteria do not cause diseases. They help in digestion and in destroying harmful microorganisms. Some intestinal bacteria produce needed vitamins. In soil and water they assist in the recycling of nutrients. Many assist in the decay, and decomposition of dead organisms and animal waste. Some bacteria convert nitrogen in the air, soil, and water into nitrogen compounds needed by plants. Fermentation, used in making alcoholic beverages by converting sugar into ethanol ("drinking alcohol"), cheese and other foods, is caused by certain bacteria. Sewage treatment plants utilize bacteria to help in the purification of water. Bacteria are also employed in the synthesis of certain drugs.

A.S. Abd-El-Aziz et al. (eds.), *Inorganic and Organometallic Macromolecules: Design and Applications.*

Many bacteria are harmful only when present in the wrong part of the body. A prime example of this is *Escherichia coli,* which is generally harmless but when located in the intestines can cause food poisoning. Other bacteria, such as anthrax, are poisonous regardless of where they are located in the body. Bacteria from the same family may exist as any number of strains with both similar and different properties. For instance, *Pseudomonas aeruginosa* strains can react toward selected drugs in a variety of ways depending on the particular strain. Unfortunately, strains found in hospitals are generally the most difficult to control and are called resistant strains. Some bacteria produce poisons—toxins that cause diseases such as tetanus, diphtheria, and scarlet fever. Others produce toxins only after their demise. Some enter the body through sores, punctures, cuts, and abrasions, whereas others are passed mistakenly from one organ to another. Bacteria also cause diseases in animals and plants.

Once in the body, harmful bacteria are met by white blood cells that destroy them. Antibodies assist in the control of these invaders. When our body is not able to effectively control bacterial invasion, a physician may prescribe a shot and/or pills that contain an antitoxin from an animal or another person or a vaccine made from dead or weakened bacteria. The vaccines stimulate antibody production in the body and some can provide long-term protection.

Bacteria are generally enclosed by a tough protective layer called the cell wall. Scientists identify strains of bacteria based on the shape of the their cell wall and their ability to retain or reject specified organic dyes (called stains). There are four major "shapes of bacteria." Cocci are round and linked together, bacilli are rod like, spirilla are spiral shaped, and vibrios appear as bent rods. Three prefixes can be added to these four major shapes to identify additional bacterial. Diplo is employed to describe paired shapes, strepto-chains, and staphlo-clusters. Thus, a staphylococci is a cluster of round bacteria. The description of the size, shape, structure, and arrangement of living objects is called the morphology of those living objects.

Certain structures, such as the mesosomes and flagellum, are not common to all bacteria. Some bacterial cells are surrounded by a vicious material which forms an envelope about the cell wall. The envelope serves to protect the bacteria from drying out through regulation of water emission, acts to block attachments and engulfment by unfriendly microorganisms, and promotes attachment of the bacteria to surfaces.

Streptococcus mutons, a bacterium associated with the production of dental caries, attaches itself to the smooth surface of teeth. The envelopes can be made of polypeptides but are generally composed of polysaccharides.

Two major groups of bacteria exist: (1)the eubacteria—which are typical of most bacteria we have about us— and (2) the archaeobacteria—which are less common and include methane-producing bacteria. This chapter focuses on eubacteria. For eubacteria, the shape-determining part of the cell wall, is largely peptidoglycan—an insoluble, porous, crosslinked giant molecule of good strength and rigidity. Whereas the peptidoglycan varies from one species to another, it is basically a polymer of *N*-acetylglucosamine, *N*-acetylmuramic acid, L-alanine, D-alaine, D-glutamate and a dipeptide. Because the peptidoglycan wall is rigid, it is constantly degraded and rebuilt to allow for growth.

Bacteria are further divided according to their ability to be stained. Thus gram-positive bacteria retain the stain from a crystal violet-iodine complex whereas gram-negative bacteria do not. The cell walls of gram-negative bacteria are generally thinner (10–15nm) than the cell walls of gram-positive species (20–25nm). The membranes connected to the cell walls may contain many features. Thus for gram-negative cells, the outer membrane is anchored to the peptidoglycan by means of a lipoprotein. The membrane is a bilayer composed mainly of phospholipids, proteins, and lipopolysaccharides (LPSs). A LPS consists of three covalently bonded units: a lipoprotein, polysaccharide, and polysaccharide *O*-antigens that extend like long arms from the membrane surface.

Although impermeable to giant molecules, the outer membrane allows smaller molecules—such as amino acids, water, small sugars, and nucleosides—to pass through using channels in special proteins called prions which span the membrane. The prions also serve other functions.

The cytoplasmic membrane lies immediately below the cell wall. It is about 7.5 nm thick and is composed largely of phospholipids and proteins. This membrane acts as a hydrophobic barrier to penetration in most water-soluble molecules.

Bacterial cells do not contain membrane-enclosed organelles corresponding to the mitochondria and chloroplasts of eucaryotic cells. Some bacteria contain mesosomes that are involved in DNA replication and cell division. The cell material bounded by the cytoplasmic membrane can be divided into: the cytoplasmic area, rich in the RNA-protein bodies called ribosomes; the chromatinic area rich in DNA; and the fluid portion.

Unlike animal or plant cells, there is no endoplasmic reticulum to which ribosomes are bound. Bacteria contain storage deposits of certain building blocks. The volutin granules serve as a reserve source of phosphate and are themselves composed of polyphosphates. Poly-β-hydroxybutyrate (PHB) a lipid-like material, serves as a carbon and energy source.

Again, in contrast to eucaryotic cells, bacterial cells contain neither a distinct membrane-enclosed nucleus nor a mitotic part. They do contain an area near the center of the cell that is called the nuclear structure, which is where DNA is found. This DNA is a single, circular, giant molecule in which all of the genes are linked.

Bacteria reproduce asexually, each cell simply dividing into two identical cells by a process called binary fission. Most bacteria reproduce rapidly, often doubling every hour. Thus in 24 hours there would be about 2^{24} or 17,000,000 (17 million) cells from one bacterial cell if adequate nutrients were available.

Binary fission results in the two bacterial cells having identical DNA. Some bacteria can exchange DNA via a single sexual process called conjugation, within which direct transfer of DNA from one type of cell—called a male—to a second type of cell—called a female. DNA may also be transferred from one cell to another through employment of viruses. Finally, some bacteria can pick up fragments of DNA from dead bacterial cells. Interestingly, traits can be transferred from one cell to another in this way. Thus, bacterial cells that are resistant to specific drugs may transfer this characteristic to nonresistant cells in this manner.

2 Antibacterial Action

It is beyond the scope of this chapter to cover antibacterial activity in any manner save an elementary and brief fashion. As noted before, bacterial cells grow and divide, eventually reaching large numbers sufficient to cause infections under certain conditions. In order to grow and divide, the bacteria must synthesize and/or procure many types of building blocks. Essentially, all antibacterial agents interfere with specific processes which allow the bacteria to grow and/or divide. Figure 17.1 contains an abbreviated general structure of a bacterium, highlighting sites of

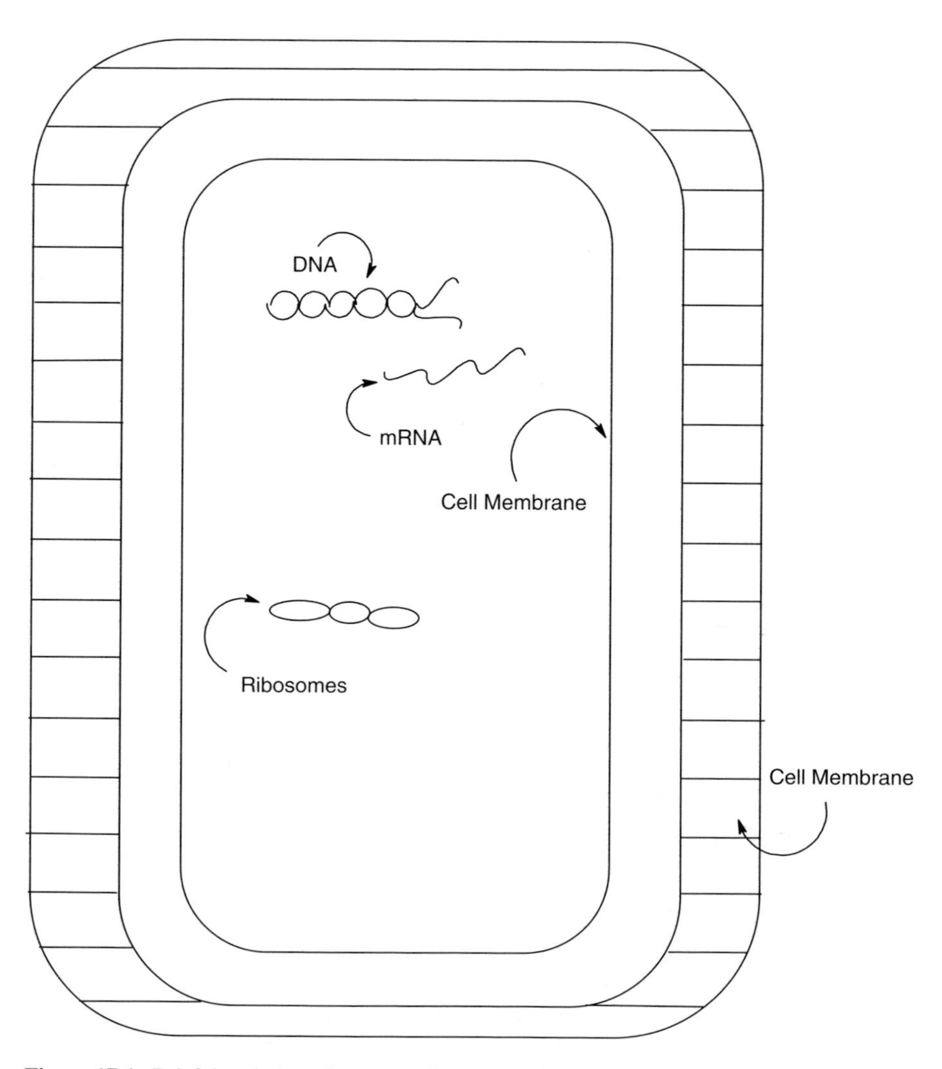

Figure 17.1 Brief description of a general bacteria highlighting sites for drug intervention

antibacterial action. The general divisions of antibacterial agents are as follows along with examples of drugs that act at that site:

1. Inhibitors of cell wall growth, including those that combine with the cell wall substrates, inhibit polymerization and attachment of new agents to the cell wall, combine with carrier molecules and those that inhibit enzymes critical to cell growth. Bacitracin, vancomycin, penicillins, cephalosporins;
2. Inhibitors of nucleic acid synthesis that include drugs that influence DNA template actions in cluding replication, inhibit RNA polymerase, and those that inhibit nucleotide metabolism (e.g., flucytosine (fungi), acyclovir (viruses), and quinolones);
3. Folate metabolism antagonists (e.g., sulfonamides, trimephoprim); and
4. Inhibition of ribosome functions (e.g., steptomycin, kanamycin, tetracyclines, fusidic acid, and erythromycin).

2.1 *Inhibition of Cell Wall Synthesis*

As noted before, bacteria have somewhat solid walls that are often in need of rebuilding as a result of growth and reproduction. Most antiwall-building antibacterial agents focus on the peptidoglycan layer. This layer protects the bacteria's hypotonic environment. Damage to this layer destroys the cell wall rigidity, leading to the demise of the bacteria. The peptidoglycan synthesis occurs in three stages. Briefly, the first stage occurs in the cytoplasm where low-chained peptides are made. Many antibacterial agents interfere with the early stages of cell wall construction. The second stage of the wall development is catalyzed by membrane-bound enzymes. The transporter is a phosphorylated undecaprenyl alcohol. Some antibacterial agents interact with this alcohol, thereby diverting it from its essential role. The third stage of wall synthesis involves the polymerization of the small units and attachment of nascent peptidoglycan to the cell wall. β-lactam antibacterial agents interfere with the cleaving of an essential peptide bond again preventing cell wall construction. Penicillin and most derivatives inhibit bacterial growth acting as β-lactam antibacterial agents.

2.2 *Inhibition of Cytoplasmic Membranes*

Lipids, lipoproteins, and proteins makeup cytoplasmic membranes. This membrane is a diffusion barrier for ions, nutrients, transport systems, and most importantly, water. It is composed of a lipid grouping with globular proteins that penetrate the lipid bilayer. Most antibacterial agents that inhibit cytoplasmic membranes do so by influencing the balance of cations, anions, or neutral compounds, thus disrupting membrane operation. Of interest is that fungal membranes contain sterols whereas bacterial membranes do not. Thus, selective inhibition is possible if sterile well-being is compromised.

2.3 Inhibition of Nucleic Acid Synthesis

Antibacterial agents can interfere with nucleic acid synthesis at a number of junctures including replication and/or transcription of DNA through preferential binding or insertion into the DNA. They may also influence the proteins involved with the replication. Other antibacterial agents interfere with the RNA synthesis, action, or proteins associated with these functions. They may also interfere with the coiling and uncoiling necessary for appropriate action.

2.4 Inhibition of Ribosome Function

Bacterial ribosomes can be divided into two subunits referred to as 50S and 30S. It is possible to limit action to just one or to both subunits. Aminoglycosides bind to ribosome subunits. These aminoglycosides are complex sugars. These agents must have a free amino or hydroxyl group(s) that binds to specific ribosomal proteins, thus inhibiting their activity. Thus, streptomycin binds to a protein in the 30S subunit causing the ribosome to misread the genetic code. In general, such drugs kill bacteria by causing misreading.

2.5 Folic Acid Inhibition

Some drugs interfere with folate metabolism whereas others block the synthesis of various components of folic acid. Unlike mammals, bacteria typically do not have a transport system to obtain preformed folic acid from their environment so folic acid must therefore be synthesized on site. Sulfonamides act through blocking the conversion of *p*-aminobenzoic acid and pteridine to dihydrofolic acid. Blocking occurs because sulfonamides have a greater activity toward pteridine synthetase as compared with *p*-aminobenzoinc acid, thus preventing the pteridine synthetase from performing its role of converting pteridine and *p*-aminoacid to dihydrofolic acid.

This section has briefly reviewed some of the sites thatantibacterial agents are known to attack. Numerous and varied, these sites show that a well planned combination of drugs within the same polymer chain have a definite possibility of intersecting. They also show that there are many combinations within which organotin compounds may inhibit bacterial growth.

3 Organotin Activities

3.1 Monomeric

Organotin compounds are the most widely used organometallic compounds with a variety of commercial uses [1]. The widespread use of organotin compounds results from a number of factors. First, tin is a relatively abundant metal and is readily available. Second, the synthesis of a number of organotin compounds is straightforward and has been widely employed for over half a century. Third, the bioactivity of organotin compounds was discovered more than 75 years ago. Because of the relatively low toxicity of certain of these organotin compounds, they offered a good window of application compared to toxicity.

Today, there is legislation that prohibits use of monomeric organotin compounds for many applications, thus there is an increased emphasis on the synthesis and application of bound organotin compounds (i.e., organotin moieties that are part of a polymer). The polymer can be part of the polymer backbone or as an appendage draping from the polymer backbone.

We have been involved in the synthesis and characterization of organotin polymers for over 30 years. The synthesis of organotin polymers was recently reviewed [2]. Much of our interest in organotin polymers is related to their antibacterial, anticancer, and antiviral activity [2–32]. The antibacterial and anticancer activity of these polymers has been recently reviewed [2,3].

Tin is a ubiquitous element and enters mammals through the food chain. It has never been found to be dysfunctional in either plant or animal tissue and has been regarded as an innocuous background material. In the inorganic state tin is generally accepted as being relatively nontoxic. The addition of one or more organic groups to the tin atom, however, has a profound effect in terms of overall biological activity. The nature and magnitude of bioactivity is determined by the nature, number, and configuration of the organic substituents. In general, the biological activity within each class of aliphatic organotins is related to the number of Sn–C bonds and the number of carbon atoms per side chain. Variation of the inorganic group, X, within any given R_3SnX series is usually found to have no significant effect on the overall bioactivity of the molecule, except that X itself may be a toxic component. In this case the biological activity of the organotin compounds may be enhanced. The bioactivity profile resulting from substitutions within R_nSnX_{4-n} may therefore be most easily appreciated by considering the effects of sequential changes in R and n. It is now well established that the progressive introduction of organic groups into the above model produces maximal toxicity when $n = 3$. However, the impact of R varies with the biological test system. The highest mammalian toxicity is seen when R is ethyl. The tributyl- and triphenyl- tins on the other hand, show highest activity against fungi and have been applied commercially without apparent adverse effects to the persons applying them. Within the class of trialkyltin compounds there are considerable variations in toxicity depending on the side-chain length of the alkyl groups. An increase in the n-alkyl chain length produces a sharp drop in biocidic activity and the long-chain species, like octyltin derivatives, are essentially nontoxic to all organisms. Tetraorganotins show a delayed toxic activity in organisms. It is suggested that only after their degradation to trisubstituted compounds will symptoms of poisoning be observed. The environmental degradation of organotin compounds, such as via ultraviolet (UV) irradiation and biological cleavage, is especially important [33]. Study showed that trisubstituted organotin compounds degraded to an inorganic tin species, via di-, and monosubstituted organotin intermediates [34]. This is one of the advantages of organotins as plant protection over conventional fungicides and pesticides, especially those based on mercury, copper, or arsenic. These latter elements are permanently toxic. Their residues remain behind on the crops and repeated spraying has raised concern over their accumulation within the environment and food chains. With organotin compounds, toxic residues are considered to be less of a problem because the toxicity of the compounds resides in the organotin radical and not the tin itself. Organotin compounds are broken down gradually over time by both UV light and weathering to SnO_2, which, like other inorganic tin compounds, is essentially without biocidal activity [35].

The general toxicity of organotins seems to follow no general rules. Toxic effects are often highly species-specific [36,37] (Table 17.1).

Some distinction can be made between the general toxic effects of the di-, tri-, and tetra- substituted compounds. It has been established that some of the trisubstituted compounds act within the central nervous system (CNS), producing cerebral edema. The disubstituted compounds, on the other hand, have no neurological effects but are powerful irritants and cause hepatic biliary inflammation following ingestion. The tetraorganotins resemble their triorganotin counterparts, although there is some delay before their effects become apparent. It has been suggested that this lag period stems from the fact that the tetraorganotin form is converted to a triorganotin derivative following metabolism in the liver [40] and/or gastrointestinal tract [41]. For the insect and plant toxicity in R_3SnX compounds, the X anion has only a minor influence on overall bioactivity, although in R_2SnX_2 compounds the X group does show an effect on bioactivity [42]. The immunotoxic effects of organotin compounds also are species-specific. The dietary administration of organotin compounds produced marked dose-dependent thymic atrophy and reduced splenic and lymph node weights. It is noteworthy that these effects do not cause any overt pathological change in other body tissues [43,44].

Table 17.1 Biological activity of triorganotin compounds as a function of species [38,39]

Species	Most active "R" group in R_3SnCl compounds
Insects	Me
Gram-negative bacteria	Et
Gram-positive bacteria	Bu
Fungi	Bu
Fish, Mollusks	Bu
Mammals	Et

Little is known about the effects of organotins in humans. There are no epidemiological studies available on the chronic low-level exposure. However, some case reports have described various health effects after acute exposure to tributyltin and triphenyltin conpounds. The patients developed acute nephropathy and mainly CNS disorders [45–47]. Some reports described the neurotoxicity of trimethyltin, triethyltin, and triphenyltin compounds in humans [48,49]. There are even some reports about cases of tributyltin, triphenyltin, and TBTO induced tumors [50–52]. However, the precise mechanism of tumor induction is not known.

3.2 Polymeric

Because of the added evaluative studies required for internal use, organotin materials are studied as materials suitable for treatment of infections (here mainly topical), contaminated sites, preventative agents, and as treatments for water sources. The organotin-containing drugs studied here were rapidly synthesized (generally within 30 seconds) with good yield employing readily available reactants. Thus, small to large scale availability of these target drugs is readily achievable. They can be used internally or topically as additives to creams, cleaning detergents and soaps, coatings (i.e., paints), plastics, and paper. They can be handled without gloves or other protective ware and have shelf-lives in excess of several years [2].

The polymers studied here have inhibited a wide variety of bacteria, fungi, and yeast. The following is a list of agents successfully inhibited by the compunds discussed in this chapter [2].

Escherichia coli	*Bacillus subtilis*	*Saccharomyces cerevisiae*
Branhamella catarrhalis	*Staphylococus epidermidis*	*Alcaligenesfaecallis*
Enterobacter aerogenes	*Neisseria mucosa*	*Staphylococcus aureus*
Klebsiella pneumoniae	*Acinetobacter calcoaceticus*	*Psudomonas aeruginosa*
Aspergillus flavus	*Aspergillus niger*	*Candida albicans*
Aspergillus fumigatus	*Penicillin sp.*	*Trichp[hyton mentagrophytes*
Trichoderma reesei	*Chaetomium globosum*	*Methicillin-resistant Stpahylococcus aureus*

Most of these products can be incorporated into paper, plastics, textiles with little loss in biological activity [2]. The products from PVA and some from lignin deserve special mention because of their ability to inhibit *Candia albicans*, the microorganism most responsible for yeast infections [53–56]. This inhibition is selective leaving much of the natural flora in tact. The organotin modified PVA products (Structure 17.1) dramatically out performed commercial preparations [56]. Products from dihaloorganotin are crosslinked whereas those from monohaloorganotin

Structure 17.1

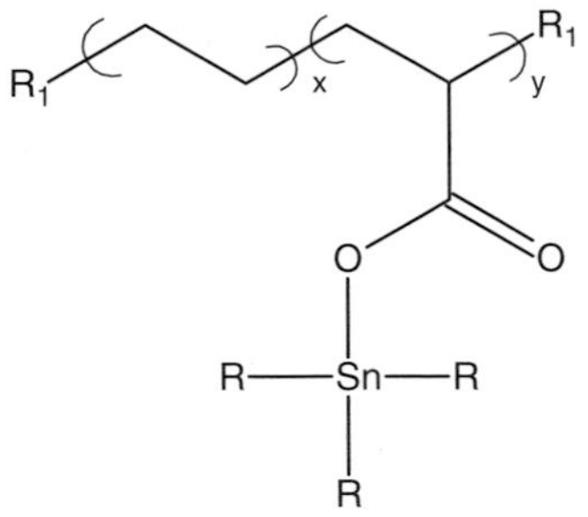

(1)

reactants are linear and soluble. Flexible films and fibers can be made from the linear products and ionomeric materials.

With the appearance of new resistant strains of common bacteria the need for new treatment rationales increases. In response to this, we have looked at particularly dangerous microorganisms where the "cure" is either extreme or unavailable. The most insidious microorganism involved in nosocomial infections is methicillin-resistant *S. aureus*MRSA (also called *Staph MRSA*). This microbe commonly colonizes those patients who are seriously ill and is prevalent in high-risk areas such as intensive care and burn units. It is also a significant risk factor in surgical wound infections. The incident of MRSA infection in hospitals is increasing at an alarming rate. It may be carried by otherwise healthy healthcare providers and transferred to the patient during routine bedside attendance. Other areas that may harbor the organism are air handling ductwork, linens, and general room contamination. We have developed a number of organotin-containing polymeric materials that inhibit MRSA and may be incorporated into soaps, cleaning agents, and coatings for the purpose of prevention and decontamination [57,58]. These include products derived from dextran, cellulose, and lignin.

The ability of these organotin-containing polymers to successfully inhibit a wide number of microorganisms, including resistant bacteria as MRSA, makes them prime candidates as anti-biological terrorism agents. They may be used to treat microorganisms such as *Bacillus anthracis*, which is not responsible for anthrax; *Yersinia bub.*, which is responsible for bubonic plague; and *Francisella tularensis*, which is responsible for tuleramia. Some of these products are also active against a number of fungus that cause mildew and rot as well as against the microorganisms responsible for ringworm and athlete's foot [2].

4 Specific Organotin Antibacterial

Because of the increased worldwide production or organotin compounds for commercial use considerable amounts of organotins have entered various ecosystems. Inorganic tin compounds are often said to be nontoxic, but their toxicity ranges from being nontoxic to moderately toxic. Organotin compounds offer varied toxicities from being mildly toxic to highly toxic with most of the compounds falling in the mildly to moderately toxic range. The toxicity–ecosystem relationship has been recently reviewed [1].

It seems strange that organotin compounds have been employed as antibacterial agents for over 50 years when relatively little is known about how they act to inhibit bacterial growth. This may be why many pharmaceutical companies steer away from organotin drugs. Although organotin compounds are widely used in industry for the prevention of bacterial infection, they are generally not prescribed by physicians to treat internal bacterial infections. Thus, drug companies have little stake in determining their biological mechanisms. They also offer a number of sites for activity and it is not simple to locate merely one site of activity, as is the case with many other antibacterial agents. It must also be noted that the variety of bacteria is so great that studying each variety of bacteria may yield different mechanistic results. Thus, alterations on growth, oxygen consumption, and physical properties of membrane lipids are stronger in *Bacillus stearothermophilus* as compared with *B. subtilis*.

Here we will look at some findings that highlight potential modes of activity as well as problems associated with looking at organotin activity. Further, essentially no mechanistic studies have been carried out on tin polymers, so results from studies of small molecule-containing organotin agents will be examined.

Because of their widespread use industrially many of the mechanistic studies have involved trialkyltin chlorides and dialkyltin chlorides. Di-*n*-butyltin dichloride (DBTC) and tri-*n*-butyltin chloride (TBTC) are the most widely used organotin compounds because of their relatively low toxicity to humans and their high toxicity to bacteria. We will also look at results from other organisms to indicate possible sites of attack.

Organotin pesticides inhibit oxidative phosphorylation intervening in the dinitrophenol uncoupling thus preventing the formation of adenosine triphosphate, a high-energy phosphate. The trialkyltins also inhibit photophosphorylation in chloroplasts and as such can serve to prevent the growth of algae.

Trimethyltin (TMT) is a powerful toxin that selectively kills cells in the CNS, lung, kidney, spleen, and in the immune system [60–62]. In fact, its toxicity has lead to its use to describe the phenotype observed after intoxication in rodents. Abnormalities after intoxication include self-mutilation, vocalization, whole-body tremors, hyper-reactivity to touching, and spontaneous seizures [63].

Similar pathological responses have been found for other mammals including humans [64,65].

TMT poisoning leads to apopotic cell death including nuclear fragmentation, chromatin condensation, membrane blebbing, caspase activation, mitochondrial dysfunction, and production of reactive oxygen compounds [60,66–69].

Davidson and coworkers found that the target for both TMT and dimethyltin dichloride (DMT) in the mitochondria is Stannin (Snn) [70].

Thus, on the cellular level, TMT poisoning results in widespread disruption of cellular machinery.

Results from five taxonomic kingdoms have been reported for TBT [71–74].

The main toxic effects reported have been endocrine disruption [75] and mitochondria function impairment [76]. Even so, the molecular mechanisms involving TBT inhibition are far from understood. Along with the sites mentioned above, biological membranes have been suggested as targets because of the high-lipophilicity of TBT [71,77].

Martins and coworkers found that there was a putative relationship between respiratory activity impairment and growth inhibition brought about by TBT suggesting that TBT influences the behavior of membrane lipids [59].

5 Combining of Antibacterial Agents

We attempted to combine antibacterial agents in the same polymer chain. One of the antibacterial agents was typically an organotin moiety. The other antibacterial agent was often a member of the penicillin family. There were reasons for combining such agents. First, by judicious choice of such agents, the combined drug can inhibit bacterial growth through several distinct routes, thus enhancing the chances for its eradication. Second, combined drugs act to prevent bacterial resistance. Third, a judicious choice of such agents can also act to inhibit mixed infections where the infecting agents contain a mixture of bacteria. Fourth, combined drugs can be employed where the precise identity of the particular bacteria is unknown—resulting in greater effectiveness of treatment. Fifth, a correct choice of combined drugs may allow synergy to occur resulting in a lower dose of drug to be utilized to successfully treat the bacterial infection.

6 Case for Polymeric Antibacterial Agents

The case for polymeric drugs is somewhat different depending on their intended use. Organotin-containing polymeric antibacterial agents can act singularly as drugs or as reservoirs for long-term release of the drug(s). Some reasons why the delivery of antibacterial drugs via polymers may be advantageous for internal use as drugs are briefly described here. First, controlled release of the drug will allow a more concentrated presentation level of the drug over a longer time period than use of small-molecule drugs. This may allow a lower dosage to be effective in curtailing the bacterial infection. Second, the polymeric drugs allow multiple contacts with the same drug-containing macromolecule. This may be important in situations such as RNA, protein, and DNA synthesis, where multiple bondings may permanently disrupt synthesis. Third, polymers can be specially designed to offer a desired solubility and hydrophobicity/hydrophilicity. The polymeric nature of such drugs make their design dominant factors in determining solubility. Fourth, polymers may be included in special delivery (so-called magic bullets) systems or have the components of the delivery system as part of their polymer. Polymers containing attached specific binding recognition agents, such as hormones or antibodies, can bind to biological receptor sites thus allowing the desired drugs, also contained in the polymer, to be delivered at appropriate concentrations to the site of infection. Sixth, an appropriate design of the chain size can greatly reduce the excretion rate, thereby increasing efficacy and reducing the required dosage. Control of chain size also can be utilized to isolate or prevent movement of the polymer drug past barriers in the body such as the blood–brain barrier. Seventh, in some cases, the body prevents rapid elimination of polymeric drugs because of their preferential attachment to various parts of the blood stream. Finally, we have found that polymers that contain unstable drugs, such as certain penicillin derivatives, are stable for two decades at room temperature.

If the material is employed externally there are additional reasons why polymers containing the organotin portion are favored over small molecules. First, regulations require use of bound tin for environmental reasons. Second, polymeric molecules are less apt to leach out remaining within the material longer offering longer term protection and activity.

7 Organotin-Containing Polymers

There are a wide variety of organotin-containing polymer structures. Some of these structures are reviewed briefly in this section in order to illustrate the types of tin-containing polymers that have been synthesized. There are a number of polymers where the organotin moiety exists as an appendage off of a polymer chain. The organotin appendage may be closely bound to the polymer, as in the case of reactions with poly(vinyl alcohol) (Structure 17.2). Here, the product may be either crosslinked if the organotin moiety has two or more reactive sites or linear when the organotin moiety has only one reactive site. The crosslinked products are typically not soluble whereas the linear products are soluble [2].

The organotin-containing moiety may also be somewhat removed from the polymer chain as in the case of poly(tributyltin-4'-vinyl-biphenylcarboxylate) (Structure 17.3) where biphenylene spacers have been introduced [78,79].

The polymers may be synthetic as in the case of poly(vinyl alcohol) or may be natural as in the case of dextrin and cellulose where the products may again be either crosslinked (Structure 17.4) or linear (Structure 17.5) depending on the functionality of the reactant.[2]

In either case, the products are composed of a variety of repeat units including unreacted units.

The products may be connected to the organotin-containing moiety through noncarbon atoms such as oxygen, sulfur, and nitrogen, or through carbon atoms (Structure 17.6).

The organotin polymers can also contain the organotin-containing moiety in the backbone (Structure 17.7). Again, the organotin moiety can be carbon or noncarbon connected [2,79].

Structure 17.2

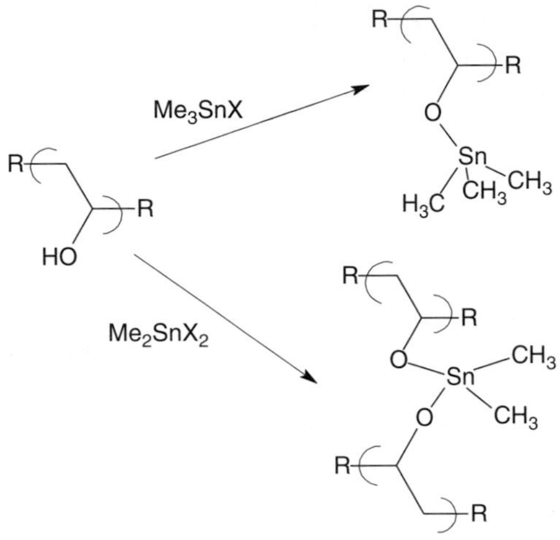

Me_3SnX

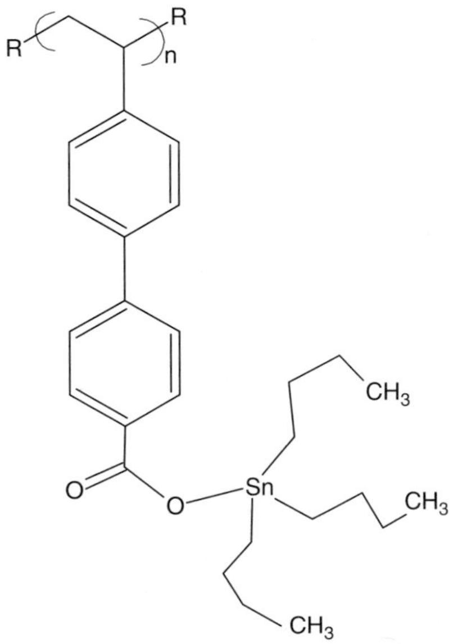

Me_2SnX_2

(2)

(3)

Structure 17.3

The polymers can also contain other metals as in the case of the ferrocene (Structure 17.8) [80,81] and cobalticinium (Structure 17.9) [82,83].

Thus, there exists a wide variety of organotin-containing structures already synthesized.

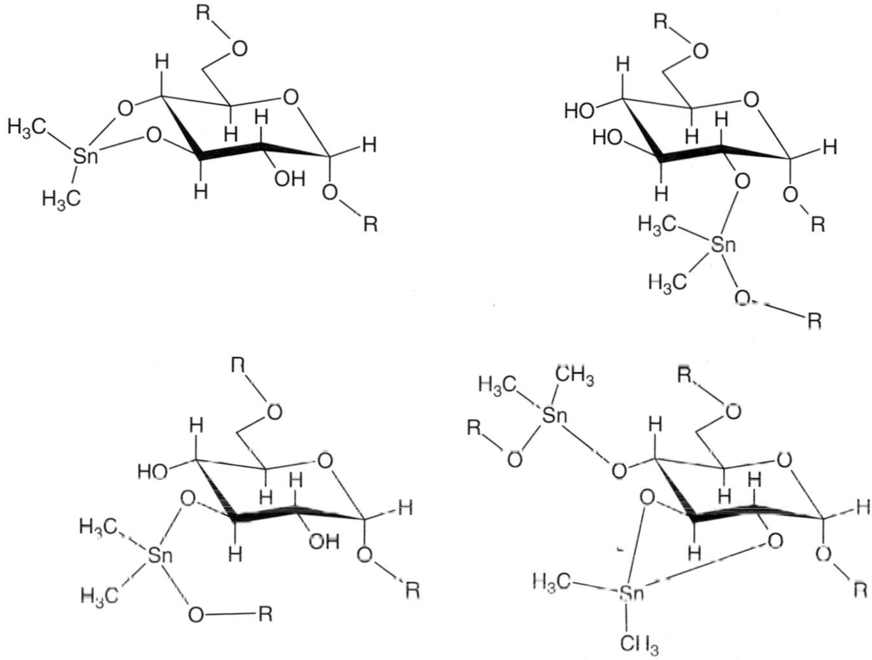

(4)

Structure 17.4

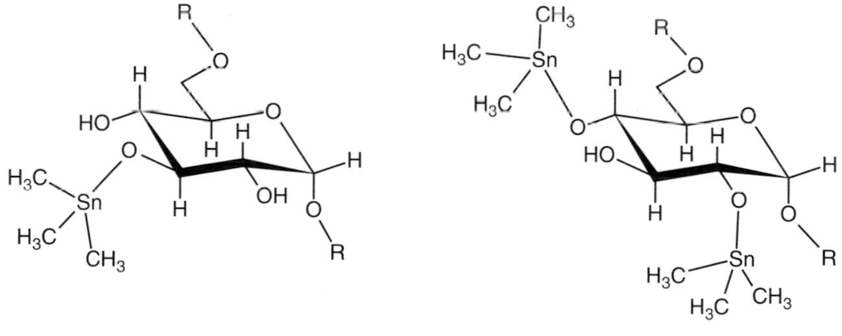

(5)

Structure 17.5

8 General Experimental

Polymers were synthesized employing the interfacial polycondensation procedure as previously described [2–14].

Structure 17.6

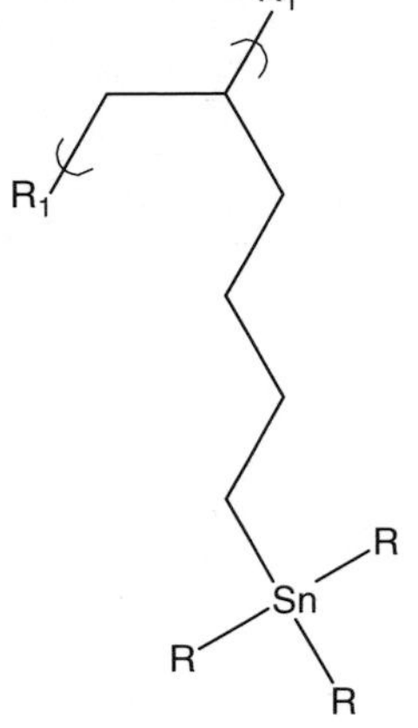

R₁ R₁ Sn R R R

(6)

CH₃ R Sn CH₃ CH₃ H₃C R n

H₃C SnH₂ CH₃ + H₂C CH₂ H₃C CH₃

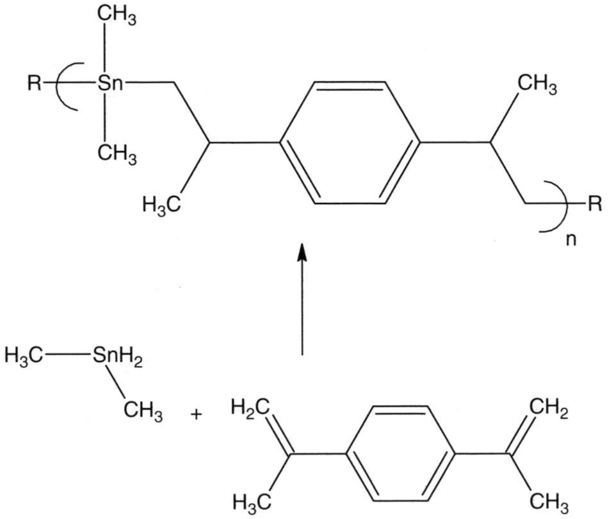

(7)

Structure 17.7

Bacterial studies were conducted in the usual manner. A suitable growth medium, such as Luri Bertani (LB) medium for prokaryote (*S. aureus* and *P. aeruginosa*) or C medium (glucose; 2% polypeptone, 0.2% yeast-extract, 0.1% potassium dihydrogenphosphate, 0.05% dipotassium hydrogenphosphate, 0.1% $MgSO_4 \times 7H_2O$,

Structure 17.8

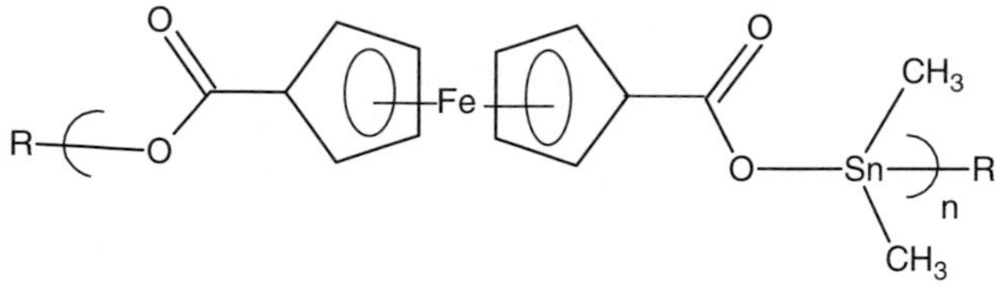

Structure 17.9

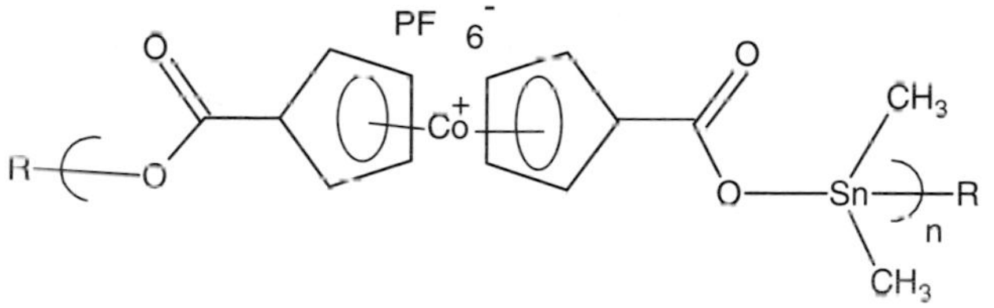

(9)

distilled water) for eukaryote (*C. albicans*) cells were used. Typically, LB agar medium plates were seeded with suspensions of the test organism and incubated at 28 °C for 24 hours. The test compounds were placed on the plates as solids in an emulsion or in solution employing paper disks. The plates were then incubated overnight at 30 °C and noted for inhibition which is indicated by a clear zone on the plate.

9 Current Test Organisms

Nine microorganisms were tested. These included both fungi and bacteria. *S. aureus* is a gram-positive cocci that causes "pimples,"' abscesses, impetigo, wound infections, pyelitis, cystitis food poisoning, pneumonia, meningitis, and enteritis. *P. aeruginosa* is a gram-positive rod that is a common inhabitant of soil and frequently found as part of the normal flora of the intestine and skin. It is an opportunistic pathogen which infects wounds; drains sinuses and decubitus ulcers; and causes urinary tract infections, eye infections, and meningitis. It is also a chief cause of death in burn patients. The family *A. faecalis* is gram-negative, aerobic, nonfermentative chemoorganotrophic, nonsporing bacteria. It often produces unusual amino acids such as D-pipecolic acid, and does not assimilate most carbohydrates as a carbon source. Some strains are being considered for industrial uses. *A. faecalis* AFK2 is a phenanthrene-degrading bacterium isolated from a soil sample as a polyaromatic hydrogen-degrading species [84]. This strain degraded phenanthrene through protocatechuate. Thus AFK2 and *Pseudomonas* species are being used as friendly scavengers that decompose a wide range of organic compounds including polyaromatic hydrocarbons. *B. subtilis* is a gram-positive straight rod that grows in long chains. They are aerobic, spore-producing bacteria found in soil and water. *B. subtilis* is often found in old sinuses or chronically infected wounds.

E. coli is found all about us. It is a popular lab organism because if grows quickly and in complex mediums. It can grow in the absence or presence of air and thus is a facultative anaerobe. There are over 70,000 cases of *E. coli* illness yearly in the United States. *S. cerevisiae* is also known as bakers' and brewers' yeast. It is a fungus that reproduces by budding. Because it is very small and unicellular, it has been widely used in genetic studies. It was the first eukaryote to have its entire genome sequenced. *C. albicans* is the major cause of yeast infections.

Thus, the tested bacteria and fungi are important pathogens and represent a somewhat broad range of microorganisms. This is a part of an ongoing study of the ability of organotin polymers to inhibit various microorganisms.

10 Monomer Results

Table 17.2 contains results for the organotin monomers tested. Whereas the majority exhibited some ability to inhibit microorganism growth, the best inhibition was by the dibutyltin dichloride and dimethyltin dichloride as expected. The ability for dibutyltin dichloride to be so active is positive as it has the lowest toxicity toward human beings. It is difficult to compare the inhibition results for the monomeric organotin dichalides with the polymers because the organotin dihalides probably have some solubility in the tested medium whereas the polymers have none.

Table 17.2 Bacterial inhibition results for organotin monomers

Test	1	2	3	4	5	6
Compounds → Organisms						
S. cerevisiae	++	0	0	++	0	++
C. albicans	++	0	0	+	0	++
Staph. aureus IAM1095	++	0	+	+++	+	++
B. subtilis IAM 1145	+++	0	+	+++	+	++
B. subtilis RM 125	++	0	+	+++	+	+
E. coli C600	+++	0	+	+++	+	+
Ps. aeruginosa PA0590	++	0	0	+++	+	+
Al. faecalis AFK2	++	0	+	+++	+	+
Al. faecalis IAM 1015	++	0	+/–	+++	+	+

Test compounds: 1. Bu_2SnCl_2; 2. $(PhCH_2)_2SnCl_2$; 3. Cy_2SnCl_2; 4. Me_2SnCl_2 (as suspension in water); 5. Oc_2SnCl_2; 6. $Ph_2Sn_2Cl_2$ where +++ = outstanding inhibition; ++ = excellent inhibition; + = good inhibition; +/– = slight inhibition; 0 = no inhibition

In general, dimethyltin and diethyltin-containing polymers are the most active at inhibiting bacterial growth [2]. This is consistent with the findings given in Table 17.1. In the present study, only dibutyltin polymers were examined because these polymers were found to be the most effective at inhibiting cell growth. Thus, diethyltin and dimethyltin derivatives should exhibit greater inhibition toward the microorganisms studied.

11 Organotin Polymers

Several groups of polymers were tested. The initial group consists of a series of organotin polyethers synthesized as part of our overall anti-cancer effort, Table 17.3 [85,86]. They have the following general structures (Structure 17.10).

The compounds are generally not active. The most active polymer is derived from 2-butyne-1,4-diol, whose synthesis was recently reported. It forms fibers on mechanical agitation so might be a candidate as a combination fiber-bactericide for use in composites.

12 Organotin Polyamines

The second series of compounds were derived from pyrimidine polyamines (Structure 17.11) and were synthesized as part of developing conducting polymers [87]. Their synthesis was also recently reported. Bacterial results are given in Table 17.4.

Table 17.3 Bacterial inhibition results for dibutyltin polyethers

Test polymers	Bu_2Sn 2-Butyne-1,4-Diol	Bu_2Sn Diethylene Glycol	Bu_2Sn Triethylene Glycol	Bu_2Sn Ethylene Glycol
Organisms				
S. cerevisiae	0	0	0	0
C. albicans	0	0	0	0
Staph. aureus	+	0	0	0
B. subtilis IAM 1145	+	0	0	±
B. subtilis RM 1015	+	0	0	0
E. coli	+	0	0	0
Ps. aeruginosa	0	0	0	0
Al. faecalis AFK2	+	0	0	0
Al. faecalis IMA 1015	+	0	±	0

+++, outstanding inhibition; ++, excellent inhibition; +, good inhibition; ±, slight inhibition; 0, no inhibition

Structure 17.10

(10)

Structure 17.11

(11)

Table 17.4 Bacterial inhibition results for dibutyltin pyrimidine polyamines

Test polymers	Bu_2Sn 1	Bu_2Sn 2	Bu_2Sn 3	Bu_2Sn 4
Organisms				
S. cerevisiae	+	0	0	0
C. albicans	+	0	0	0
Staph. aureus	±	0	0	0
B. subtilis IAM 1145	0	0	±	±
B. subtilis RM 1015	±	0	0	0
E. coli	±	±	0	0
Ps. aeruginosa	0	0	0	0
Al. faecalis AFK2	±	0	±	±
Al. faecalis IMA 1015	±	0	±	±

Where the diamine 1. 4,6-diamino-5-isoamyl-2-(3-phenylpropylamino)pyrimidine; 2. 4,6-diamino-5-nitropyrimidine; 3. 4,6-diamino-2-methyl-mercaptopyrimidine; and 4. 4,6-diamino-2-methyl-5-nitrosopyrimidine; and where +++ = outstanding inhibition; ++ = excellent inhibition; + = good inhibition; ± = slight inhibition; and 0 = no inhibition

As with the polyethers, inhibition was selective occurring for most of the organisms for the 4,6-diamino-5-isoamyl-2-(3-phenylpropyl-amino)-pyrimidine product.

The low activities for the polyethers and polyamines are not unexpected because they are all dibutyltin derivatives. As noted before, the most active organotin derivatives against bacteria are the methyl and ethyltin derivatives with the butyltin derivatives exhibiting markedly less activities. Even so, some of the polymers inhibit selected bacteria.

13 Combined Antibacterial Agents

We have synthesized and tested a number of organotin polymers containing known drugs, in particular known antibacterial agents. These include ampicillin, cephalexin, norfloxacin, and ticarcillin [2]. We have also synthesized organotin-containing polymers incorporating ciprofloxacin. Some results for these compounds, along with a comparison to monomeric model compounds containing the ciprofloxacin moiety, are discussed in this section.

Ciprofloxacin recently received much press because it was the drug of choice for combating anthrax. Ciprofloxacin hydrochloride, $C_{17}H_{18}FN_3O_3{\cdot}HCl{\cdot}H_2O$, is a synthetic broad-spectrum antibacterial agent. Ciprofloxacin differs from other quinolones in that it has a fluorine atom at the 6-position, a piperazine moiety at the 7-position, and a cyclopropyl ring at the 1-position.

We have already reported on the synthesis and structural characterization (Structure 17.12) of organotin-ciprofloxacin polymers [88–91]. They are moderate antiviral agents inhibiting such viruses as reovirus ST3, vaccinia WR, herpes simplex virus (HSV)-1, and varicella zoster virus (VZV) [92]. These viruses are responsible for smallpox, chickenpox, herpes, and other related illncsses.

The ciprofloxacin polymers also exhibit good inhibition of Balb 3T3 cells and the dibutyltin derivative is a candidate for further study as an anticancer drug. The

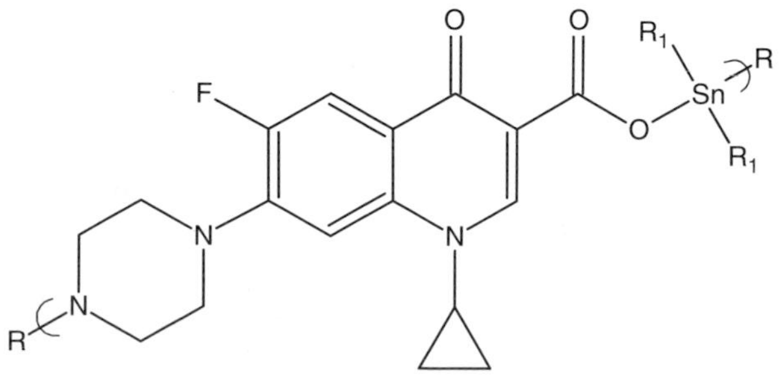

O O R$_1$ R F Sn O N N R N R$_1$

(12)

Structure 17.12

dimethyltin-ciprofloxacin product has a GI_{50} of about 25 µg/mL; the diphenyl-ciprofloxacin has a GI_{50} of about 0.5 µg/mL and the dibutyltin-ciprofloxacin has a GI_{50} of about 0.1 µg/mL, well within the range found for cisplatin. Again, methyl groups result in a much less active polymer as compared with butyl groups, consistent with the general trend previously found with other organotin polymers [93].

Table 17.5 presents data related to the ability of the organotin-ciprofloxacin polymers and model compounds to inhibit a variety of bacteria and yeast.

The most active polymer was the divinyltin-ciprofloxacin. It inhibited all of the tested microorganisms, with diethyltin-ciprofloxacin the second most active, as expected. Of interest was the similar, but varied activity of the di-*n*-butyltin and di-*t*-butyltin products. This is consistent with small structural changes resulting in changes in a products ability to inhibit microorganism growth.

As part of our structure-property study related to the inhibition of viral, cancer cell, and bacterial growth, we synthesized various model "monomeric" or "dimeric" small model compounds. Here, the diester compound was synthesized from enrofloxacin and dibutyltin dichloride (Table 17.6, compound 1) (Structure 17.13). The corresponding diamine was also formed from the diester of ciprofloxacin (Table 17.6, compound 2 (Structure 17.14).

Table 17.5 Bacterial inhibition results for organotin polymers of ciprofloxacin

Test Polymers (Organotin)	Diallyl	Divinyl	Diphenyl	Dibenzyl	Dicyclo
Organisms					
S. cerevisiae	0	+++	0	0	0
C. albicans	0	+++	0	0	0
Staph. aureus	+	+++	+	+++	+
B. subtilis IAM 1145	+	+++	++	+++	++
B. subtilis RM 1015	+++	+++	+/−	+++	+
E. coli	++	+++	+	+	+
Ps. aeruginosa	+++	+++	+++	+	+++
Al. faecalis AFK2	++	+++	+	+	+
Al. faecalis IMA 1015	+++	+++	+	+	+
Test Polymers (Organotin)	Diethyl	Dipropyl	Di-*n*-butyl	Di-*t*-Butyl	Dioctyl
Organisms					
S. cerevisiae	0	0	0	0	0
C. albicans	0	0	0	0	0
Staph. aureus	+++	++	+	+	+
B. subtilis IAM 1145	+++	+	+	++	++
B. subtilis RM 1015	++	+	+/−	++	+/−
E. coli	+	+++	+	+	+
Ps. aeruginosa	+++	+	+++	+	+++
Al. faecalis AFK2	++	+++	++	++	+
Al. faecalis IMA 1015	+++	+++	++	+++	+

Where +++ = outstanding inhibition; ++= excellent inhibition; += good inhibition; +/−= slight inhibition; and 0 = no inhibition

Table 17.6 Bacterial inhibition results for organotin model compound of ciprofloxacin and other small organotin compounds

Test compounds	1	2	3	4
Organisms				
S. cerevisiae	0	0	0	0
C. albicans	0	+/–	+	0
Staph. aureus	+++	+	+++	+++
B. subtilis IAM 1145	+++	+	+++	+++
B. subtilis RM 1015	++	+++	+++	+++
E. coli	+++	+	++	+++
Ps. aeruginosa	+++	+++	+++	+++
Al. faecalis AFK2	+++	++	+++	+++
Al. faecalis IMA 1015	+++	+	+++	+++

1 = enrofloxacin ciprofloxacin-dibutyltin; 2 = diester-ciprofloxacin-dibutyltin; 3 = ciprofloxacin tributyltin, 4 = ciprofloxacin

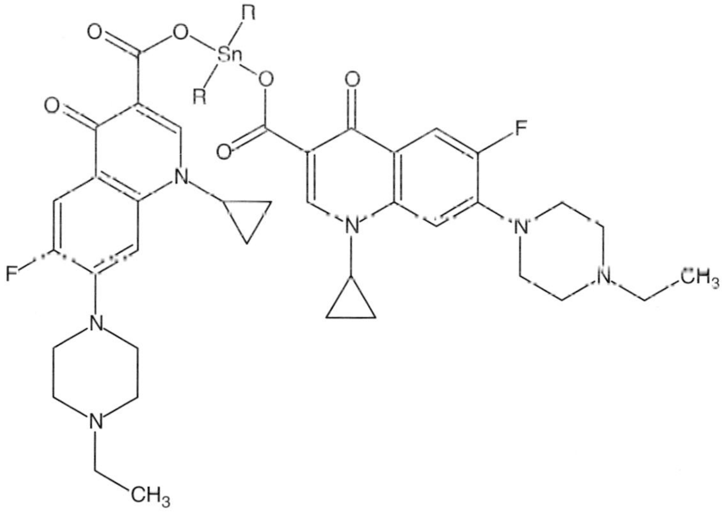

Structure 17.13

A third model compound was made from reacting tributyltin chloride with ciprofloxacin (Table 17.6, compund 3). It is believed to have Structure 17.15.

The ability to inhibit the various microorganisms by the three model small molecules is similar to that of the dibutyltin-ciprofloxacin polymer.

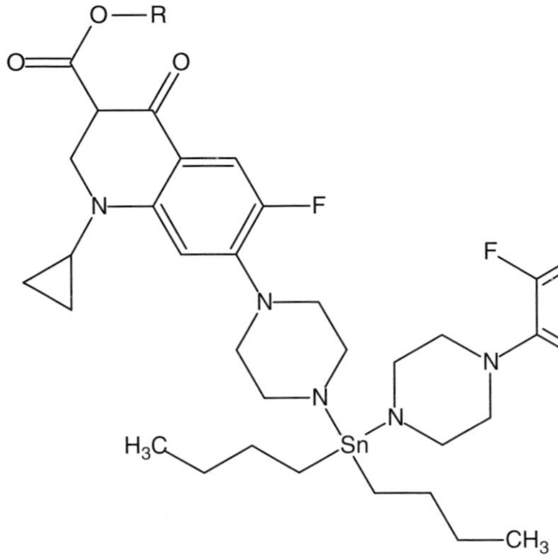

(14)

Structure 17.14
Structure 17.15

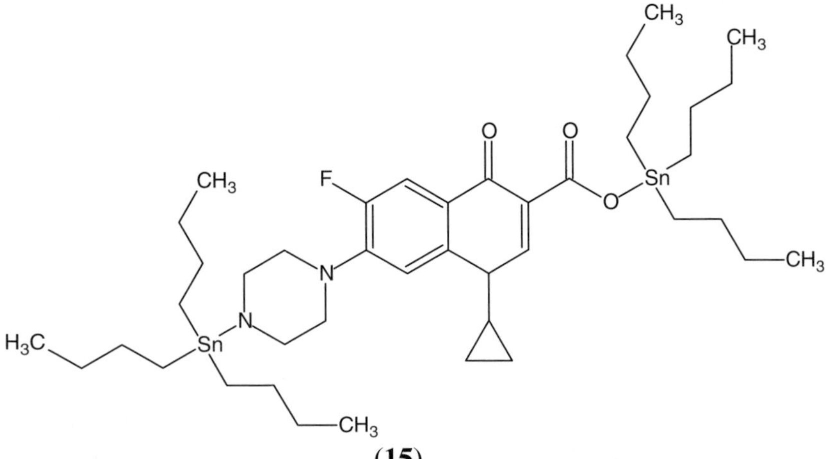

(15)

Whereas widespread inhibition is important for some applications—such as in caulks, sealants, and coatings—selective inhibition is important for other applications such as the treatment of yeast infections and protection from unwanted bacteria such as *Al. faecalis*.

References

1. Hoch M. (2001) Appl Geochem. 1:719.
2. Carraher C. (2005) In:. Group IVB polymers. Macromolecules containing metal and metal-like elements, vol. 3. Wiley, Hoboken.
3. Carraher C, Siegmann-Louda D. (2004) In: Biomedical applications. Macromolecules ontaining metal and metal-like elements, vol 4. Wiley, Hoboken.
4. Roner M, Carraher C, Zhao A, Roehr J, Bassett K, Siegmann-Louda D. (2004) Polym Mater Sci Eng. 90:515.
5. Siegmann-Louda D, Carraher C, Pflueger F, Coleman J, Harless S, Luing H. (2000) Polym Mat Sci Eng. 82:83.
6. Carraher C, Li F, Butler C. (2000) J Polym Mater. 17:377.
7. Carraher C, Stewart H, Carraher S, Nagata M, Miao S. (2001) J Polym Mater. 18:111.
8. Carraher C, Lanz L. (2003) J Polym Mater. 20:91.
9. Carraher C, Morie K. (2004) Polym Mater Sci Eng. 91:556.
10. Carraher C, Siegmann-Louda D. (2004) In: Biochemical applications. Macromolecules containing metal and metal-like elements, vol 3. Wiley, Hoboken, NJ
11. Siegmann-Louda D, Carraher C, Pflueger F, Coleman J, Harless S, Luing H. (2000) Polym Mat Sci Eng. 82:83
12. Siegmann-Louda D, Carraher C, Ross J, Li F, Mannke K, Harless S. (1999) Polym Mat Sci Eng. 81:151.
13. Siegmann-Louda D, Carraher C, Pfueger F, Nagy D. (2001) Polym Mat Sci Eng. 84:658.
14. Siegmann-Louda D, Carraher C, Chamely D, Cardoso A, Snedden D, (2002)Polym Mat Sci Eng. 86:293.
15. Siegmann-Louda D, Carraher C, Graham M, Doucetter R, Lanz L. (2002)Polym Mat Sci Eng. 87:247.
16. Siegmann-Louda D, Carraher C, Snedden D, Komulainen A. (2004) Polym Mater Sci Eng. 90:512.
17. Doucette R, Siegmann-Louda D, Carraher C, Cardoso A. (2004) Polym Mater Sci Eng. 91:564.
18. Bleicher R, Carraher C. (2002) Polym Mater Sci Eng. 86:289.
19. Roner M, Carraher C, Zhao A, Roehr J, Bassett K, Siegmann-Louda D. (2003) Polym Mater Sci Eng. 89:525.
20. Roner M, Carraher C, Zhao A, Roehr J, Bassett K, Siegmann-Louda D. (2004) Polym Mater Sci Eng. 90:515.
21. Roner M, Carraher C, Roehr J, Bassett K, Siegmann-Louda D. (2004) Polym Mater Sci Eng. 91:744.
22. Roner M, Carraher C, Dhanji S. (2005) Polym Mater Sci Eng. 93:410.
23. Sabir T, Carraher C. (In Press) J Polym Mater.
24. Siegmann-Louda D, Carraher C, Gordon O, Battin A. (2005) Polym Mater Sci Eng. 93:407.
25. Carraher C, Lanz L. (2005) J Polym Mater. 21:51.
26. Doucette R, Siegmann-Louda D, Carraher C. (2004) Polym Mater Sci Eng. 91:567, 569.
27. Carraher C, Lee J. (2004) Polym Mater Sci Eng. 90:408.
28. Carraher C, Scherubel G. (1972) Makromolekular Chemie. 152:61.
29. Carraher C, Scherubel G. (1972) Makromolekular Chemie. 160:259.
30. Carraher C, Scherubel G. (1971) J Polym Sci. A-1, 9:983.
31. Siegmann-Louda D, Carraher C, Nagy D, Snedden D, Rosa J. (2003) J Polym Mater Sci Eng. 89:487.
32. Barot G, Carraher C, Siegmann-Louda D. (2005) Polym Mater Sci Eng. 92:339.
33. Lyman T. (1969) Metals handbook, 8th ed., vol. 1. American Society For Metals, p1131.
34. Blunden S. (1983) J Organometallic Chem 248:149.
35. World Health Organization. (1980) Tin and organotin compounds: a preliminary review, environmental health criteria ser., no 15. Task Group on Environmental Health Aspects of Tin and Organotin Compounds, Geneva.

36. Tauberger G, Klimmer O. (1961) Arch Exper Path Pharmakol. 242:370.
37. Tauberger G. (1963) Ned Exp. 9:393.
38. Hamman I. (1978) Nachrichten Bayer 31:61.
39. Rosenburg B. (1973) Naturwissenchaften 60:399.
40. Cremer JE. (1958) Biochem J. 68:685.
41. Iwai H, Wada O.. (1981) Ind Health 19:247.
42. Nicklin S, Robson MW.. (1988) App Organometallic Chem 2(6):487.
43. Seinen W, Vos J, Spanje I, Snoek M, Brands R, Hooykaas H. (1977) Toxicol Appl Pharmacol 42:197.
44. Seinen W, Vos J, Krieken R, Penninks A, Brands R, Hooykaas H. (1977) Toxicol Appl Pharmacol 42:213.
45. Colosio C, Tomasini M, Cairoli S, Foa V, Minoia C, Marinovich M, Galli C. (1991) Br J Ind Med. 48:136.
46. Grace C, Ng S, Cheong L. (1991) Contact Dermatitis 25:250.
47 Lin J, Hsueh S. (1993) Am J Nephrol. 13:124.
48. Krinke G. (2000) Experimental and clinical neurotoxicology. Oxford University Press, p 1206.
49. Lin T, Hung D, Kao C, Hu W, Yang D. (1998) Human Exp Toxicol. 17:403.
50. World Health Organization. (1992) Pesticide residues in food: evaluations 1991 Part II Toxicology. Geneva, p 173.
51. Tennekes H, Luetkemeier H, Vogel W, et al. (1989) Unpublished report 046980 (A40468) of RCC research and Consulting Company AG, Itingen.
52. Wester P, Kranjnc E, van Leeuwen F, et al. (1990) Food Chem. Toxicol. 28:179.
53. Carraher C, Reckleben L, Butler C. (1991) Polym Mater Sci Eng.63:704.
54. Carraher C, Butler C. (1991) US Patent No. 5,043,463.
55. Carraher C, Butler C, Reckleben L. (1991) Cosmetic and pharmaceutical applications of polymers. Plenum Press, New York, NY.
56. Carraher C, Butler C, Reckleben L, Taylor A, Saurinao V. (1992) Polymer P. 33:539.
57. Carraher C, Butler C. (1998) US Patent No. 5,840,760.
58. Butler C, Carraher C. (1999) Poly Mater Sci Eng. 80:365.
59. Martins J, Jurado A, Moreno A, Madeira V. (2005) Toxicol in Vitro. 19:943.
60. Brown W, Aldridge W. (1979) Am J Pathol. 97:59.
61. Philbert MA, Billingsley M, Reuhl K. (2000) Toxicol Pathol. 28:43.
62. Snoeij N, van Iersel A, Penninks A, Seinen W. (1985) Toxicol Appl Pharmacol. 81:274.
63. Dyer RS, Walsh T, Wonderlin W, Bercegeay M. (1982) Neurobehav Toxicol Teratol. 4:127.
64. Ross WD, Emmett E, Steiner J, Tureen R. (1981) Am J Psychiatry. 138: 1092.
65. Feldman GR, White RF, Ikechukwu E. (1993) Arch Neurol. 50:1320.
66. Geloso M, Vercilli A, Corvino V, et al. (2002) Exp Neurol. 175:152.
67. Jenkins SM, Barone S. (2004) Toxicol Lett. 147:63.
68. LeBel C, Ali S, McKee M, Bondy S. (1990) Toxicol Appl Pharmacol. 104:17.
69. Stein KE, Reiter L, Lemasters J. (1988) Toxicol Appl Pharmacol. 94:394.
70. Davidson C, Reese B, Billingsley M, Yun J. (2004) Mol Pharm. 66:855.
71. White JS, Tobin J, Cooney J. (1999) Can J Microbiol. 45:541.
72. White JS, Tobin J. (2004) Appl Micro Biotech. 63:445.
73. Smith LD, Negri A, Philipp E, Webster N, Heyward A. (2003) Marine Biology.15:651.
74. Jensen HF, Holmer M, Dahllof I. (In press) Marine Pollution Bull.
75. Schuylte-Oehlmann U, Ochlmann J, Fioroni P, Bauer B. (1997) Marine Biology. 12:128.
76. Jurkiewicz M, Averill-Bates D, Marion M, Denizeau F. (2004) Biophysica Acta. 15:1693.
77. Gadd GM. (2000) Sci Total Environment. 258:119.
78. Adrova N, Koton M, Moskvina E. (1962) Ivz Alad Nauk SSSR Otdel Kim Nauk. 1804.
79. Adrova N, Koton M, Klages V. (1961) Vysokomol Soden. 3:1041.
80. Carraher C, Jorgensen S, Lessek P. (1976) J Appl Polym Sci. 20:2255.
81. Carraher C, Lessek P. (1974) Angew Makromol Chem. 38:57.
82. Carraher C, Peterson G, Sheats J, Kirsch T. (1974) Makromol Chem. 175:3089.
83. Carraher C, Peterson G, Sheats J, Kirsch T. (1974) J Macromol Sci Chem. A8:1009.

84. Kiyohara H, Nagao K, Kouno K, Yano K. (1982) App Environmental Microbiol. 43:458.
85. Barot G, Carraher C, Siegmann-Louda D. (2005) Polym Mater Sci Eng. 92:339.
86. Barot G, Carraher C, Siegmann-Louda D. (2005) Polym Mater Sci Eng. 93:417.
87. Battin A, Carraher C. (2005) Polym Mater Sci Eng. 92:496.
88. Zhao A, Carraher C. (2003) Polym Mater Sci Eng. 89:367.
89. Zhao A, Carraher C, Siegmann-Louda D. (2004) Polym Mater Sci Eng. 90:472.
90. Zhao A, Carraher C, Scopelliti M, Pellerito L. (2005) Polym Mater Sci Eng. 92:347.
91. Zhao A, Carraher C, Barone G, Pellerito C, Scopelliti M, Pellerito L. (2005) Polym Mater Sci Eng. 93:414.
92. Roner M, Zhao A, Carraher C, Roehr J, Bassett K, Siegmann-Louda D. (2003) Polym Mater Sci Eng. 89: 525.
93. Siegmann-Louda D, Zhao A, Carraher C, Fletcher A, Herrera Y. (2005) Polym Mater Sci Eng. 92:433.

Polymeric Organotin Fibers

Girish Barot and Charles E. Carraher Jr.

1 Background

It is unusual for most metal-containing polymers to form fibers because of a lack of appropriate solubility of the polymers and subsequently fiber formation. It is more unusual for polymers themselves to spontaneously form fibers. We initially reviewed this tendency for a number of Group IVB polyester more than three decades ago [1,2]. Briefly, the polymers were formed through employment of the interfacial polymerization process. The product is precipated from reaction, collected on filter paper in a Buchner filter with suction, washed with the organic solvent and water to remove unreacted materials and salts, and subsequently washed in a glass Petri dish with acetone to remove the product from the filter paper. The product is allowed to dry. When first scraped from glass Petri dishes, many of these products produced fibers. Generally, no fibers were observed from visual or microscopic observation. As the product was recovered from the dish, fibers were spontaneously formed. In some cases the fibers were present on evaporation of the liquid. More typically, fiber was formed as the polymer was scraped, using a flat-ended steel spathula. It appears that the mechanical agitation is sufficient to induce fiber formation.

Most of the fiber-forming structures contain rigid backbones as is the case with the terephthalic acid-derived products shown in (**1**).

Others contain semi-rigid backbone structures as in the case of the itaconic acid (methylene succinic acid) (**2**).

whereas others have relatively flexible backbones as those from azelatic acid (nonanediolic acid) (**3**).

A.S. Abd-El-Aziz et al. (eds.), *Inorganic and Organometallic Macromolecules: Design and Applications.*

Although the structures are all drawn for *M = Ti*, products from Zr and Hf also gave fibers.

The structures and physical properties of these fibers were studied and compared with analogous nonfibrous portions of the product. The phase transitions via differential scanning calorimetry were at essentially the same locations and magnitudes and were the same in both air or nitrogen. Degradation was similar. Whereas DSC noted transitions occurring in the 90 to 250 °C range, weight loss occurred for many of the products to about 1,000 °C with less weight reduction occurring in nitrogen as opposed to to air. Infrared spectra of the solids prior to heating were similar to those taken after heating. Finally, there were no noticeable differences in the infrared spectra between the fibrous and nonfibrous portions of the same polymer in the range of 3,000 to 200 cm^{-1}. The nonfibrous products typically exhibited a sharp

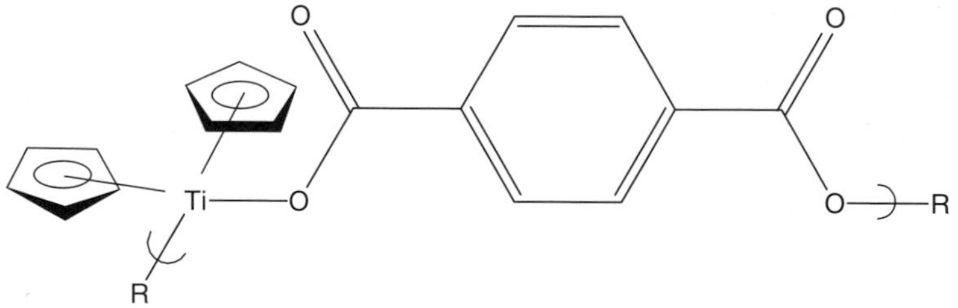

(1)

(1)

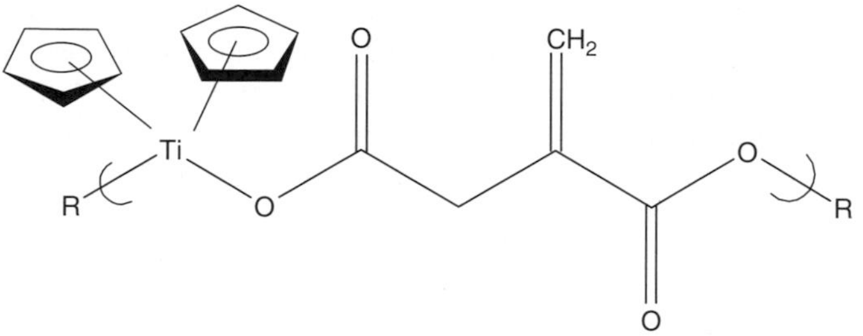

(2)

(2)

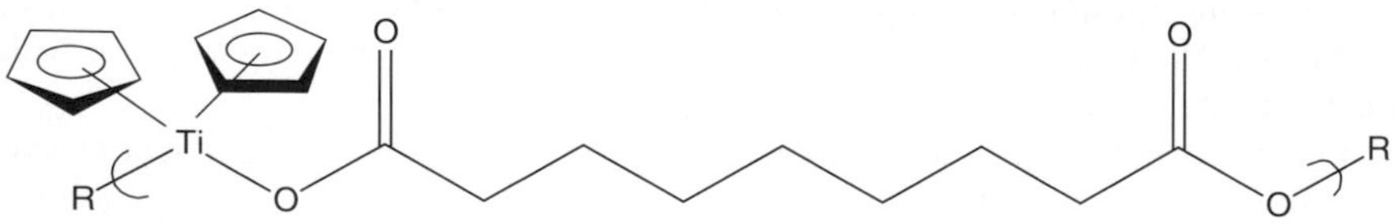

(3)

(3)

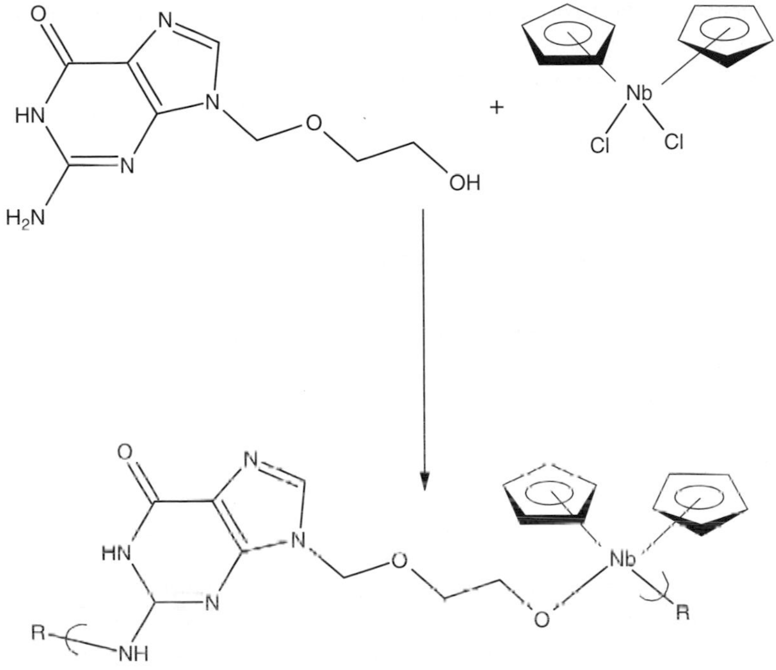

(4)

(4)

band 3,130 cm^{-1} associated with the Cp and terphthalate moieties for the terephthalic acid-derived product. This peak was less prominent in the fibrous product being somewhat covered by bands between 3,480 and 3,100 cm^{-1}. These bands were more intense for the fibrous products as compared with the nonfibrous products. Increased intensity of these bands may result from increased hydrogen bonding.

The fibers were flexible and some retained some flexibility up to 500 °C. The fibers retained their original flexibility and other properties for more than 30 years.

Since that time we have noticed fiber formation for only a few products and have generally only noted it in passing when presenting the synthesis of new polymers.

Recently, we noticed a number of fiber-forming products, some derived from acyclovir and various metallocene dichlorides such as vanadocene [3], and niobocene (**4**) [4]. The formation of fibers from a simple organotin polyether and from other organotin products is described here.

As a side note, the method of synthesis may be critical for fiber formation. The interfacial polymerization system is believed to offer not only a rapid alternative method for polymer synthesis when employing "high-energy" Lewis acids, such as acid chlorides, but it also offers some orientation to the forming polymer chains. This orientation is probably a result of polymerization occurring in a somewhat two-dimensional, layered environment similar to that of lignin. For lignin, this results in a somewhat layered, two-dimensional structure. Similar structures were found for other polymers when formed within cavities that resemble two-dimensional templates. Similar constraints are also found for most of the self-assembly polymerizing systems. Thus, such orientation is not unusual.

2 Organotin Polyethers-General

We have synthesized a number of organotin condensation polymers for several reasons, including study of their biological activity. The topic of organotin polymers was recently reviewed [5], as was their use as anticancer agents [6]. In general, the order of ability to inhibit cell growth with respect to the alkyl chain on the organotin is Bu > Pr > Et > Me with the methyl, octyl, and lauryl groups all essentially inactive [5,8–14].

Some time ago, we synthesized a number of organotin polyethers of the general form (**5**) [16–20].

Recently, we tested the ability of some of these to inhibit Balb 3T3 cell growth as a measure of their potential as anticancer drugs. We will focus on only the dibutyltin products here [21]. The GI50 (growth inhibition of 50%) for the 1,6-hexanediol product was 5 μg/ml. For the 1,4-butanediol dibutyltin dichloride product the GI50 was 0.25 μg/ml. And for the 1,4-butenediol product the GI50 was 0.025 μg/ml , the lowest GI50 thus far found for the organotin polymers. By comparison, the GI50 for cisplatin, the most widely employed anticancer drug, is 0.4 μg/ml.

These results suggested two structural windows to be investigated. First, as the distance between the oxygen atoms decreased, activity increased. Second, that unsaturation—the presence of π bonds—may contribute to the ability of the organotin polyethers to inhibit cell growth. In investigating these windows the product of dibutyltin dichloride and 1,4-butynediol was synthesized (**6**). This product is described in this chapter.

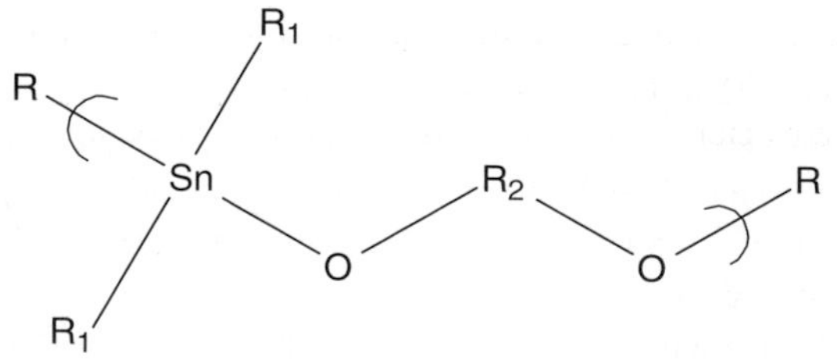

R$_1$ R Sn R$_2$ R O O R$_1$

(**5**)

(**5**)

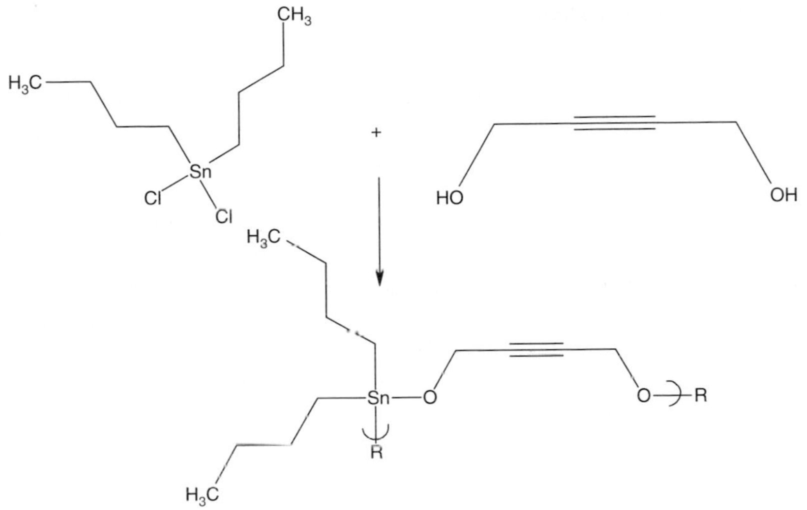

(6)

(6)

3 Experimental

Dibutyltin dichloride was used as obtained from Aldrich as was 1,4-butynediol. Reactions were carried out using the interfacial polycondensation technique. Briefly, an aqueous solution (30 ml) containing the diol (0.00300 M) and sodium hydroxide (0.0060 M) was transferred to a one quart Kimax emulsifying jar fitted on top of a Waring Blender (model 1120; no load speed of about 18,000 rpm; reactions were carried out at about 25 °C). Stirring was begun and a hexane solution (30 ml) containing dibutyltin dichloride (0.00300 M) was rapidly added (over 3–4 seconds) through a hole in the jar lid using a powder funnel. The resulting solution was blended for 15 seconds. The precipitate was recovered using vacuum filtration and washed several times with deionized water and chloroform to remove unreacted materials and unwanted by-products. The solid was washed onto a glass Petri dish and allowed to dry at room temperature.

Solubilities were determined by placing between 1 and 10 mg of polymer in 3 ml of liquid. The solid-liquid combinations were observed over a period of 2 to 4 weeks.

Light scattering was carried out employing a Brice-Phoenix BP 3000 Universal Light Scattering Photometer. Refractive indices were obtained using a Bauch & Lomb Abbe Model 3-L refractometer.

Infrared spectra were obtained employing KBr pellets using a Mattson Instruments galaxy Series 4020 FTIR using 32 scans and an instrumental resolution of 4 1/cm.

The fibers were examined using an Olympus CH30 Microscope with variable magnifications (×10–100).

4 1,4-Butynediol Product

We noticed an important similarity with respect to structures that are apt to form these fibers, namely that rigid structures are more likely to form fibers. In the present study, several of the created polyethers did not produce fibers upon recovery from the Petri dish. These structures included products from ethylene glycol, 1,6-hexanediol, 1,4-hexanediol, and 1,4-butenediol. When we made the analogous polyether using 1,4-butynediol, fibers were present. Of the polyethers made in this series, the 1,4-butynediol product should be the most rigid to be consistent with the former observation. We repeated the reaction several times with varying results as a result of the delicate balance required in forming these fibers. In all cases, some fibers were made but in different amounts. The fibers are described in greater detail in a subsequent section.

The product was polymeric with a molecular weight in HMPA via light scattering photometry of 12,0000 corresponding to an average chain length of 380. The molecular weight was taken weekly for 5 weeks and did not change. Thus, the polymer is reasonably stable in solution for at least a month.

Infrared spectral analysis was consistent with the presence of units from each of the reactants and with the formation of the Sn–O unit (Table 18.1). The product showed bands at about 2955 (all bands are given in cm^{-1}) from the C–H structure for the butyl groups from the dibutyltin moiety. Bands around 2921 and a doublet near 2856 are from the C–H structure of the diol-methylene moiety. The band at 3400 was believed to result from the presence of a Sn–OH unit end group. It was relatively small, consistent with a low population of Sn–OH groups relative to other groups, and with the long chain length of the product. The Sn–Cl band found in dibutyltin tin at about 630 was missing, consistent with a lack of detectable Sn–Cl end groups. The Sn–C assymetric structure was found at 594 and the symmetric Sn–C structure was found at about 566. The CC triple bond appeared in the 2260 to 2190 region. It is a "forbidden" IR band but appears weakly for many compounds. For the 1,4-butynediol it appeared at about 2214 and for the product it was a very weak band at about 2240.

The Sn–O stretch was assigned to be within a wide range of 400 to 700. For the dibutyltin product of acyclovir the Sn–O was found at 420. The presence of a new band at about 426 was tentatively assigned to the Sn–O moiety. Bands attributed to

Table 18.1 Selected IR locations for 1,4-butynediol and the derived organotin polymers

Assignment	1,4-Butynediol	Organotin polymer
C–H Stretch	2864	2856
CH_2 Scissor	1443	1466
CH_2 Wag	1411; 1363	1421; 1377
CH_2 Twist	1250	1287
C–C,C–O St., CH_2 Twist	1130	1128
C–C,C–OSt., CH_2Rock	ca850	861

the Sn–O–C, C–O stretch were found at 1012 and 861. The aliphatic C–O–C ether asymmetrical stretching vibration band was found in the range of 1150 to 1060. This band was found at 1068 for the product. Bands characteristic of methylene deformation were present at 1466 and 1150. The CH_2 scissoring band was assigned to 1377. The CH_2 twist was found at 1287.

The product was tested for ability to inhibit Balb 3T3 cells. Preliminary results showed that they exhibited a GI_{50} value of 0.05 µg/ml. By comparison, cisplatin, the most widely used anticancer drug, shows a GI_{50} value of 0.50 µg/m; —about ten times the concentration found for the dibutyltin/2-butyne-1,4-diol product. This was consistent with our idea of a structural window including a low number of carbons between oxygen and present of unsaturation being positive features in developing anticancer drugs that inhibit cell growth at low concentrations.

5 Organotin Polymer Fibers

As part of an effort to better understand the structural requirements for forming fibers, we surveyed a number of organotin polymers. Table 18.2 contains the results of this survey.

As seen in Table 18.2, fiber formation has a complex structural relationship and is not usual. For the approximately 50 polymers tested, only 5 formed fibers.

As previously noted, the product of dibutyltin dichloride and 2-butynene-1,4-diol formed fibers. The fibers were generally clear and colorless. They were generally smooth and linear, with some having small branches coming off the main fiber. These branches were clearly seen on inspection at greater magnifications. Some of the fibers were 1 mm in length and 0.007 mm diameter corresponding with aspect ratios (length to diameter) generally greater than 100. Using a needle, the fibers can be bent without breaking. The nonfiber portion was present as somewhat flat plates. The percentage by bulk that were fibers varied depending on the particular synthesis from about 1% to more than 15%.

The synthetic procedures appeared to be the same but obviously small changes affect the amount of fiber formation.

Following are pictures of these fibers. Figure 18.1 contains a picture showing both the fibers and nonfiber plates under 10-fold magnification.

Figure 18.2 contains a similar picture except emphasizing one of the longer fibers.

Figure 18.3 contains a magnification of the fiber to 60-fold.

Figure 18.4 contains the fiber magnified 100-fold.

Figure 18.5 shows a fiber with branch points at 60-fold magnification.

In summary, the product from dibutyltin dichloride and 1,4-butynediol forms fibers on mechanical agitation.

Two of the products from 1,1'-dicarboxylic ferrocene formed fibers (**7**). The fibers from dibutyltin dichloride were light brown in color with the product containing only about 5% fiber mass. Again, the fibers had some short spikes coming off a central fiber (Figure 18.6). Fiber lengths were about 200 mm with a diameter of about 8 mm for an aspect ratio of about 25. The product from dimethyltin dichloride

Table 18.2 Fiber forming capability of selected organotin polymers

Organotin dichloride	Lewis base	Fiber formation
Me_2SnCl_2	Norfloxacin	No
Et_2SnCl_2	Norfloxacin	No
Pr_2SnCl_2	Norfloxacin	No
Bu_2SnCl_2	Norfloxacin	No
Oc_2SnCl_2	Norfloxacin	No
Me_2SnCl_2	Ampicillin	No
Et_2SnCl_2	Ampicillin	No
Bu_2SnCl_2	Ampicillin	No
Bz_2SnCl_2	Ampicillin	No
Me_2SnCl_2	Cephalexin	No
Et_2SnCl_2	Cephalexin	No
Bu_2SnCl_2	Cephalexin	No
Me_2SnCl_2	Kinetin	No
Et_2SnCl_2	Kinetin	No
Pr_2SnCl_2	Kinetin	No
Oc_2SnCl_2	Kinetin	No
Et_2SnCl_2	2-Chloro-*p*-phenylenediamine	Yes
Pr_2SnCl_2	2-Chloro-*p*-phenylenediamine	Yes
n-Bu_2SnCl_2	2-Chloro-*p*-phenylenediamine	No
t-Bu_2SnCl_2	2-Chloro-*p*-phenylenediamine	No
Ph_2SnCl_2	2-Chloro-*p*-phenylenediamine	No
Bu_2SnCl_2	Ethylene glycol	No
Bu_2SnCl_2	Diethylene glycol	No
Bu_2SnCl_2	Triethylene glycol	No
Bu_2SnCl_2	Tetraethylene glycol	No
Bu_2SnCl_2	Pentaethylene glycol	No
Bu_2SnCl_2	1,3-Propanediol	No
Bu_2SnCl_2	1,4-Butanediol	No
Bu_2SnCl_2	1,6-Hexanediol	No
Ph_2SnCl_2	Ethylene Glycol	No
Bu_2SnCl_2	1,4-Benzenediamine	No
Bu_2SnCl_2	Polyethylene glycol (MW 400)	No
Bu_2SnCl_2	2-Butynene-1,4-diol	Yes
Me_2SnCl_2	Tricarcillin	No
Et_2SnCl_2	Tricarcillin	No
Pr_2SnCl_2	Tricarcillin	No
Bu_2SnCl_2	Tricarcillin	No
Oc_2SnCl_2	Tricarcillin	No
Cy_2SnCl_2	Tricarcillin	No
Ph_2SnCl_2	Tricarcillin	No
Bu_2SnCl_2	Bisphenol A	No
Bu_2SnCl_2	Adipic Acid	No
Me_2SnCl_2	Adipic Acid	No
Me_2SnCl_2	1,1'-Dicarboxylic Ferrocene	Yes
Et_2SnCl_2	1,1'-Dicarboxylic Ferrocene	No
Bu_2SnCl_2	1,1'-Dicarboxylic Ferrocene	Yes
Oc_2SnCl_2	1,1'-Dicarboxylic Ferrocene	No
Cy_2SnCl_2	1,1'-Dicarboxylic Ferrocene	No
$Benzyl_2SnCl_2$	1,1'-Dicarboxylic Ferrocene	No
Ph_2SnCl_2	1,1'-Dicarboxylic Ferrocene	No

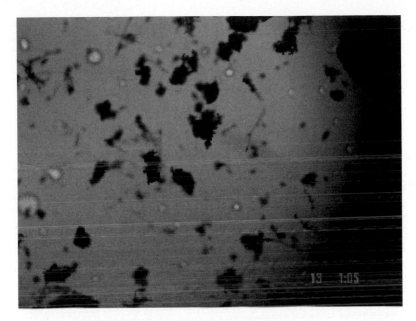

Figure 18.1 Fiber and nonfiber plates under 10 fold magnification

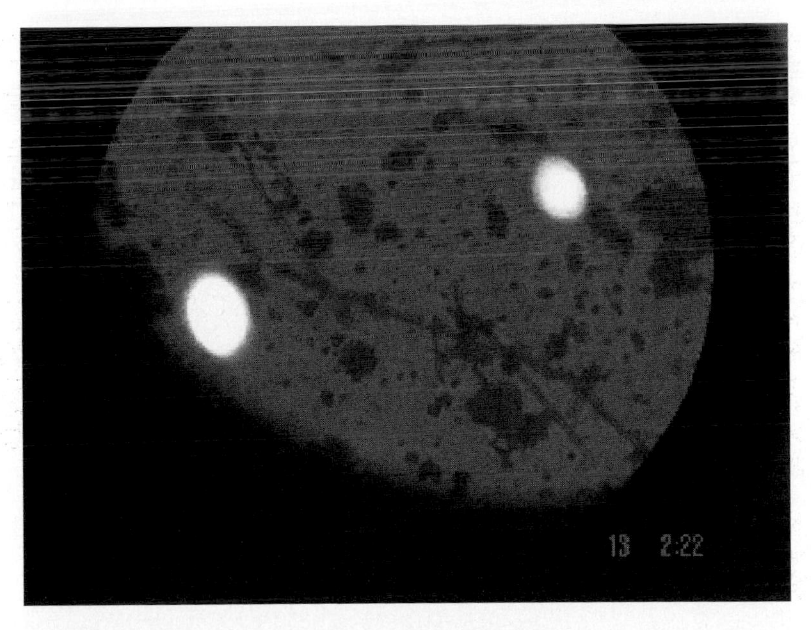

Figure 18.2 Picture of longer fiber under 10-fold magnification

gave about 50% fibers that bunched together forming balls (Figure 18.7). They were also light brown in color with a few spikes coming off of the central fiber with an average length of about 100 mm and diameter of 0.001 mm giving an aspect ratio of about 10^5, which is high.

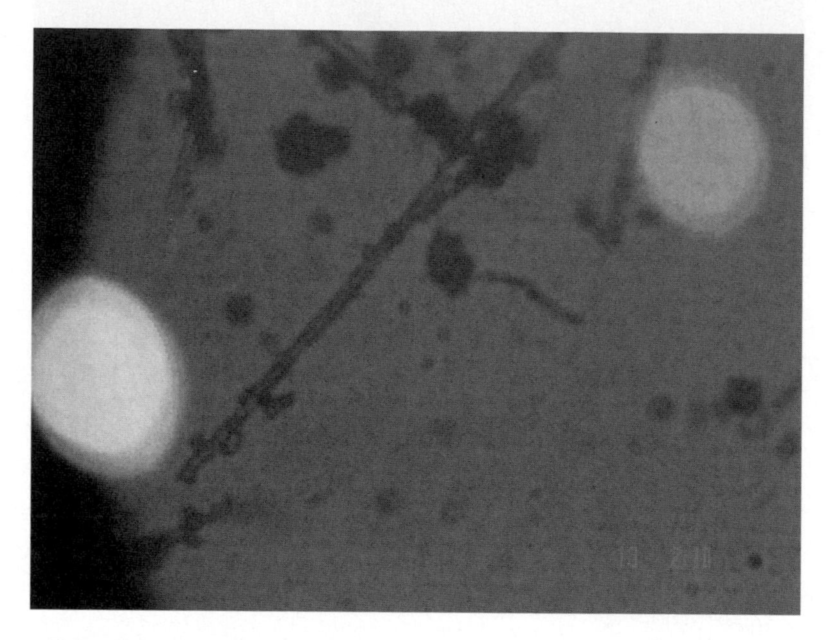

Figure 18.3 60-fold magnification of fiber

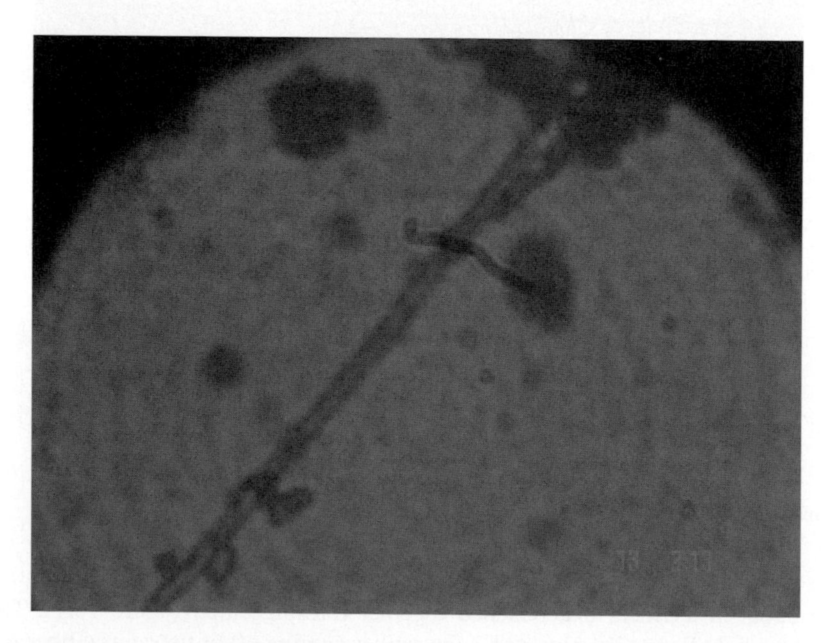

Figure 18.4 Fiber magnified 100-fold

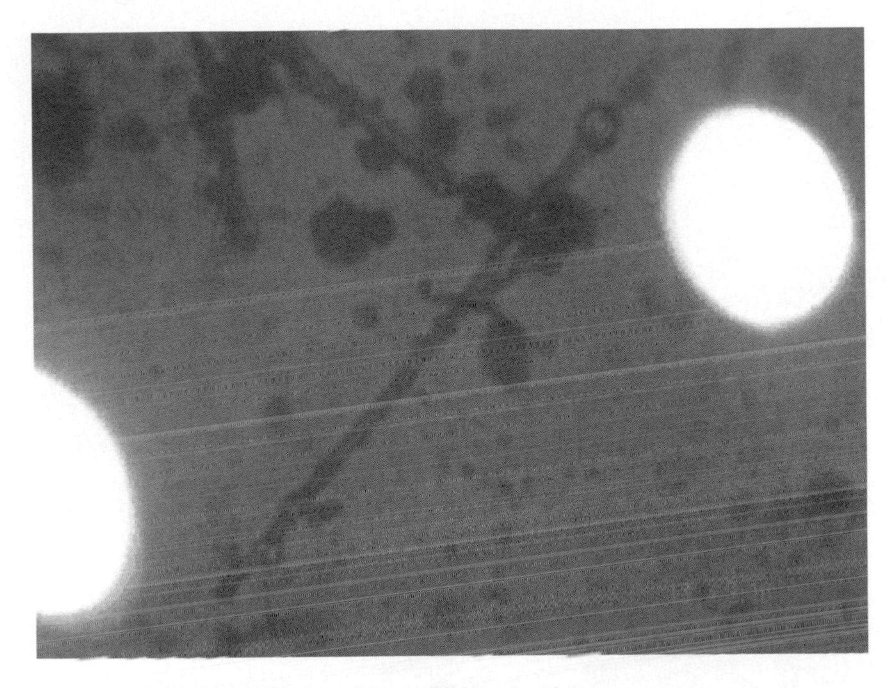

Figure 18.5 60-fold magnification of fiber with branch points

Figure 18.6 Product from dibutyltin dichloride and 1,1'-dicarboxylic ferrocene

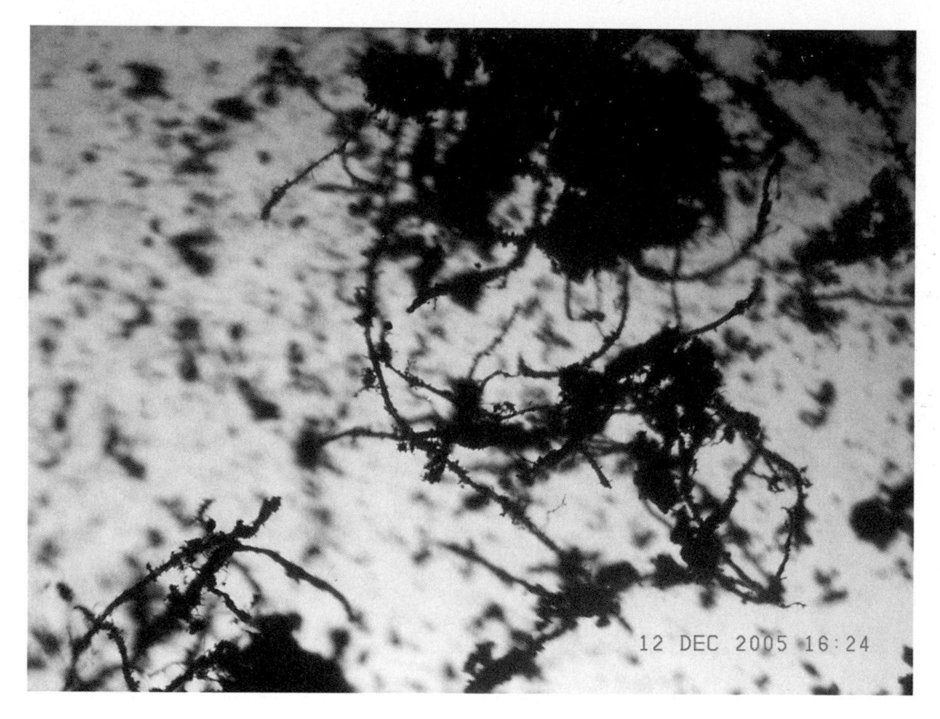

Figure 18.7 Fibers from the product of dimethyltin dichloride and 1,1 -dicarboxylic ferrocene

Two products derived from 2-chloro-p-phenylenediamine also gave fibers (**8**). The dipropyltin dichloride fibers were also light brown containing a few spikes off a central main fiber (Figure 18.8). They were formed in about 5% yield with fiber lengths of 200 mm and diameters of about 10 mm giving an aspect ratio of about 20. The diethyltin dichloride derived products gave about 1% production of light brown fibers (Figure 18.9). The fibers were about 1 mm by 0.05 mm giving an aspect ratio of about 20.

It is too early to describe the structural requirements for fiber formation. Further work is needed to see if there are structural requirements for fiber formation and what those structural requirements are.

6 Application

The most obvious area of application are as fibers for composites. There are a number of metallic whiskers that are employed as high-strength fibers in composites. These composites are among the strongest known. The mechanical and electrical properties of the fibers also need to be studied. It is possible that the fibers are semiconductors to near conductors, allowing their use as directional electrical wires.

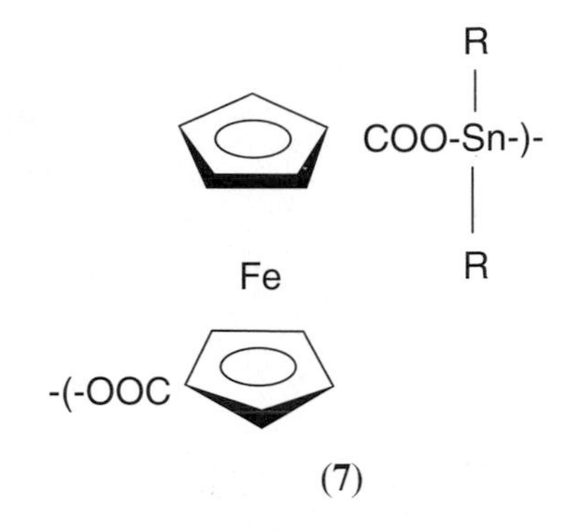

(7)

(8)

(8)

Figure 18.8 Fibers from the product of dipropyltin dichloride and 2-chloro-p-phenylenediamine

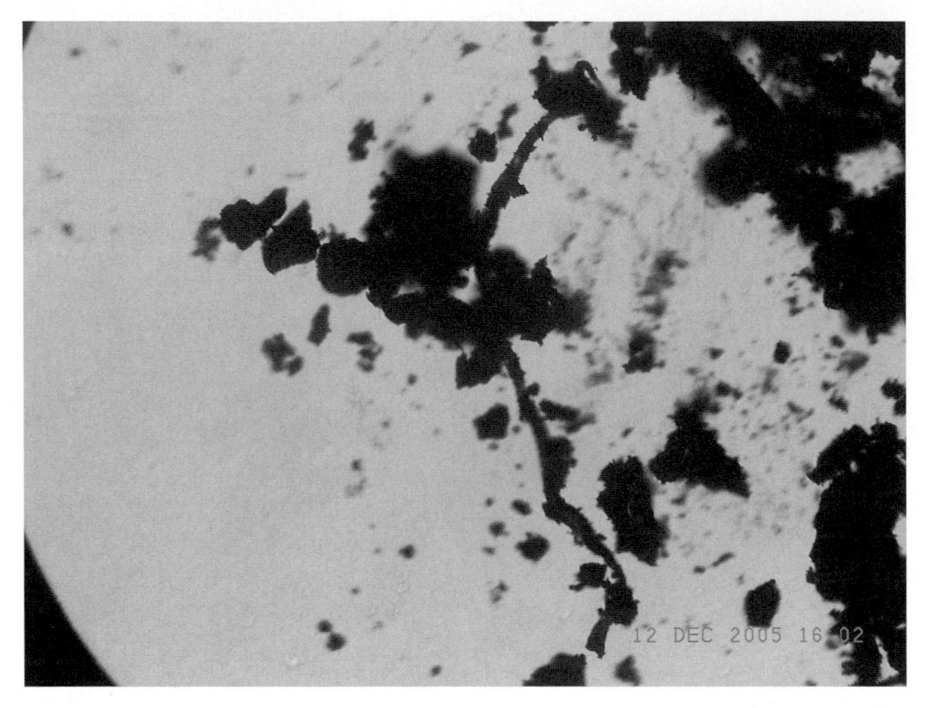

Figure 18.9 Fibers from the product of diethyltin dichloride and 2-chloro-p-phenylenediamine

References

1. Carraher C. (1971) ChemTech,.
2. Carraher C. Unpublished results.
3. Sabir T, Carraher C. (2005) Polym Mater Sci Eng. 93:396
4. Sabir T, Carraher C. Polym Mater Sci Eng. in press.
5. Carraher C. (2005) Macromolecules containing metal and metal-like elements, vol. 3. Group IVB Polymers, Wiley, Hoboken, New Jersey.
6. Carraher C, Siegmann-Louda D. (2004) Macromolecules containing metal and metal-like elements, vol 4, biomedical applications. Wiley, Hoboken, New Jersey.
7. Siegmann-Louda D, Carraher C, Pflueger F, Coleman J, Harless S, Luing H. (2000) Polym Mat Sci Eng. 82:83.
8. Siegmann-Louda D, Carraher C, Ross J, Li F, Mannke K, Harless S. (1999) Polym Mat Sci Eng. 81:151.
9. Siegmann-Louda D, Carraher C, Pfueger F, Nagy D. (2001) Polym Mat Sci Eng. 84:658.
10. Siegmann-Louda D, Carraher C, Chamely D, Cardoso A, Snedden D. (2002) Polym Mat Sci Eng. 86:293.
11. Siegmann-Louda D, Carraher C, Graham M, Doucetter R, Lanz L. (2002) Polym Mat Sci Eng. 87:247.
12. Siegmann-Louda D, Carraher C, Snedden D, Komulainen A. (2004) Polym Mater Sci Eng. 90:512.
13. Doucette R, Siegmann-Louda D, Carraher C, Cardoso A. (2004) Polym Mater Sci Eng. 91:564.

14. Doucette R, Siegmann-Louda D, Carraher C. (2004) Polym Mater Sci Eng. 91:567 and 569.
15. Doucette R, Siegmann-Louda D, Carraher C. (2004) Polym Mater Sci Eng. 91:567 and 569.
16. Carraher C, Lee JL. (2004) Polym Mater Sci Eng. 90:408.
17. Carraher C, Scherubel G. (1972) Makromolekular Chemie. 152:61.
18. Carraher C, Scherubel G. (1972) Makromolekular Chemie. 160:259.
19. Carraher C, Scherubel G. (1971) J Polym Sci. A-1, 9:983.
20. Barot G, Carraher C, Siegmann-Louda D. Polym Mater Sci Eng. in press.
21. Siegmann-Louda D, Carraher C, Nagy D, Snedden D, Rosa J. (2003) J Polym Mater Sci Eng. 89:487.

Index

C

D

Printed in the United States of America